PROBABILITY AND PROBABILITY DISTRIBUTIONS

Probability   Bayes' theorem

$$P(B_i|A) = \frac{P(B_i) \cdot P(A|B_i)}{P(B_1) \cdot P(A|B_1) + P(B_2) \cdot P(A|B_2) + \cdots + P(B_k) \cdot P(A|B_k)}$$

Conditional probability

$$P(A|B) = \frac{P(A \cap B)}{P(B)}$$

General addition rule

$$P(A \cup B) = P(A) + P(B) - P(A \cap B)$$

General multiplication rule

$$P(A \cap B) = P(B) \cdot P(A|B) \quad \text{or} \quad P(A \cap B) = P(A) \cdot P(B|A)$$

Mathematical expectation

$$E = a_1 p_1 + a_2 p_2 + \cdots + a_k p_k$$

Probability   Binomial distribution
Distributions

$$f(x) = \binom{n}{x} p^x (1 - p)^{n-x}$$

Mean of probability distribution

$$\mu = \sum x \cdot f(x)$$

Standard deviation of probability distribution

$$\sigma = \sqrt{\sum (x - \mu)^2 \cdot f(x)}$$

Continued inside back cover

MODERN
ELEMENTARY
STATISTICS

Prentice-Hall, Inc., Englewood Cliffs, New Jersey 07632

6th edition

John E. Freund

Arizona State University

MODERN ELEMENTARY STATISTICS

Library of Congress Cataloging in Publication Data

Freund, John E.
  Modern elementary statistics.

  Includes bibliographies and index.
  1. Statistics.   I. Title.
QA276.12.F738   1984   519.5   83-3099
ISBN 0-13-593525-3

The following figures are taken from Freund and Williams, *Elementary Business Statistics*, 4th ed. Copyright © 1982 by Prentice-Hall, Inc.: 2.12 and 12.4. Used by permission.

The following figures are taken from Freund, *Statistics: A First Course*, 3rd ed. Copyright © 1981 by Prentice-Hall, Inc.: 2.6, 3.2, 3.4, 5.1, 5.8, 5.10 to 5.12, R.1, 7.8, 7.11, unnumbered figures on pp. 422, 424, 426, and 428, and Tables I, III, V, VII, VIII, X, XI, XII, and XIII. Used by permission.

MODERN ELEMENTARY STATISTICS, Sixth Edition by John E. Freund

Editorial/production supervision: Karen J. Clemments
Interior/cover design: Suzanne Behnke
Manufacturing buyer: John Hall

Printed in the United States of America

10   9   8   7   6   5

ISBN 0-13-593525-3

Prentice-Hall International, Inc., *London*
Prentice-Hall of Australia Pty. Limited, *Sydney*
Editora Prentice-Hall do Brasil, Ltda., *Rio de Janeiro*
Prentice-Hall Canada Inc., *Toronto*
Prentice-Hall of India Private Limited, *New Delhi*
Prentice-Hall of Japan, Inc., *Tokyo*
Prentice-Hall of Southeast Asia Pte. Ltd., *Singapore*
Whitehall Books Limited, *Wellington, New Zealand*

CONTENTS

All sections marked ★ are optional.

8 THE NORMAL DISTRIBUTION   *204*

9 SAMPLING AND SAMPLING DISTRIBUTIONS   *233*

10 INFERENCES ABOUT MEANS   *264*

11 INFERENCES ABOUT STANDARD DEVIATIONS   310

12 INFERENCES ABOUT PROPORTIONS   323

13 ANALYSIS OF VARIANCE   365

14 REGRESSION   401

PREFACE

The foremost objective of this edition, like that of all previous editions, is to acquaint beginning students in the various sciences with the fundamentals of modern statistics. Although the basic organization has remained the same, there are substantial changes in content and in format.

Many of the examples and exercises are new, and as in previous editions they are distributed over a wide variety of applications. New also are the review exercises at the end of each chapter. Altogether, there are almost 1,000 exercises, numbered consecutively throughout each chapter. Most of them are drawn from actual problems, but many have been modified and scaled down somewhat to simplify the computational load.

Most changes in the text are based on suggestions of colleagues and students, who generously took the time to share their experiences and thoughts. For instance, there is some new material on stem-and-leaf plots; the chapter on nonparametric tests now includes small-sample tests with appropriate tables; statistical terminology and notation are updated to conform with current usage; the basic discussion of statistical inference includes also null hypotheses which are not simple hypotheses about population parameters; the order of the chapters on analysis of variance and nonparametric tests is reversed; cross-referencing among exercises is held to a minimum; the binomial probability table has been expanded; and each chapter has a checklist of key terms.

As in the previous editions, the author shall resist the temptation to tell anyone specifically what chapters, sections, or subjects to study or teach, but it should be observed that there are topics which may be omitted without loss of continuity. Upon the urging of many colleagues, and our publishers, some sections and related material are thus marked ★, meaning that they are optional. This posed all sorts of difficulties, for what one person may consider optional another person may well consider essential. For instance, many instructors will not cover the material on the description of grouped data, yet a person who has to work with published government data may well consider it essential. Although quite a few colleagues indicated that they do not cover the material on subjective probability, to mark it optional would be an affront to those who hold the subjectivist point of view. However, risking their wrath, the sections dealing with Bayes' theorem and Bayesian estimation are marked optional.

As indicated in previous editions, the study of statistics may not only be directed toward applications in various specialized fields of inquiry, but it may also be presented at various levels of mathematical difficulty and in almost any balance between theory and application. As it is more important, in the author's opinion, to understand the meaning and implications of basic ideas than it is to memorize an impressive list of formulas, some of the details that are sometimes included in introductory courses in statistics have been sacrificed. This may be unfortunate in some respects, but it should prevent the reader from getting lost in an excessive amount of detail which could easily obscure the more important issues. It is hoped that this will avoid some of the unfortunate consequences which often result from the indiscriminate application of so-called standard techniques without a thorough understanding of the basic ideas that are involved.

It cannot be denied that a limited amount of mathematics is a prerequisite for any course in statistics, and that a thorough study of the theoretical principles of statistics would require a knowledge of mathematical subjects taught ordinarily only on the graduate level. Since this book is designed for students with relatively little background in mathematics, the aims, and, therefore, also the prerequisites needed here are considerably more modest. Actually, the mathematical background needed for this study of statistics is amply covered in a course of college algebra; in fact, even a good knowledge of high school algebra provides a sufficient foundation.

The author would like to express his appreciation to the many colleagues and students whose helpful suggestions and criticisms contributed greatly to previous editions of this text as well as to this sixth edition. In particular, the author would like to thank his son, John, and Rita Ewer for reading various drafts of the manuscript and helping with the proofreading, and Victor Romano for working on the answers to the exercises. The author would also like to express his appreciation to Robert Sickles, mathematics editor, and to Karen J. Clemments, production editor, for their courteous cooperation in the production of this book.

The author is also indebted to the Macmillan Publishing Co. for their permission to reproduce the material in Table II; to Professor E. S. Pearson and the *Biometrika* trustees to reproduce the material in Tables III and IV; to the Addison-Wesley Publishing Company to base Table VI on Table 11.4 of D. B. Owen's *Handbook of Statistical Tables*; to the American Cynamid Co. to reproduce the material in Table VII; to the editor of the *Annals of Mathematical Statistics* to reproduce the material in Table VIII; and to the RAND Corporation to reproduce the sample pages of random normal numbers shown in Table XII.

JOHN E. FREUND
Scottsdale, Arizona

MODERN ELEMENTARY STATISTICS

Everything dealing with the collection, processing, analysis, and interpretation of numerical data belongs to the domain of statistics. This includes such diversified tasks as calculating a baseball player's batting average, collecting and presenting data on marriages and divorces, evaluating the effectiveness of commercial products, forecasting the weather, or studying the vibrations of airplane wings.

The word "statistics" itself is used in several ways. In one connection it means a collection of data such as those found in the financial pages of newspapers or in the *Statistical Abstract of the United States*. In another connection it means the totality of methods used in the collection, processing, analysis, or interpretation of any kind of data. In this latter sense, statistics is a branch of applied mathematics, and it is this field of mathematics which is the subject matter of this book. In yet another sense, statistics are particular descriptions of numerical data.

In Sections 1.1 and 1.2 we discuss the recent growth of statistics and its ever widening range of applications. In Section 1.3 we explain the distinction between descriptive statistics and statistical inference, and in Section 1.4 we warn the reader against the indiscriminate mathematical treatment (addition, multiplication, etc.) of statistical data.

1 INTRODUCTION

1.1
THE GROWTH OF MODERN STATISTICS

There are two reasons why the scope of statistics and the need to study statistics have grown enormously in the last few decades. One reason is the increasingly quantitative approach employed in all the sciences, as well as in business and many other activities which directly affect our lives. This includes the use of mathematical techniques in the evaluation of anti-pollution controls, in inventory planning, in the analysis of traffic patterns, in the study of the effects of various kinds of medications, in the evaluation of teaching techniques, in the analysis of competitive behavior of business-men and governments, in the study of diet and longevity, and so forth.

The other reason is that the amount of data that is collected, processed, and disseminated to the public for one reason or another has increased almost beyond comprehension. To act as watchdogs, more and more persons with some knowledge of statistics are needed to take an active part in the collection of the data, in the analysis of the data, and, what is equally important, in all of the preliminary planning. Without the latter, it is frightening to think of all the things that can go wrong in the compilation of statistical data. The results of costly surveys can be useless if questions are ambiguous or asked in the wrong way, if they are asked of the wrong persons, in the wrong place, or at the wrong time.

EXAMPLE   To determine public sentiment about the continuation of a certain government program, an interviewer asks: "Do you feel that this wasteful program should be stopped?" Explain why this will probably not yield the desired information.

Solution   This is called "begging the question" and may well yield misleading results, because the interviewer suggests that the program is, in fact, wasteful.

EXAMPLE   To study consumer reaction to a new convenience food, a house-to-house survey is conducted during weekday mornings, with no provisions for return visits in case no one is home. Explain why this may well yield misleading information.

Solution   This survey will fail to reach most of the persons who are most likely to use the product: single persons and husbands and wives who are both employed.

1.2
THE STUDY OF STATISTICS

The subject of statistics may be presented at various levels of mathematical difficulty, and it may be directed toward applications in various fields of inquiry. Accordingly, many textbooks have been written on business statistics, educational statistics, medical statistics, psychological statistics, ..., and even on statistics for historians. Although problems arising in these various disciplines will sometimes require special statistical techniques, none of the basic methods discussed in this text is restricted to any particular field of application. In the same way in which $2 + 2 = 4$ regardless of whether we are adding dollar amounts, horses, or trees, the methods we shall present provide **statistical models** which apply regardless of whether the data are IQ's, tax payments, reaction times, humidity readings, test scores, and so on. To illustrate this further, consider Exercise 12.70 on page 361, made up by the author.

12.70 In a random sample of 200 retired persons, 137 stated that they prefer living in an apartment to living in a one-family home. At the 0.05 level of significance, does this refute the claim that 60 percent of all retired persons prefer living in an apartment to living in a one-family home?

Except for the reference to the "level of significance," a technical term, the question asked here should be clear, and it should also be clear that the answer would be of interest mainly to social scientists or to persons in the construction industry. However, if we wanted to cater to the special interests of students of biology, engineering, education, or ecology, we might rephrase the exercise as follows:

12.70 In a random sample of 200 citrus trees exposed to a 20° frost, 137 showed some damage to their fruit. At the 0.05 level of significance, does this refute the claim that 60 percent of all citrus trees exposed to a 20° frost will show some damage to their fruit?

12.70 In a random sample of 200 transistors made by a given manufacturer, 137 passed an accelerated performance test. At the 0.05 level of significance, does this refute the claim that 60 percent of all the transistors made by the manufacturer will pass the test?

12.70 In a random sample of 200 high school seniors in a large city, 137 said that they will go on to college. At the 0.05 level of significance, does this refute the claim that 60 percent of all the high school seniors in this city will go on to college?

12.70 In a random sample of 200 cars tested for the emission of pollutants, 137 failed to meet a state's legal standards. At the 0.05 level of significance, does this refute the claim that 60 percent of all cars tested in this state will fail to meet its legal emission standards?

So far as the work in this book is concerned, the statistical treatment of all these versions of Exercise 12.70 is the same, and with some imagination the reader should be able to rephrase it for almost any field of specialization. We could present, and so designate, special problems for readers with special interests, but this would defeat our main goal of impressing upon the reader the importance of statistics in all of science, business, and everyday life. To attain this goal, the examples and the exercises in this text cover a wide spectrum of interests.

To avoid giving a wrong impression with the above example, let us make it clear that all statistical problems cannot be squeezed into the same mold. Although the methods we shall study in this book are widely applicable, it is important always to make sure that we are using the right kind of statistical model.

EXERCISES

1.1 Rephrase the exercise referred to on page 3 so that it would be of special interest to
   (a) a cosmetics salesperson;
   (b) a musician;
   (c) a traffic engineer.

1.2 "Bad" statistics may well result from asking questions in the wrong way or of the wrong persons. Explain why the following may lead to useless data:
   (a) To study executives' reaction to its copying machines, Xerox Corporation hires a research organization to ask executives the question: "How do you like using Xerox copiers?"
   (b) To determine what the average person spends on a vacation, a researcher interviews passengers on a luxury cruise.

1.3 "Bad" statistics may well result from asking questions in the wrong place or at the wrong time. Explain why the following may lead to useless data:
   (a) To predict an election, a poll taker interviews persons coming out of the building which houses the national headquarters of a political party.
   (b) To study the spending patterns of families in a certain income group, a survey is conducted during the first three weeks of December.

1.4 Explain why each of the following studies may fail to yield the desired information:
   (a) To determine the average income of its graduates ten years after graduation, a university alumni office sent questionnaires in 1982 to all members of the class of 1972, and the estimate is based on the questionnaires returned.
   (b) To determine the proportion of improperly sealed cans of coffee, a quality control engineer checks every 50th can coming off an assembly line.

1.3
DESCRIPTIVE STATISTICS
AND STATISTICAL INFERENCE

The origin of modern statistics may be traced to two areas of interest which, on the surface, have very little in common: government (political science) and games of chance.

Governments have long used censuses to count persons and property, and the problem of describing, summarizing, and analyzing census data has led to the development of methods which, until recently, constituted about all there was to the subject of statistics. These methods, which originally consisted mainly of presenting data in the form of tables and charts, make up what we now call **descriptive statistics**. This includes anything done to data which is designed to summarize, or describe, them without going any further; that is, without attempting to infer anything that goes beyond the data themselves. For instance, if tests performed on six small cars imported in 1980 showed that they were able to accelerate from 0 to 60 miles per hour in 18.7, 19.2, 16.2, 12.3, 17.5, and 13.9 seconds, and we report that half of them accelerated from 0 to 60 mph in less than 17.0 seconds, our work belongs to the domain of descriptive statistics. This would also be the case if we claim that these six cars averaged

$$\frac{18.7 + 19.2 + 16.2 + 12.3 + 17.5 + 13.9}{6} = 16.3 \text{ seconds}$$

but not if we conclude that half of *all* cars imported that year could accelerate from 0 to 60 mph in less than 17.0 seconds.

Although descriptive statistics is an important branch of statistics and it continues to be widely used, statistical information usually arises from samples (from observations made on only part of a large set of items), and this means that its analysis will require generalizations which go beyond the data. As a result, the most important feature of the recent growth of statistics has been a shift in emphasis from methods which merely describe to methods which serve to make generalizations; that is, a shift in emphasis from descriptive statistics to the methods of **statistical inference**.

Such methods are required, for instance, to predict the operating life span of a hand-held calculator (on the basis of the performance of several such calculators); to estimate the 1990 assessed value of all property in Pima County, Arizona (on the basis of business trends, population projections, and so forth); to compare the effectiveness of two reducing diets (on the basis of the weight losses of persons who have been on the diets); to determine the most effective dose of a new medication (on the basis of tests performed with volunteer patients from selected hospitals); or to predict the flow of traffic on a freeway which has not yet been built (on the basis of past traffic counts on alternative routes).

In each of the situations described in the preceding paragraph, there are uncertainties because there is only partial, incomplete, or indirect information, and it is with the use of the methods of statistical inference that we judge the merits of the results and, perhaps, suggest a "most profitable" choice, a "most promising" prediction, or a "most reasonable" course of action.

In view of the uncertainties, we handle problems like these with statistical methods which find their origin in games of chance. Although the mathematical study of games of chance dates back to the seventeenth century, it was not until the early part of the nineteenth century that the theory developed for "heads or tails," for example, or "red or black" or "even or odd," was applied also to real-life situations where the outcomes were "boy or girl," "life or death," "pass or fail," and so forth. Thus, **probability theory** was applied to many problems in the behavioral, natural, and social sciences, and nowadays it provides an important tool for the analysis of any situation (in science, in business, or in everyday life) which in some way involves an element of uncertainty or chance. In particular, it provides the basis for the methods which we use when we generalize from observed data, namely, when we use the methods of statistical inference.

1.4
THE NATURE OF STATISTICAL DATA ★†

Statistical data are the raw material of statistical investigations, and they arise whenever measurements are made or observations are classified. They may be weights of animals, measurements of personality traits, or earthquake intensities, and they may be simple "yes or no" answers or descriptions of persons' marital status as single, married, widowed, or divorced. Since we said on page 1 that statistics deals with numerical data, this requires some explanation, because "yes or no" answers and descriptions of marital status would hardly seem to qualify as being numerical. Observe, however, that we can record "yes or no" answers to a question as 0 and 1 (or as 1 and 2, or perhaps as 29 and 30 if we are referring to the 15th "yes or no" question of a test), and that we can record a person's marital status as 1, 2, 3, or 4, depending on whether the person is single, married, widowed, or divorced. In this artificial, or nominal way, categorical (qualitative, or descriptive) data can be made into numerical data, and if we thus code the various categories, we refer to the numbers we record as **nominal data.**

---

† As is explained in the Preface, all sections marked ★ are optional. Although the material in this section is meant to serve as a warning against the indiscriminate mathematical treatment of statistical data, it is most relevant to students of the behavioral and social sciences, where artificial scales serve to measure, say, neurotic tendencies, happiness, or conformity to social standards.

Nominal data are numerical in name only, because they do not share any of the properties of the numbers we deal with in ordinary arithmetic. For instance, if we record marital status as 1, 2, 3, or 4 as suggested above, we cannot write $3 > 1$ or $2 < 4$, and we cannot write $2 - 1 = 4 - 3$, $1 + 3 = 4$, or $4 \div 2 = 2$. It is important, therefore, always to check whether mathematical calculations performed in a statistical analysis are really legitimate.

Let us now consider some examples where data share some, but not necessarily all, of the properties of the numbers we deal with in ordinary arithmetic. For instance, in mineralogy the hardness of solids is sometimes determined by observing "what scratches what." If one mineral can scratch another it receives a higher hardness number, and on Mohs' scale the numbers from 1 to 10 are assigned, respectively, to talc, gypsum, calcite, fluorite, apatite, feldspar, quartz, topaz, sapphire, and diamond. With these numbers we can write $6 > 3$, for example, or $7 < 9$, since feldspar is harder than calcite and quartz is softer than sapphire. On the other hand, we cannot write $10 - 9 = 2 - 1$, for example, because the difference in hardness between diamond and sapphire is actually much greater than that between gypsum and talc. Also, it would be meaningless to say that topaz is twice as hard as fluorite simply because their respective hardness numbers on Mohs' scale are 8 and 4.

If we cannot do anything except set up inequalities, as was the case in the preceding example, we refer to the data as **ordinal data.** In connection with ordinal data, $>$ does not necessarily mean "greater than;" it may be used to denote "happier than," "preferred to," "more difficult than," "tastier than," and so forth.

If we can also form differences, but not multiply or divide, we refer to the data as **interval data.** To give an example, suppose we are given the following temperature readings in degrees Fahrenheit: 63°, 68°, 91°, 107°, 126°, and 131°. Here we can write $107° > 68°$ or $91° < 131°$, which simply means that 107° is warmer than 68° and that 91° is colder than 131°. Also, we can write $68° - 63° = 131° - 126°$, since equal temperature differences are equal in the sense that the same amount of heat is required to raise the temperature of an object from 63° to 68° as from 126° to 131°. On the other hand, it would not mean much if we say that 126° is twice as hot as 63°, even though $126 \div 63 = 2$. To show why, we have only to change to the Celsius scale, where the first temperature becomes $\frac{5}{9}(126 - 32) = 52.2°$, the second temperature becomes $\frac{5}{9}(63 - 32) = 17.2°$, and the first figure is now more than three times the second. This difficulty arises because the Fahrenheit and Celsius scales both have artificial origins (zeros); in other words, the number 0 of neither scale is indicative of the absence of whatever quantity we are trying to measure.

If we can also form quotients, we refer to the data as **ratio data,** and such data are not difficult to find. They include all the usual measurements (or determinations) of length, height, money amounts, weight, volume, area, pressure, elapsed time (though not calendar time), sound intensity, density, brightness, velocity, and so on.

The distinction we have made here between nominal, ordinal, interval, and ratio data is important, for as we shall see, the nature of a set of data may suggest the use of particular statistical techniques.

EXERCISES†

1.5 In five biology tests a student received grades of 46, 61, 74, 79, and 88. Which of the following conclusions can be obtained from these figures by purely descriptive methods and which require generalizations? Explain your answers.
(a) Only two of the grades exceeded 75.
(b) The student's grades increased from each test to the next.
(c) The student must have studied harder for each successive test.
(d) The difference between the highest and lowest grades is 42.

1.6 Jean and Tom are real estate salespersons. In the first three months of 1982 Jean sold 1, 4, and 2 one-family homes and Tom sold 2, 0, and 3. Which of the following conclusions can be obtained from these figures by purely descriptive methods and which require generalizations? Explain your answers.
(a) During the three months Jean sold more cars than Tom.
(b) Jean is a better real estate salesperson than Tom.
(c) Jean sold at least one one-family home in each of the three months.
(d) Tom probably took a vacation, or was ill, during the second month.

1.7 On three consecutive days, a traffic policeman issued 9, 14, and 10 speeding tickets, and 5, 10, and 12 tickets for going through red lights. Which of the following conclusions can be obtained from these data by purely descriptive methods and which require generalizations? Explain your answers.
(a) Altogether on these three days, the policeman issued more speeding tickets than tickets for going through red lights.
(b) On two of the three days, the policeman issued more speeding tickets than tickets for going through red lights.
(c) The policeman issued the smallest number of tickets on the first day because he was new on the job.
(d) This policeman will seldom give more than 15 speeding tickets on any one day.

1.8 The three oranges which a person bought at a supermarket weighed 9, 8, and 13 ounces. Which of the following conclusions can be obtained from these data by purely descriptive methods and which require generalizations? Explain your answers.
(a) The average weight of the three oranges is 10 ounces.
(b) The average weight of oranges sold at that supermarket is 10 ounces.

★1.9 Will we get nominal data or ordinal data if
(a) mechanics have to say whether changing the spark plugs on a new model car is very difficult, difficult, easy, or very easy;
(b) the religion of persons attempting suicide is coded 1, 2, 3, 4, or 5, representing Protestant, Catholic, Jewish, other, and none;

----

† Exercises marked ★ pertain to optional material.

(c) consumers must say whether they prefer brand *A* to brand *B*, like them equally, or prefer brand *B* to brand *A*;

(d) consumers must say whether they prefer brand *A* to brand *B*, like them equally, prefer brand *B* to brand *A*, or have no opinion.

★1.10 Are the following nominal, ordinal, interval, or ratio data? Explain your answers.

(a) Social Security numbers.

(b) The number of passengers on buses from Los Angeles to San Diego.

(c) Vocational interest scores consisting of the total number of "yes" answers given to a set of questions, if it can be assumed that each "yes" answer represents the same increment of vocational interest.

(d) Military ranks.

★1.11 IQ scores are sometimes looked upon as interval data. What assumption would this entail about the differences in intelligence of three persons with IQ's of 95, 105, and 135? Is this assumption reasonable?

★1.12 On page 8 we indicated that data pertaining to calendar time (for instance, the years in which Army beat Navy in football) are not ratio data. Explain why. What kind of time measurements do constitute ratio data?

1.5
CHECKLIST OF KEY TERMS
(with page references to their definitions)†

Descriptive statistics, 5
★ Interval data, 7
★ Nominal data, 6
★ Ordinal data, 7
★ Ratio data, 7
Statistical inference, 5
Statistical model, 3

1.6
REVIEW EXERCISES‡

1.13 The paid attendance at a small college's home football games was 12,305, 10,984, 6,850, 11,733, and 10,641. Which of the following conclusions can be obtained from these figures by purely descriptive methods and which require generalizations? Explain your answers.

(a) The attendance at the third home game was so low because it rained.

(b) Among the five games, the paid attendance was highest at the first game.

(c) The paid attendance exceeded 11,000 at two of the five games.

(d) The paid attendance increased from the third home game to the fourth home game because the college's football team had been winning.

---

† Terms marked ★ pertain to optional material.
‡ Review exercises marked ★ pertain to optional material.

★1.14 Are the following nominal, ordinal, interval, or ratio data? Explain your answers.
   (a) Elevations above sea level.
   (b) Responses to the question whether (in the downtown area of a large city) living conditions are "getting much worse," "getting a little worse," "staying the same," "getting a little better," or "getting much better."
   (c) Ages of secondhand cars.
   (d) Responses on drivers' licenses with regard to eye color.

1.15 Explain why each of the following data may well fail to yield the desired information:
   (a) To predict a municipal election, a public opinion poll telephones persons selected haphazardly from the city's telephone directory.
   (b) To determine public sentiment about certain import restrictions, an interviewer asks voters: "Do you feel that this unfair practice should be stopped?"

★1.16 If students calculate their grade-point indexes (that is, average their grades) by counting A, B, C, D, and F as 1, 2, 3, 4, and 5, what does this assume about the nature of the grades?

1.17 Explain why each of the following data may well fail to yield the desired information:
   (a) To see how the public feels about imports from India, selected persons are asked whether they like Indian art.
   (b) To ascertain facts about their bathing habits, a sample of the citizens of a European country are asked how many times on the average they bathe per week.

1.18 Using the same model car, five drivers averaged 23.4, 22.5, 24.0, 23.4, and 22.7 miles per gallon. Which of the following conclusions can be obtained by purely descriptive methods and which require generalizations? Explain your answers.
   (a) More often than any of the other figures, the drivers averaged 23.4 miles per gallon.
   (b) More often than any of the other figures, drivers of this kind of car will average 23.4 miles per gallon.
   (c) None of the averages differs from 23.5 by more than 1.0 mile.
   (d) If the whole experiment is repeated, none of the drivers will average less than 22.5 or more than 24.5 miles per hour.

★1.19 In two major golf tournaments one professional golfer finished second and ninth, while another finished sixth and fifth. Comment on the argument that since $2 + 9 = 6 + 5$, the overall performance of the two golfers in these two tournaments was equally good.

1.20 Rephrase the exercise referred to on page 3 so that it would be of special interest to
   (a) a lawyer;
   (b) a travel agent;
   (c) an author.

1.7
REFERENCES

A brief and informal discussion of what statistics is and what statisticians do may be found in a pamphlet titled *Careers in Statistics*, which is published by the American Statistical Association. It may be obtained by writing to this organization at 806 15th Street, N.W., Washington, D.C., 20005. Among the few books on the history of statistics, there is on the elementary level

WALKER H. M., *Studies in the History of Statistical Method*. Baltimore: The Williams & Wilkins Company, 1929.

and on the more advanced level

PEARSON, E. S., and KENDALL, M. G., eds., *Studies in the History of Statistics and Probability*. New York: Hafner Press, 1970.

KENDALL, M. G., and PLACKETT, R. L., eds., *Studies in the History of Statistics and Probability, Vol. II*. New York: Macmillan Publishing Co., Inc., 1977.

A more detailed discussion of the nature of statistical data and the general problem of scaling (namely, the problem of constructing scales of measurement) may be found in

HILDEBRAND, D. K., LAING, J. D., and ROSENTHAL, H., *Analysis of Ordinal Data*. Beverly Hills, Sage Publications, Inc., 1977.

REYNOLDS, H. T., *Analysis of Nominal Data*. Beverly Hills, Sage Publications, Inc., 1977.

SIEGEL, S., *Nonparametric Statistics for the Behavioral Sciences*. New York: McGraw-Hill Book Company, 1956.

In recent years the collection of statistical data has grown at such a rate that it would be impossible to keep up with even a small part of the things which directly affect our lives unless this information is disseminated in "pre-digested" or summarized form. Of course, we do not always deal with very large sets of data (in many cases they are prohibitively costly and hard to collect), but the problem of putting mass data into a suitable form is so important that it requires special attention.

The most common method of summarizing data is to present them in condensed form in tables or charts, and the study of this at one time took up the better part of elementary courses in statistics. Nowadays, the scope of statistics has expanded to such an extent that much less time is devoted to this kind of work—in fact, we shall talk about it only in this chapter.

Sections 2.1 and 2.2 deal with problems related to the grouping of data and the presentation of such groupings in graphical form; in Section 2.3 we discuss a relatively new way of presenting grouped data.

2 SUMMARIZING DATA: FREQUENCY DISTRIBUTIONS

2.1
FREQUENCY DISTRIBUTIONS

When we deal with large sets of data, a good overall picture and sufficient information can often be conveyed by grouping the data into a number of classes. For instance, the total billings, to the nearest dollar, of 8,644 selected law firms may be summarized as follows:

Total billings	Number of firms
Less than $100,000	1,406
$100,000–$249,999	4,352
$250,000–$499,999	1,833
$500,000–$749,999	489
$750,000–$999,999	163
$1,000,000 or more	401
Total	8,644

Also, 2,439 complaints about comfort-related characteristics of an airline's planes may be presented as follows:

Nature of complaint	Number of complaints
Inadequate leg room	719
Uncomfortable seats	914
Narrow aisles	146
Insufficient carry-on facilities	218
Insufficient restrooms	58
Miscellaneous other complaints	384
Total	2,439

Tables like these are called **frequency distributions** (or simply **distributions**). The first one shows how the law firms' billings are distributed among the chosen intervals, and the second one shows how the complaints are distributed among the different categories. If data are grouped according to numerical size, as in the first table, the resulting table is called a **numerical** or **quantitative distribution**; if data are grouped into nonnumerical categories, as in the second table, the resulting table is called a **categorical** or **qualitative distribution**.

Frequency distributions present data in a relatively compact form, give a good overall picture, and contain information that is adequate for many purposes, but some things which can be determined from the original data cannot be determined from a distribution. For instance, from the first table on page 13 we can find neither the exact size of the smallest and the largest of the total billings, nor the exact total or average of the total billings of the 8,644 firms. Similarly, from the second table on page 13 we cannot tell how many of the complaints about uncomfortable seats were over the width of the seats, how many of the complaints about insufficient carry-on facilities were over space for suit carriers, and so forth. Nevertheless, frequency distributions present **raw** (unprocessed) **data** in a more readily usable form, and the price we pay for this—the loss of certain information—is usually a fair exchange.

The construction of frequency distributions consists essentially of three steps: (1) choosing the **classes** (intervals or categories), (2) tallying the data into these classes, and (3) counting the number of items (tallies) in each class. Since the last two steps are purely mechanical, we shall concentrate on the first, namely, the problem of choosing a suitable classification.

The two things we must consider in choosing a classification scheme for a numerical distribution are how many classes we should use and the range of values each class should cover, that is, from where to where each class should go. Both choices are essentially arbitrary, but the following rules are usually observed:

> **We seldom use fewer than 6 or more than 15 classes; the exact number we use in a given situation will depend mainly on the number of measurements or observations we have to group.**

Clearly, we would lose more than we gain if we group 5 observations into 12 classes with most of them empty, and we would lose a great deal of information if we group 10,000 measurements into 3 classes.

> **We always make sure that each item (measurement or observation) will go into one and only one class.**

To this end we must make sure that the smallest and largest values fall within the classification, that none of the values can fall into gaps between successive classes, and that successive classes do not overlap, namely, that successive classes have no values in common.

> **Whenever possible, we make the classes the same length; that is, we make them cover equal ranges of values.**

If we can, we also make these ranges multiples of numbers that are easy to work with, such as 5, 10, or 100, for this facilitates constructing, reading, and using the distribution.

Since the law firm billings of the example on page 13 were rounded to the nearest dollar, only the last of these rules was violated in constructing the distribution on page 13. (Had the billings been given to the nearest cent, however, a billing of, say, $249,999.53 would fall between the second class and the third class, and we would also have violated the second rule.) Actually, the third rule was violated in several ways: the intervals $100,000–$249,999 and $250,000–$499,999 cover unequal ranges of values, the first class has no specific lower limit, and the last has no specific upper limit.

In general, we refer to any class of the "lens than," "or less," "more than," or "or more" type as an **open class**. If a set of data contains a few values which are much greater than or much smaller than the rest, open classes are quite useful in reducing the number of classes required to accommodate the data. However, we usually avoid open classes because they make it impossible to calculate certain values of interest, such as an average or a total.

As we have suggested, the appropriateness of a classification may depend on whether the data are rounded to the nearest dollar or to the nearest cent. Similarly, it may depend on whether the data are given to the nearest inch or to the nearest hundredth of an inch, whether they are given to the nearest second or to the nearest microsecond, whether they are rounded to the nearest percent or to the nearest tenth of a percent, and so on.

For instance, if we want to group the heights of children, we could use the first of the following three classifications if the heights are given to the nearest inch, the second if the heights are given to the nearest tenth of an inch, and the third if the heights are given to the nearest hundredth of an inch:

Height (inches)	Height (inches)	Height (inches)
25–29	25.0–29.9	25.00–29.99
30–34	30.0–34.9	30.00–34.99
35–39	35.0–39.9	35.00–39.99
40–44	40.0–44.9	40.00–44.99
45–49	45.0–49.9	45.00–49.99
etc.	etc.	etc.

To illustrate the construction of a numerical frequency distribution, let us go through the actual steps of grouping a set of data.

EXAMPLE   Construct a distribution of the following amounts of sulfur oxides (in tons) emitted by an industrial plant on 80 days:

15.8	26.4	17.3	11.2	23.9	24.8	18.7	13.9	9.0	13.2
22.7	9.8	6.2	14.7	17.5	26.1	12.8	28.6	17.6	23.7
26.8	22.7	18.0	20.5	11.0	20.9	15.5	19.4	16.7	10.7
19.1	15.2	22.9	26.6	20.4	21.4	19.2	21.6	16.9	19.0
18.5	23.0	24.6	20.1	16.2	18.0	7.7	13.5	23.5	14.5
14.4	29.6	19.4	17.0	20.8	24.3	22.5	24.6	18.4	18.1
8.3	21.9	12.3	22.3	13.3	11.8	19.3	20.0	25.7	31.8
25.9	10.5	15.9	27.5	18.1	17.9	9.4	24.1	20.1	28.5

Solution   Since the smallest value is 6.2 and the largest is 31.8, we might choose the six classes 5.0–9.9, 10.0–14.9, ..., and 30.0–34.9, we might choose the seven classes 5.0–8.9, 9.0–12.9, ..., and 29.0–32.9, or we might choose the nine classes 5.0–7.9, 8.0–10.9, ..., and 29.0–31.9. Note that in each case the classes accommodate all the data, they do not overlap, and they are all of the same size.

Deciding upon the second of these classifications, we now tally the 80 measurements and obtain the results shown in the following table:

Tons of sulfur oxides	Tally	Frequency
5.0– 8.9	///	3
9.0–12.9	//// ////	10
13.0–16.9	//// //// ////	14
17.0–20.9	//// //// //// //// ////	25
21.0–24.9	//// //// //// //	17
25.0–28.9	//// ////	9
29.0–32.9	//	2
	Total	80

The numbers given in the right-hand column of this table, which show how many items fall into each class, are called the **class frequencies**. The smallest and largest values that can go into any given class are called its **class limits,** and for the distribution of the emission data they are 5.0 and 8.9, 9.0 and 12.9, 13.0 and 16.9, ..., and 29.0 and 32.9. More specifically, 5.0, 9.0, 13.0, ..., and 29.0 are called the **lower class limits,** and 8.9, 12.9, 16.9, ..., and 32.9 are called the **upper class limits.**

Using this terminology, we can now say that the choice of the class limits depends on the extent to which the numbers we want to group are rounded. If our data are weights to the nearest pound, the class 150–159 actually contains all weights between 149.5 and 159.5; and if our data are lengths rounded to the nearest tenth of an inch, the class 5.0–7.4 actually contains all lengths between 4.95 and 7.45. Such pairs of values are usually called **class boundaries** or **real class limits.**

For the distribution of the emission data, the class boundaries are 4.95, 8.95, 12.95, . . . , and 32.95, namely, the midpoints between the respective class limits. Of course, these values must be, by their very nature, impossible values which cannot occur among the data being grouped. For instance, for the distribution of the law firm billings on page 13, the class boundaries, where they exist, are the impossible values $99,999.50, $249,999.50, $499,999.50, $749,999.50, and $999,999.50. They are impossible values because the figures are rounded to the nearest dollar.

Numerical distributions also have what we call **class marks** and **class intervals.** Class marks are simply the midpoints of the classes, and they are found by adding the lower and upper limits of a class (or its lower and upper boundaries) and dividing by 2. A class interval is merely the length of a class, or the range of values it can contain, and it is given by the difference between its boundaries. If the classes of a distribution are all equal in length, their common class interval, which we call the **class interval of the distribution,** is also given by the difference between any two successive class marks.

EXAMPLE   Find the class marks and the class interval of the distribution of the emission data.

Solution   The class marks are $\dfrac{5.0 + 8.9}{2} = 6.95$, $\dfrac{9.0 + 12.9}{2} = 10.95, \ldots$, and

$\dfrac{29.0 + 32.9}{2} = 30.95$, and the class interval of the distribution is $10.95 - 6.95 = 4$. Note that the class interval is not given by the difference between the upper and lower limits of a class, which here is 3.9, not 4.

There are essentially two ways in which frequency distributions can be modified to suit particular needs. One way is to convert a distribution into a **percentage distribution** by dividing each class frequency by the total number of items grouped, and then multiplying by 100.

EXAMPLE   Convert the distribution of the emission data into a percentage distribution.

Solution   The first class contains $\frac{3}{80} \cdot 100 = 3.75$ percent of the data, the second class contains $\frac{10}{80} \cdot 100 = 12.50$ percent of the data, . . . , and the seventh class contains $\frac{2}{80} \cdot 100 = 2.50$ percent of the data. These results are shown in the following table:

Tons of sulfur oxides	Percentage
5.0– 8.9	3.75
9.0–12.9	12.50
13.0–16.9	17.50
17.0–20.9	31.25
21.0–24.9	21.25
25.0–28.9	11.25
29.0–32.9	2.50
	100.00

Percentage distributions are often used when it is desired to compare two or more distributions; for instance, if we wanted to compare the emission of sulfur oxides of the plant of our example with that of a plant at a different location.

The other way of modifying a frequency distribution is to convert it into a "less than," "or less," "more than," or "or more" **cumulative distribution.** To this end we simply add the class frequencies, starting either at the top or at the bottom of the distribution.

EXAMPLE Convert the original distribution of the emission data into a "less than" cumulative distribution.

Solution Since none of the values is less than 5.0, 3 are less than 9.0, 3 + 10 = 13 are less than 13.0, 3 + 10 + 14 = 27 are less than 17.0, ..., the results are as shown in the following table:

Tons of sulfur oxides	Cumulative frequency
Less than 5.0	0
Less than 9.0	3
Less than 13.0	13
Less than 17.0	27
Less than 21.0	52
Less than 25.0	69
Less than 29.0	78
Less than 33.0	80

Note that instead of "less than 5.0" we could have written "4.9 or less" or "less than 4.95," instead of "less than 9.0" we could have written "8.9 or less" or "less than 8.95," and so on.

In the same way we can also convert a percentage distribution into a **cumulative percentage distribution.** We simply add the percentages starting either at the top or at the bottom of the distribution.

So far we have discussed only the construction of numerical distributions, but the general problem of constructing categorical (or qualitative) distributions is about the same. Here again we must decide how many categories (classes) to use and what kind of items each category is to contain, making sure that all the items are accommodated and that there are no ambiguities. Since the categories must often be chosen before any data are actually collected, it is prudent to include a category labeled "others" or "miscellaneous."

For categorical distributions, we do not have to worry about such mathematical details as class limits, class boundaries, and class marks. On the other hand, there is often a serious problem with ambiguities and we must be very careful and explicit in defining what each category is to contain. For instance, if we had to classify items sold at a supermarket into "meats," "frozen foods," "baked goods," and so forth, it would be difficult to decide, for example, where to put frozen beef pies. Similarly, if we had to classify occupations, it would be difficult to decide where to put a farm manager, if our table contained (without qualification) the two categories "farmers" and "managers." For this reason, it is advisable, where possible, to use standard categories developed by the Bureau of the Census and other government agencies. References to lists of such categories may be found in the book by P. M. Hauser and W. R. Leonard mentioned on page 35.

EXERCISES

2.1 The weights of the members of a football team (to the nearest pound) vary from 177 to 265 pounds. Indicate the limits of ten classes into which these weights might be grouped.

2.2 The weekly wages paid to the piecework employees of an electronics firm varied from $185.20 to $283.25. Indicate the limits of six classes into which these wages might be grouped.

2.3 The number of empty seats on flights from Atlanta to St. Louis are grouped into a table having the classes 0–4, 5–9, 10–14, 15–19, 20–24, 25–29, and 30 or more. Is it possible to determine from this distribution the number of flights on which there were

    (a) fewer than 20 empty seats;
    (b) more than 20 empty seats;
    (c) at least 9 empty seats;
    (d) at most 9 empty seats;
    (e) exactly 5 empty seats;
    (f) anywhere from 10 through 25 empty seats?

2.4 The following is the distribution of the weights of 125 mineral specimens collected on a field trip:

Weight (grams)	Number of specimens
0– 19.9	19
20.0– 39.9	38
40.0– 59.9	35
60.0– 79.9	17
80.0– 99.9	11
100.0–119.9	3
120.0–139.9	2
Total	125

If possible, find how many of the specimens weigh
  (a) at most 59.9 grams;
  (b) less than 40.0 grams;
  (c) more than 100.0 grams;
  (d) 80.0 grams or less;
  (e) exactly 20.0 grams;
  (f) anywhere from 40.0 to 80.0 grams.

2.5 The number of congressmen absent each day during a session of Congress are grouped into a distribution having the classes 0–19, 20–39, 40–59, 60–79, 80–99, 100–119, 120–139, and 140–159. Determine
  (a) the lower class limits;
  (b) the upper class limits;
  (c) the class marks;
  (d) the class interval of the distribution.

2.6 The number of nurses on duty each day at a hospital are grouped into a distribution having the classes 20–34, 35–49, 50–64, 65–79, and 80–94. Find
  (a) the class limits;
  (b) the class boundaries;
  (c) the class marks;
  (d) the class interval of the distribution.

2.7 The class marks of a distribution of temperature readings (given to the nearest degree Celsius) are 16, 25, 34, 43, 52, and 61. Find
  (a) the class boundaries;
  (b) the class limits.

2.8 To group data on the number of rainy days reported by a weather station for the month of July during the last fifty years, a meteorologist uses the classes 0–5, 6–11, 12–16, 18–24, and 24–30. Explain where difficulties might arise.

2.9 The following are the grades which 50 students obtained in a psychology test:

75	89	66	52	90	68	83	94	77	60
38	47	87	65	97	49	65	70	73	81
85	77	83	56	63	79	69	82	84	70
62	75	29	88	74	37	81	76	74	63
69	73	91	87	76	58	63	60	71	82

Group these grades into a distribution having the classes 20–29, 30–39, 40–49, . . . , and 90–99.

2.10 Convert the distribution of the preceding exercise into a percentage distribution.

2.11 Convert the distribution of Exercise 2.9 into a cumulative "less than" distribution, beginning with "less than 20."

2.12 The following are the body weights (in grams) of 60 rats used in a study of vitamin deficiencies:

125	128	106	111	116	123	119	114	117	143
136	92	115	118	121	137	132	120	104	125
119	115	101	129	87	108	110	133	135	126
127	103	110	126	118	82	104	137	120	95
146	126	119	119	105	132	126	118	100	113
106	125	117	102	146	129	124	113	95	148

Group these weights into a distribution having the classes 80–89, 90–99, 100–109, . . . , and 140–149.

2.13 Convert the distribution of the preceding exercise into a percentage distribution, rounding the percentages to one decimal.

2.14 Convert the distribution of Exercise 2.12 into an "or more" cumulative percentage distribution.

2.15 The following are the miles per gallon obtained with 40 tankfuls of gas:

24.5	23.6	24.1	25.0	22.9	24.7	23.8	25.2	24.9	24.1
23.7	24.4	24.7	23.9	25.1	24.6	23.3	24.3	24.8	22.8
24.6	23.9	24.1	24.4	24.5	25.7	23.6	24.0	24.7	23.1
23.9	24.2	24.7	24.9	25.0	24.8	24.5	23.4	24.6	25.3

Group these figures into a distribution having the classes 22.5–22.9, 23.0–23.4, 23.5–23.9, 24.0–24.4, 24.5–24.9, 25.0–25.4, and 25.5–25.9.

2.16 Convert the distribution of the preceding exercise into a cumulative "or less" distribution.

2.17 A survey made at a resort city showed that 50 tourists arrived by the following means of transportation: car, train, plane, plane, plane, bus, train, car, car, car, plane, car, plane, train, car, car, bus, car, plane, plane, train, train, plane, plane, car, car, train, car, car, plane, car, car, plane, bus, plane, bus, car, plane, car, car, train, train, car, plane, bus, plane, car, car, train, and bus. Construct a categorical distribution showing the frequencies corresponding to the different means of transportation.

2.18 Asked to rate the maneuverability of a car as excellent, very good, good, fair, poor, or very poor, forty drivers responded as follows: very good, good, good, fair, excellent, good, good, good, very good, poor, good, good, good, good, very good, good, fair, good, good, very poor, very good, fair, good, good, excellent, very good, good, good, good, fair, fair, very good, good, very good, excellent, very good, fair, good, good, and very good. Construct a categorical distribution showing the frequencies corresponding to these ratings of the maneuverability of the car.

2.2
GRAPHICAL PRESENTATIONS

When frequency distributions are constructed primarily to condense large sets of data and display them in an "easy to digest" form, it is usually advisable to present them graphically. The most common form of graphical presentation of statistical data is the **histogram,** an example of which is shown in Figure 2.1. A histogram is constructed by representing the measurements or observations that are grouped (in Figure 2.1, the sulfur oxides emission data) on a horizontal scale, the class frequencies on a vertical scale,

FIGURE 2.1 *Histogram of the distribution of the emission data.*

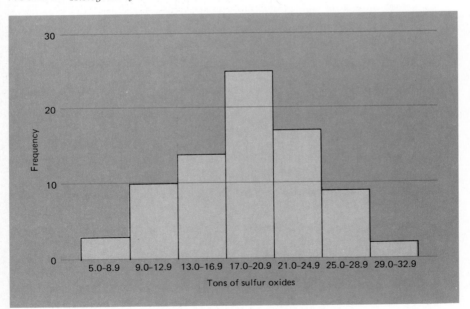

and drawing rectangles whose bases equal the class interval and whose heights are determined by the corresponding class frequencies. The markings on the horizontal scale can be the class limits as in Figure 2.1, the class boundaries, the class marks, or arbitrary key values. For easy readability, it is usually better to indicate the class limits, although the rectangles actually go from one class boundary to the next. Histograms cannot be used in connection with frequency distributions having open classes, and they must be used with extreme care if the class intervals are not all equal (see page 26).

Similar to histograms are **bar charts,** such as the one shown in Figure 2.2. The heights of the rectangles, or bars, again represent the class frequencies, but there is no pretense of having a continuous horizontal scale.

Another, less widely used form of graphical presentation is the **frequency polygon** (see Figure 2.3). Here the class frequencies are plotted at the class marks and the successive points are connected by means of straight lines. Note that we added classes with zero frequencies at both ends of the distribution to "tie down" the graph to the horizontal scale. If we apply the same technique to a cumulative distribution, we obtain what is called an **ogive.** However, the cumulative frequencies are plotted at the class boundaries instead of the class marks—it stands to reason that the cumulative frequency corresponding, say, to "less than 13.0" should be plotted at 12.95, the class boundary, since "less than 13.0" actually includes everything up to 12.95.

FIGURE 2.2 *Bar chart of the distribution of the emission data.*

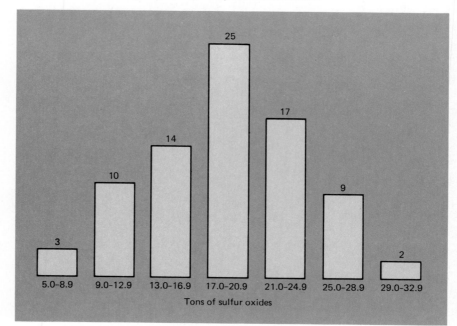

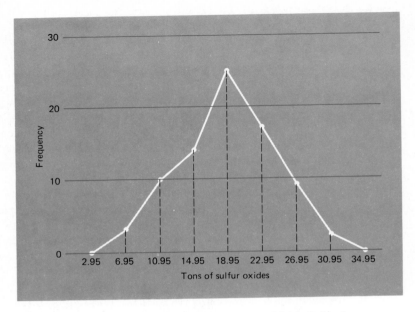

FIGURE 2.3 *Frequency polygon of the distribution of the emission data.*

FIGURE 2.4 *Ogive of the distribution of the emission data.*

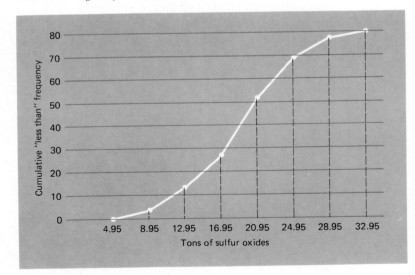

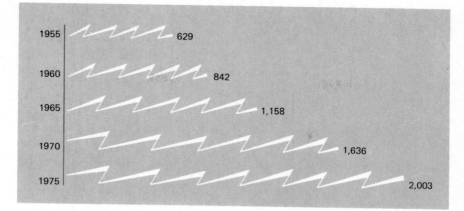

FIGURE 2.5 *Electric energy production in the United States (billions of kilowatt-hours).*

Figure 2.4 shows an ogive corresponding to the "less than" distribution of the emission data.

Although the visual appeal of histograms, frequency polygons, and ogives exceeds that of frequency tables, there are ways in which distributions can be presented even more dramatically and often more effectively. Two kinds of such pictorial presentations (often seen in newspapers, magazines, and reports of various sorts) are illustrated by the **pictograms** of Figures 2.5 and 2.6.

FIGURE 2.6 *Population of the United States.*

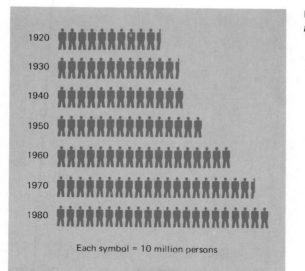

Categorical (or qualitative) distributions are often presented graphically as **pie charts** such as the one shown in Figure 2.7, where a circle is divided into sectors (pie-shaped pieces) which are proportional in size to the corresponding frequencies or percentages. To construct a pie chart we first convert the distribution into a percentage distribution. Then, since a complete circle corresponds to 360 degrees, we obtain the central angles of the various sectors by multiplying the percentages by 3.6.

Intentionally or unintentionally, frequency tables, histograms, and other pictorial presentations are sometimes very misleading. Suppose, for instance, that when we grouped the emission data we combined the two classes 17.0–20.9 and 21.0–24.9 into one class, the class 17.0–24.9. This new class has a frequency of 25 + 17 = 42, but in Figure 2.8, where we still use the heights of the rectangles to represent the class frequencies, we get the erroneous impression that this class contains about two-thirds of the data (instead of slightly more than one half). This is due to the fact that when we compare the sizes of rectangles, triangles, or other plane figures, we instinctively compare their areas and not their sides. This does not matter when the class intervals are all equal, but in Figure 2.8 the class 17.0–24.9 is twice as wide as the others, and we should compensate for this by dividing the height of the rectangle by 2. Figure 2.9 (where the vertical scale, which has lost its significance, has been omitted) shows the result of this adjustment. Now we get the correct impression that the class 17.0–24.9 contains just about half the data; 42 out of 80 to be exact.

FIGURE 2.7 *Federal budget receipts: 1978.* (*Source: Statistical Abstract of the United States, 100th ed.*)

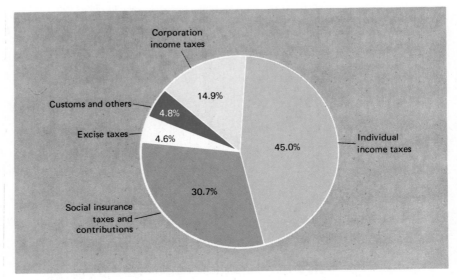

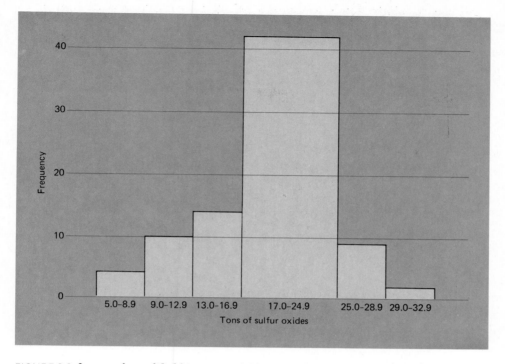

FIGURE 2.8 *Incorrectly modified histogram of the distribution of the emission data.*

FIGURE 2.9 *Correctly modified histogram of the distribution of the emission data.*

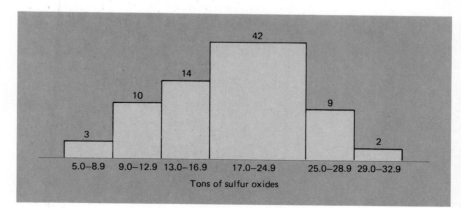

2.19 The following is the distribution of the total finance charges, to the nearest dollar, which 240 customers paid on their budget accounts at a department store:

Amount (dollars)	Frequency
0–19	16
20–39	78
40–59	77
60–79	54
80–99	15

Draw a histogram of this distribution.

2.20 Convert the distribution of the preceding exercise into a "less than" cumulative distribution and draw its ogive.

2.21 The following is the distribution of the numbers of mistakes 150 students made in translating a certain passage from French to English:

Number of mistakes	Number of students
10–14	5
15–19	57
20–24	42
25–29	28
30–34	17
35–39	1

Draw a bar chart of this distribution.

2.22 Convert the distribution of the preceding exercise into a "less than" cumulative percentage distribution and draw its ogive.

2.23 Draw a histogram of the weight distribution of Exercise 2.4 on page 20.

2.24 Convert the distribution of Exercise 2.4 on page 20 into a cumulative "or more" distribution and draw its ogive.

2.25 With reference to Exercise 2.21, modify the distribution so that it has the three classes 10–19, 20–24, and 25–39, and draw its histogram.

2.26 The following table shows how workers in Phoenix, Arizona, get to work:

Means of transportation	Percentage
Ride alone	81
Car pool	14
Ride bus	3
Varies or work at home	2

Construct a pie chart of this percentage distribution.

2.27 Draw a pie chart of the distribution obtained in Exercise 2.18 on page 22.

2.28 The pictogram of Figure 2.10 is intended to illustrate that in a certain region average family income has doubled from $7,000 in 1960 to $14,000 in 1975. Explain why this pictogram conveys a misleading impression and suggest how it might be modified.

$7,000 in 1960

$14,000 in 1975

FIGURE 2.10 *Pictogram for Exercise 2.28.*

2.3
STEM-AND-LEAF PLOTS

In the preceding sections we directed our attention to the grouping of mass data, with the objective of putting such data into a manageable form. As we saw, this entailed some loss of information. In recent years, similar techniques have been proposed for the preliminary exploration of relatively small sets of data which yield a good overall picture of the data without any loss of information.

To illustrate, consider the following scores which twenty students got in a history test:

| 69 | 84 | 52 | 93 | 61 | 74 | 79 | 65 | 88 | 63 |
| 57 | 64 | 67 | 72 | 74 | 55 | 82 | 61 | 68 | 77 |

Proceeding as in Section 2.1, we might group these data into the following distribution:

Scores	Tally	Frequency
50–59	///	3
60–69	##// ///	8
70–79	##//	5
80–89	///	3
90–99	/	1

where the tally pictures the overall pattern of the data like a histogram (or bar chart) lying on its side.

If we wanted to avoid the loss of information inherent in the above table, we could replace the tally marks with the last digits of the corresponding scores, getting

```
50–59 | 2  7  5
60–69 | 9  1  5  3  4  7  1  8
70–79 | 4  9  2  4  7
80–89 | 4  8  2
90–99 | 3
```

This can also be written as

```
5* | 2  7  5
6* | 9  1  5  3  4  7  1  8
7* | 4  9  2  4  7
8* | 4  8  2
9* | 3
```

where * is a placeholder for 0, 1, 2, 3, 4, 5, 6, 7, 8, or 9, or simply as

```
5 | 2  7  5
6 | 9  1  5  3  4  7  1  8
7 | 4  9  2  4  7
8 | 4  8  2
9 | 3
```

In either of these final forms, the table is referred to as a **stem-and-leaf plot** — each line is a **stem** and each digit on a stem to the right of the vertical line is a **leaf**. To the left of the vertical line are the **stem labels,** which, in our example, are 5*, 6*, ..., and 9*, or 5, 6, ..., and 9.

Essentially, a stem-and-leaf plot presents the same picture as the corresponding tally, yet it retains all the original information. For instance, if a stem-and-leaf plot has the stem

$$12* \ | \ 3 \quad 5 \quad 2 \quad 0 \quad 8$$

the corresponding data are 123, 125, 122, 120, and 128, and if a stem-and-leaf plot has the stem

$$3** \quad | \quad 17, 03, 55, 89$$

with two-digit leaves, the corresponding data are 317, 303, 355, and 389.

There are various ways in which stem-and-leaf plots can be modified to meet particular needs (see Exercises 2.33 and 2.43), but we shall not go into this here in any detail as it has been our objective only to present one of the relatively new techniques which come under the general heading of **exploratory data analysis.** Let us point out, though, that in a stem-and-leaf plot the data are partially ordered—they are ordered with respect to the stem labels—and as we shall see in Chapter 3, this simplifies the determination of certain further descriptions.

EXERCISES

2.29   The following are the IQ's of sixteen high school students: 120, 105, 112, 108, 102, 117, 100, 108, 103, 107, 115, 143, 98, 126, 103, and 114. Construct a stem-and-leaf plot with the stem labels 9, 10, . . . , and 14.

2.30   The following are the weights of twenty-four applicants for jobs with a city's fire department: 216, 170, 194, 212, 194, 205, 186, 190, 181, 198, 204, 223, 169, 226, 196, 175, 207, 183, 199, 187, 203, 218, 187, and 192. Construct a stem-and-leaf plot with one-digit leaves.

2.31   The following are the weekly earnings (in dollars) of fifteen salespersons: 305, 255, 319, 167, 270, 291, 512, 283, 334, 362, 188, 217, 440, 195, and 408. Construct a stem-and-leaf plot with the stem labels 1, 2, 3, 4, and 5 (and, hence, with two-digit leaves).

2.32   List the data which correspond to the following stems of stem-and-leaf plots:
(a) 1* | 0   2   7   5   1   1   8
(b) 12 | 5   3   3   0   2
(c) 3** | 45, 18, 66, 01
(d) 1.5 | 0   7   2   2   9

2.33   If we want to construct a stem-and-leaf plot with more stems than there would be otherwise, we might use * as a placeholder for 0, 1, 2, 3, and 4, and · as a placeholder for 5, 6, 7, 8, and 9. For the data on page 29 we would thus get the **double-stem plot**

5*	2
5·	7 5
6*	1 3 4 1
6·	9 5 7 8
7*	4 2 4
7·	9 7
8*	4 2
8·	8
9*	3

where we doubled the number of stems by cutting the interval covered by each stem in half.

(a) Construct a double-stem plot with one-digit leaves for the data of Exercise 2.30.

(b) The following are the ages of thirty heads of household in a retirement community: 68, 81, 61, 62, 76, 65, 69, 73, 78, 60, 64, 74, 57, 70, 68, 66, 83, 71, 59, 66, 61, 60, 85, 72, 76, 65, 67, 73, 72, and 67. Construct a double-stem plot with one-digit leaves.

2.4
CHECKLIST OF KEY TERMS
(with page references to their definitions)

2.5
REVIEW EXERCISES

2.34 The class marks of a distribution of the daily number of calls received by a small cab company are 22, 27, 32, 37, 42, 47, and 52. If the class intervals are all equal, what are the class limits?

2.35 The following are the numbers of deer observed in 72 sections of land in a wildlife count:

18	8	9	22	12	16	20	33	15	21	18	13
13	19	0	2	14	17	11	18	16	13	12	6
8	12	13	21	8	11	19	1	14	4	19	16
2	16	11	18	10	28	15	24	8	20	6	7
21	0	16	12	20	17	13	20	10	16	5	10
15	10	16	14	29	17	4	18	21	10	16	9

Group these data into a distribution having the classes 0–4, 5–9, 10–14; 15–19, 20–24, 25–29, and 30–34.

2.36 Draw a histogram of the distribution of the preceding exercise.

2.37 Convert the distribution of Exercise 2.35 into a cumulative "less than" distribution and draw its ogive.

2.38 The following are the systolic blood pressures of twenty hospital patients: 165, 135, 151, 153, 155, 182, 142, 158, 146, 149, 124, 162, 173, 204, 159, 130, 177, 162, 141, and 156. Construct a stem-and-leaf plot with the stem labels 12, 13, ..., and 20.

2.39 The following is the distribution of the number of meals which 60 real estate salespersons charged as business expenses in a given week:

Number of meals	Frequency
0–1	16
2–3	25
4–5	13
6–7	4
8–9	2

Find

   (a) the class marks;
   (b) the class boundaries;
   (c) the class interval of the distribution.

2.40 Convert the distribution of the preceding exercise into an "or less" cumulative percentage distribution and draw its ogive.

2.41 Asked whether they ever accept social invitations from their students, 40 tennis pros replied: occasionally, rarely, rarely, never, rarely, frequently, occasionally, occasionally, never, rarely, rarely, never, occasionally, occasionally, frequently, occasionally, rarely, rarely, occasionally, never, occasionally, occasionally, never, never, rarely, rarely, occasionally, occasionally, frequently, occasionally, never, occasionally, rarely, rarely, never, frequently, occasionally, rarely, rarely, and occasionally. Construct a categorical distribution and draw its pie chart.

2.42 The pictogram of Figure 2.11 is intended to illustrate the fact that the total value of corporation stock held by U.S. life insurance companies tripled from 1960 to 1970. Explain why this pictogram conveys a misleading impression and suggest how it might be modified.

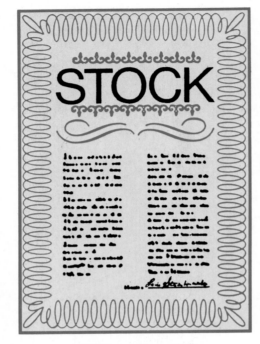

5 billion dollars in 1960   15 billion dollars in 1970

FIGURE 2.11 *Value of corporation stock held by U.S. life insurance companies.*

2.43 If we want to construct a stem-and-leaf plot equivalent to a distribution with a class interval of 2, we can use * as a placeholder for 0 and 1, *t* for 2 and 3, *f* for 4 and 5, *s* for 6 and 7, and · for 8 and 9. The resulting stem-and-leaf plot is called a **five-stem plot**.

(a) The following are the heights (in inches) of a sample of 18 college freshmen: 69, 71, 66, 66, 68, 75, 69, 67, 69, 68, 70, 72, 64, 70, 72, 66, 71, and 68. Construct a five-stem plot with the stem labels $6f$, $6s$, $6·$, $7*$, $7t$, and $7f$.

(b) The following is part of a five-stem plot:

$23f$	5 4 4 4 5 4
$23s$	6 7 6 6
$23·$	9 8
$24*$	1

List the corresponding measurements.

2.44 The daily number of persons attending an art exhibit are grouped into a distribution with the classes 0–39, 40–79, 80–119, and 120–159. Is it possible to determine from this distribution on how many days

    (a) at least 79 persons attended the exhibit;

    (b) more than 79 persons attended the exhibit;

    (c) 40 or more persons attended the exhibit;

    (d) at most 120 persons attended the exhibit?

2.45 The ages of a company's employees are to be grouped into the following classes: under 19, 20–24, 25–29, 30–34, 34–39, and over 39. Explain where difficulties might arise.

2.46 In 1978 there were 78 major opera companies in the United States, 458 community opera companies, and 420 opera workshops. Present this information in the form of a

    (a) bar chart;

    (b) pie chart.

2.47 The annual salaries paid to teachers in a certain school district vary from \$13,200 to \$27,700. Indicate the limits of five classes into which these figures might be grouped.

★2.48 Among histograms, bar charts, and pie charts, which ones can be used to represent

    (a) nominal data;

    (b) ordinal data;

    (c) interval data?

2.49 Measurements of the lengths of fish to the nearest tenth of an inch are grouped into a table whose classes have the boundaries 5.95, 7.95, 9.95, 11.95, 13.95, and 15.95. What are the lower and upper limits of each class?

2.50 List the data which correspond to the following stems of stem-and-leaf plots:

    (a) 125 | 3 0 4 8 7 6 6 5

    (b) 34 | 67, 05, 19, 48

2.6
REFERENCES

Discussions of what not to do in the presentation of statistical data may be found in

CAMPBELL, S. K., *Flaws and Fallacies in Statistical Thinking*. Englewood Cliffs, N.J.: Prentice-Hall, Inc., 1974.

HUFF, D., *How to Lie with Statistics*. New York: W. W. Norton & Company, Inc., 1954.

REICHMAN, W. J., *Use and Abuse of Statistics*. New York: Penguin Books, 1971.

Useful references to lists of standard categories may be found in

HAUSER, P. M., and LEONARD, W. R., *Government Statistics for Business Use, 2nd ed.* New York: John Wiley & Sons, Inc., 1956.

For further information about exploratory data analysis, and stem-and-leaf plots in particular, see

HARTWIG, F., and DEARING, B. E., *Exploratory Data Analysis.* Beverly Hills: Sage Publications, Inc., 1979.

KOOPMANS, L. H., *An Introduction to Contemporary Statistics.* Boston: Duxbury Press, 1981.

TUKEY, J. W., *Exploratory Data Analysis.* Reading, Mass.: Addison-Wesley Publishing Company, Inc., 1977.

When we describe sets of data we try to say neither too little nor too much. So, depending on the purpose they are to serve, statistical descriptions can be very brief or very elaborate. Sometimes it may be satisfactory to present data just as they are, in raw form, and let them speak for themselves; on other occasions it may be necessary only to group the data and present their distribution in tabular or graphical form. Usually, though, data have to be summarized further, and in this chapter we shall concentrate on the two most widely used kinds of statistical descriptions, called measures of location and measures of variation. The former are discussed in Sections 3.2 through 3.4, and the latter in Sections 3.5 through 3.8. The description of grouped data is treated in Section 3.9, and some further kinds of descriptions are presented in Section 3.10.

3 SUMMARIZING DATA: STATISTICAL DESCRIPTIONS

3.1
POPULATIONS AND SAMPLES

Before we study particular statistical descriptions, let us make the following distinction: if a set of data consists of all conceivably possible (or hypothetically possible) observations of a certain phenomenon, we call it a **population;** if a set of data contains only a part of these observations, we call it a **sample.** We added the phrase "hypothetically possible" to take care of such clearly hypothetical situations as where we look at the outcomes (heads or tails) of 12 flips of a coin as a sample from the potentially unlimited number of flips of the coin, where we look at the weights of ten 30-day-old lambs as a sample of the weights of all (past, present, or future) 30-day-old lambs, or where we look at four determinations of the uranium content of an ore as a sample of all possible determinations of the uranium content of the ore. In fact, we often look at the results of an experiment as a sample of what we might obtain if the experiment were repeated over and over again.

On page 5 we said that statistics dealt originally with the description of human populations (political units), but as it grew in scope, the term "population" took on the much wider connotation given to it above. Whether or not it sounds strange to refer to the heights of all the trees in a forest or the speeds of all the cars passing a checkpoint as populations is beside the point—in statistics, "population" is a technical term with a meaning of its own.

Although we are free to call any group of items a population, what we do in practice depends on the context in which the items are to be viewed. Suppose, for instance, that we are offered a lot of 400 ceramic tiles, which we may or may not buy depending on their strength. If we measure the breaking strength of 20 of these tiles in order to estimate the average breaking strength of all the tiles, these 20 measurements are a sample from the population which consists of the breaking strengths of the 400 tiles. In another context, however, if we consider entering into a long-term contract calling for the delivery of tens of thousands of such tiles, we would look upon the breaking strengths of the original 400 tiles only as a sample. Similarly, the complete figures for a recent year, giving the elapsed times between the filing and disposition of divorce suits in San Diego County, can be looked upon as either a population or a sample. If we are interested only in San Diego County and that particular year, we would look upon the data as a population; on the other hand, if we want to generalize about the time that is required for the disposition of divorce suits in the entire United States, in some other county, or in some other year, we would look upon the data as a sample.

As we have used it here, the word "sample" has very much the same meaning as it has in everyday language. A newspaper considers the attitudes of 150 readers toward a proposed school bond to be a sample of the attitudes of all its readers toward the bond; and a consumer considers a box of Mrs. See's candy a sample of the firm's product. Later, we shall use the word "sample" only when referring to data which can reasonably serve as the basis for valid generalizations about the populations from which they came; in this more technical sense, many sets of data which are popularly called samples are not samples at all.

In this chapter we shall describe things statistically but not make any generalizations. For future reference, though, it is important even here to distinguish between populations and samples. Thus, we shall use different symbols depending on whether we are describing populations or samples, and sometimes even different formulas.

3.2
MEASURES OF LOCATION

It is often necessary to represent a set of data by means of a single number which, in its way, is descriptive of the entire set. Exactly what sort of number we choose depends on the particular characteristic we want to describe. In one study we may be interested in an extreme (smallest or largest) value among the data; in another, in the value which is exceeded by only 10 percent of the data; and in still another, in the total of all the values. In the next section, we shall consider several measures which somehow describe the center or middle of a set of data—appropriately, they are called **measures of location,** or more specifically, **measures of central location.**

3.3
THE MEAN, THE MEDIAN, AND THE MODE

Among the different measures of central location, by far the best known and the most widely used is the **arithmetic mean,** or simply the **mean,** which we define as follows:

The mean of a set of values is the sum of the values divided by their number.

In everyday language, the mean is often called the "average," and on occasion we shall call it that ourselves. As we shall see, however, there are other "averages" in statistics, and we cannot afford to speak loosely when there is any danger of ambiguity.

EXAMPLE For the 12 months of 1982, a police department reported 3, 2, 4, 4, 9, 7, 8, 5, 2, 3, 7, and 6 armed robberies. Find the mean, namely, the average number of armed robberies per month.

Solution The total for the 12 months is $3 + 2 + 4 + 4 + 9 + 7 + 8 + 5 + 2 + 3 + 7 + 6 = 60$, and therefore

$$\text{mean} = \frac{60}{12} = 5$$

EXAMPLE Given that the 89th through 93rd Congresses of the United States enacted 1,283, 1,002, 941, 768, and 295 measures (bills, acts, or resolutions), find the mean.

Solution Altogether, these five congresses enacted $1,283 + 1,002 + 941 + 768 + 295 = 4,289$ measures, so that "on the average" they passed

$$\text{mean} = \frac{4,289}{5} = 857.8$$

Since we shall have occasion to calculate the means of many different sets of sample data, it will be convenient to have a simple formula which is always applicable. This requires that we represent the figures to be averaged by some general symbol such as x, y, or z; the number of values in a sample, the **sample size,** is usually denoted by the letter n. Choosing the letter x, we can refer to the n values in a sample as x_1 (which is read "x sub-one"), $x_2, x_3, \ldots$, and x_n, and write

$$\text{sample mean} = \frac{x_1 + x_2 + x_3 + \cdots + x_n}{n}$$

This formula is perfectly general and it will take care of any set of sample data, but it can be made more compact by assigning the sample mean the symbol $\bar{x}$ (which is read "x bar") and using the $\sum$ notation. The symbol $\sum$ is capital *sigma*, the Greek letter for S. In this notation, we let $\sum x$ stand for "the sum of the x's" (that is, for the sum $x_1 + x_2 + x_3 + \cdots + x_n$), and we can write

Sample mean

$$\bar{x} = \frac{\sum x}{n}$$

If we refer to the measurements as y's or z's, we write their mean as $\bar{y}$ or $\bar{z}$. In the formula for $\bar{x}$, the term $\sum x$ does not state explicitly which values of x are to be added; let it be understood, however, that $\sum x$ always refers to the sum of all the x's under consideration in a given situation. In Section 3.11 on page 74, the use of the sigma notation is discussed in more detail.

The mean of a population of N items is defined in the same way. It is the sum of the N items, $x_1 + x_2 + x_3 + \cdots + x_N$, or $\sum x$, divided by the **population size** N. Assigning the population mean the symbol μ (*mu*, the Greek letter for lowercase *m*), we write

Population mean

$$\mu = \frac{\sum x}{N}$$

with the reminder that $\sum x$ is now the sum of all N values of x which constitute the population.

Also, to distinguish between descriptions of populations and descriptions of samples, we not only use different symbols such as μ and $\bar{x}$, but we refer to descriptions of populations as **parameters** and descriptions of samples as **statistics**. Parameters are usually denoted by Greek letters.

To illustrate the terminology and notation introduced in this section, suppose we are interested in the mean lifetime of a production lot (considered to be a population) of $N = 40{,}000$ light bulbs. Obviously, we cannot test all of the light bulbs for there would be none left to use or sell, so we take a sample, calculate $\bar{x}$, and use this quantity to estimate μ. If $n = 5$ and the light bulbs in the sample last 967, 949, 940, 952, and 922 hours, we have

$$\bar{x} = \frac{967 + 949 + 940 + 952 + 922}{5} = 946 \text{ hours}$$

If these lifetimes constitute a sample in the technical sense (that is, a set of data from which valid generalizations can be made), we can estimate the mean lifetime μ of all the 40,000 light bulbs as 946 hours.

For non-negative data, the mean not only describes the middle of a set of data, but it also puts some limitation on their size. If we multiply by n on both sides of the equation $\bar{x} = \dfrac{\sum x}{n}$, it follows that $\sum x$ equals $n \cdot \bar{x}$ and hence cannot exceed it.

EXAMPLE If the mean annual salary paid to the top three executives of a firm is \$96,000, can one of them receive an annual salary of \$300,000?

Solution Since $n = 3$ and $\bar{x} = \$96{,}000$, we get $\sum x = 3 \cdot 96{,}000 = \$288{,}000$, and it is impossible for any one of the executives to receive more than that.

EXAMPLE If nine high school juniors averaged 41 on the verbal part of the PSAT/NMSQT test, at most how many of them can have scored 65 or more?

Solution Since $n = 9$ and $\bar{x} = 41$, we get $\sum x = 9 \cdot 41 = 369$, and since 65 goes into 369 at most five times ($369 = 5 \cdot 65 + 44$), it follows that at most five of these high school juniors can have scored 65 or more.

The popularity of the mean as a measure of the "middle" or "center" of a set of data is not accidental. Any time we use a single number to describe some aspect of a set of data, there are certain requirements, or desirable features, that should be kept in mind. Aside from the fact that the mean is a simple and familiar measure, the following are some of its noteworthy properties:

It can be calculated for any set of numerical data, so it always exists.

A set of numerical data has one and only one mean, so it is always unique.

It lends itself to further statistical treatment; as we shall see, for example, the means of several sets of data can always be combined into the overall mean of all the data.

It is relatively reliable in the sense that means of many samples drawn from the same population usually do not fluctuate, or vary, as widely as other statistical measures used to estimate the mean of a population.

The last of these properties is of fundamental importance in statistical inference, and we shall study it in some detail in Chapter 9.

There is another property of the mean which, on the surface, seems desirable:

It takes into account every item of a set of data.

However, sometimes samples contain very small or very large values which are so far removed from the main body of the data that the appropriateness of including them in a sample is questionable. Such values may be due to chance, or they may be due to gross errors in recording the data, gross errors in calculations, malfunctioning of equipment, or other identifiable sources of contamination. In any case, when such values are averaged in with the other values, they can affect the mean to such an extent that it is debatable whether it really provides a useful description of the "middle" of the data.

EXAMPLE With reference to the illustrations dealing with the light bulbs on page 41, suppose that the second value is recorded incorrectly as 499 instead of 949. Find the error this would cause in the mean lifetime of the five light bulbs.

Solution The mean of 967, 499, 940, 952, and 922 is

$$\bar{x} = \frac{967 + 499 + 940 + 952 + 922}{5} = 856$$

and this differs from 946, the mean we got on page 41, by $946 - 856 = 90$ hours.

EXAMPLE The ages of six students who went on a geology field trip are 18, 19, 20, 17, 19, and 18, and the age of the instructor who went with them is 50. Find the mean age of these seven persons.

Solution The mean is

$$\bar{x} = \frac{18 + 19 + 20 + 17 + 19 + 18 + 50}{7} = 23$$

but any statement to the effect that the mean age of the group is 23 could easily be misinterpreted.

To avoid the possibility of being misled by very small or very large values, we sometimes describe the "middle" or "center" of a set of data with statistical measures other than the mean. One of these, the **median**, is defined as follows:

The median of a set of data is the value of the middle item, or the mean of the values of the two middle items, when the data are arranged according to size.

The symbol we use for the median of n sample values $x_1, x_2, x_3, \ldots$, and x_n is $\tilde{x}$; if a set of data constitutes a population, we denote its median by $\tilde{\mu}$.

EXAMPLE Nine corporations reported that in 1982 they made cash donations to 9, 16, 11, 10, 13, 12, 6, 9, and 12 colleges. Find the median number of donations.

Solution Arranging these figures according to size, we get

6   9   9   10   11   12   12   13   16

and it can be seen that the median is 11.

EXAMPLE Among groups of 40 students interviewed at each of ten different colleges, 18, 13, 15, 12, 8, 3, 7, 14, 16, and 3 said that they jog regularly. Find the median.

Solution Arranging these figures according to size, we get

$$3 \quad 3 \quad 7 \quad 8 \quad 12 \quad 13 \quad 14 \quad 15 \quad 16 \quad 18$$

and it can be seen that the median is $\dfrac{12 + 13}{2} = 12.5$, the mean of 12 and 13.

In general, if there are n items (measurements or observations) and n is odd, the median is the value of the $\dfrac{n + 1}{2}$ th largest item. For instance, the median of $n = 25$ items is the value of the $\dfrac{25 + 1}{2} = 13$th largest item, and the median of $n = 47$ items is the value of the $\dfrac{47 + 1}{2} = 24$th largest item.

If n is even, there is no single middle item and the median is defined as the mean of the two middle items, but we can still use the formula $\dfrac{n + 1}{2}$ to locate the median. For instance, if $n = 16$, then $\dfrac{16 + 1}{2} = 8.5$ and the median is the mean of the values of the 8th and 9th largest items; also, if $n = 50$, then $\dfrac{50 + 1}{2} = 25.5$, and the median is the mean of the values of the 25th and 26th largest items.

It is important to remember that the formula $\dfrac{n + 1}{2}$ is not a formula for the median itself; it merely tells us how many of the ordered values we have to count until we reach the item whose value is the median (or the two items whose values we have to average).

If we calculate the mean and the median of a set of data, their values will usually not be the same. For instance, the mean of 8, 6, 3, 15, 10, and 13 is $\dfrac{8 + 6 + 3 + 15 + 10 + 13}{6} = 9\frac{1}{6}$, and the median is 9. Each of these averages describes the "middle" or "center" of the data in its own way. The median is average in the sense that it splits the data into two parts so that, unless there are duplicates, as many of the values are less than the median as are greater than the median. The mean, on the other hand, is typical in the sense of a center of gravity, as is illustrated by Figure 3.1. To put it differently, the mean is typical in the sense that if each value in a set of

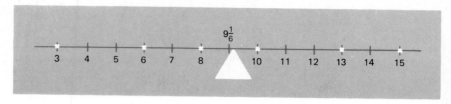

FIGURE 3.1 *The mean as a center of gravity.*

data is replaced by the same number while the total remains unchanged, this number will have to be the mean. This follows directly from the fact that $n \cdot \bar{x} = \sum x$.

Like the mean, the median always exists and is unique for any set of data. Also like the mean, the median is simple enough to find once the data have been arranged according to size, but ordering large sets of data manually can be very tedious. To simplify this task, it may help to utilize the partial ordering of a stem-and-leaf plot.

EXAMPLE   The following are the numbers of passengers on 25 runs of a ferry-boat: 52, 84, 40, 57, 61, 65, 77, 64, 62, 35, 82, 58, 50, 78, 103, 71, 75, 41, 53, 66, 60, 95, 58, 49, and 89. Find the median.

Solution   First we construct a stem-and-leaf plot with one-digit leaves, getting

$$
\begin{array}{r|llllll}
3 & 5 \\
4 & 0 & 1 & 9 \\
5 & 2 & 7 & 8 & 0 & 3 & 8 \\
6 & 1 & 5 & 4 & 2 & 6 & 0 \\
7 & 7 & 8 & 1 & 5 \\
8 & 4 & 2 & 9 \\
9 & 5 \\
10 & 3
\end{array}
$$

Since ten of the values fall on the first three stems, the $\dfrac{25 + 1}{2} = 13$th value, the median, is the third smallest value on the fourth stem. As can be seen by inspection, it is 62.

Unlike the mean, the median is not so easily affected by extreme values. For instance, on page 41 we showed that the mean of 967, 949, 940, 952, and 922 (the lifetimes of five light bulbs) is 946, and on page 43 we showed that if we misread 949 as 499, the mean becomes 856, which is off by 946 − 856 = 90 hours. Now, if we use the median instead of the mean, we find that for the original data the median is 949, when 949 is misread as 499 the median is 940, so that the error is only $949 - 940 = 9$.

Also unlike the mean, the median can be used to define the middle of a number of objects, properties, or qualities which can be ranked, namely, when we deal with ordinal data. For instance, we might rank a number of tasks according to their difficulty and then describe the middle (or median) one as being of "average" difficulty; also, we might rank samples of chocolate fudge according to their consistency and then describe the middle (or median) one as having "average" consistency.

Perhaps the most important difference between the median and the mean is that in problems of statistical inference the mean is usually more reliable than the median. This is meant to say that the medians of many samples drawn from the same population will usually vary more widely than the corresponding sample means (see Exercise 3.19 on page 51, Exercise 3.40 on page 62, and Exercise 9.42 on page 62).

Another measure that is sometimes used to describe the "middle" of a set of data is the **mode,** which is defined simply as the value which occurs with the highest frequency. Its two main advantages are that it requires no calculation, only counting, and it can be determined even for qualitative, or nominal, data.

EXAMPLE   The 20 meetings of a square dance club were attended by 26, 25, 28, 23, 25, 24, 24, 21, 23, 26, 28, 26, 24, 32, 25, 27, 24, 23, 24, and 22 of its members. Find the mode.

Solution   Among the twenty numbers, 21, 22, 27, and 32 each occurs once; 28 occurs twice; 23, 25, and 26 each occurs three times; and 24 occurs five times. Thus, 24 is the **modal attendance.**

Also, if more visitors to California want to see Disneyland than any other tourist attraction, we say that Disneyland is their **modal choice.**

Aside from the fact that the mode is seldom of any use in statistical inference, it also has the disadvantage that it may not exist (which is the case when no two values are alike) or that it may not be unique.

EXAMPLE   A sample of the records of a motor vehicle bureau shows that 18 drivers in a certain age group received 3, 2, 0, 0, 2, 3, 3, 1, 0, 1, 0, 3, 4, 0, 3, 2, 3, and 0 traffic tickets during the last three years. Find the mode.

Solution   As can be seen, the number 4 occurs once, the number 1 occurs twice, the number 2 occurs three times, and the numbers 0 and 3 each occur six times. So, there are two modes 0 and 3.

The presence of more than one mode is sometimes indicative of the fact that the data are not homogeneous, namely, that they can be looked upon as a combination of several distinct sets of data. Thus, in the preceding example we might infer that there are many very good drivers and many very poor drivers, while fewer drivers fall into the categories between these two extremes.

There are many other measures of central location besides the mean, the median, and the mode, and some of them will be introduced in the exercises that follow Section 3.4. The question of what particular "average" should be used in a given situation is not always easily answered, and the fact that there is a good deal of arbitrariness in the selection of statistical descriptions has led some persons to believe that the magic of statistics can be used to prove almost anything. A famous nineteenth-century British statesman said that there are three kinds of lies: lies, damned lies, and statistics, and Exercises 3.17 and 3.18 on pages 50 and 51 describe a situation where this kind of criticism would be justified.

3.4
THE WEIGHTED MEAN ★

When averaging quantities, it is often necessary to account for the fact that not all of them are equally important in the phenomenon being described. For instance, in 1978, clams, crabs, lobsters, scallops, and shrimp brought commercial fishermen 124.3, 20.6, 177.1, 253.5, and 210.9 cents per pound. The mean of these figures is 157.28 cents, but we cannot very well say that this is the average price per pound which commercial fishermen received for these kinds of shellfish—the five figures do not carry equal weight because there are vast differences in the size of the catch.

In order to give quantities being averaged their proper degree of importance, it is necessary to assign them (relative importance) **weights,** and then calculate a **weighted mean.** In general, the weighted mean $\bar{x}_w$ of a set of numbers $x_1, x_2, x_3, \ldots,$ and x_n, whose relative importance is expressed numerically by a corresponding set of numbers $w_1, w_2, w_3, \ldots,$ and w_n, is given by

Weighted mean

$$\bar{x}_w = \frac{w_1 x_1 + w_2 x_2 + \cdots + w_n x_n}{w_1 + w_2 + \cdots + w_n} = \frac{\sum w \cdot x}{\sum w}$$

If all the weights are equal, this formula reduces to that of the ordinary (arithmetic) mean.

EXAMPLE   Given that in 1978 commercial fishermen caught 87.7 million pounds of clams, 449.1 million pounds of crabs, 39.0 million pounds of lobsters, 33.3 million pounds of scallops, and 434.8 million pounds of shrimp, use these weights and the prices in the text above to determine the average price per pound which commercial fishermen received for the five kinds of shellfish.

Solution   Substituting $x_1 = 124.3$, $x_2 = 20.6$, $x_3 = 177.1$, $x_4 = 253.5$, $x_5 = 210.9$, $w_1 = 87.7$, $w_2 = 449.1$, $w_3 = 39.0$, $w_4 = 33.3$, and $w_5 = 434.8$ into the formula for $\bar{x}_w$, we get

$$\bar{x}_w = \frac{(87.7)(124.3) + (449.1)(20.6) + (39.0)(177.1) + (33.3)(253.5) + (434.8)(210.9)}{87.7 + 449.1 + 39.0 + 33.3 + 434.8}$$

$$= \frac{127{,}200.34}{1{,}043.9}$$

$$= 121.85 \text{ cents per pound}$$

The figure in the denominator, 1,043.9, is the total catch in millions of pounds, and the figure in the numerator, 127,200.34, is the total value of the catch in millions of cents, that is, in units of $10,000. Note also that the value we obtained for $\bar{x}_w$ is much less than that of $\bar{x}$, 121.85 compared to 157.28, and this is due largely to the big catch of relatively less expensive crabs.

A special application of the formula for the weighted mean arises when we must find the overall mean, or **grand mean,** of k sets of data having the means $\bar{x}_1, \bar{x}_2, \bar{x}_3, \ldots,$ and $\bar{x}_k$, and consisting of $n_1, n_2, n_3, \ldots,$ and n_k measurements or observations. The result is given by

Grand mean of combined data

$$\bar{\bar{x}} = \frac{n_1\bar{x}_1 + n_2\bar{x}_2 + \cdots + n_k\bar{x}_k}{n_1 + n_2 + \cdots + n_k} = \frac{\sum n \cdot \bar{x}}{\sum n}$$

where the weights are the sizes of the samples, the numerator is the total of all the measurements or observations, and the denominator is the number of items in the combined samples.

EXAMPLE   In a psychology class there are 9 freshmen, 14 sophomores, 22 juniors, and 5 seniors. If the freshmen averaged 68 in the final examination, the sophomores averaged 75, the juniors averaged 83, and the seniors averaged 81, what is the mean grade for the entire class?

Solution   Substituting $\bar{x}_1 = 68$, $\bar{x}_2 = 75$, $\bar{x}_3 = 83$, $\bar{x}_4 = 81$, $n_1 = 9$, $n_2 = 14$, $n_3 = 22$, and $n_4 = 5$ into the formula for the grand mean of combined data, we get

$$\bar{\bar{x}} = \frac{9 \cdot 68 + 14 \cdot 75 + 22 \cdot 83 + 5 \cdot 81}{9 + 14 + 22 + 5}$$

$$= \frac{3,893}{50}$$

$$= 77.86$$

or 78 rounded to the nearest integer.

EXERCISES

3.1   The dean of a college has complete records on how many failing grades each faculty member gave to his or her students during the academic year 1981–1982. Give one example each of a situation in which these data would be looked upon as
  (a) a population;
  (b) a sample.

3.2   Suppose that the final election returns from a county show that the two candidates for a certain office received 14,283 and 12,695 votes. What offices might these candidates be running for so that these figures would constitute
  (a) a population;
  (b) a sample?

3.3   At their first inaugurations, the first ten presidents of the United States were 57, 61, 57, 57, 58, 57, 61, 54, 68, and 51 years old. Find the mean age of these presidents at their first inauguration.

3.4   The following are the ages of twenty persons empaneled for jury duty by a court: 49, 61, 53, 38, 47, 30, 63, 52, 34, 41, 58, 48, 42, 33, 31, 57, 25, 52, 60, and 46. Find the mean age.

3.5   Trying to determine the calories per serving of lasagna in a laboratory assignment in nutrition, fifteen students obtained 329, 335, 347, 318, 322, 330, 351, 362, 315, 342, 346, 353, 316, 327, and 333.
  (a) Find the mean.
  (b) Recalculate the mean by first subtracting 300 from each value, finding the mean of the numbers thus obtained, and then adding 300 to the result. What general simplification does this suggest for the calculation of a mean?

3.6   By mistake, an instructor erased the grade which one of ten students received. If the nine other students have grades of 48, 71, 79, 95, 45, 57, 75, 83, and 97, and the mean of all ten grades is 72, what grade did the instructor erase?

3.7   An elevator in a department store is designated to carry a maximum load of 3,000 pounds. Is it overloaded if at one time it carries
  (a) 18 passengers whose mean weight is 135 pounds;
  (b) 12 women whose mean weight is 123 pounds and 9 men whose mean weight is 175 pounds?

3.8 The mean weight of the 46 members of a football team is 212 pounds. If none of the players weighs less than 165 pounds, at most how many of them can weigh 250 pounds or more?

★3.9 Generalizing the argument of the examples on page 41, it can be shown that for any set of non-negative data with the mean $\bar{x}$, the fraction of the data that are greater than or equal to any positive constant k cannot exceed $\bar{x}/k$. Use this result, called **Markov's theorem**, in the following problems:
(a) If the mean breaking strength of certain linen threads is 32.5 ounces, at most what fraction of the threads can have a breaking strength of 40.0 ounces or more?
(b) If the diameters of the citrus trees in an orchard have a mean of 15.8 cm, at most what fraction of the trees can have a diameter of 25.0 cm or more?

3.10 Records show that in Phoenix, Arizona, the normal daily maximum temperature for each month is 65, 69, 74, 84, 93, 102, 105, 102, 98, 88, 74, and 66 degrees Fahrenheit. Verify that the mean of these figures is 85 and comment on the claim that, in Phoenix, the average daily maximum temperature is a very comfortable 85 degrees.

3.11 The following are the amounts of time (in minutes) which a person had to wait for the bus to work on fifteen working days: 15, 10, 2, 17, 6, 8, 3, 10, 2, 9, 5, 9, 13, 1, and 10. Determine
(a) the mean;
(b) the median;
(c) the mode.

3.12 The following are the weight losses (in pounds) of ten persons following a prescribed diet for two weeks: 4.3, 3.4, 3.8, 5.2, 4.4, 2.9, 3.7, 5.6, 4.1, and 3.6. Calculate
(a) the mean;
(b) the median.

3.13 The grades of ten students in a history test are 65, 77, 92, 71, 53, 73, 96, 51, 65, and 86. Find
(a) the median;
(b) the mode.

3.14 Twenty-five NBA games lasted, respectively, 138, 142, 113, 126, 135, 142, 159, 157, 140, 157, 121, 128, 142, 164, 155, 139, 143, 158, 140, 118, 142, 146, 123, 130, and 137 minutes. Determine the median
(a) directly;
(b) by first constructing a stem-and-leaf plot.

3.15 Find the mode (if it exists) of each of the following sets of blood pressure readings:
(a) 144, 145, 146, 146, 148, 146, 146, 145, 147, 145, 144;
(b) 146, 149, 146, 141, 146, 149, 147, 146, 149, 149, 145;
(c) 167, 151, 175, 144, 152, 148, 156, 169, 143, 177, 161.

3.16 Asked for their favorite color, thirty persons said: red, blue, blue, green, yellow, blue, brown, red, red, red, green, white, blue, red, yellow, blue, blue, red, green, yellow, blue, blue, orange, green, blue, blue, green, red, purple, and blue. What is their modal choice?

3.17 A consumer testing service obtained the following miles per gallon in five test runs performed with each of three compact cars:

$$Car\ A:\quad 27.9,\ 30.4,\ 30.6,\ 31.4,\ 31.7$$

$$Car\ B:\quad 31.2,\ 28.7,\ 31.3,\ 28.7,\ 31.3$$

$$Car\ C:\quad 28.6,\ 29.1,\ 28.5,\ 32.1,\ 29.7$$

    (a) If the manufacturers of car A want to advertize that their car performed best in this test, which of the "averages" discussed in this text could they use to substantiate their claim?

    (b) If the manufacturers of car B want to advertize that their car performed best in this test, which of the "averages" discussed in this text could they use to substantiate their claim?

★3.18 Suppose that the manufacturers of car C of the preceding exercise hire an unscrupulous statistician and instruct him to find some kind of "average" which will show that their car performed best in the test. Show that the **midrange**, the mean of the smallest and largest values, will serve their purpose.

3.19 To verify the claim that the mean is generally more reliable than the median (namely, that it is subject to smaller chance fluctuations), a student conducted an experiment consisting of 12 tosses of three dice. The following are his results: 2, 4, and 6; 5, 3, and 5; 4, 5, and 3; 5, 2, and 3; 6, 1, and 5; 3, 2, and 1; 3, 1, and 4; 5, 5, and 2; 3, 3, and 4; 1, 6, and 2; 3, 3, and 3; 4, 5, and 3.

    (a) Calculate the twelve medians and the twelve means.

    (b) Group the medians and the means obtained in (a) into separate distributions having the classes 1.5–2.5, 2.5–3.5, 3.5–4.5, and 4.5–5.5. (Note that there will be no ambiguities since the medians of three whole numbers and the means of three whole numbers cannot equal 2.5, 3.5, or 4.5.)

    (c) Draw histograms of the two distributions obtained in part (b) and explain how they illustrate the claim that the mean is generally more reliable than the median.

3.20 Repeat Exercise 3.19 with your own data by repeatedly rolling three dice (or one die three times) and construct corresponding distributions of the twelve medians and means. (If no dice are available, simulate the experiment mentally or by drawing numbered slips of paper out of a hat.)

★3.21 The **geometric mean** of n positive numbers is the nth root of their product. For example, the geometric mean of 3 and 12 is $\sqrt{3 \cdot 12} = \sqrt{36} = 6$ and the geometric mean of $\frac{1}{3}$, 1, and 81 is $\sqrt[3]{\frac{1}{3} \cdot 1 \cdot 81} = \sqrt[3]{27} = 3$. The geometric mean is used mainly to average ratios, rates of change, economic indexes, and the like.

    (a) Find the geometric mean of 8 and 32.

    (b) Find the geometric mean of 1, 2, 8, and 16.

    (c) During a recent flu epidemic, 12 cases were reported on the first day, 18 on the second day, and 48 on the third day. Thus, from the first day to the second day the number of cases reported was multiplied by $\frac{18}{12}$, and from the second day to the third day the number of cases was multiplied by $\frac{48}{18}$. Find the geometric mean of these two growth rates and (assuming that the growth pattern continues) predict the number of cases that will be reported on the fourth and fifth days.

In actual practice, geometric means are usually calculated by making use of the fact that the logarithm of the geometric mean of a set of positive numbers equals the arithmetic mean of their logarithms.

★3.22 The **harmonic mean** of n numbers $x_1, x_2, x_3, \ldots,$ and x_n is defined as n divided by the sum of the reciprocals of the n numbers, or $\dfrac{n}{\sum 1/x}$. The harmonic mean has limited usefulness, but it is appropriate in some special situations. For instance, if a commuter drives 10 miles on a freeway at 60 miles per hour and the next 10 miles off the freeway at 30 miles per hour, he will not have averaged $\dfrac{60 + 30}{2} = 45$ miles per hour. He will have driven 20 miles in a total of 30 minutes, so his average speed is 40 miles per hour.

(a) Verify that the harmonic mean of 60 and 30 is 40, so that it gives the appropriate "average" for this example.

(b) If an investor buys $9,000 worth of a company's stock at $45 a share and $9,000 worth at $36 a share, calculate the average price which the investor pays per share, and verify that it is the harmonic mean of $45 and $36.

(c) If a bakery buys $36 worth of an ingredient at 60 cents per pound, $36 worth at 72 cents per pound, and $36 worth at 90 cents per pound, what is the average cost per pound?

★3.23 An instructor counts the final examination in a course four times as much as each of the four one-hour examinations. What is the average grade of a student who received grades of 74, 80, 61, and 77 in the four one-hour examinations and 83 in the final examination?

★3.24 From 1970 to 1975 the cost of food increased by 53 percent in a certain city, the cost of housing increased by 40 percent, and the cost of transportation increased by 34 percent. If the average salaried worker spent 28 percent of his or her income on food, 25 percent on housing, and 14 percent on transportation, what is the combined percentage increase in the cost of these items?

★3.25 If a person invests $3,000 at 7 percent, $5,000 at 8 percent, and $12,000 at 10 percent, what is the average return on these investments?

★3.26 In three high schools, the average IQ's of the students are 105, 109, and 101. If there are 824, 486, and 590 students in the three schools, what is the average IQ of all the students in the three schools?

★3.27 In 1980 a college paid its 52 instructors a mean salary of $13,200, its 96 assistant professors a mean salary of $15,800, its 67 associate professors a mean salary of $18,900, and its 35 full professors a mean salary of $23,500. What was the mean salary paid to the 250 members of this faculty?

★3.28 In 1977, a sample survey yielded the following data on the average number of times per year that 200 persons in various age groups visit a doctor:

Age group	Number of persons in the sample	Mean number of visits
Under 6	20	6.5
6–16	55	2.9
17–44	68	5.0
45 and over	57	6.2

What is the mean for all 200 persons in the sample?

3.5
MEASURES OF VARIATION

One of the most important characteristics of almost any set of data is that the values are not all alike; indeed, the extent to which they are unalike, or vary among themselves, is of basic importance in statistics. Measures of central location describe one important aspect of a set of data—their "middle" or their "average"—but they tell us nothing about this other basic characteristic. Consequently, we require ways of measuring the extent to which data are dispersed, or spread out, and the statistical measures which provide this information are called **measures of variation.** The following are some examples which will illustrate the importance of measuring the variability of statistical data.

Suppose that in a hospital each patient's pulse rate is taken in the morning, at noon, and in the evening, and that on a certain day the pulse rate of patient A is 72, 76, and 74, while that of patient B is 72, 91, and 59. The mean pulse rates of the two patients are the same, 74, but observe the difference in variability. Whereas patient A's pulse rate is stable, that of patient B fluctuates widely.

Suppose that the manager of a supermarket is interested in stocking certain 1-pound bags of mixed nuts. The knowledge that on the average there are 12 almonds per bag leaves room for various possibilities. If most of the bags contain anywhere from 10 to 14 almonds, we would say that the product is quite consistent. The situation would be altogether different, however, if some of the bags have no almonds at all while others have 20 or more.

The concept of variability is of special importance in statistical inference. Suppose, for example, that we have a coin which is slightly bent and we wonder whether it is still balanced or "fair;" that is, whether it will still come up heads 50 percent of the time. So, we toss the coin 100 times and get 28 heads and 72 tails. Is there anything out of the ordinary about this result? Specifically, does the shortage of heads—only 28 where we might have expected 50—imply that the coin is not fair?

To answer this question, we must have some idea about the magnitude of the fluctuations, or variations, brought about by chance in the number of times a fair coin comes up heads when it is tossed 100 times. Suppose, thus, that we take a coin in mint condition, repeatedly toss it 100 times, and that in 10 sets of 100 tosses each we get 44, 59, 50, 53, 40, 46, 51, 48, 54, and 56 heads. Since the number of heads varies from 40 to 59, we might conclude that a shortage or excess of 10 heads from the expected 50 heads is not unusual, but that the shortage of 22 heads, which we got with the bent coin, is so large that we would be reluctant to attribute it to chance. In other words, it would seem reasonable to conclude that the bent coin is not fair.

We have given these three examples to show that the concept of variability plays an important role in the analysis of statistical data. Next, we shall see how it can actually be measured.

3.6
THE RANGE

To introduce one way of measuring variability, let us refer to the first of the three examples of the preceding section, and let us observe that the pulse rate of patient A varied from 72 to 76, while that of patient B varied from 59 to 91. These extreme (smallest and largest) values give an indication of the variability of the two sets of data, and just about the same information is conveyed if we take the differences between the respective extremes. For patient A we obtain a **range** of $76 - 72 = 4$ and for patient B we obtain a range of $91 - 59 = 32$. Also, for the air pollution data on page 16 the smallest value is 6.2, the largest value is 31.8, so that the range of the data is $31.8 - 6.2 = 25.6$; and for the lifetimes of the five light bulbs on page 41 the smallest value is 922, the largest value is 967, so that the range of the data is $967 - 922 \doteq 45$.

The range is easy to calculate and easy to understand, but despite these advantages it is generally not a very useful measure of variation. Its main shortcoming is that it tells us nothing about the dispersion of the values which fall between the two extremes. For instance, each of the following sets of data

Sample 1:	6	18	18	18	18	18	18	18	18	18
Sample 2:	6	6	6	6	6	18	18	18	18	18
Sample 3:	6	7	9	11	12	14	15	16	17	18

has a range of $18 - 6 = 12$, but the dispersion is quite different in each case.

In some cases, when the sample size is quite small, the range can be an adequate measure of variation. For instance, it is used widely in industrial quality control to keep a close check on the consistency of raw materials or products, or on the uniformity of a process, on the basis of small samples taken at regular intervals of time.

3.7
THE VARIANCE
AND THE STANDARD DEVIATION

To define the **standard deviation,** by far the most generally useful measure of variation, let us observe that the dispersion of a set of data is small if the values are closely bunched about their mean, and that it is large if the values are scattered widely about their mean. It would seem reasonable, therefore, to measure the variation of a set of data in terms of the amounts by which the values deviate from their mean. If a set of numbers $x_1, x_2, x_3, \ldots,$ and x_n, constituting a sample, has the mean $\bar{x}$, the differences $x_1 - \bar{x}$, $x_2 - \bar{x}$, $x_3 - \bar{x}, \ldots$, and $x_n - \bar{x}$ are called the **deviations from the mean**, and it suggests itself that we might use their average (namely, their mean) as a measure of the variation of the sample. Unfortunately, this will not do. Unless the x's are all equal, some of the deviations will be positive, some will be negative, and as the reader will be asked to show in Exercise 3.71 on page 76, the sum of the deviations from the mean, $\sum (x - \bar{x})$, and consequently also their mean, is always zero.

Since we are really interested in the magnitude of the deviations, and not in whether they are positive or negative, we might simply ignore the signs and define a measure of variation in terms of the absolute values of the deviations from the mean. Indeed, if we add the deviations from the mean as if they were all positive or zero and divide by n, we obtain the statistical measure which is called the **mean deviation**. This measure has intuitive appeal, but because of the absolute values it leads to serious difficulties, theoretical ones, in problems of inference.

An alternative approach is to work with the squares of the deviations from the mean, as this will also eliminate the effect of the signs. Squares of real numbers cannot be negative; in fact, squares of the deviations from a mean are all positive unless a value happens to coincide with the mean. Then, if we average the squared deviations from the mean and take the square root of the result (to compensate for the fact that the deviations were squared), we get

$$\sqrt{\frac{\sum (x - \bar{x})^2}{n}}$$

and this is how, traditionally, the standard deviation used to be defined. Expressing literally what we have done here mathematically, it is also called the **root-mean-square deviation**.

Nowadays, it is customary to modify this formula by dividing the sum of the squared deviations from the mean by $n - 1$ instead of n. Following this practice, which will be explained below, let us define the **sample standard deviation**, denoted by s, as

Sample standard deviation

$$s = \sqrt{\frac{\sum (x - \bar{x})^2}{n - 1}}$$

and its square, the **sample variance**, as

Sample variance

$$s^2 = \frac{\sum (x - \bar{x})^2}{n - 1}$$

To facilitate the calculation of standard deviations, a table of square roots is given at the end of the book.

These formulas for the standard deviation and the variance apply to samples, but if we substitute μ for $\bar{x}$ and N for n, we obtain analogous formulas for the standard deviation and the variance of a population. It has become fairly general practice to denote the population standard deviation by σ (*sigma*, the Greek letter for lowercase s) when dividing by N, and by S when dividing by $N - 1$. Thus, for σ we write

Population standard deviation

$$\sigma = \sqrt{\frac{\sum (x - \mu)^2}{N}}$$

Ordinarily, the purpose of calculating a sample statistic (such as the mean, the standard deviation, or the variance) is to estimate the corresponding population parameter. If we actually took many samples from a population which has the mean μ, calculated the sample means $\bar{x}$, and then averaged all these estimates of μ, we should find that their average is very close to μ. However, if we calculated the variance of each sample by means of the formula $\dfrac{\sum (x - \bar{x})^2}{n}$, and then averaged all these supposed estimates of σ^2, we would probably find that their average is less than σ^2. Theoretically, it can be shown that we can compensate for this by dividing by $n - 1$ instead of n in the formula for s^2. Estimators which have the desirable property that their values will on the average equal the quantity they are supposed to estimate are said to be **unbiased**; otherwise, they are said to be **biased**. So, we say that $\bar{x}$ is an unbiased estimator of the population mean μ, and that s^2 is an unbiased estimator of the population variance σ^2. It does not follow from this that s is also an unbiased estimator of σ, but when n is large the bias is small and can usually be ignored.

In calculating the sample standard deviation using the formula by which it is defined, we must (1) find $\bar{x}$, (2) determine the n deviations from the mean $x - \bar{x}$, (3) square these deviations, (4) add all the squared deviations, (5) divide by $n - 1$, and (6) take the square root of the result arrived at in step 5.

EXAMPLE   A bacteriologist found 6, 12, 9, 10, 6, and 8 microorganisms of a certain kind in six cultures. Calculate s.

Solution   First we calculate the mean

$$\bar{x} = \frac{6 + 12 + 9 + 10 + 6 + 8}{6} = 8.5$$

and then we set up the work required to find $\sum (x - \bar{x})^2$ in the following table:[†]

x	$x - \bar{x}$	$(x - \bar{x})^2$
6	−2.5	6.25
12	3.5	12.25
9	0.5	0.25
10	1.5	2.25
6	−2.5	6.25
8	−0.5	0.25
	0.0	27.50

Then we divide $\sum (x - \bar{x})^2 = 27.50$ by $6 - 1 = 5$ and take the square root. We thus get

$$s = \sqrt{\frac{27.50}{5}} = \sqrt{5.5} = 2.3$$

rounded to one decimal.

It was easy to calculate s in this example because the data were whole numbers and the mean was exact to one decimal. However, the calculations required by the formula defining s and s^2 can be quite tedious, and it is usually preferable to use the following computing formula, which can be derived by applying the rules for summations given in Section 3.11:

Computing formula for the sample standard deviation

$$s = \sqrt{\frac{n(\sum x^2) - (\sum x)^2}{n(n - 1)}}$$

EXAMPLE   Use the computing formula to rework the preceding example.

Solution   First we calculate the two sums

$$\sum x = 6 + 12 + 9 + 10 + 6 + 8 = 51$$

and

$$\sum x^2 = 6^2 + 12^2 + 9^2 + 10^2 + 6^2 + 8^2 = 461$$

[†] Note that the sum of the entries in the middle column is equal to zero. Since this must always be the case, it provides a check on the calculations.

Then, substituting these sums and $n = 6$ into the formula, we find that

$$s = \sqrt{\frac{6(461) - (51)^2}{6 \cdot 5}} = \sqrt{\frac{165}{30}} = \sqrt{5.5} = 2.3$$

rounded to one decimal.

The result which we obtained here is the same as before. Indeed, the computing formula for s gives the exact value of s, not an approximation, and its main advantage is that we do not have to work with the deviations from the mean—all we need is the sum of the x's and the sum of their squares. Aside from its advantage in manual calculations, the computing formula for s (or a slight modification of it) is the one usually preprogrammed into electronic statistical calculators.

Note that the computing formula for s can also be used to find σ, provided that we substitute n for the factor $n - 1$ in the denominator before we replace s and n with σ and N.

3.8
SOME APPLICATIONS
OF THE STANDARD DEVIATION

In subsequent chapters, sample standard deviations will be used mainly to estimate population standard deviations in problems of inference. To get more of a feeling for what a standard deviation really measures, we shall devote this section to some applications.

In the argument which led to the definition of the standard deviation, we observed that the dispersion of a set of data is small if the values are bunched closely about their mean, and that it is large if the values are scattered widely about their mean. Correspondingly, we can now say that if the standard deviation of a set of data is small, the values are concentrated near the mean, and if the standard deviation is large, the values are scattered widely about the mean. This idea is expressed more formally by the following theorem, called **Chebyshev's theorem** after the Russian mathematician P. L. Chebyshev (1821–1894):

Chebyshev's theorem

> *For any set of data (population or sample) and any constant k greater than 1, at least $1 - 1/k^2$ of the data must lie within k standard deviations on either side of the mean.*

Thus, we can be sure that at least $1 - \dfrac{1}{2^2} = \dfrac{3}{4}$, or 75 percent, of the values in any set of data must lie within two standard deviations on either side of the mean; at least $1 - \dfrac{1}{3^2} = \dfrac{8}{9}$, or 88.9 percent, must lie within three standard deviations on either side of the mean; and at least $1 - \dfrac{1}{5^2} = \dfrac{24}{25}$, or 96 percent, must lie within five standard deviations on either side of the mean. Here we arbitrarily let $k = 2$, 3, and 5.

EXAMPLE If all the 1-pound cans of coffee filled by a food processor have a mean weight of 16.00 ounces with a standard deviation of 0.02 ounce, at least what percentage of the cans must contain between 15.80 and 16.20 ounces of coffee?

Solution Since k standard deviations, or $k(0.02)$, equals $16.20 - 16.00 = 16.00 - 15.80 = 0.20$, we find that $k = \dfrac{0.20}{0.02} = 10$. Thus, at least $1 - \dfrac{1}{10^2} = 0.99$, or 99 percent, of the cans must contain between 15.80 and 16.20 ounces of coffee.

Chebyshev's theorem applies to any kind of data, but it tells us only "at least what percentage" must lie between certain limits. For distributions having the general shape of the cross section of a bell (see Figure 3.2), we can make the much stronger statement that

About 68 percent of the values will lie within one standard deviation of the mean, about 95 percent will lie within two standard deviations of the mean, and about 99.7 percent will lie within three standard deviations of the mean.

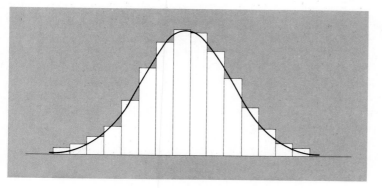

FIGURE 3.2 *Bell-shaped distribution.*

This result is sometimes referred to as the **empirical rule,** presumably because such percentages are observed in practice; actually, it is a theoretical result based on the normal distribution, which we shall study in Chapter 8.

In Section 3.5 we showed that there are many ways in which information about the variability of a set of data can be of importance. Another application arises in the comparison of numbers belonging to different sets of data. To illustrate, suppose that the final examination in a French course consists of two parts, vocabulary and grammar, and that a certain student scored 66 points in the vocabulary part and 80 points in the grammar part. At first glance it would seem that the student did much better in grammar than in vocabulary, but suppose that all the students in the class averaged 51 points in the vocabulary part with a standard deviation of 12, and 72 points in the grammar part with a standard deviation of 16. Thus, we can argue that the student's score in the vocabulary part is $\dfrac{66 - 51}{12} = 1.25$ standard deviations above the average for the class, while her score in the grammar part is only $\dfrac{80 - 72}{16} = 0.50$ standard deviation above the average for the class. Whereas the original scores cannot be meaningfully compared, these new scores, expressed in terms of standard deviations, can. Clearly, the given student rates much higher on her command of French vocabulary than on her knowledge of French grammar, compared to the rest of the class.

What we have done here consisted of converting the grades into **standard units** or **z-scores.** In general, if x is a measurement belonging to a set of data having the mean $\bar{x}$ (or μ) and the standard deviation s (or σ), then its value in standard units, denoted by z, is

Formula for converting to standard units

$$z = \frac{x - \bar{x}}{s} \qquad or \qquad z = \frac{x - \mu}{\sigma}$$

depending on whether the data constitute a sample or a population. In these units, z tells us how many standard deviations a value lies above or below the mean of the set of data to which it belongs. Standard units will be used frequently in later chapters.

One disadvantage of the standard deviation as a measure of variation is that it depends on the units of measurement. For instance, the weights of certain objects may have a standard deviation of 0.1 ounce or 2,835 milligrams, which is the same, but neither value really tells us whether it reflects a great deal of variation or very little variation. If the objects we are weighing are the eggs of small birds, either figure would reflect a great deal of variation, but this would not be the case if the objects we are weighing are 100-pound bags of potatoes. What we need in a situation like this is a **measure of relative variation,** such as the **coefficient of variation**

Coefficient of variation

$$V = \frac{s}{\bar{x}} \cdot 100 \qquad or \qquad V = \frac{\sigma}{\mu} \cdot 100$$

which expresses the standard deviation as a percentage of what is being measured, at least on the average.

EXAMPLE In recent months, the price of sirloin steak averaged $2.87 with a standard deviation of $0.13, and the price of T-bone steak averaged $3.90 with a standard deviation of $0.16. For which of these two cuts of beef is the price relatively more variable?

Solution The two coefficients of variation are

$$\frac{0.13}{2.87} \cdot 100 = 4.5 \text{ percent} \qquad and \qquad \frac{0.16}{3.90} \cdot 100 = 4.1 \text{ percent}$$

Thus, the price of sirloin steak is relatively more variable than the price of T-bone steak, even though the second standard deviation is greater than the first.

EXERCISES

3.29 According to the most recent figures available, 40, 11, 47, 22, and 6 percent of the population of the five Pacific states used fluoridated water. What is the range of these percentages?

3.30 The following are the closing prices of two stocks on five consecutive Fridays:

$$Stock\ A: \quad 18\tfrac{1}{4}, 18, 17\tfrac{3}{8}, 17\tfrac{5}{8}, 18\tfrac{1}{8}$$

$$Stock\ B: \quad 20, 20\tfrac{1}{4}, 19\tfrac{7}{8}, 20\tfrac{1}{2}, 20\tfrac{1}{8}$$

Calculate the range for each stock and judge which one is less variable.

3.31 The 25 teachers of an elementary school are given an intensive course in first aid. Find the range of the following numbers of correct answers they gave in a test administered after the completion of the course: 18, 12, 15, 9, 11, 16, 20, 15, 14, 18, 18, 15, 10, 17, 13, 17, 19, 8, 19, 20, 16, 12, 18, 11, and 14.

3.32 Calculate the standard deviation σ of the population which consists of the integers 1, 2, 3, 4, 5, and 6.

3.33 In five attempts, it took a person 10, 16, 12, 9, and 13 minutes to change a tire on a car. Calculate the standard deviation of this sample using
(a) the formula which defines s;
(b) the computing formula for s.

3.34 The following are the wind velocities reported at an airport at 6 P.M. on six consecutive days: 13, 8, 15, 11, 3, and 10. Find the variance of these figures using
(a) the formula by which s^2 is defined;
(b) the computing formula for s^2.

3.35 On four days it took a person 17, 12, 15, and 21 minutes to drive to work.
  (a) Use the computing formula for s to calculate the standard deviation of these data.
  (b) Subtract 10 from each of the values and then use the computing formula for s to calculate the standard deviation of the resulting data. What general rule does this suggest for simplifying the calculation of s?

3.36 It has been claimed that for samples of size $n = 4$, the range should be roughly twice as large as the standard deviation. Check this claim with reference to the following data, representing the numbers of emergency surgeries that were performed at a hospital on four days: 3, 6, 2, and 6.

3.37 It has been claimed that for samples of size $n = 10$, the range should be roughly three times as large as the standard deviation. Check this claim with reference to the following data, representing the speeds at which ten cars were timed at a checkpoint: 61, 50, 55, 39, 59, 66, 46, 47, 65, and 52.

3.38 The following are the numbers of commercial television stations that were licensed in 1978 in the New England states: 5, 7, 12, 1, 2, and 2. Use the computing formula, suitably modified, to calculate σ.

3.39 With reference to Exercise 3.4 on page 49, find the standard deviation of the ages of the sample of persons empaneled for jury duty.

3.40 Find the variance of the twelve means and also that of the twelve medians obtained in part (a) of Exercise 3.19 on page 51. What is illustrated by the difference in the size of these two variances?

3.41 According to Chebyshev's theorem, what can we assert about the percentage of any set of data that must lie within k standard deviations on either side of the mean when
  (a) $k = 8$;
  (b) $k = 12$?

3.42 An airline's records show that its flights between two cities arrive on the average 4.6 minutes late with a standard deviation of 1.4 minutes. At least what percentage of its flights between these two cities arrive anywhere between
  (a) 1.8 minutes late and 7.4 minutes late;
  (b) 1.0 minutes early and 10.2 minutes late;
  (c) 3.8 minutes early and 13.0 minutes late?

3.43 A study of the nutritional value of a certain kind of bread shows that on the average one slice contains 0.260 milligram of thiamine (vitamin B_1) with a standard deviation of 0.005 milligram. Between what values must be the thiamine content of
  (a) at least $\frac{35}{36}$ of all slices of this bread;
  (b) at least $\frac{63}{64}$ of all slices of this bread?

3.44 With reference to the preceding exercise, at least what percentage of the slices of the given kind of bread must have a thiamine content between 0.245 and 0.275 milligram? What can we say about this percentage if it can be assumed that the distribution of the thiamine contents of the slices of bread is bell-shaped?

3.45 In a city in the Southwest, restaurants charge on the average $8.65 for a steak dinner (with a standard deviation of $0.40), $5.65 for a chicken dinner (with a standard deviation of $0.25), and $12.95 for a lobster dinner (with a standard deviation of $0.30). If a restaurant in this city charges $9.25 for a steak dinner,

$6.15 for a chicken dinner, and $13.95 for a lobster dinner, which of the three dinners is relatively most overpriced?

3.46 Among two persons on a reducing diet, the first belongs to an age group for which the mean weight is 146 pounds with a standard deviation of 14 pounds, and the second belongs to an age group for which the mean weight is 160 pounds with a standard deviation of 17 pounds. If their respective weights are 178 pounds and 193 pounds, which of the two is more seriously overweight for his or her age group?

3.47 The applicants to one state university have an average ACT mathematics score of 20.4 with a standard deviation of 3.1, while the applicants to another state university have an average ACT mathematics score of 21.1 with a standard deviation of 2.8. With respect to which of these two universities is a student in a relatively better position, if he or she scores
   (a) 25 on this test;
   (b) 30 on this test?

3.48 To compare the precision of two micrometers, a laboratory technician studies recent measurements made with both instruments. The first micrometer was recently used to measure the diameter of a ball bearing and several measurements had a mean of 4.98 mm and a standard deviation of 0.018 mm; the second was recently used to measure the unstretched length of a spring and several measurements had a mean of 2.56 in. with a standard deviation of 0.012 in. Which of the two micrometers is relatively more precise?

3.49 In five tests, one student averaged 63.2 with a standard deviation of 3.3, while another student averaged 78.8 with a standard deviation of 5.3. Which student is relatively more consistent?

3.50 One patient's blood pressure, measured daily over several weeks, averaged 182 with a standard deviation of 12.6, while that of another patient averaged 124 with a standard deviation of 9.4. Which patient's blood pressure is relatively more variable?

3.9
THE DESCRIPTION OF GROUPED DATA ⋆

Since published data are often available only in the form of a frequency distribution, we shall discuss briefly the calculation of statistical descriptions of grouped data.

As we have already seen, the grouping of data entails some loss of information. Each item loses its identity, so to speak; we only know how many items there are in each class, so we must be satisfied with approximations. In the case of the mean and the standard deviation, we can usually get good approximations by assigning to each item falling into a class the value of the class mark. For instance, to calculate the mean or the standard deviation of the grouped sulfur oxides emission data on page 16, we treat the three values falling into the class 5.0–8.9 as if they were all 6.95, the ten values falling into the class 9.0–12.9 as if they were all 10.95,..., and the two values falling into the class 29.0–32.9 as if they were both 30.95. This procedure is usually quite satisfactory, since the errors which are thus introduced into the calculations will more or less "average out."

To write general formulas for the mean and the standard deviation of a distribution with k classes, let us denote the successive class marks by $x_1, x_2, \ldots,$ and x_k, and the corresponding class frequencies by $f_1, f_2, f_3, \ldots,$ and f_k. Then, the sum of all the measurements or observations is given by $x_1 f_1 + x_2 f_2 + x_3 f_3 + \cdots + x_k f_k = \sum x \cdot f$, the sum of their squares is given by $x_1^2 f_1 + x_2^2 f_2 + x_3^2 f_3 + \cdots + x_k^2 f_k = \sum x^2 \cdot f$, and the formula for $\bar{x}$ and the computing formula for s can be written as[†]

$$\bar{x} = \frac{\sum x \cdot f}{n} \quad \text{and} \quad s = \sqrt{\frac{n(\sum x^2 \cdot f) - (\sum x \cdot f)^2}{n(n-1)}}$$

EXAMPLE   Calculate the mean and the standard deviation of the distribution of the sulfur oxides emission data on page 16.

Solution   To get $\sum x \cdot f$ and $\sum x^2 \cdot f$, we perform the calculations shown in the following table, where the first column contains the class marks, the second column is copied from the original distribution, and the third and fourth columns contain the products $x \cdot f$ and $x^2 \cdot f$:

Class marks x	Frequencies f	$x \cdot f$	$x^2 \cdot f$
6.95	3	20.85	144.9075
10.95	10	109.50	1,199.0250
14.95	14	209.30	3,129.0350
18.95	25	473.75	8,977.5625
22.95	17	390.15	8,953.9425
26.95	9	242.55	6,536.7225
30.95	2	61.90	1,915.8050
	80	1,508.00	30,857.0000

Then, substituting the appropriate totals into the formulas, we get

$$\bar{x} = \frac{1,508.00}{80} = 18.85$$

[†] To get corresponding formulas for the mean μ and the standard deviation σ of a population, we substitute N for n in the formula for the mean, and also in the formula for the standard deviation, after replacing the factor $n-1$ by n.

and

$$s = \sqrt{\frac{80(30,857) - (1,508)^2}{80 \cdot 79}} = 5.55$$

These calculations were rather tedious, and we went through them only to dramatize the simplification that can be attained by **coding** the class marks so that we have smaller numbers to work with. When the class intervals are all equal, this coding consists of assigning the value 0 to one of the class marks (preferably at or near the center of the distribution), and representing all the class marks by means of successive integers. For instance, if a distribution has nine classes and the class mark of the middle class is assigned the value 0, the successive class marks of the distribution are assigned the values $-4, -3, -2, -1, 0, 1, 2, 3,$ and 4.

Of course, when we code the class marks in this way, we must account for it in the formulas for the mean and the standard deviation. Referring to the new (coded) class marks as u's, we write

Mean of grouped data (with coding)

$$\bar{x} = x_0 + \frac{\sum u \cdot f}{n} \cdot c$$

where x_0 is the class mark in the original scale to which we assign 0 in the new scale, c is the class interval, n is the number of items grouped, and $\sum u \cdot f$ is the sum of the products obtained by multiplying each of the new class marks by the corresponding class frequency. Similarly, we write

Standard deviation of grouped data (with coding)

$$s = c \sqrt{\frac{n(\sum u^2 \cdot f) - (\sum u \cdot f)^2}{n(n-1)}}$$

where $\sum u^2 \cdot f$ is the sum of the products obtained by multiplying the squares of the new class marks by the corresponding class frequencies. If a set of data constitutes a population rather than a sample, we make the same modifications in the formulas as before.

EXAMPLE To demonstrate the simplification brought about by coding, recalculate the mean and the standard deviation of the distribution of the sulfur oxides emission data.

Solution Arranging the work, as before, in a table, we get

Original class marks x	New class marks u	f	$u \cdot f$	$u^2 \cdot f$
6.95	-3	3	-9	27
10.95	-2	10	-20	40
14.95	-1	14	-14	14
18.95	0	25	0	0
22.95	1	17	17	17
26.95	2	9	18	36
30.95	3	2	6	18
		80	-2	152

where the class mark 18.95 is taken to be 0 in the new scale. (Of course, we could have used any class mark as the zero for the new scale, but the objective is to make the numbers, and the arithmetic, as simple as possible.)

Substituting $c = 4$, $x_0 = 18.95$, $n = 80$, $\sum u \cdot f = -2$, and $\sum u^2 \cdot f = 152$ into the above formulas for $\bar{x}$ and s, we get

$$\bar{x} = 18.95 + \frac{-2}{80} \cdot 4 = 18.85$$

and

$$s = 4 \sqrt{\frac{80(152) - (-2)^2}{80 \cdot 79}} = 5.55$$

These results are, as they should be, identical with the ones which we obtained earlier without coding.

Once a set of data has been grouped, we cannot find the exact value of the median because of the loss of information which results from the act of grouping. So, we define the median of a distribution in a special way, as follows:

The median of a distribution is the number which is such that half the total area of the rectangles of the histogram of the distribution lies to its left and the other half lies to its right.

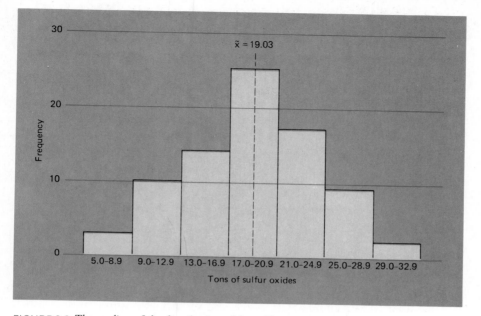

FIGURE 3.3 *The median of the distribution of the sulfur oxides emission data.*

This definition, which is illustrated by Figure 3.3, is equivalent to the assumption that the values in the class containing the median of the ungrouped data are distributed evenly (that is, spread out evenly) throughout that class.

To find the dividing line between the two halves of a histogram (each of which represents $\frac{n}{2}$ of the items grouped), we must count $\frac{n}{2}$ of the items starting at one end of the distribution. How this is done is illustrated by the following example:

EXAMPLE Find the median of the distribution of the sulfur oxides emission data.

Solution Since $\frac{n}{2} = \frac{80}{2} = 40$, we must count 40 items starting at either end.

Counting from the bottom (that is, beginning with the smallest values), we find that $3 + 10 + 14 = 27$ of the values fall into the first three classes, and that $3 + 10 + 14 + 25 = 52$ of the values fall into the first four classes. Therefore, we must count $40 - 27 = 13$ more values beyond the 27 which fall into the first three classes, and on the assumption that the 25 values in the fourth class are spread evenly throughout that class, we can do this by adding $\frac{13}{25}$

of the class interval of 4 to 16.95, the lower boundary of the fourth class. This gives us

$$\tilde{x} = 16.95 + \frac{13}{25} \cdot 4 = 19.03$$

for the median of this distribution.

In general, if L is the lower boundary of the class into which the median must fall, f is its frequency, c is its class interval, and j is the number of items we still lack when we reach L, then the median of the distribution is given by

Median of grouped data

$$\tilde{x} = L + \frac{j}{f} \cdot c$$

If we prefer, we can find the median of a distribution by starting to count at the other end (beginning with the largest values) and subtracting an appropriate fraction of the class interval from the upper boundary U of the class into which the median must fall. The corresponding formula is

Alternate formula for the median of grouped data

$$\tilde{x} = U - \frac{j'}{f} \cdot c$$

where j' is the number of items we still lack when we reach U.

EXAMPLE   Use the alternate formula to find the median of the distribution of the sulfur oxides emission data.

Solution   Since $2 + 9 + 17 = 28$ of the values fall above 20.95, we need $40 - 28 = 12$ of the 25 values which fall into the next class to reach the median, and we write

$$\tilde{x} = 20.95 - \frac{12}{25} \cdot 4 = 19.03$$

The result is, of course, the same.

Note that the median of a distribution can be found regardless of whether the class intervals are all equal; in fact, it can usually be found even when either or both of the classes at the top and at the bottom of a distribution are open (see Exercise 3.60).

The method by which we found the median of a distribution can also be used to determine more general measures of location called **fractiles** or **quantiles.** By definition, a fractile, or quantile, is a value at or below which a given fraction of the data must lie. There are, for instance, the three **quartiles** Q_1, Q_2, and Q_3, which are such that 25 percent of the data are less than or equal to Q_1, 50 percent are less than or equal to Q_2, and 75 percent are less than or equal to Q_3. Also, there are the nine **deciles, $D_1, D_2, \ldots$,** and D_9, which are such that 10 percent of the data are less than or equal to D_1, 20 percent are less than or equal to D_2, and so on; and there are the 99 **percentiles,** $P_1, P_2, \ldots$, and P_{99}, which are such that 1 percent of the data are less than or equal to P_1, 2 percent are less than or equal to P_2, and so on. It should be clear from this that Q_2, D_5, and P_{50} are all equal to the median, and that P_{25} equals Q_1 and P_{75} equals Q_3.

EXAMPLE   Find Q_1, D_8, and P_5 for the distribution of the sulfur oxides emission data.

Solution   Using the formulas for the median and counting in each case the appropriate fraction of the number of items grouped in the distribution, we find that

$$Q_1 = 12.95 + \frac{7}{14} \cdot 4 = 14.95$$

$$D_8 = 24.95 - \frac{5}{17} \cdot 4 = 23.77$$

and

$$P_5 = 8.95 + \frac{1}{10} \cdot 4 = 9.35$$

3.10
SOME FURTHER DESCRIPTIONS ⋆

So far we have discussed statistical descriptions which come under the general heading of "measures of location" and "measures of variation." Actually, there is no limit to the number of ways in which statistical data can be described, and statisticians continually develop new methods of describing characteristics of numerical data that are of interest in particular problems. In this section we shall consider briefly the problem of describing the overall shape of a distribution.

Although frequency distributions can take on almost any shape or form, most of the distributions we meet in practice can be described fairly well by one or another of a few standard types. Among these, foremost in importance is the aptly described symmetrical **bell-shaped distribution** shown in Figure 3.4. Indeed, there are theoretical reasons why, in many cases, distributions of actual data can be expected to follow its pattern. The other two distributions of Figure 3.4 can still, by a stretch of the imagination, be called bell-shaped, but they are not symmetrical. Distributions like these, having a "tail" on one side or the other, are said to be **skewed**; if the tail is on the left we say that they are **negatively skewed** and if the tail is on the right we say that they are **positively skewed.** Distributions of incomes or wages are often positively skewed because of the presence of some relatively high values that are not offset by correspondingly low values.

There are several ways of measuring the extent to which a distribution is skewed. A relatively easy one is based on the fact that for a perfectly symmetrical bell-shaped distribution such as the one of Figure 3.4, the values of the median and the mean coincide. Since the presence of some relatively high values that are not offset by correspondingly low values will tend to make the mean greater than the median (and the presence of

FIGURE 3.4 *Bell-shaped distributions.*

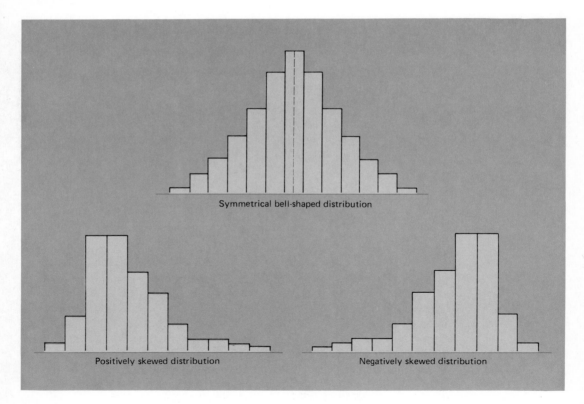

Symmetrical bell-shaped distribution

Positively skewed distribution

Negatively skewed distribution

some relatively low values that are not offset by correspondingly high values will tend to make the mean less than the median), we can use this relationship between the mean and the median to define a relatively simple measure of the extent to which a distribution is skewed. It is called the **Pearsonian coefficient of skewness**, and it is given by

Pearsonian coefficient of skewness

$$SK = \frac{3(mean - median)}{standard\ deviation}$$

For a perfectly symmetrical distribution the value of SK is 0, and in general its values must fall between -3 and 3.

EXAMPLE Find the Pearsonian coefficient of skewness for the distribution of the sulfur oxides emission data.

Solution Substituting into the formula the values of the mean, $\bar{x} = 18.85$, the median, $\tilde{x} = 19.03$, and the standard deviation, $s = 5.55$, we get

$$SK = \frac{3(18.85 - 19.03)}{5.55} = -0.01$$

Since this value is very close to 0, we can say that the distribution is nearly symmetrical.

Two other kinds of distributions which sometimes arise in practice are the **reverse J-shaped** and **U-shaped distributions** shown in Figure 3.5. As can be seen from this figure, the names of these distributions quite literally describe their shape. Examples of such distributions may be found in Exercises 3.62 and 3.63 on pages 73 and 74.

FIGURE 3.5 *Reverse J-shaped and U-shaped distributions.*

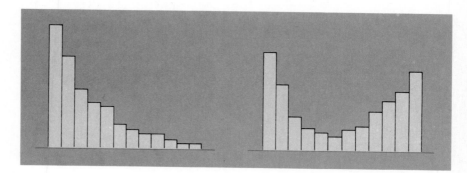

★3.51 Find $\bar{x}$ and s for the following distribution of the percentages of the students belonging to a certain ethnic group in a sample of 50 elementary schools:

Percentage	Number of schools
0– 4	18
5– 9	15
10–14	9
15–19	7
20–24	1

★3.52 Find the median of the distribution of the preceding exercise and, using the results of that exercise, calculate the Pearsonian coefficient of skewness.

★3.53 The following is the distribution of the grades obtained by 500 students in a geography test:

Grade	Number of students
10–24	44
25–39	70
40–54	92
55–69	147
70–84	115
85–99	32

Find $\bar{x}$ and s.

★3.54 With reference to the distribution of the preceding exercise, find
 (a) the median;
 (b) the quartiles Q_1 and Q_3;
 (c) the deciles D_1 and D_9;
 (d) the percentiles P_5 and P_{95}.

★3.55 Use the results of the two preceding exercises to find the Pearsonian coefficient of skewness for the distribution of the grades.

★3.56 Find $\bar{x}$ and s for the following distribution of the ages of the members of a union:

Age (years)	Frequency
15–19	16
20–24	35
25–29	44
30–34	27
35–39	17
40–44	8
45–49	2
50–54	1

★3.57 With reference to the distribution of the preceding exercise, find
  (a) the median;
  (b) the quartiles Q_1 and Q_3;
  (c) the deciles D_2 and D_8;
  (d) the percentiles P_1 and P_5.

★3.58 Use the results of the two preceding exercises to calculate the Pearsonian co-efficient of skewness for the distribution of the ages of the members of the union.

★3.59 Sometimes we use the **midquartile** $\frac{1}{2}(Q_1 + Q_3)$ as a measure of central location instead of the median or the mean, the **interquartile range** $Q_3 - Q_1$, or the **semi-interquartile range** $\frac{1}{2}(Q_3 - Q_1)$, as a measure of variation instead of the range or the standard deviation, and the **coefficient of quartile variation** $\dfrac{Q_3 - Q_1}{Q_3 + Q_1} \cdot 100$

as a measure of relative variation instead of the coefficient of variation.
  (a) Use the results of part (b) of Exercise 3.54 to calculate the midquartile and the interquartile range for the grade distribution of Exercise 3.53.
  (b) Use the results of part (b) of Exercise 3.57 to calculate the semi-interquartile range and the coefficient of quartile variation for the age distribution of Exercise 3.56.

★3.60 For each of the following distributions determine whether it is possible to find the mean, the median, or both:

(a)

Grade	Frequency
40–49	5
50–59	18
60–69	27
70–79	15
80–89	6

(b)

IQ	Frequency
Less than 90	3
90– 99	14
100–109	22
110–119	19
More than 119	7

(c)

Weight	Frequency
100 or less	41
101–110	13
111–120	8
121–130	3
131–140	1

★3.61 To study the effect of grouping on the calculation of the mean and the standard deviation, calculate
  (a) the mean of the raw (ungrouped) sulfur oxides emission data on page 16,
  (b) their standard deviation.

For part (b) it will be helpful to use a calculator on which sums of squares can be accumulated, or a statistical calculator which gives the value of s directly. Compare the values obtained here with the corresponding values obtained for the grouped data on pages 64 and 65, namely, $\bar{x} = 18.85$ and $s = 5.55$.

★3.62 Roll three dice 150 times and construct a distribution showing how many times there were zero 6's, how many times there was one 6, how many times there were two 6's, and how many times there were three 6's. Draw a histogram of this distribution and describe its shape.

★3.63   If a coin is flipped five times, the result may be represented by means of a sequence of H's and T's (for example, HHTTH), where H stands for *heads* and T for *tails*. Having obtained such a sequence of H's and T's, we can then check after each successive flip whether the number of heads exceeds the number of tails. For example, for the sequence HHTTH, heads is ahead after the first flip, after the second flip, after the third flip, not after the fourth flip, but again after the fifth flip; altogether, it is ahead four times. Repeat this experiment 50 times, and construct a histogram showing in how many cases heads was ahead altogether 0 times, 1 time, 2 times, ..., and 5 times. Explain why the resulting distribution should be U-shaped.

3.11
TECHNICAL NOTE (Summations)

In the notation introduced on page 40, $\sum x$ does not tell us which, or how many, values of x we must add. This is taken care of by the more explicit notation

$$\sum_{i=1}^{n} x_i = x_1 + x_2 + \cdots + x_n$$

where it is made clear that we are adding the x's whose subscripts i are $1, 2, \ldots, n$. We did not use this notation in the text, in order to simplify the overall appearance of the various formulas, assuming that it is clear in each case what x's we are referring to and how many there are.

Using the $\sum$ notation, we shall also have occasion to write such expressions as $\sum x^2, \sum xy, \sum x^2f, \ldots$, which (more explicitly) represent the sums

$$\sum_{i=1}^{n} x_i^2 = x_1^2 + x_2^2 + x_3^2 + \cdots + x_n^2$$

$$\sum_{j=1}^{m} x_j y_j = x_1 y_1 + x_2 y_2 + \cdots + x_m y_m$$

$$\sum_{i=1}^{n} x_i^2 f_i = x_1^2 f_1 + x_2^2 f_2 + \cdots + x_n^2 f_n$$

Working with two subscripts, we shall also have the occasion to evaluate **double summations** such as

$$\sum_{j=1}^{3} \sum_{i=1}^{4} x_{ij} = \sum_{j=1}^{3} (x_{1j} + x_{2j} + x_{3j} + x_{4j})$$

$$= x_{11} + x_{21} + x_{31} + x_{41} + x_{12} + x_{22} + x_{32} + x_{42}$$

$$+ x_{13} + x_{23} + x_{33} + x_{43}$$

To verify some of the formulas involving summations that are stated but not proved in the text, we need the following rules:

Rules for summations

$$\text{Rule } A: \quad \sum_{i=1}^{n} (x_i \pm y_i) = \sum_{i=1}^{n} x_i \pm \sum_{i=1}^{n} y_i$$

$$\text{Rule } B: \quad \sum_{i=1}^{n} k \cdot x_i = k \cdot \sum_{i=1}^{n} x_i$$

$$\text{Rule } C: \quad \sum_{i=1}^{n} k = k \cdot n$$

The first of these rules states that the summation of the sum (or difference) of two terms equals the sum (or difference) of the individual summations, and it can be extended to the sum or difference of more than two terms. The second rule states that we can, so to speak, factor a constant out of a summation, and the third rule states that the summation of a constant is simply n times that constant. All of these rules can be proved by actually writing out in full what each of the summations represents.

EXERCISES

3.64 Write each of the following in full; that is, without summation signs:

(a) $\sum_{i=1}^{6} x_i$;

(b) $\sum_{i=1}^{5} y_i$;

(c) $\sum_{i=1}^{3} x_i y_i$;

(d) $\sum_{j=1}^{8} x_j f_j$;

(e) $\sum_{i=3}^{7} x_i^2$;

(f) $\sum_{j=1}^{4} (x_j + y_j)$.

3.65 Write each of the following as summations:

(a) $z_1 + z_2 + z_3 + z_4 + z_5$;

(b) $x_5 + x_6 + x_7 + x_8 + x_9 + x_{10} + x_{11} + x_{12}$;

(c) $x_1 f_1 + x_2 f_2 + x_3 f_3 + x_4 f_4 + x_5 f_5 + x_6 f_6$;

(d) $y_1^2 + y_2^2 + y_3^2$;

(e) $2x_1 + 2x_2 + 2x_3 + 2x_4 + 2x_5 + 2x_6 + 2x_7$;

(f) $(x_2 - y_2) + (x_3 - y_3) + (x_4 - y_4)$;

(g) $(z_2 + 3) + (z_3 + 3) + (z_4 + 3) + (z_5 + 3)$;

(h) $x_1 y_1 f_1 + x_2 y_2 f_2 + x_3 y_3 f_3 + x_4 y_4 f_4$.

3.66 Given $x_1 = 1$, $x_2 = 3$, $x_3 = -2$, $x_4 = 4$, $x_5 = -1$, $x_6 = 2$, $x_7 = 1$, and $x_8 = 2$, find

(a) $\sum_{i=1}^{8} x_i$;

(b) $\sum_{i=1}^{8} x_i^2$.

3.67 Given $x_1 = 3$, $x_2 = 4$, $x_3 = 5$, $x_4 = 6$, $x_5 = 7$, $f_1 = 3$, $f_2 = 7$, $f_3 = 10$, $f_4 = 5$, and $f_5 = 2$, find

(a) $\sum_{i=1}^{5} x_i$;

(c) $\sum_{i=1}^{5} x_i f_i$;

(b) $\sum_{i=1}^{5} f_i$;

(d) $\sum_{i=1}^{5} x_i^2 f_i$.

3.68 Given $x_1 = 2$, $x_2 = -3$, $x_3 = 4$, $x_4 = -2$, $y_1 = 5$, $y_2 = -3$, $y_3 = 2$, and $y_4 = -1$, find

(a) $\sum_{i=1}^{4} x_i$;

(d) $\sum_{i=1}^{4} y_i^2$;

(b) $\sum_{i=1}^{4} y_i$;

(e) $\sum_{i=1}^{4} x_i y_i$.

(c) $\sum_{i=1}^{4} x_i^2$;

3.69 Given $x_{11} = 3$, $x_{12} = 1$, $x_{13} = -2$, $x_{14} = 2$, $x_{21} = 1$, $x_{22} = 4$, $x_{23} = -2$, $x_{24} = 5$, $x_{31} = 3$, $x_{32} = -1$, $x_{33} = 2$, and $x_{34} = 3$, find

(a) $\sum_{i=1}^{3} x_{ij}$ separately for $j = 1, 2, 3$, and 4;

(b) $\sum_{j=1}^{4} x_{ij}$ separately for $i = 1, 2$, and 3.

3.70 With reference to the preceding exercise, evaluate the double summation $\sum_{i=1}^{3} \sum_{j=1}^{4} x_{ij}$ using

(a) the results of part (a) of that exercise;

(b) the results of part (b) of that exercise.

3.71 Show that $\sum_{i=1}^{n} (x - \bar{x}) = 0$ for any set of x's whose mean is $\bar{x}$.

3.72 Is it true in general that $\left(\sum_{i=1}^{n} x_i \right)^2 = \sum_{i=1}^{n} x_i^2$? (*Hint:* Check whether the equation holds for $n = 2$.)

3.12
CHECKLIST OF KEY TERMS
(with page references to their definitions)

Arithmetic mean, 39
★ Bell-shaped distribution, 70
Biased estimator, 56
Chebyshev's theorem, 58
★ Coding, 65
★ Coefficient of quartile variation, 73
Coefficient of variation, 60
Decile, 69
Deviation from mean, 55

3.13
REVIEW EXERCISES

3.73 The following are the television audience ratings which a new comedy series received during the first five weeks of the 1982–1983 season: 13.9, 18.3, 15.5, 16.2, and 14.1. Find
    (a) the mean;
    (b) the median.

3.74 According to Chebyshev's theorem, what can we assert about the percentage of any set of data that must lie within k standard deviations on either side of the mean when

(a) $k = 4$;

(b) $k = 7$?

★3.75 In a benefit sale, a service organization sold 220 books at an average price of $2.10, 80 cakes at an average price of $2.75, and 50 craft items at an average price of $6.55. What is the average price per item sold?

★3.76 The following is the distribution of the number of days it rained in Seattle in 60 months:

Number of days	Frequency
5– 7	5
8–10	9
11–13	12
14–16	18
17–19	13
20–22	3

Calculate $\bar{x}$, $\tilde{x}$, and s.

★3.77 Use the results of the preceding exercise to calculate the Pearsonian coefficient of skewness for the given distribution.

★3.78 With reference to the distribution of Exercise 3.76, find

(a) the quartiles Q_1 and Q_3;

(b) the deciles D_8 and D_9;

(c) the percentiles P_5 and P_{15}.

3.79 If $\sum x = 16$ and $\sum x^2 = 39$ for a sample of size $n = 8$, find

(a) the mean;

(b) the standard deviation.

3.80 A producer of television commercials knows exactly how much money was spent on the production of each of ten one-minute commercials. Give one example each of a problem in which these data would be looked upon as

(a) a population;

(b) a sample.

3.81 Six women, working in industrial sales, earned the following amounts (in thousands of dollars) in a given year: 16.3, 24.5, 32.2, 21.8, 12.6, and 20.8. Calculate the mean, the median, the range, and the standard deviation of these earnings.

★3.82 In a certain class, a student scored 78, 83, and 88 on the first test, the second test, and the final. If the instructor considers the second test to be twice as important as the first test and the final to be three times as important as the second test, what is the student's average score on the three tests?

3.83 In a sample survey based on 100 families in Phoenix, Arizona, it was found that they had on the average 14.8 medications on hand. If the coefficient of variation is 15 percent, what is the standard deviation of the sample?

3.84 A current events test is given to a large group of students in a metropolitan school district. If the mean score is 40 points and the standard deviation is 5 points, what is the minimum fraction of the scores which must lie between 15 and 65 points?

3.85 During the last three months, nine persons shopped on the average in 5.9 clothing stores. Is it possible that at least six of them shopped in seven or more clothing stores?

3.86 Forty registered voters were asked whether they considered themselves Democrats, Republicans, or Independents. Use the following results to determine their modal choice: Democrat, Republican, Independent, Independent, Democrat, Independent, Republican, Republican, Independent, Democrat, Democrat, Independent, Democrat, Independent, Republican, Independent, Independent, Independent, Democrat, Democrat, Republican, Independent, Independent, Republican, Republican, Democrat, Republican, Democrat, Independent, Independent, Democrat, Democrat, Independent, Republican, Independent, Independent, Democrat, Independent, Republican, Democrat.

★3.87 The following is the distribution of the sizes of a sample of 50 orders received by the mail-order department of a department store:

Size of order	Number of orders
$20.00– 39.99	4
$40.00– 59.99	13
$60.00– 79.99	17
$80.00– 99.99	14
$100.00–119.99	2

Calculate
   (a) the mean and the median;
   (b) the standard deviation;
   (c) the quartiles Q_1 and Q_3;
   (d) the coefficient of variation and the coefficient of quartile variation.

★3.88 The following are the numbers of passengers on 56 sight-seeing buses in Boston, Massachusetts:

Number of persons	Frequency
15 or less	15
16–20	8
21–25	11
26–30	13
31–35	6
36–40	3

If possible, find
   (a) the median;
   (b) the mean;
   (c) the quartile Q_1;
   (d) the quartile Q_3.

3.89 In tests of the storage performance of certain frozen meat pies, a sample of ten specimens withstood 31, 28, 25, 34, 29, 33, 28, 31, 30, and 28 freeze–thaw cycles before damage appeared. Find the mean, median, and mode of these data.

3.90 For a large group of students the mean score on a scholastic aptitude test is 185 points and the standard deviation is 12 points. At least what percentage of the scores must lie between
      (a) 161 and 209 points;
      (b) 137 and 233 points;
      (c) 167 and 203 points?

★3.91 The following are the 1978 prices (in cents per can) and quantities produced (in millions of standard cases of No. 303 cans) of selected fruits and vegetables:

	Price	Quantity
Fruit cocktail	48.8	11.1
Peas	37.7	25.3
Tomatoes	38.0	49.2

Find the mean price (in cents per can) paid for these food items.

3.92 In standard units, a grade of 90 in a large economics class is $z = 2$. Given that the coefficient of variation for the entire class is 12.5 percent, find the mean and the standard deviation of the grades.

3.93 With reference to Exercise 2.9 on page 21, find the median of the data by first constructing a stem-and-leaf plot.

3.14
REFERENCES

An informal discussion of the ethics involved in choosing among measures of location is given in

> HUFF, D., *How to Lie with Statistics*. New York: W. W. Norton & Company, Inc., 1954.

A proof that division by $n - 1$ makes the sample variance an unbiased estimator of the population variance may be found in most textbooks on mathematical statistics; for instance, in

> FREUND, J. E., and WALPOLE, R. E., *Mathematical Statistics, 3rd ed.* Englewood Cliffs, N.J.: Prentice-Hall, Inc., 1980.

Some information about the effects of grouping on the calculation of x and s may be found in many of the older textbooks on statistics; for example, in

> MILLS, F. C., *Introduction to Statistics*. New York: Holt, Rinehart and Winston, 1956.

We can hardly predict the outcome of a football game unless we know what teams are going to play, and we cannot very well predict which television program will get the highest rating in a given week unless we know at least what shows will be on the air. More generally, we cannot make intelligent predictions or decisions unless we know at least what is possible—in other words, we must know what is possible before we can judge what is probable.

Thus, Sections 4.1, 4.2, and 4.3 will be devoted to "what is possible" in given situations, and then, in Section 4.4, we shall learn how to judge also "what is probable."

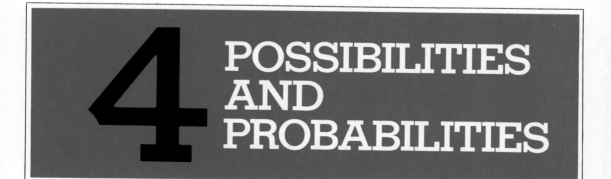

4 POSSIBILITIES AND PROBABILITIES

4.1
COUNTING

In contrast to the complex methods used nowadays in science, in business, and even in everday life, the simple process of counting still plays an important role. We still have to count 1, 2, 3, 4, 5, . . . , for example, to determine how many persons take part in a demonstration, the size of the response to a questionnaire, the number of damaged cases in a shipment of wines from Portugal, or when preparing a report showing how many times the temperature in Phoenix, Arizona, went over 100 degrees in a given month. Sometimes, the process of counting can be simplified by using mechanical devices (for instance, when counting spectators passing through turnstiles), or by performing counts indirectly (for instance, by subtracting the serial numbers of invoices to determine the total number of sales). At other times, the process of counting can be simplified greatly by means of special mathematical techniques, such as the ones given below.

In the study of "what is possible," there are essentially two kinds of problems. There is the problem of listing everything that can happen in a given situation, and then there is the problem of determining how many different things can happen (without actually constructing a complete list). The second kind of problem is especially important, because there are many situations in which we do not need a complete list, and hence, can save ourselves a great deal of work. Although the first kind of problem may seem straightforward and easy, this is not always the case.

EXAMPLE To meet a graduation requirement, each of three students must study a foreign language, including, among others, French, German and Spanish. In how many ways can they make their choice, if we are interested only in how many of them will study French, how many of them will study German, and how many of them will study Spanish?

Solution Clearly, there are many possibilities: all three students may decide to study German; one of them may decide to study French while the other two decide to study Spanish; one of them may decide to study German while the other two decide to study languages other than French, German, or Spanish; all three may decide to study languages other than French, German, or Spanish; and so forth.

Continuing this way, we may be able to list all twenty possibilities, but the chances are that we will omit at least one or two.

To handle this kind of problem systematically, it helps to draw a **tree diagram** like that of Figure 4.1. This diagram shows that first there are four possibilities (four branches) corresponding to 0, 1, 2, or 3 of the students deciding to study French. Then, for German there are four branches emanating from the top branch, three from the second branch, two from the third branch, and one from the bottom branch. Evidently, there are again four possibilities, 0, 1, 2, or 3, when none of the students decides to study French; three possibilities, 0, 1, or 2, when two of the students decides to study French; two possibilities, 0 or 1, when two of the students decide to study French; and one possibility, 0, when all three decide to study French. For Spanish the reasoning is the same, and we find that (going from left to right) there are altogether twenty different paths along the "branches" of the tree. In other words, there are twenty distinct possibilities.

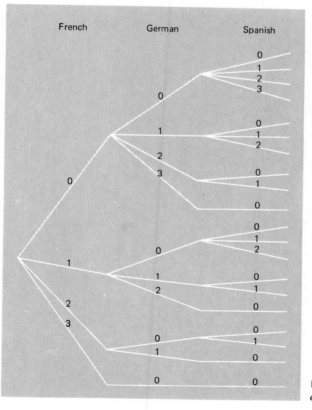

FIGURE 4.1 *Tree diagram for choice of language example.*

EXAMPLE In a medical study, patients are classified according to whether they have blood type A, B, AB, or O, and also according to whether their blood pressure is low, normal, or high. In how many different ways can a patient thus be classified?

Solution As is apparent from the tree diagram of Figure 4.2, the answer is 12. Starting at the top, the first path along the "branches" corresponds to a patient having blood type A and low blood pressure, the second path corresponds to a patient having blood type A and normal blood pressure, . . . , and the twelfth path corresponds to a patient having blood type O and high blood pressure.

The answer we got in the second example is $4 \cdot 3 = 12$, the product of the number of ways in which a patient can be classified according to blood type and the number of ways in which a patient can be classified according to

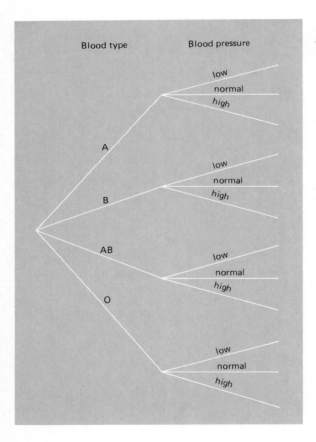

FIGURE 4.2 *Tree diagram for classification of patients in medical study.*

blood pressure. Generalizing from this example, let us state the following rule:

Multiplication of choices

> *If a choice consists of two steps, of which the first can be made in m ways and for each of these the second can be made in n ways, then the whole choice can be made in m · n ways.*

To prove this, we have only to draw a tree diagram similar to that of Figure 4.2. First there are m branches corresponding to the possibilities in the first step, and then there are n branches emanating from each of these branches to represent the possibilities in the second step. This leads to $m \cdot n$ paths along the branches of the tree diagram, and hence to $m \cdot n$ possibilities.

EXAMPLE   If a travel agency offers special weekend trips to 12 different cities, by air, rail, or bus, in how many different ways can such a trip be arranged?

Solution   Since $m = 12$ and $n = 3$, there are $12 \cdot 3 = 36$ ways.

EXAMPLE   If a history department schedules four lecture sections and sixteen discussion groups for a course in modern European history, in how many different ways can a student choose a lecture section and a discussion group?

Solution   Since $m = 4$ and $n = 16$, there are $4 \cdot 16 = 64$ ways.

By using appropriate tree diagrams, we can easily generalize the above rule so that it will apply to choices involving more than two steps. For k steps, where k is a positive integer, we get the following rule:

Multiplication of choices (generalized)

> *If a choice consists of k steps, of which the first can be made in n_1 ways, for each of these the second can be made in n_2 ways, ..., and for each of these the kth can be made in n_k ways, then the whole choice can be made in $n_1 \cdot n_2 \cdot \ldots \cdot n_k$ ways.*

We simply multiply the numbers of ways in which the different steps can be made.

EXAMPLE   If a new-car buyer has the choice of four body styles, three engines, and ten colors, in how many different ways can he or she order one of these cars?

Solution   Since $n_1 = 4, n_2 = 3$, and $n_3 = 10$, there are $4 \cdot 3 \cdot 10 = 120$ different ways.

EXAMPLE   If the new-car buyer of the preceding example also has the option to order the car with or without air conditioning and with or without automatic transmission, how many different choices does he or she have?

Solution   Since $n_1 = 4$, $n_2 = 3$, $n_3 = 10$, $n_4 = 2$, and $n_5 = 2$, there are $4 \cdot 3 \cdot 10 \cdot 2 \cdot 2 = 480$ different choices.

EXAMPLE   A test consists of twelve multiple-choice questions, each permitting four possible answers. In how many different ways can a student mark the test paper with one answer to each question?

Solution   Since $n_1 = n_2 = \cdots = n_{12} = 4$, there are altogether $4 \cdot 4 \cdot 4 \cdot 4 \cdot 4 \cdot 4 \cdot 4 \cdot 4 \cdot 4 \cdot 4 \cdot 4 \cdot 4 = 16{,}777{,}216$ different ways. Among these there is only one case where all the answers are correct and $3 \cdot 3 \cdot 3 \cdot 3 \cdot 3 \cdot 3 \cdot 3 \cdot 3 \cdot 3 \cdot 3 \cdot 3 \cdot 3 = 531{,}441$ cases where all the answers are wrong.

4.2
PERMUTATIONS

The rule for the multiplication of choices and its generalization are often applied when several choices are made from one set and we are concerned with the order in which they are made.

EXAMPLE   If twenty paintings are entered in an art show, in how many different ways can the judges award a first prize and a second prize?

Solution   Since the first prize can be awarded in $m = 20$ ways and the second prize must be awarded to one of the other $n = 19$ paintings, there are $20 \cdot 19 = 380$ ways.

EXAMPLE   In how many different ways can the 52 members of a labor union choose a president, a vice-president, a secretary, and a treasurer?

Solution   Since $n_1 = 52, n_2 = 51, n_3 = 50$, and $n_4 = 49$ (regardless of which officer is chosen first, second, third, or fourth), there are $52 \cdot 51 \cdot 50 \cdot 49 = 6{,}497{,}400$ ways.

In general, if r objects are selected from a set of n distinct objects, any particular arrangement (order) of these objects is called a **permutation.** For instance, 4 1 2 3 is a permutation of the first four positive integers;

Maine, Vermont, and Connecticut is a permutation (a particular ordered arrangement) of three of the six New England states; and

Cubs, Phillies, Pirates, Cardinals

Mets, Expos, Cardinals, Phillies

are two different permutations (ordered arrangements) of four of the six baseball teams in the Eastern Division of the National League.

EXAMPLE Determine the number of different permutations of two of the five vowels a, e, i, o, and u, and list them all.

Solution Since $m = 5$ and $n = 4$, there are $5 \cdot 4 = 20$ different permutations; they are

ae	*ai*	*ao*	*au*	*ei*	*eo*	*eu*	*io*	*iu*	*ou*
ea	*ia*	*oa*	*ua*	*ie*	*oe*	*ue*	*oi*	*ui*	*uo*

To find a formula for the total number of permutations of r objects selected from a set of n distinct objects, such as the six baseball teams or the five vowels, we observe that the first selection is made from the whole set of n objects, the second selection is made from the $n - 1$ objects which remain after the first selection has been made, the third selection is made from the $n - 2$ objects which remain after the first two selections have been made, ..., and the rth and final selection is made from the $n - (r - 1) = n - r + 1$ objects which remain after the first $r - 1$ selections have been made. Therefore, direct application of the generalized rule for the multiplication of choices yields the result that the total number of permutations of r objects selected from a set of n distinct objects, which we shall denote $_nP_r$, is

$$n(n - 1)(n - 2) \cdot \ldots \cdot (n - r + 1)$$

Since products of consecutive integers arise in many problems relating to permutations and other kinds of special arrangements or selections, it is convenient to introduce here the **factorial notation.** In this notation, the product of all positive integers less than or equal to the positive integer n is called "n factorial" and denoted by $n!$. Thus,

$$1! = 1$$
$$2! = 2 \cdot 1 = 2$$
$$3! = 3 \cdot 2 \cdot 1 = 6$$
$$4! = 4 \cdot 3 \cdot 2 \cdot 1 = 24$$
$$5! = 5 \cdot 4 \cdot 3 \cdot 2 \cdot 1 = 120$$
$$6! = 6 \cdot 5 \cdot 4 \cdot 3 \cdot 2 \cdot 1 = 720$$

$$\cdot \quad \cdot \quad \cdot \quad \cdot \quad \cdot$$

and in general $n! = n(n-1)(n-2) \cdot \ldots \cdot 3 \cdot 2 \cdot 1$. Also, to make various formulas more generally applicable, we let $0! = 1$ by definition.

To express the formula for $_nP_r$ in terms of factorials, we note, for example, that $12 \cdot 11 \cdot 10! = 12!$, $9 \cdot 8 \cdot 7 \cdot 6! = 9!$, and $37 \cdot 36 \cdot 35 \cdot 34 \cdot 33! = 37!$. Similarly,

$$_nP_r \cdot (n-r)! = n(n-1)(n-2) \cdot \ldots \cdot (n-r+1) \cdot (n-r)!$$
$$= n!$$

so that $_nP_r = \dfrac{n!}{(n-r)!}$. To summarize:

Number of permutations of n objects taken r at a time

The number of permutations of r objects selected from a set of n distinct objects is

$$_nP_r = n(n-1)(n-2) \cdot \ldots \cdot (n-r+1)$$

or, in factorial notation,

$$_nP_r = \frac{n!}{(n-r)!}$$

The first formula is generally easier to use because it requires fewer steps, but many students find the one in factorial notation easier to remember.

EXAMPLE Find the number of ways in which four of ten new movies can be ranked first, second, third, and fourth according to their attendance figures for the first six months.

Solution For $n = 10$ and $r = 4$ the first formula yields

$$_{10}P_4 = 10 \cdot 9 \cdot 8 \cdot 7 = 5{,}040$$

and the second formula yields

$$_{10}P_4 = \frac{10!}{(10-4)!} = \frac{10!}{6!} = \frac{10 \cdot 9 \cdot 8 \cdot 7 \cdot 6!}{6!} = 5{,}040$$

EXAMPLE Find the number of permutations of zero objects selected from a set of 25 objects.

Solution   Here we cannot use the first formula, but substitution of $n = 25$ and $r = 0$ into the second formula yields

$$_{25}P_0 = \frac{25!}{(25 - 0)!} = \frac{25!}{25!} = 1$$

This result may be trivial, but it shows that the factorial notation makes the formula for the number of permutations more generally applicable.

To find the formula for the number of permutations of n distinct objects taken all together, we substitute $r = n$ into either formula for $_nP_r$ and get

Number of
permutations
of n objects
taken all together

$$_nP_n = n!$$

EXAMPLE   In how many ways can eight teaching assistants be assigned to eight sections of a course in college algebra.

Solution   Substituting $n = 8$, we get

$$_8P_8 = 8! = 40,320$$

When the n objects are not all distinct, the formula for $_nP_n$ must be modified, as is shown in Exercises 4.22 and 4.23 on page 94.

4.3
COMBINATIONS

There are many problems in which we want to know the number of ways in which r objects can be selected from a set of n objects, but we do not care about the order in which the selection is made. For instance, we may want to know in how many ways a committee of four can be selected from among the 45 members of a college fraternity, or the number of ways in which the IRS can choose five of 36 tax returns for a special audit. To derive a formula which applies to problems like these, let us first examine the following 24 permutations of three of the first four letters of the alphabet:

abc	*acb*	*bac*	*bca*	*cab*	*cba*
abd	*adb*	*bad*	*bda*	*dab*	*dba*
acd	*adc*	*cad*	*cda*	*dac*	*dca*
bcd	*bdc*	*cbd*	*cdb*	*dbc*	*dcb*

If we are not concerned with the order in which three letters are chosen from the four letters a, b, c, and d, there are only four ways in which the selection can be made. They are abc, abd, acd, and bcd; namely, the letters shown in the first column. Each row of the table merely contains the $3! = 6$ different permutations of the letters shown in the first column.

In general, there are $r!$ permutations of any r objects we select from a set of n distinct objects, and hence the $_nP_r$ permutations of r objects selected from a set of n objects contain each set of r objects $r!$ times. (In our example, the 24 permutations of three letters selected from among the first four letters of the alphabet contain each set of three letters $3! = 6$ times.) Therefore, to find the number of ways in which r objects can be selected from a set of n distinct objects, also called the number of **combinations** of n objects taken r at a time and denoted by $\binom{n}{r}$, we divide $_nP_r$ by $r!$, and we get[†]

Number of combinations of n objects taken r at a time

> The number of ways in which r objects can be selected from a set of n distinct objects is
>
> $$\binom{n}{r} = \frac{n(n-1)(n-2) \cdot \ldots \cdot (n-r+1)}{r!}$$
>
> or, in factorial notation,
>
> $$\binom{n}{r} = \frac{n!}{r!(n-r)!}$$

---

where only the second formula can be used when $r = 0$. For $n = 0$ to $n = 20$, the values of $\binom{n}{r}$ may be read from Table X at the end of the book, where these quantities are called **binomial coefficients** (see Exercise 4.33).

EXAMPLE In how many ways can a person choose three books from a list of eight best-sellers?

Solution For $n = 8$ and $r = 3$ the first formula yields

$$\binom{8}{3} = \frac{8 \cdot 7 \cdot 6}{3!} = 8 \cdot 7 = 56$$

[†] Instead of $\binom{n}{r}$ we sometimes use the symbol $_nC_r$.

and the second formula yields

$$\binom{8}{3} = \frac{8!}{3!5!} = \frac{8 \cdot 7 \cdot 6 \cdot \cancel{5!}}{3!\cancel{5!}} = \frac{8 \cdot 7 \cdot 6}{3 \cdot 2 \cdot 1} = 56$$

Basically, the work is the same, but the first formula required fewer steps.

EXAMPLE In how many different ways can a committee of four be selected from among the 64 staff members of a hospital?

Solution For $n = 64$ and $r = 4$, the first formula yields

$$\binom{64}{4} = \frac{64 \cdot 63 \cdot 62 \cdot 61}{4!} = 635,376$$

The result of the first example, but not that of the second, can be verified in Table X.

EXAMPLE In how many different ways can the director of a research laboratory choose two chemists from among seven applicants and three physicists from among nine applicants?

Solution The two chemists can be selected in $\binom{7}{2}$ ways, the three physicists can be selected in $\binom{9}{3}$ ways, and by the multiplication of choices we have

$$\binom{7}{2} \cdot \binom{9}{3} = 21 \cdot 84 = 1,764$$

In this case we looked up the binomial coefficients in Table X.

When r objects are selected from a set of n distinct objects, $n - r$ of the objects are left, and consequently, there are as many ways of leaving (or selecting) $n - r$ objects from a set of n distinct objects as there are ways of selecting r objects. Symbolically, we write

Rule for binomial coefficients

$$\binom{n}{r} = \binom{n}{n - r} \qquad \text{for } r = 0, 1, 2, \ldots, n$$

Sometimes this rule serves to simplify details and sometimes it is used in connection with Table X.

EXAMPLE Determine the value of $\binom{75}{72}$.

Solution To avoid having to write down the product $75 \cdot 74 \cdot 73 \cdot \ldots \cdot 4$ and cancel $72 \cdot 71 \cdot 70 \cdot \ldots \cdot 4$, we write directly

$$\binom{75}{72} = \binom{75}{3} = \frac{75 \cdot 74 \cdot 73}{3!} = 67{,}525$$

EXAMPLE Find the value of $\binom{19}{13}$.

Solution $\binom{19}{13}$ cannot be looked up directly in Table X, but $\binom{19}{19-13} = \binom{19}{6} = 27{,}132$ can.

EXERCISES

4.1 Suppose that in a baseball World Series (in which the winner is the first team to win four games) the National League champion leads the American League champion three games to two. Construct a tree diagram to show the number of ways in which these teams may win or lose the remaining game or games.

4.2 A person with $2 in his pocket bets $1, even money, on the flip of a coin, and he continues to bet $1 so long as he has any money. Draw a tree diagram to show the various things that can happen during the first three flips of the coin. In how many of the cases will he be
   (a) exactly $1 ahead;
   (b) exactly $1 behind?

4.3 A student can study 0, 1, or 2 hours for a statistics test on any given night. Draw a tree diagram to show that there are six different ways in which she can study altogether 4 hours for the test on three consecutive nights.

4.4 There are four routes, A, B, C, and D, between a person's home and the place where he works, but route A is one-way so that he cannot take it on the way to work, and route D is one-way so that he cannot take it on the way home.
   (a) Draw a tree diagram showing the various ways he can go to and from work.
   (b) Draw a tree diagram showing the various ways he can go to and from work, but does not go by the same route both ways.

4.5 A student has to take two different courses in the humanities, one each during the two semesters of her freshman year. Draw a tree diagram to show the various ways in which she can make her choice, if during the first semester they offer introductory courses in philosophy, fine arts, history, and literature, and during the second semester they offer the same courses in philosophy and history, but not those in the other subjects.

4.6 If the NCAA has applications from four universities for hosting its intercollegiate wrestling championships in 1986 and 1987, in how many ways can they select the sites for these championship meets
   (a) if they are not to be held at the same university;
   (b) if they may be held at the same university?

4.7 In a political science survey voters are classified into six categories according to income and into five categories according to education. In how many different ways can a voter thus be classified?

4.8 In an optics kit there are six concave lenses, four convex lenses, two prisms, and two mirrors. In how many different ways can one choose a concave lens, a convex lens, a prism, and a mirror from this kit?

4.9 A psychologist preparing three-letter nonsense words for use in a memory test chooses the first letter from among the consonants q, w, x, and z; the second letter from among the vowels e, i, and u; and the third letter from among the consonants c, f, p, and v.
   (a) How many different three-letter nonsense words can he construct?
   (b) How many of these nonsense words will begin with the letter w?
   (c) How many of these nonsense words will end either with the letter f or the letter p?

4.10 In a doctor's office there are eight recent issues of *Newsweek*, six issues of the *New Yorker* magazine, and five issues of the *Reader's Digest*. In how many different ways can a patient waiting to see the doctor glance at one of each kind of magazine, if the order does not matter.

4.11 A true–false test consists of 15 questions. In how many different ways can a student mark one answer to each question?

4.12 Determine whether each of the following is true or false:
   (a) $20! = 20 \cdot 19 \cdot 18 \cdot 17!$;
   (b) $4! \cdot 3! = 12!$;
   (c) $3! + 4! = 7!$;
   (d) $16! = \dfrac{17!}{17}$.

4.13 Determine whether each of the following is true or false:

   (a) $\dfrac{1}{2!} + \dfrac{1}{2!} = 1$;

   (b) $\dfrac{15!}{13!} = 15 \cdot 14$;

   (c) $\dfrac{9!}{7!2!} = 72$;

   (d) $4! + 0! = 25$.

4.14 How many different signals can be made by arranging five of eight differently colored flags on a vertical pole?

4.15 On a vacation, a person wants to visit four of the nation's 22 historical parks. If the order of the visits matters, in how many ways can this person plan the trip?

4.16 In an English class, the students are given the choice of eight different essay topics. In how many different ways can four students each choose a topic
   (a) if no two students may choose the same topic;
   (b) if there is no restriction on the choice of topics?

4.17 If the drama club of a college has to choose four of ten half-hour skits to present on one evening from 8:00 to 10:00, in how many different ways can it arrange the schedule?

4.18 If there are nine cars in a race, in how many different ways can they place first, second, and third?

4.19 In how many different ways can a person arrange four paintings next to each other horizontally on a wall?

4.20 In how many different ways can the manager of a baseball team arrange the batting order of the nine players in the starting lineup?

4.21 In how many ways can a television director schedule six different commercials during the six time slots allocated to commercials during the telecast of the first period of a hockey game?

★4.22 If among n objects r are alike, and the others are all distinct, the number of permutations of these n objects taken all together is $\dfrac{n!}{r!}$.

    (a) How many permutations are there of the letters in the word "class"?

    (b) In how many ways (according to manufacturer only) can five cars place in a stock-car race, if three of the cars are Fords, one is a Chevrolet, and one is a Dodge?

    (c) In how many ways can the television director of Exercise 4.21 fill the six time slots allocated to commercials, if she has four different commercials, of which a given one is to be shown three times, while each of the others is to be shown once?

    (d) Present an argument to justify the formula given in this exercise.

★4.23 If among n objects r_1 are identical, another r_2 are identical, and the rest (if any) are all distinct, the number of permutations of these n objects taken all together is $\dfrac{n!}{r_1! \cdot r_2!}$.

    (a) How many permutations are there of the letters in the word "greater"?

    (b) In how many ways can the television director of Exercise 4.21 fill the six time slots allocated to commercials, if she has only two different commercials, each of which is to be shown three times?

    (c) Generalize the formula so that it applies if among n objects r_1 are identical, another r_2 are identical, another r_3 are identical, and the rest (if any) are all distinct. In how many ways can the television director of Exercise 4.21 fill the six time slots allocated to commercials, if she has three different commercials, each of which is to be shown twice?

    (d) In its cookbook section, a bookstore has four copies of *The New York Times Cookbook*, two copies of *The Joy of Cooking*, five copies of the *Better Homes and Gardens Cookbook*, and one copy of *The Secret of Cooking for Dogs*. If these books are sold one at a time, in how many different sequences can they be sold?

★4.24 The number of ways in which n distinct objects can be arranged in a circle is $(n - 1)!$.

    (a) Present an argument to justify this formula.

    (b) In how many ways can six persons be seated at a round table (if we care only who sits on whose left or right side)?

    (c) In how many ways can a window dresser display four tennis rackets in a circular arrangement?

4.25 Calculate the number of ways in which a restaurant chain can choose 2 of 13 locations for new franchises.

4.26 Calculate the number of ways in which five of 36 tax returns can be chosen for a special audit.

4.27 A true–false test consists of 15 questions. Calculate the numbers of ways in which a student can mark one answer to each question and get
    (a) 8 right and 7 wrong;
    (b) 10 right and 5 wrong.

4.28 A carton of 12 transistor batteries contains one that is defective. In how many ways can an inspector choose three of the batteries and
    (a) get the one that is defective;
    (b) not get the one that is defective?

4.29 With reference to the preceding exercise, suppose that two of the batteries are defective. In how many ways can the inspector choose three of the batteries and get
    (a) none of the defective batteries;
    (b) one of the defective batteries;
    (c) both of the defective batteries?

4.30 Among the eight nominees for two vacancies on a school board are four men and four women. In how many ways can these vacancies be filled
    (a) with any two of the eight nominees;
    (b) with any two of the female nominees;
    (c) with one of the male nominees and one of the female nominees?

4.31 To fill a number of vacancies, the personnel manager of a company has to choose three secretaries from among ten applicants and two bookkeepers from among five applicants. In how many different ways can the personnel manager fill the five vacancies?

4.32 A men's clothing store carries eight kinds of sweaters, six kinds of slacks, and nine kinds of shirts. In how many different ways can two of each kind be chosen for a special sale?

4.33 The quantity $\binom{n}{r}$ is called a binomial coefficient because it is, in fact, the coeffi-cient of $a^{n-r}b^{r}$ in the binomial expansion of $(a + b)^n$. Verify this for $n = 2$, 3, and 4 by expanding $(a + b)^2$, $(a + b)^3$, and $(a + b)^4$, and comparing the coeffi-cients with the corresponding entries for $n = 2$, $n = 3$, and $n = 4$ in Table X.

★4.34 A table of binomial coefficients is easy to construct by following the pattern shown below, which is called **Pascal's triangle.**

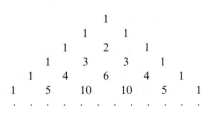

In this arrangement, each row begins with a 1, ends with a 1, and each other entry is the sum of the nearest two values in the row immediately above.

(a) Use Table X to verify that the third row of the triangle contains the values of $\binom{2}{r}$ for $r = 0, 1,$ and 2; the fourth row contains the values of $\binom{3}{r}$ for $r = 0, 1, 2,$ and 3; and the fifth row contains the values of $\binom{4}{r}$ for $r = 0, 1, 2, 3,$ and 4.

(b) Construct the next two rows of the triangle and use Table X to verify the results.

4.35 Verify the identity $\binom{n+1}{r} = \binom{n}{r} + \binom{n}{r-1}$ by expressing each of the binomial coefficients in terms of factorials. Explain why this identity justifies the method used in the construction of Pascal's triangle in the preceding exercise.

4.4
PROBABILITY

So far we have studied only what is possible in a given situation. In some instances we listed all possibilities and in others we merely determined how many different possibilities there are. Now we shall go one step further and judge also what is probable and what is improbable.

The most common way of measuring the uncertainties connected with events (say, the outcome of a presidential election, the side effects of a new medication, the durability of an exterior paint, or the total number of points we may roll with a pair of dice) is to assign them **probabilities** or to specify the **odds** at which it would be fair to bet that the events will occur. In this section we shall learn how probabilities are interpreted and how their numerical values are determined; odds will be discussed in Section 5.3.

Historically, the oldest way of measuring uncertainties is the **classical probability concept**. It was developed originally in connection with games of chance, and it lends itself most readily to bridging the gap between possibilities and probabilities. The classical probability concept applies only when all possible outcomes are equally likely, in which case we say that

The classical probability concept

> *If there are n equally likely possibilities, of which one must occur and s are regarded as favorable, or as a "success," then the probability of a "success" is given by the ratio $\dfrac{s}{n}$.*

In the application of this rule, the terms "favorable" and "success" are used rather loosely—what is favorable to one player is unfavorable to his opponent, and what is a success from one point of view is a failure from another.

Thus, the terms "favorable" and "success" can be applied to any particular kind of outcome, even if "favorable" means that a television set does not work, or "success" means that someone catches the flu. This usage dates back to the days when probabilities were quoted only in connection with games of chance.

EXAMPLE What is the probability of drawing an ace from a well-shuffled deck of 52 playing cards?

Solution There are $s = 4$ aces among the $n = 52$ cards, so we get

$$\frac{s}{n} = \frac{4}{52} = \frac{1}{13}$$

EXAMPLE What is the probability of rolling a 3 or a 4 with a balanced die?

Solution In this case, $s = 2$ and $n = 6$, and the probability is

$$\frac{s}{n} = \frac{2}{6} = \frac{1}{3}$$

Although equally likely possibilities are found mostly in games of chance, the classical probability concept applies also in a great variety of situations where gambling devices are used to make **random selections**— say, when offices are assigned to research assistants by lot, when laboratory animals are chosen for an experiment so that each one has the same chance of being selected, when each family in a township has the same chance of being included in a survey, or when machine parts are chosen for inspection so that each part produced has the same chance of being selected.

EXAMPLE If three of twenty tires are defective and four of them are randomly chosen for inspection, what is the probability that one of the defective tires will be included?

Solution There are $\binom{20}{4} = 4{,}845$ equally likely ways of choosing four of the twenty tires because of the random selection. The number of favorable outcomes is the number of ways in which one of the defective tires and three of the nondefective tires can be selected, or $\binom{3}{1}\binom{17}{3} =$ $= 3 \cdot 680 = 2{,}040$. It follows that the probability is

$$\frac{s}{n} = \frac{2{,}040}{4{,}845} = \frac{8}{19}$$

or approximately 0.42. The values of the binomial coefficients $\binom{20}{4}$, $\binom{3}{1}$, and $\binom{17}{3}$ were read directly from Table X.

A major shortcoming of the classical probability concept is its limited applicability, for there are many situations in which the various possibilities cannot all be regarded as equally likely. This would be the case, for instance, if we are concerned with the question of whether it will rain on a certain day; when we wonder whether a person will get a raise; when we want to predict the outcome of an election or the score of a baseball game; or when we want to judge whether a stock market index will go up or down.

Among the various probability concepts, most widely held is the **frequency interpretation**, according to which

<table>
<tr>
<td>The frequency
interpretation
of probability</td>
<td>The probability of an event (happening or outcome) is the proportion of the time that events of the same kind will occur in the long run.</td>
</tr>
</table>

If we say that that the probability is 0.78 that a jet from San Francisco to Phoenix will arrive on time, we mean that such flights arrive on time 78 percent of the time. Also, if the Weather Service predicts that there is a 40 percent chance for rain (that the probability is 0.40 that it will rain), they mean that under the same weather conditions it will rain 40 percent of the time. More generally, we say that an event has a probability of, say, 0.90, in the same sense in which we might say that our car will start in cold weather 90 percent of the time. We cannot guarantee what will happen on any particular occasion—the car may start and then it may not—but if we kept records over a long period of time, we should find that the proportion of "successes" is very close to 0.90.

In accordance with the frequency interpretation of probability, we estimate the probability of an event by observing what fraction of the time similar events have occurred in the past.

EXAMPLE   If records show that (over a period of time) 468 of 600 jets from San Francisco to Phoenix arrived on time, what is the probability that any one jet from San Francisco to Phoenix will arrive on time?

Solution   In the past $\frac{468}{600} = 0.78$ of the flights arrived on time, and we use this figure as an estimate of the probability.

EXAMPLE   If records show that 504 of 813 automatic dishwashers sold by a large retailer required repairs within the warranty year, what is the probability that an automatic dishwasher sold by the retailer will not require repairs within the warranty year?

Solution   Since $813 - 504 = 309$ of the dishwashers did not require repairs, we estimate the probability as $\frac{309}{813} = 0.38$.

When probabilities are estimated in this way, it is only reasonable to ask just how good the estimates are. In Chapter 12 we shall answer this question in some detail, but for now let us refer to an important theorem called the **Law of Large Numbers.** Informally, this theorem may be stated as follows:

The Law of Large Numbers

If a situation, trial, or experiment is repeated again and again, the proportion of successes will tend to approach the probability that any one outcome will be a success.

To illustrate this law, we repeatedly flipped a balanced coin and recorded the accumulated proportion of heads after every fifth flip. The results are shown in Figure 4.3, where the proportion of heads can be seen to fluctuate, but come closer and closer to $\frac{1}{2}$, the probability of heads for each flip of the coin (see also Exercise 7.50 on page 195).

In the frequency interpretation, the probability of an event is defined in terms of what happens to similar events in the long run, so let us examine briefly whether it is at all meaningful to talk about the probability of an event which can occur only once. For instance, can we assign a probability to the event that Ms. Bertha Jones will be able to leave the hospital within four days after having an appendectomy, or to the event that a certain major-party candidate will win an upcoming gubernatorial election? If we put ourselves

FIGURE 4.3 *Graph illustrating the Law of Large Numbers.*

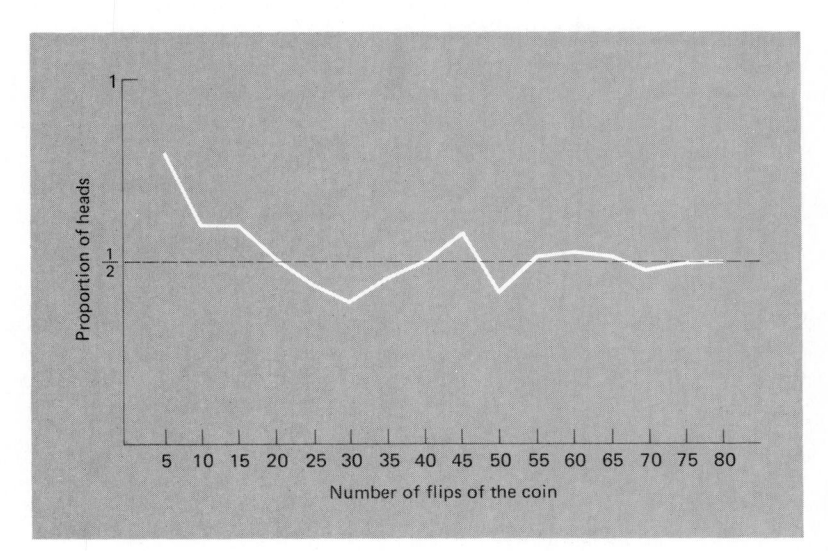

in the position of Ms. Jones' doctor, we might check medical records, discover that patients left the hospital within four days after an appendectomy in, say, 34 percent of hundreds of cases, and apply this figure to Ms. Jones. This may not be of much comfort to Ms. Jones, but it does provide a meaning for a probability statement about her leaving the hospital within four days—the probability is 0.34.

This illustrates that when we make a probability statement about a specific (nonrepeatable) event, the frequency interpretation of probability leaves us no choice but to refer to a set of similar events. As can well be imagined, however, this can easily lead to complications, since the choice of "similar" events is generally neither obvious nor straightforward. With reference to Ms. Jones' appendectomy, we might consider as "similar" only cases in which the patients were of the same sex, only cases in which the patients were also of the same age as Ms. Jones, or only cases in which the patients were also of the same height and weight as Ms. Jones. Ultimately, the choice of "similar" events is a matter of personal judgment, and it is by no means contradictory that we can arrive at different probability estimates, all valid, concerning the same event.

With regard to the question whether a certain major-party candidate will win an upcoming gubernational election, suppose that we ask the persons who have conducted a poll "how sure" they are that the candidate will win. If they say they are "95 percent sure" (that is, if they assign a probability of 0.95 to the candidate's winning the election), this is not meant to imply that he would win 95 percent of the time if he ran for office a great number of times. Rather, it means that the pollsters' prediction is based on a method which "works" 95 percent of the time. It is in this way that we must interpret many of the probabilities attached to statistical results.

An alternative point of view, which is currently gaining in favor, is to interpret probabilities as **personal** or **subjective** evaluations. Such probabilities express the strength of one's belief with regard to the uncertainties that are involved, and they apply especially when there is little or no direct evidence, so that there really is no choice but to consider collateral (indirect) information, "educated guesses," and perhaps intuition and other subjective factors. Subjective probabilities are sometimes determined by putting the issues in question on a "put up or shut up" basis, as will be explained in Chapter 6.

EXERCISES

4.36 When one card is drawn from a well-shuffled deck of 52 playing cards, what are the probabilities of getting
    (a) a black queen;
    (b) a queen, king, or ace of any suit;
    (c) a red card;
    (d) a 5, 6, 7, or 8?

4.37 If H stands for heads and T for tails, the four possible outcomes for two flips of a coin are HH, HT, TH, and TT. If it can be assumed that these four possibilities are equally likely, what are the probabilities of getting 0, 1, or 2 heads?

4.38 A bowl contains 12 red beads, 20 white beads, 15 blue beads, and 3 black beads. If one of the beads is drawn at random, what are the probabilities that it will be
  (a) red;
  (b) white or blue;
  (c) black;
  (d) neither white nor black?

4.39 If we roll a balanced die, what are the probabilities of getting
  (a) a 1 or a 6;
  (b) an even number?

4.40 Among the 15 applicants for three positions at a newspaper, 10 are college graduates. If the selection is random, what are the probabilities that the positions will be filled with three applicants
  (a) with college degrees;
  (b) without college degrees?

4.41 A carton of 24 light bulbs includes two that are defective. If two of the bulbs are chosen at random, what are the probabilities that
  (a) neither bulb will be defective;
  (b) one of the bulbs will be defective;
  (c) both bulbs will be defective?

4.42 If two cards are drawn from a well-shuffled deck of 52 playing cards, what are the probabilities of getting
  (a) two spades;
  (b) two kings;
  (c) a king and an ace?

4.43 A car rental agency has 18 compact cars and 12 intermediate-size cars. If four of the cars are randomly selected for a safety check, what is the probability that two of them will be compact cars?

4.44 Among 842 armed robberies in a certain city, 143 were never solved. Estimate the probability that an armed robbery in this city will not be solved.

4.45 In a sample of 300 students attending a large university, 78 expressed the opinion that out-of-state tuition is too high. Estimate the probability that a student attending that university, selected at random, will share this opinion.

4.46 In a sample of 446 cars stopped at a roadblock, only 67 of the drivers had their seat belts fastened. Estimate the probability that a driver stopped on that road will have his or her seat belt fastened.

4.47 A department store's records show that 782 of 920 women who entered the store on a Saturday afternoon made at least one purchase. Estimate the probability that a woman who enters the store on a Saturday afternoon will make at least one purchase.

4.5
CHECKLIST OF KEY TERMS
(with page references to their definitions)

4.6
REVIEW EXERCISES

4.48 Certain government employees are classified into six categories according to age and four categories according to marital status. In how many ways can one of these employees be classified?

4.49 In a union election, Mr. Brown, Ms. Green, and Ms. Jones are running for president, and Mr. Adams, Ms. Roberts, and Mr. Smith are running for vice-president. Construct a tree diagram showing the nine possible outcomes, and use it to determine the number of ways in which the two union officials will not be of the same sex.

4.50 In how many ways can three students, A, B, and C, divide up ten assignments, so that A gets two of the assignments, B gets five, and C gets three?

4.51 If 448 of 700 television viewers interviewed in a certain area feel that local news coverage is inadequate, estimate the probability that a television viewer randomly selected in that area will feel this way.

4.52 In how many different ways can eight proposed titles for a new television series be ranked first, second, and third by a group of studio executives?

4.53 Determine whether each of the following is true or false:

(a) $5! = \dfrac{7!}{42}$;

(b) $3! + 2! = 5!$;

(c) $\dfrac{20!}{16!} = 20 \cdot 19 \cdot 18$;

(d) $3! + 0! = 7$.

4.54 What is the probability of rolling a 5 with a pair of balanced dice?

4.55 The five finalists in the Miss Arizona contest are Miss Cochise, Miss Gila, Miss Pima, Miss Maricopa, and Miss Yuma. Draw a tree diagram showing the different ways in which the judges can choose the winner and the first runner-up.

4.56 In how many different ways can a person arrange eight books on a shelf?

4.57 A personality inventory consists of ten questions, each with five different answers. In how many different ways can a person choose one answer for each question?

4.58 Among the classified ads of a newspaper listing foreign-made cars, there are listings for eight Japanese cars, six German cars, and two Italian cars. In how many different ways can a person choose two of the Japanese cars, three of the German cars, and one of the Italian cars to inspect?

4.59 If 768 of 1,200 letters mailed by a government agency were delivered within 48 hours, estimate the probability that any one letter mailed by the agency will be delivered within 48 hours.

4.60 If an auditorium schedules six concerts, eight lectures, and five plays, in how many different ways can one choose
   (a) one of the concerts and one of the plays;
   (b) one of the concerts, one of the lectures, and one of the plays?

4.61 In how many ways can a person buy a pound each of three of the twelve kinds of cheese carried by a gourmet food shop?

4.62 For weekday afternoons, a television station schedules four soap operas, two situation comedies, and two game shows. In how many different ways can a viewer choose two of the soap operas, one of the situation comedies, and one of the game shows?

4.63 A business employs three persons named Jones: Harry Jones, Norma Jones, and Richard Jones. Draw a tree diagram to show the different ways in which the payroll department can distribute their paychecks so that each of them receives a check made out to a Jones. In how many of the possibilities will
   (a) none of them get the right check;
   (b) only one of them get the right check?

4.64 Suppose that someone flips a coin 100 times and gets 30 heads, which is short of the number of heads she might expect. Then she flips the coin another 100 times and gets 44 heads, which is again short of the number of heads she might expect. Can she accuse the Law of Large Numbers of "letting her down?" Explain.

4.7
REFERENCES

Informal introductions to probability, written primarily for the layman, may be found in

GARVIN, A. D., *Probability in Your Life*. Portland, Maine: J. Weston Walch Publisher, 1978.

HUFF, D., and GEIS, I., *How to Take a Chance*. New York: W. W. Norton & Company, Inc., 1959.

LEVINSON, H. C., *Chance, Luck, and Statistics*. New York: Dover Publications, Inc., 1963.

MOSTELLER, F., KRUSKAL, W. H., LINK, R. F., PIETERS, R. S., and RISING, G. R., *Statistics by Example: Weighing Chances*. Reading, Mass.: Addison-Wesley Publishing Company, Inc., 1973.

In the study of probability there are three fundamental kinds of questions: (1) What do we mean when we say that the probability of an event is, say, 0.50, 0.78, or 0.04? (2) How are the numbers we call probabilities determined, or measured in actual practice? (3) What are the mathematical rules which probabilities must obey?

The first two kinds of questions were already studied in Chapter 4. In the classical probability concept we are concerned with equally likely possibilities and count "favorable" outcomes; in the frequency interpretation we are concerned with proportions of "successes" in the long run and base our estimates on what has happened in the past; and when it comes to subjective probabilities we are concerned with a measure of a person's belief and in Section 5.3 we shall see how such probabilities can actually be determined.

In this chapter, after some preliminaries in Section 5.1, we shall concentrate on the rules which probabilities must obey. This includes the basic postulates in Section 5.2, the relationship between probabilities and odds in Section 5.3, addition rules in Section 5.4, the definition of conditional probability in Section 5.5, multiplication rules in Section 5.6, and finally Bayes' theorem in Section 5.7.

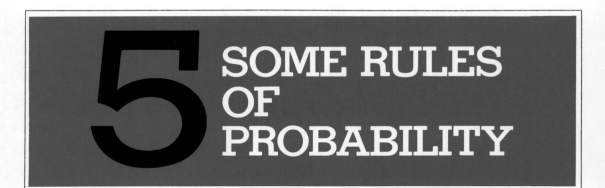

5 SOME RULES OF PROBABILITY

5.1
SAMPLE SPACES AND EVENTS

In statistics, a set of all possible outcomes of an experiment is called a **sample space** and it is usually denoted by the letter S. For instance, if a zoologist must choose three of 24 guinea pigs for an experiment, the sample space consists of the $\binom{24}{3} = 2{,}024$ ways in which the selection can be made; if the dean of a college must assign two of 84 faculty members as advisors to a political science club, the sample space consists of the $\binom{84}{2} = 3{,}486$ ways in which this can be done. Also, if we are concerned with the number of days it rains in Chicago during the month of January, the sample space is the set

$$S = \{0, 1, 2, 3, 4, \ldots, 30, 31\}$$

To avoid misunderstandings about the terms "outcome" and "experiment," let us make it clear that statisticians use them in a very wide, and unconventional, sense. For lack of a better term, "experiment" refers to any process of observation or measurement. Thus, an **experiment** may consist of counting how many times a student has been absent; it may consist of the simple process of noting whether a light is on or off, or whether a person is single or married; or it may consist of the very complicated process of obtaining and evaluating data to predict trends in the economy, to find the source of social unrest, or to study the cause of a disease. The results one obtains from an experiment, whether they are instrument readings, counts, "yes or no" answers, or values obtained through extensive calculations, are called the **outcomes** of the experiment.

When we study the outcomes of an experiment, we usually identify the various possibilities with numbers, points, or some other kinds of symbols, so that we can treat all questions about them mathematically, without having to go through long verbal descriptions of what has taken place, is taking place, or will take place. For instance, if there are eight candidates for a scholarship and we let a, b, c, d, e, f, g, and h denote that it is awarded to Ms. Adam, Mr. Bean, Miss Clark, and so on, then the sample space for this experiment is the set

$$S = \{a, b, c, d, e, f, g, h\}$$

The use of points rather than letters or numbers has the advantage that it makes it easier to visualize the various possibilities, and perhaps discover some special features which several of the outcomes may have in common. For instance, if a used-car dealer has two 1980 Chevettes in stock and we are interested in how many of them each of two salespersons sells in a given week, we could write the outcomes as (0, 0), (1, 0), (0, 1), (2, 0), (1, 1), and (0, 2). Here (0, 1) represents the outcome that the first salesperson sells neither of the Chevettes and the second salesperson sells one, (1, 1) represents the outcome that each of the two salespersons sells one of the Chevettes, and (2, 0) represents the outcome that the first salesperson sells both. Geometrically, the sample space may be pictured as in Figure 5.1, from which it is apparent, for instance, that they sell both cars in three of the six possibilities.

Usually, we classify sample spaces according to the number of elements, or points, which they contain. The ones we have mentioned so far in this section contained 2,024, 3,486, 31, 8, and 6 elements, and we refer to them all as **finite**; in each case, the number of possibilities is finite, or fixed. In this Chapter we shall consider only finite sample spaces, but in later chapters we shall work also with sample spaces that are **infinite.** For instance, an infinite sample space arises when we throw a dart at a target and there is a continuum of points we may hit.

In statistics, any subset of a sample space is called an **event.** By subset we mean any part of a set, including the set as a whole and, trivially, a set called the **empty set** and denoted by $\varnothing$, which has no elements at all. For

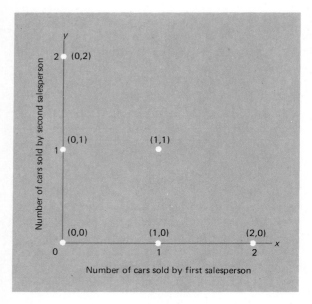

FIGURE 5.1 *Sample space for two-salesperson example.*

instance, for the example dealing with the number of days it rains in Chicago during the month of January,

$$A = \{18, 19, 20, 21, 22, 23, 24\}$$

is the event that there will be from 18 to 24 rainy days, and

$$B = \{20, 21, 22, \ldots, 30, 31\}$$

is the event that there will be at least 20 rainy days. Also, with reference to Figure 5.1,

$$C = \{(1, 0), (0, 1)\}$$

is the event that, between them, the two salespersons sell only one of the 1980 Chevettes,

$$D = \{(0, 0), (1, 0), (0, 1), (1, 1)\}$$

is the event that neither salesperson sells both, and

$$E = \{(1, 1)\}$$

is the event that each salesperson sells one.

Note that events C and E have no elements in common. Such events are referred to as **mutually exclusive,** which means that they cannot both occur at the same time. Evidently, if, between them, the two salespersons sell only one of the two cars, they cannot each sell one. On the other hand, events D and E are not mutually exclusive since they both contain the outcome $(1, 1)$, where each salesperson sells one of the two Chevettes.

In many probability problems we are interested in events that can be expressed in terms of two or more events by forming **unions, intersections,** and **complements.** In general, the union of two events X and Y, denoted by $X \cup Y$, is the event which consists of all the elements (outcomes) contained in event X, in event Y, or in both; the intersection of two events X and Y, denoted by $X \cap Y$, is the event which consists of all the elements (outcomes) contained in both X and Y; and the complement of X, denoted by X', is the event which consists of all the elements (outcomes) of the sample space that are not contained in X. We usually read $\cup$ as "or," $\cap$ as "and," and X' as "not X."

EXAMPLE For the example about the number of days that it rains in Chicago in January and the events A and B defined on page 105, list the outcomes comprising each of the following events, and also express the events in words:

(a) $A \cup B$; (c) B';

(b) $A \cap B$; (d) $A' \cap B'$.

Solution **(a)** Since $A \cup B$ contains all the elements that are in A, in B, or in both, we find that

$$A \cup B = \{18, 19, 20, \ldots, 30, 31\}$$

and this is the event that there will be at least 18 rainy days; **(b)** since $A \cap B$ contains all the elements that are in both A and B, we find that

$$A \cap B = \{20, 21, 22, 23, 24\}$$

and this is the event that there will be from 20 to 24 rainy days; **(c)** since B' contains all the elements of the sample space that are not in B, we find that

$$B' = \{0, 1, 2, \ldots, 18, 19\}$$

and this is the event that there will be at most 19 rainy days; **(d)** since $A' \cap B'$ contains all the elements of the sample space that are neither in A nor in B, we find that

$$A' \cap B' = \{0, 1, 2, \ldots, 16, 17\}$$

and this is the event that there will be at most 17 rainy days.

Sample spaces and events, particularly relationships among events, are often pictured by means of **Venn diagrams** such as those of Figures 5.2 and

FIGURE 5.2 *Venn diagrams.*

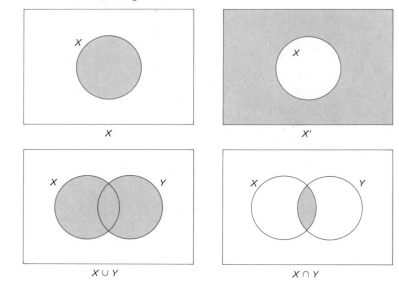

5.3. In each case, the sample space is represented by a rectangle, and events by circles or parts of circles within the rectangle. The tinted regions of the four Venn diagrams of Figure 5.2 represent event X, the complement of event X, the union of events X and Y, and the intersection of events X and Y.

EXAMPLE If X is the event that Mr. Green is a naturalized citizen and Y is the event that his wife is a naturalized citizen, what events are represented by the tinted regions of the four Venn diagrams of Figure 5.2?

Solution The tinted region of the first diagram represents the event that Mr. Green is a naturalized citizen; the tinted region of the second diagram represents the event that Mr. Green is not a naturalized citizen; the tinted region of the third diagram represents the event that Mr. Green, his wife, or both are naturalized citizens; and the tinted region of the fourth diagram represents the event that Mr. Green and his wife are both naturalized citizens.

When we deal with three events, we draw the circles as in Figure 5.3. In this diagram, the circles divide the sample space into eight regions, numbered 1 through 8, and it is easy to determine whether the corresponding events are in X or X', Y or Y', and Z or Z'.

EXAMPLE If X is the event that unemployment will go up, Y is the event that stock prices will go up, and Z is the event that interest rates will go up, express in words what events are represented by the following regions of the Venn diagram of Figure 5.3:
(a) region 4;
(b) regions 1 and 3 together;
(c) regions 3, 5, 6, and 8 together.

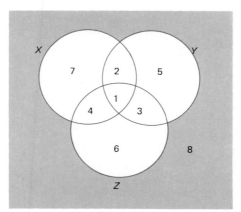

FIGURE 5.3 *Venn diagram.*

Solution **(a)** Since this region is contained in X and Z but not Y, it represents the event that unemployment and interest rates will go up, but stock prices will not go up; **(b)** since this is the region common to Y and Z, it represents the event that stock prices and interest rates will go up; **(c)** since this is the entire region outside X, it represents the event that unemployment will not go up.

EXERCISES

5.1 With reference to the illustration on page 105, suppose that $a, b, c, d, e, f, g,$ and h denote the events that Ms. Adam, Mr. Bean, Miss Clark, Mrs. Daly, Mr. Earl, Ms. Fuentes, Ms. Gardner, and Mr. Hall is awarded the scholarship, and that $A = \{b, e, h\}$, $B = \{a, b, e\}$, and $C = \{a, f, h\}$. Express each of the following symbolically and also in words:

    (a) A';
    (b) $A \cup B$;
    (c) $A \cap B$;
    (d) C';
    (e) $B \cup C$;
    (f) $A' \cap C'$.

5.2 In an experiment, persons are asked to pick a number from 1 to 10, so that for each person the sample space is the set $S = \{1, 2, \ldots, 9, 10\}$. If $C = \{1, 2, 3, 4, 5, 6\}$, $D = \{5, 6, 7, 8, 9, 10\}$, and $E = \{4, 5, 6, 7\}$, express each of the following symbolically and also in words:

    (a) C';
    (b) $C \cup D$;
    (c) $C \cap D$;
    (d) $D \cup E$;
    (e) $D \cap E'$;
    (f) $D' \cap E'$.

5.3 To construct sample spaces for experiments in which we deal with categorical data, we often code the various alternatives by assigning them numbers. For instance, if persons are asked whether their favorite color is red, yellow, blue, green, brown, white, purple, or some other color, we might assign these alternatives the codes 1, 2, 3, 4, 5, 6, 7, and 8. If $A = \{3, 4\}$, $B = \{1, 2, 3, 4, 5, 6, 7\}$, and $C = \{6, 7, 8\}$, express each of the following symbolically:

    (a) B';
    (b) $A \cup B$;
    (c) $A \cap B$;
    (d) C';
    (e) $B \cup C$;
    (f) $B \cap C'$.

5.4 With reference to the preceding exercise, which of the pairs of events, A and B, A and C, and B and C, are mutually exclusive?

5.5 With reference to the two salespersons and Figure 5.1, describe each of the following events in words:

    (a) $F = \{(0, 0), (1, 1)\}$;
    (b) $G = \{(1, 0), (1, 1)\}$;
    (c) $H = \{(0, 0), (0, 1), (0, 2)\}$.

5.6 With reference to the preceding exercise, which of the pairs of events, F and G, F and H, and G and H, are mutually exclusive?

5.7 With reference to the two salespersons and Figure 5.1, list the points of the sample space which comprise the following events:

    (a) One of the salespersons sells both cars.
    (b) The second salesperson sells one of the cars.
    (c) The first salesperson sells at least one car.

5.8 A movie critic has two days in which to view some of the pictures that have recently been released. She wants to see at least three of the movies but not more than three on either day.

    (a) Using two coordinates so that (3, 1), for example, represents the event that she will see three of the movies on the first day and one on the second day, draw a diagram similar to that of Figure 5.1 showing the ten points of the corresponding sample space.

    (b) If T is the event that altogether she will see three of the movies, U is the event that she will see more of the movies on the second day than on the first, V is the event that she will see three of the movies on the first day, and W is the event that she will see equally many movies on both days, express each of these events symbolically by listing its elements.

5.9 With reference to the preceding exercise, which of the six pairs of events, T and U, T and V, T and W, U and V, U and W, and V and W, are mutually exclusive?

5.10 A small marina has three fishing boats which are sometimes in dry dock for repairs.

    (a) Using two coordinates so that (2, 1), for example, represents the event that two of the fishing boats are in dry dock and one is rented out for the day, and (0, 2) represents the event that none of the boats is in dry dock and two are rented out for the day, draw a diagram similar to that of Figure 5.1 showing the ten points of the corresponding sample space.

    (b) If K is the event that at least two of the boats are rented out for the day, L is the event that more boats are in dry dock than are rented out for the day, and M is the event that all the boats that are not in dry dock are rented out for the day, express each of these events symbolically by listing its elements.

    (c) With reference to part (b), list the elements of the sample space that are in K' and in $L \cap M$, and describe in words the corresponding events.

    (d) With reference to part (b), which of the three pairs of events, K and L, K and M, and L and M, are mutually exclusive?

5.11 Among six applicants for an executive job, A is a college graduate, foreign born, and single; B is not a college graduate, foreign born, and married; C is a college graduate, native born, and married; D is not a college graduate, native born, and single; E is a college graduate, native born, and married; and F is not a college graduate, native born, and married. One of these applicants is to get the job, and the event that the job is given to a college graduate, for example, is denoted $\{A, C, E\}$. State in a similar manner the event that the job is given to

    (a) a single person;

    (b) a native-born college graduate;

    (c) a married person who is foreign born.

5.12 If we code tails and heads as 0 and 1, we could let (1, 0, 0) represent the event that we get heads, tails, tails (in that order) in three flips of a coin. Use this notation to list the eight possible outcomes for three flips of a coin, and draw the corresponding three-dimensional sample space.

5.13 Which of the following pairs of events are mutually exclusive? Explain your answers.

    (a) A driver getting a ticket for speeding and a ticket for going through a red light.

(b) Being foreign-born and being President of the United States.

(c) A baseball player getting a walk and hitting a home run in the same at bat.

(d) A baseball player getting a walk and hitting a home run in the same game.

(e) Having rain and sunshine on July 4, 1988.

5.14 In Figure 5.4, L is the event that a driver has liability insurance and C is the event that he or she has collision insurance. Explain in words what events are represented by regions 1, 2, 3, and 4.

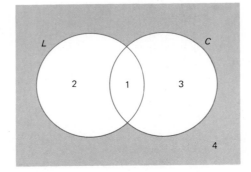

FIGURE 5.4 *Venn diagram for Exercise 5.14.*

5.15 With reference to the preceding exercise, what events are represented by

(a) regions 1 and 3 together;

(b) regions 3 and 4 together;

(c) regions 1, 2, and 3 together?

5.16 In Figure 5.5, W is the event that a new novel is well written and F is the event that it is a financial success. Explain in words what events are represented by regions 1, 2, 3, and 4.

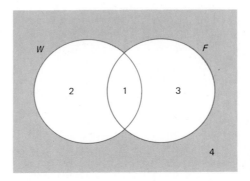

FIGURE 5.5 *Venn diagram for Exercise 5.16.*

5.17 With reference to the preceding exercise, what events are represented by

(a) regions 1 and 2 together;

(b) regions 2 and 4 together;

(c) regions 2, 3, and 4 together?

5.18 With reference to the example on page 109 and Figure 5.3, what regions or combinations of regions represent the following events?

    (a) Unemployment, stock prices, and interest rates will go up.

    (b) Stock prices will go up, but unemployment and interest rates will not go up.

    (c) Interest rates will go up, but unemployment will not go up.

    (d) Unemployment or stock prices, but not interest rates, will go up.

5.19 In Figure 5.6, E, T, and N are the events that a car brought to a garage needs an engine overhaul, transmission repairs, or new tires. Express in words what events are represented by

    (a) region 1;   (d) regions 1 and 4 together;

    (b) region 3;   (e) regions 2 and 5 together;

    (c) region 7;   (f) regions 3, 5, 6, and 8 together.

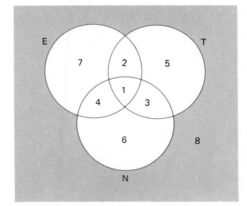

FIGURE 5.6 *Venn diagram for Exercise 5.19.*

5.20 With reference to the preceding exercise and the Venn diagram of Figure 5.6, list the regions or combinations of regions which represent the events that a car brought to the garage needs

    (a) transmission repairs, but neither an engine overhaul nor new tires;

    (b) an engine overhaul and transmission repairs;

    (c) transmission repairs or new tires, but not an engine overhaul;

    (d) new tires.

5.21 Venn diagrams are often used to verify relationships among sets, subsets, or events, without requiring formal proofs based on the algebra of sets. We simply check whether the expressions which are supposed to be equal are represented by the same region of a Venn diagram. Use Venn diagrams to show that

    (a) $A \cup (A \cap B) = A$;

    (b) $(A \cap B) \cup (A \cap B)' = A$;

    (c) $(A \cap B)' = A' \cup B'$ and also $(A \cup B)' = A' \cap B'$;

    (d) $A \cup B = (A \cap B) \cup (A \cap B') \cup (A' \cap B)$;

    (e) $A \cap (B \cup C) = (A \cap B) \cup (A \cap C)$.

5.2
THE POSTULATES OF PROBABILITY

We turn now to the question of how probabilities must "behave," and we begin by stating the three basic postulates. To formulate these postulates and some of their immediate consequences, we shall continue the practice of denoting events by capital letters, and we shall write the probability of event A as $P(A)$, the probability of event B as $P(B)$, and so forth. As before, we shall denote the set of all possible outcomes, the sample space, by the letter S. As we shall formulate them here, the three postulates of probability apply when the sample space S is finite.

*First two
postulates of
probability*

> 1. *Probabilities are positive real numbers or zero; symbolically, $P(A) \geq 0$ for any event A.*
> 2. *Every sample space has probability 1; symbolically, $P(S) = 1$ for any sample space S.*

Let us justify these two postulates, as well as the one which follows with reference to the classical probability concept and the frequency interpretation; in Section 5.3 we shall see to what extent the postulates are compatible also with subjective probabilities.

So far as the first postulate is concerned, the fraction $\frac{s}{n}$ is always positive or zero, and so are percentages or proportions. The second postulate states indirectly that certainty is identified with a probability of 1; after all, one of the possibilities included in S must occur, and it is to this certain event that we assign a probability of 1. For equally likely outcomes, $s = n$ for the whole sample space and $\frac{s}{n} = \frac{n}{n} = 1$; and in the frequency interpretation, a probability of 1 implies that the event will occur 100 percent of the time, or in other words, that it is certain to occur.

In actual practice, we also assign a probability of 1 to events which are "practically certain" to occur. For instance, we would assign a probability of 1 to the event that at least one person will vote in the next presidential election, and that among all new cars sold during any one model year at least one will be involved in an accident before it has been driven 12,000 miles.

The third postulate of probability is especially important, but it is not quite so obvious as the other two.

3. *If two events are mutually exclusive, the probability that one or the other will occur equals the sum of their probabilities. Symbolically,*

$$P(A \cup B) = P(A) + P(B)$$

for any two mutually exclusive events A and B.

For instance, if the probability that weather conditions will improve during a certain week is 0.62 and the probability that they will remain unchanged is 0.23, then the probability that they will either improve or remain unchanged is $0.62 + 0.23 = 0.85$. Similarly, if the probabilities that a student will get an A or a B in a course are 0.13 and 0.29, then the probability that he or she will get either an A or a B is $0.13 + 0.29 = 0.42$.

This postulate is also compatible with the classical probability concept and the frequency interpretation. In the classical concept, if s_1 of n equally likely possibilities constitute event A and s_2 others constitute event B, then these $s_1 + s_2$ equally likely possibilities constitute event $A \cup B$, and we have

$$P(A) = \frac{s_1}{n}, \qquad P(B) = \frac{s_2}{n}, \qquad P(A \cup B) = \frac{s_1 + s_2}{n},$$

and $P(A) + P(B) = P(A \cup B)$. In accordance with the frequency interpretation, if one event occurs, say, 36 percent of the time, another event occurs 41 percent of the time, and they cannot both occur at the same time (that is, they are mutually exclusive), then one or the other will occur $36 + 41 = 77$ percent of the time; this satisfies the third postulate.

By using the three postulates of probability, we can derive many further rules according to which probabilities must "behave"—some of them are easy to prove and some are not, but they all have important applications. Among the immediate consequences of the three postulates we find that probabilities can never be greater than 1, that an event which cannot occur has probability 0, and that the probabilities that an event will occur and that it will not occur always add up to 1. Symbolically,

$$P(A) \leq 1 \qquad \text{for any event } A$$
$$P(\varnothing) = 0$$
$$P(A) + P(A') = 1 \qquad \text{for any event } A$$

The first of these results simply expresses the fact that there cannot be more favorable outcomes than there are outcomes, and that an event cannot occur more than 100 percent of the time. The second result expresses the fact that when an event cannot occur there are $s = 0$ favorable outcomes, and that such an event occurs zero percent of the time. In actual practice, we also assign 0 probability to events which are so unlikely that we are "practically certain" they will not occur. For instance, we would assign 0 probability to the event that a monkey set loose on a typewriter will by chance type Plato's *Republic* word for word without a single mistake.

The third result can also be derived from the postulates of probability, and it can easily be seen that it is compatible with the classical probability concept and the frequency interpretation. In the classical concept, if there are s "successes" there are $n - s$ "failures," the corresponding probabilities are $\frac{s}{n}$ and $\frac{n - s}{n}$, and their sum is $\frac{s}{n} + \frac{n - s}{n} = 1$. In accordance with the frequency interpretation, if certain investments are successful 22 percent of the time, then they are not successful 78 percent of the time, the corresponding probabilities are 0.22 and 0.78, and their sum is 1.

The examples which follow show how the postulates and the further rules we gave on page 115 are put to use in actual practice.

EXAMPLE   If A and B are the events that a consumer testing service will rate a given stereo system very good or good, $P(A) = 0.22$ and $P(B) = 0.35$, find
    (a) $P(A')$;
    (b) $P(A \cup B)$;
    (c) $P(A \cap B)$.

Solution   **(a)** From the third of the further rules, we find that $P(A') = 1 - P(A) = 1 - 0.22 = 0.78$; **(b)** since A and B are mutually exclusive, it follows from the third postulate that $P(A \cup B) = P(A) + P(B) = 0.22 + 0.35 = 0.57$; **(c)** since A and B are mutually exclusive, $A \cap B = \varnothing$ and it follows that $P(A \cap B) = 0$ in accordance with the second of the further rules.

In problems like this, it often helps to draw a Venn diagram, fill in the probabilities associated with the various regions, and then read the answers directly off the diagram.

EXAMPLE   If U and W are the events that a person will take a United Airlines flight or a Western Airlines flight between two cities, $P(U) = 0.48$ and $P(W) = 0.27$, find $P(U' \cap W')$, the probability that the person will take an airline other than United or Western.

Solution   Drawing the Venn diagram as in Figure 5.7, we first put a 0 into region 1 because U and W are mutually exclusive events. It follows

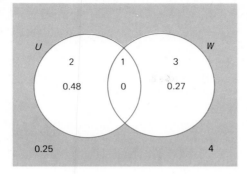

FIGURE 5.7 *Venn diagram.*

that the 0.48 probability of event U must go into region 2, the 0.27 probability of event W must go into region 3, and since the probability for the entire sample space equals 1, we put $1 - (0.48 + 0.27) = 0.25$ into region 4. Since $U' \cap W'$ is represented by region 4, the region outside both circles, we find that $P(U' \cap W') = 0.25$.

5.3
PROBABILITIES AND ODDS⋆

If an event is twice as likely to occur as not to occur, we say that the **odds** are 2 to 1 that it will occur; if an event is three times as likely to occur as not to occur, we say that the odds are 3 to 1; if an event is ten times as likely to occur as not to occur, we say that the odds are 10 to 1; and so forth. In general,

> **The odds that an event will occur are given by the ratio of the probability that the event will occur to the probability that it will not occur.**

Symbolically,

Formula relating odds to probabilities

> *If the probability of an event is p, the odds for its occurrence are a to b, where a and b are positive values such that*
>
> $$\frac{a}{b} = \frac{p}{1 - p}$$

It is customary to express odds as a ratio of two positive integers having no common factor, and if an event is more likely not to occur than to occur, to give the odds that it will not occur rather than the odds that it will occur.

EXAMPLE   What are the odds for the occurrence of an event if its probability is
(a) $\frac{4}{7}$;
(b) 0.95;
(c) 0.20?

Solution   (a) By definition, the odds are $\frac{4}{7}$ to $1 - \frac{4}{7} = \frac{3}{7}$, or 4 to 3; (b) by definition, the odds are 0.95 to $1 - 0.95 = 0.05$, 95 to 5, or 19 to 1; (c) by definition, the odds are 0.20 to $1 - 0.20 = 0.80$, 20 to 80, or 1 to 4, but we say instead that the odds against the occurrence of the event are 4 to 1.

In gambling, the word "odds" is also used to denote the ratio of the wager of one party to that of another. For instance, if a gambler says that he will give 3 to 1 odds on the occurrence of an event, he means that he is willing to bet $3 against $1 (or perhaps $30 against $10 or $1,500 against $500) that the event will occur. If such **betting odds** actually equal the odds that the event will occur, we say that the betting odds are **fair.**

EXAMPLE   Suppose that $\frac{1}{9}$ of the mail between two cities is not delivered within 48 hours. If someone offers to bet $30 against $4 that a certain letter between the two cities will be delivered within 48 hours, are these betting odds fair?

Solution   Since $1 - \frac{1}{9} = \frac{8}{9}$ of the mail between the two cities is delivered on time, the odds are 8 to 1, and the bet would be fair if the person offered to bet $32 against $4 that the letter will be delivered on time. Thus, the original bet of $30 against $4 favors the person offering the bet; it is not fair.

The preceding discussion provides the groundwork for a way of measuring **subjective probabilities.** If a businessman feels that the odds for the success of a new clothing store are 3 to 2, this means that he is willing to bet (or considers it fair to bet), say, $300 against $200 that the new store will be a success. In this way he expresses his belief regarding the uncertainties connected with the success of the store, and to convert it into a probability we first solve the equation $\dfrac{a}{b} = \dfrac{p}{1 - p}$ for p and then substitute $a = 3$ and $b = 2$. Leaving the details to the reader in Exercise 5.42, let us merely state here that the first step is taken care of by the following result:

Formula relating probabilities to odds

> *If the odds are a to b that an event will occur, the probability of its occurrence is*
>
> $$p = \frac{a}{a + b}$$

Then, substituting $a = 3$ and $b = 2$, we find that, according to the business-man, the probability for the store's success is $p = \dfrac{3}{3 + 2} = 0.60$.

EXAMPLE If an applicant for a teaching position feels that the odds are 7 to 4 that she will get the job, what probability is she thus assigning to her getting the job?

Solution Substituting $a = 7$ and $b = 4$ into the formula for p, we get

$$p = \frac{7}{7 + 4} = \frac{7}{11}$$

or approximately 0.64.

Let us now see whether subjective probabilities, determined in this way, "behave" in accordance with the postulates of probability. Since a and b are positive quantities, $\dfrac{a}{a + b}$ cannot be negative and this satisfies the first postulate. As for the second postulate, we note that the surer we are that an event will occur, the "better" odds we should be willing to give—say, 100 to 1, 1,000 to 1, or perhaps even 1,000,000 to 1. The corresponding prob-abilities are $\dfrac{100}{100 + 1} = 0.99$, $\dfrac{1,000}{1,000 + 1} = 0.999$, and $\dfrac{1,000,000}{1,000,000 + 1} = 0.999999$, and it can be seen that the surer we are that an event will occur, the closer its probability will be to 1.

The third postulate—$P(A \cup B) = P(A) + P(B)$ for any two mutually exclusive events A and B—does not necessarily apply to subjective prob-abilities, but proponents of the subjectivist point of view impose it as a **consistency criterion**. In other words, if a person's subjective probabilities "behave" in accordance with the third postulate, he or she is said to be consistent; otherwise, the person's probability judgments ought not to be taken seriously.

EXAMPLE A student feels that the odds are 9 to 1 that he will not get an A in statistics, 4 to 1 that he will not get a B, and 7 to 3 that he will get neither an A nor a B. Are the corresponding probabilities consistent?

Solution The probabilities that the student will get an A, a B, or an A or a B are $\dfrac{1}{1 + 9} = \dfrac{1}{10}$, $\dfrac{1}{1 + 4} = \dfrac{1}{5}$, and $\dfrac{3}{3 + 7} = \dfrac{3}{10}$. Since $\dfrac{1}{10} + \dfrac{1}{5} = \dfrac{3}{10}$, the probabilities are consistent.

5.22 In a study of the adequacy of fuel supplies, C stands for the event that a power plant uses coal and I is the event that it provides inexpensive electricity. State in words what probabilities are expressed by

(a) $P(C')$;
(b) $P(I')$;
(c) $P(C \cup I)$;

(d) $P(C \cap I)$;
(e) $P(C \cup I')$;
(f) $P(C' \cap I')$.

5.23 If F is the event that a dishonest stockbroker is in financial difficulties, T is the event that he has tax problems, and Q is the event that he uses questionable sales practices, write in symbolic form the probabilities that a dishonest stockbroker

(a) is in financial difficulties and uses questionable sales practices;
(b) does not have tax problems but is in financial difficulties;
(c) uses questionable sales practices and/or is in financial difficulties;
(d) has neither financial difficulties nor tax problems.

5.24 Explain why there must be a mistake in each of the following statements:

(a) The probability that a mineral sample will contain copper is 0.28 and the probability that it will not contain copper is 0.62.
(b) The probability that Dan will pass the bar examination is 0.34 and the probability that he will not pass is -0.66.
(c) The probability that the home team will win an upcoming football game is 0.77, the probability that it will tie the game is 0.08, and the probability that it will win or tie the game is 0.95.
(d) The probability that a chemistry experiment will succeed is 0.73 and the probability that it will not succeed is 0.47.

★5.25 Express symbolically what general rule is violated in each of the following assertions:

(a) The probability that an ambulance service will receive more than ten calls on a certain day is 0.53 and the probability that it will receive more than twelve calls is 0.60.
(b) The probability that a student will get a passing grade in English is 0.72 and the probability that she will get a passing grade in English and French is 0.85.

5.26 Given the mutually exclusive events C and D for which $P(C) = 0.29$ and $P(D) = 0.43$, find

(a) $P(C')$;
(b) $P(D')$;
(c) $P(C \cap D)$;

(d) $P(C \cup D)$;
(e) $P(C' \cup D')$;
(f) $P(C' \cap D')$.

5.27 The probabilities that a typist will make at most three mistakes when typing a report, or make from four to eight mistakes, are 0.58 and 0.26. Find the probabilities that the typist will make

(a) at least four mistakes;
(b) at most eight mistakes;
(c) more than eight mistakes.

5.28 If the probabilities that a certain missile will explode during lift-off or have its guidance system fail in flight are, respectively, 0.0002 and 0.0005, find the probabilities that such a missile will

(a) not explode during lift-off;

(b) either explode during lift-off or have its guidance system fail in flight;

(c) not explode during lift-off nor have its guidance system fail in flight.

★5.29 Convert each of the following probabilities to odds:

(a) The probability of rolling "7 or 11" with a pair of balanced dice is $\frac{2}{9}$.

(b) The probability of getting three heads and three tails in six flips of a coin is $\frac{5}{16}$.

(c) The probability that the last digit of a car's license plate is 2, 3, 4, 5, 6, or 7 is $\frac{6}{10}$.

★5.30 If the probability is 0.36 that next year's inflation rate will exceed this year's, what are the corresponding odds?

★5.31 If the probability that a shipment of laboratory supplies will arrive on time is $\frac{4}{13}$, what are the odds that it will not arrive on time?

★5.32 Convert each of the following odds to probabilities:

(a) If three eggs are randomly chosen from a carton of twelve eggs of which three are cracked, the odds are 34 to 21 that at least one of them will be cracked.

(b) If a person has eight $1 bills, five $5 bills, and one $20, and randomly selects three of them, the odds are 11 to 2 that they will not all be $1 bills.

(c) If we arbitrarily arrange the letters in the word "nest," the odds are 5 to 1 that we will not get a meaningful word in the English language.

★5.33 A football coach claims that the odds are 2 to 1 that his team will win an upcoming game, and that the odds against his team's losing or tieing are, respectively, 4 to 1 and 9 to 1. Can these odds be right? Explain.

★5.34 A stockbroker is unwilling to bet $40 against $120 that the price of a certain stock will go up within a week. What does this tell us about the subjective probability he assigns to the stock's price going up within a week? (*Hint:* The answer should read "less than....")

★5.35 If a student is anxious to bet $25 against $5 that she will pass a certain course, what does this tell us about the personal probability she assigns to her passing the course? (*Hint:* The answer should read "greater than....")

★5.36 A television producer is willing to bet $1,500 against $1,000, but not $2,000 against $1000, that a new game show will be a success. What does this tell us about the probability which the producer assigns to the show's success?

★5.37 A high school principal feels that the odds are 7 to 5 against her getting a $1,000 raise and 11 to 1 against her getting a $2,000 raise. Furthermore, she feels that it is an even-money bet that she will get one of these raises or the other. Discuss the consistency of the corresponding subjective probabilities.

★5.38 Asked about his political future, a party official replies that the odds are 2 to 1 that he will not run for the House of Representatives and 4 to 1 that he will not run for the Senate. Furthermore, he feels that the odds are 7 to 5 that he will run for one or the other. Are the corresponding probabilities consistent?

★5.39 There are two Porsches in a race, and a reporter feels that the odds against their winning are, respectively, 3 to 1 and 4 to 1. To be consistent, what odds should he assign to the event that neither car will win?

5.40 The following is a proof of the fact that $P(A) \leq 1$ for any event A: By definition A and A' represent mutually exclusive events and $A \cup A' = S$ (since A and A'

together constitute all the points of the sample space S). So, we can write $P(A \cup A') = P(S)$, and it follows that

$$P(A) + P(A') = P(S) \qquad \text{step 1}$$

$$P(A) + P(A') = 1 \qquad \text{step 2}$$

$$P(A) = 1 - P(A') \qquad \text{step 3}$$

$$P(A) \leq 1 \qquad \text{step 4}$$

State which of the three postulates justify the first, second, and fourth steps of this proof; the third step is simple arithmetic. Note also that in step 2 we actually proved the third of the three rules on page 115.

5.41 Making use of the fact that $S \cup \varnothing = S$ for any sample space S, and S and $\varnothing$ are mutually exclusive (by default), prove the second of the three rules on page 115, namely, $P(\varnothing) = 0$.

★5.42 Verify algebraically that the formula $\dfrac{a}{b} = \dfrac{p}{1-p}$, solved for p, yields $p = \dfrac{a}{a+b}$.

5.4
ADDITION RULES

The third postulate of probability applies only to two mutually exclusive events, but it can easily be generalized; repeatedly using this postulate, it can be shown that

Generalization of Postulate 3

> *If k events are mutually exclusive, the probability that one of them will occur equals the sum of their individual probabilities; symbolically,*
>
> $$P(A_1 \cup A_2 \cup \cdots \cup A_k) = P(A_1) + P(A_2) + \cdots + P(A_k)$$
>
> *for any mutually exclusive events $A_1, A_2, \ldots,$ and A_k.*

where, again, $\cup$ is usually read "or."

EXAMPLE The probabilities that a woman will buy a new dress for a party at Bullock's, the Broadway Southwest department store, or the May Co., are 0.22, 0.18, and 0.35. What is the probability that she will buy the new dress at one of these stores?

Solution Since the three possibilities are mutually exclusive, direct substitution into the formula yields

$$0.22 + 0.18 + 0.35 = 0.75$$

EXAMPLE The probabilities that a consumer testing service will rate a new antipollution device for cars very poor, poor, fair, good, very good, or excellent are 0.07, 0.12, 0.17, 0.32, 0.21, and 0.11. What are the probabilities that it will rate the device

(a) very poor, poor, fair, or good;

(b) good, very good, or excellent?

Solution Since the possibilities are all mutually exclusive, direct substitution into the formula yields

$$0.07 + 0.12 + 0.17 + 0.32 = 0.68$$

for part **(a)** and

$$0.32 + 0.21 + 0.11 = 64$$

for part **(b)**.

The job of assigning probabilities to all possible events connected with a given situation can be very tedious. For instance, if there are as few as four outcomes (points) in a sample space S, there is $\binom{4}{0} = 1$ subset (the empty set) which contains no outcomes at all, there are $\binom{4}{1} = 4$ subsets which contain one outcome, $\binom{4}{2} = 6$ subsets which contain two outcomes, $\binom{4}{3} = 4$ subsets which contain three outcomes, and there is $\binom{4}{4} = 1$ subset (the whole sample space S) which contains all four outcomes. Thus, there are $1 + 4 + 6 + 4 + 1 = 16$ different subsets of a sample space with four outcomes.

In general, if we let $a = 1$ and $b = 1$ in the binomial expansion of $(a + b)^n$, it follows directly that a sample space with n outcomes has 2^n different subsets, namely, that

$$\binom{n}{0} + \binom{n}{1} + \binom{n}{2} + \cdots + \binom{n}{n-1} + \binom{n}{n} = 2^n$$

where $\binom{n}{0}, \binom{n}{1}, \binom{n}{2}, \ldots, \binom{n}{n-1}$, and $\binom{n}{n}$ are the numbers of subsets containing $0, 1, 2, \ldots, n - 1$, and n outcomes. As we saw in the preceding paragraph, a sample space with $n = 4$ outcomes has $2^4 = 16$ different subsets. As the number of outcomes increases, the number of subsets increases very rapidly; for instance, a sample space with $n = 20$ outcomes has $2^{20} = 1,048,576$ different subsets.

Fortunately, it is seldom necessary to assign probabilities to all possible subsets or events, and the following rule (which is a direct application of the above generalization of the third postulate) makes it relatively easy to determine the probability of any event on the basis of the probabilities of the individual outcomes (points) of the sample space:

Rule for calculating the probability of an event

> *The probability of any event A is given by the sum of the probabilities of the individual outcomes comprising A.*

In the special case where the outcomes are all equiprobable, this rule leads to the formula $P(A) = \dfrac{s}{n}$, which we used earlier in connection with the classical probability concept. Here, n is the total number of outcomes in the sample space and s is the number of "successes," namely, the number of outcomes in event A.

EXAMPLE   Referring again to the two-salesperson example on page 106, suppose that the six points of the sample space have the probabilities shown in Figure 5.8. Find the probabilities that
    (a) the first salesperson will not sell either car;
    (b) between them, they will sell both cars;
    (c) the second salesperson will sell at least one of the cars.

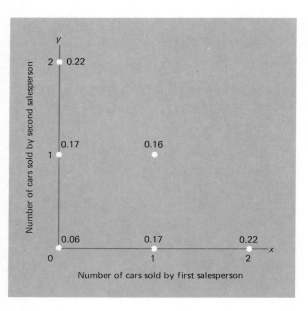

FIGURE 5.8 *Sample space with probabilities assigned to the individual outcomes.*

Solution **(a)** Adding the probabilities associated with the points (0, 0), (0, 1) and (0, 2), we get 0.06 + 0.17 + 0.22 = 0.45; **(b)** adding the probabilities associated with the points (2, 0), (1, 1), and (0, 2), we get 0.22 + 0.16 + 0.22 = 0.60; **(c)** adding the probabilities associated with the points (0, 1), (1, 1), and (0, 2), we get 0.17 + 0.16 + 0.22 = 0.55.

EXAMPLE Assuming that the 44 points (outcomes) of the sample space of Figure 5.9 are all equiprobable, find $P(A)$.

Solution Since there are $s = 10$ outcomes in A and the $n = 44$ outcomes of the sample space are all equiprobable, it follows that $P(A) = \frac{10}{44} = \frac{5}{22}$.

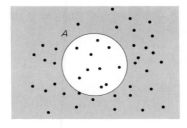

FIGURE 5.9 *Sample space with 44 outcomes.*

Since the third postulate of probability applies only to mutually exclusive events, it cannot be used, for example, to find the probability that at least one of two roommates will pass a final examination in Freshman English, the probability that a bird watcher will spot a roadrunner or a cactus wren, or the probability that a customer will buy a shirt or a sweater at a J. C. Penny store. Both roommates can pass the examination, a bird watcher can spot both kinds of birds, and a customer can buy both a shirt and a sweater at the same store.

To find a formula for $P(A \cup B)$ which holds regardless of whether A and B are mutually exclusive events, let us consider the Venn diagram of Figure 5.10. It concerns the job applications of a recent business school graduate,

FIGURE 5.10 *Venn diagram.*

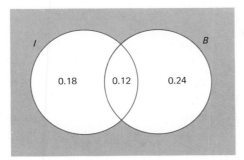

and the letters I and B stand for his getting a job offer from an insurance company or a bank. It follow from the figures in the Venn diagram that

$$P(I) = 0.18 + 0.12 = 0.30$$

$$P(B) = 0.12 + 0.24 = 0.36$$

and

$$P(I \cup B) = 0.18 + 0.12 + 0.24 = 0.54$$

where we could add the probabilities in each case because they represent mutually exclusive events (nonoverlapping regions of the Venn diagram).

If we erroneously use the third postulate of probability to calculate $P(I \cup B)$, we get $P(I) + P(B) = 0.30 + 0.36 = 0.66$, which exceeds the correct value by 0.12. This error results from adding $P(I \cap B) = 0.12$ in twice, once in $P(I) = 0.30$ and once in $P(B) = 0.36$, and we could correct for it by subtracting 0.12 from 0.66. Thus, we could write

$$P(I \cup B) = P(I) + P(B) - P(I \cap B)$$

$$= 0.30 + 0.36 - 0.12$$

$$= 0.54.$$

and this agrees, as it should, with the result obtained before.

Since the argument used in this example holds for any two events A and B, we can now state the following **general addition rule**, which applies regardless of whether A and B are mutually exclusive events:

General addition rule

$$P(A \cup B) = P(A) + P(B) - P(A \cap B)$$

When A and B are mutually exclusive, $P(A \cap B) = 0$ (since by definition the two events cannot both occur at the same time), and the above formula reduces to that of the third postulate of probability. In this connection, the third postulate of probability is also referred to as the **special addition rule.**

EXAMPLE   If the probabilities are 0.20, 0.15, and 0.03 that a student will get a failing grade in chemistry, in English, or in both, what is the probability that the student will get a failing grade in at least one of these subjects?

Solution Direct substitution into the above formula yields

$$0.20 + 0.15 - 0.03 = 0.32$$

EXAMPLE If the probabilities are 0.87, 0.36, and 0.29 that a family, randomly chosen as part of a sample survey in a large metropolitan area, owns a color television set, a black-and-white set, or both, what is the probability that a family in this area will own one or the other or both kinds of sets?

Solution Substituting these values into the formula for the general addition rule, we get

$$0.87 + 0.36 - 0.29 = 0.94$$

The general addition rule can be generalized further so that it applies to more than two events (see Exercise 5.54 on page 129).

EXERCISES

5.43 A police department needs new tires for its patrol cars and the probabilities are 0.17, 0.22, 0.03, 0.29, 0.21, and 0.08 that it will buy Uniroyal tires, Goodyear tires, Michelin tires, General tires, Goodrich tires, or Armstrong tires. Find the probabilities that it will buy
    (a) Goodyear or Goodrich tires;
    (b) Uniroyal, General, or Goodrich tires;
    (c) Michelin or Armstrong tires;
    (d) Goodyear, General, or Armstrong tires.

5.44 The probabilities that a student will get an A, a B, or a C in a history course are 0.09, 0.15, and 0.53. What is the probability that the student will get a grade lower than C?

5.45 The probabilities that a TV station will receive 0, 1, 2, 3, . . . , 8, or at least 9 complaints after showing a controversial program are, respectively, 0.01, 0.03, 0.07, 0.15, 0.19, 0.18, 0.14, 0.12, 0.09, and 0.02. What are the probabilities that after showing such a program the station will receive
    (a) at most 4 complaints;
    (b) at least 6 complaints;
    (c) from 5 to 8 complaints?

5.46 With reference to Figure 5.8, find the probabilities that
    (a) the first salesperson will sell one of the cars;
    (b) between them, they will sell at most one of the cars;
    (c) neither salesperson will sell both cars.

5.47 Figure 5.11 pertains to the number of persons who are invited to a conference and the number of persons who attend. If each of the 35 points of the sample space has the probability $\frac{1}{35}$, what are the probabilities that

(a) at most three persons will attend;

(b) at least six persons will be invited;

(c) one invited person will not attend?

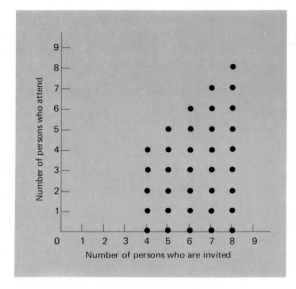

FIGURE 5.11 *Sample space for Exercise 5.47.*

5.48 With reference to Figure 5.9, suppose that each outcome in A is twice as likely as each outcome in A'. Find $P(A)$.

5.49 If H stands for heads and T for tails, the 16 possible outcomes for four flips of a coin are HHHH, HHHT, HHTH, HTHH, THHH, HHTT, HTHT, HTTH, THHT, THTH, TTHH, HTTT, THTT, TTHT, TTTH, and TTTT. Assuming that these 16 possibilities are all equally likely, what are the probabilities of getting 0, 1, 2, 3, or 4 heads in four flips of a balanced coin?

5.50 The probabilities that a person convicted of reckless driving will be fined, have his license revoked, or both are 0.88, 0.62, and 0.55. What is the probability he will be fined or have his license revoked?

5.51 A geology professor has two graduate assistants helping her with her research. The probability that the older of the two will be absent on any given day is 0.08, the probability that the younger of the two will be absent on any given day is 0.06, and the probability that they will both be absent on any given day is 0.02. Find the probability that either or both of the graduate assistants will be absent on any given day.

5.52 The probabilities that a person stopping at a gas station will ask to have his tires checked is 0.14, the probability that he will ask to have his oil checked is

0.27, and the probability that he will ask to have them both checked is 0.09. What are the probabilities that a person stopping at this gas station will have

(a) his tires, his oil, or both checked;

(b) neither his tires nor his oil checked?

5.53 Among the 64 doctors on the staff of a hospital, 58 carry malpractice insurance, 33 are surgeons, and 31 of the surgeons carry malpractice insurance. If one of these doctors is chosen by lot to represent the hospital staff at an AMA convention (that is, each of the doctors has a probability of $\frac{1}{64}$ of being selected), what is the probability that the one chosen is not a surgeon and does not carry malpractice insurance?

★5.54 It can be shown that for any three events A, B, and C, the probability that at least one of them will occur is given by $P(A \cup B \cup C) = P(A) + P(B) + P(C) - P(A \cap B) - P(A \cap C) - P(B \cap C) + P(A \cap B \cap C)$. Use this formula in the following problems:

(a) The probabilities that a person visiting his dentist will have his teeth cleaned, a cavity filled, a tooth extracted, his teeth cleaned and a cavity filled, his teeth cleaned and a tooth extracted, a cavity filled and a tooth extracted, or his teeth cleaned, a cavity filled, and a tooth extracted are 0.47, 0.29, 0.22, 0.08, 0.06, 0.07, and 0.03. What is the probability that the person will have at least one of these things done?

(b) Suppose that if a person visits Disneyland, the probabilities that he will go on the Jungle Cruise, the Monorail, the Matterhorn ride, the Jungle Cruise and the Monorail, the Jungle Cruise and the Matterhorn ride, the Monorail and the Matterhorn ride, or the Jungle Cruise, the Monorail, and the Matterhorn ride are 0.74, 0.70, 0.62, 0.52, 0.46, 0.44, and 0.34. What is the probability that a person visiting Disneyland will go on at least one of these three rides?

5.5
CONDITIONAL PROBABILITY

Difficulties can easily arise when we quote probabilities without specifying the sample space. For instance, if we ask for the probability that a lawyer makes more than $80,000 per year, we may well get many different answers, and they can all be correct. One of them might apply to all lawyers in the United States, another to corporation lawyers, a third to lawyers employed by the federal government, another to lawyers handling only divorces, and so forth. Since the choice of the sample space (that is, the set of all possibilities under consideration) is by no means always self-evident, it is helpful to use the symbol $P(A|S)$ to denote the **conditional probability** of event A relative to the sample space S, or as we often call it "the probability of A given S." The symbol $P(A|S)$ makes it explicit that we are referring to a particular sample space S, and it is generally preferable to the abbreviated notation $P(A)$ unless the tacit choice of S is clearly understood. It is also preferable when we have to refer to different sample spaces in the same problem.

To elaborate on the idea of a conditional probability, suppose that a consumer research organization has studied the service under warranty provided by the 200 tire dealers in a large city, and that their findings are summarized in the following table:

	Good service under warranty	*Poor service under warranty*
Name-brand tire dealers	64	16
Off-brand tire dealers	42	78

Suppose, further, that a person randomly selects one of these tire dealers, where "randomly" means that each of the dealers has the same chance, a probability of $\frac{1}{200}$, of being selected. Now, if we let N denote the selection of a name-brand dealer and G the selection of a dealer who provides good service under warranty, we find that

$$P(N \cap G) = \frac{64}{200} = 0.32$$

$$P(N) = \frac{64 + 16}{200} = 0.40$$

and

$$P(G) = \frac{64 + 42}{200} = 0.53$$

where all these probabilities were calculated by means of the formula $\frac{s}{n}$ for equally likely possibilities.

Since the third of these probabilities is particularly disconcerting—there is almost a fifty–fifty chance of choosing a tire dealer who provides poor service under warranty—let us see what will happen if we limit the choice to name-brand dealers. This reduces the number of equally likely choices to $64 + 16 = 80$, and the probability of choosing a dealer who provides good service under warranty given that he is a name-brand dealer is

$$P(G|N) = \frac{64}{80} = 0.80$$

This is quite an improvement over $P(G) = 0.53$, as might have been expected.

Observe that the conditional probability $P(G|N)$ can also be written as

$$P(G|N) = \frac{64/200}{80/200} = \frac{P(N \cap G)}{P(N)}$$

which is the ratio of the probability of choosing a name-brand dealer who provides good service under warranty to the probability of choosing a name-brand dealer.

Generalizing from this example, let us now make the following definition of **conditional probability,** which applies to any two events A and B belonging to a given sample space S:

Definition of conditional probability

> *If $P(B)$ is not equal to zero, then the conditional probability of A relative to B, namely, the probability of A given B, is*
>
> $$P(A|B) = \frac{P(A \cap B)}{P(B)}$$

EXAMPLE   With reference to the tire dealers of the preceding illustration, what is the probability that an off-brand tire dealer will give good service under warranty?

Solution   As can be seen from the table, $P(G \cap N') = \dfrac{42}{200}$ and $P(N') = \dfrac{42 + 78}{200} = \dfrac{120}{200}$, so that substitution into the formula yields

$$P(G|N') = \frac{P(G \cap N')}{P(N')} = \frac{42/200}{120/200} = 0.35$$

Of course, the fraction $\dfrac{42}{42 + 78} = \dfrac{42}{120} = 0.35$ could have been obtained directly from the second row of the table.

Although we introduced the formula for $P(A|B)$ by means of an example in which the possibilities were all equally likely, this is not a requirement for its use. The only restriction is that $P(B)$ must not equal zero.

EXAMPLE   If the probability that a research project will be well planned is 0.60 and the probability that it will be well planned and well executed is 0.54, what is the probability that a well-planned research project will be well executed?

Solution   Substituting into the formula which defines conditional probability, we get $\dfrac{0.54}{0.60} = 0.90$.

EXAMPLE   With reference to the sample space of Figure 5.8 on page 124, what is the probability that the first salesperson will sell one of the cars given that the second salesperson will sell neither?

Solution   Since the probability that the first salesperson will sell one of the cars and the second salesperson will sell neither is 0.17, and the probability that the second salesperson will sell neither car is $0.06 + 0.17 + 0.22 = 0.45$, substitution into the formula yields $\dfrac{0.17}{0.45} = 0.38$ rounded to two decimals.

To introduce another concept which is important in the study of probability, let us consider the following problem:

EXAMPLE   The probabilities that a student will fail mathematics, geology, or both are $P(M) = 0.30$, $P(G) = 0.20$, and $P(M \cap G) = 0.06$. What is the probability that he will fail geology given that he will fail mathematics?

Solution   Substituting into the formula which defines conditional probability, we get

$$P(G|M) = \frac{P(G \cap M)}{P(M)} = \frac{0.06}{0.30} = 0.20$$

What is special, and interesting, about this result is that $P(G|M) = P(G) = 0.20$; that is, the probability of event G is the same regardless of whether event M has occured (occurs, or will occur).

In general, if $P(A|B) = P(A)$, we say that event A is **independent** of event B, and since it can be shown that event B is independent of event A whenever event A is independent of event B, we say simply that A **and** B **are independent** whenever one is independent of the other. If two events A and B are not independent, we say that they are **dependent.**

5.6
MULTIPLICATION RULES

So far we have used the formula $P(A|B) = \dfrac{P(A \cap B)}{P(B)}$ only to calculate conditional probabilities, but if we multiply on both sides of the equation by $P(B)$, we get the following formula, called the **general multiplication rule,** which enables us to calculate the probability that two events will both occur:

General
multiplication rule

$$P(A \cap B) = P(B) \cdot P(A \mid B)$$

In words, this formula states that the probability that two events will both occur is the product of the probability that one of the events will occur and the conditional probability that the other event will occur given that the first event has occurred (occurs, or will occur). As it does not matter which event is referred to as A and which is referred to as B, the above formula can also be written as

General
multiplication rule

$$P(A \cap B) = P(A) \cdot P(B \mid A)$$

EXAMPLE   If we randomly select two shirts, one after the other, from a carton containing twelve shirts, three of which have blemishes, what is the probability that both of them will have blemishes?

Solution   Since the selection is random, the probability that the first shirt we pick will have blemishes is $\frac{3}{12}$, and the probability that the second shirt we pick will have blemishes given that the first one has blemishes is $\frac{2}{11}$. Clearly, there are only two shirts with blemishes among the eleven which remain after one shirt with blemishes has been picked. Hence, the probability of getting two shirts with blemishes is

$$\frac{3}{12} \cdot \frac{2}{11} = \frac{1}{22}$$

The same kind of argument leads to the result that the probability of getting two shirts without blemishes is

$$\frac{9}{12} \cdot \frac{8}{11} = \frac{12}{22}$$

and it follows, by subtraction, that the probability of getting one shirt with blemishes and one shirt without blemishes is

$$1 - \frac{1}{22} - \frac{12}{22} = \frac{9}{22}$$

SEC. 5.6/ MULTIPLICATION RULES   **133**

When A and B are independent events, we can substitute $P(A)$ for $P(A|B)$ in the first of the two formulas for $P(A \cap B)$, or $P(B)$ for $P(B|A)$ in the second, and we obtain

Special multiplication rule

$$P(A \cap B) = P(A) \cdot P(B)$$

Thus, the probability that two independent events will both occur is simply the product of their probabilities. This rule is sometimes used as the definition of independence; in any case, it may be used to determine whether two given events are independent.

EXAMPLE   What is the probability of getting two heads in two flips of a balanced coin?

Solution   Since the probability of heads is $\frac{1}{2}$ for each flip and the two flips are independent, the probability is $\frac{1}{2} \cdot \frac{1}{2} = \frac{1}{4}$.

EXAMPLE   Two cards are drawn at random from an ordinary deck of 52 playing cards. What is the probability of getting two aces if
(a) the first card is replaced before the second card is drawn;
(b) the first card is not replaced before the second card is drawn?

Solution   **(a)** Since there are four aces among the 52 cards, we get $\frac{4}{52} \cdot \frac{4}{52} = \frac{1}{169}$;
(b) since there are only three aces among the 51 cards which remain after one ace has been removed from the deck, we get $\frac{4}{52} \cdot \frac{3}{51} = \frac{1}{221}$. The distinction between the two parts of this exercise is important in statistics, where we sometimes **sample with replacement** and sometimes **sample without replacement.**

EXAMPLE   If $P(C) = 0.65$, $P(D) = 0.40$, and $P(C \cap D) = 0.24$, are the events C and D independent?

Solution   Since $P(C) \cdot P(D) = (0.65)(0.40) = 0.26$ and not 0.24, the two events are not independent.

The special multiplication rule can easily be generalized so that it applies to the occurrence of three or more independent events—again, we multiply together all the individual probabilities.

EXAMPLE   What is the probability of getting three heads in three flips of a balanced coin?

Solution   The flips of the coin are independent, and we get $\frac{1}{2} \cdot \frac{1}{2} \cdot \frac{1}{2} = \frac{1}{8}$.

EXAMPLE   What is the probability of first rolling four 3's and then a 2 or a 5 in five rolls of a balanced die?

Solution   Multiplying the five probabilities, we get

$$\frac{1}{6} \cdot \frac{1}{6} \cdot \frac{1}{6} \cdot \frac{1}{6} \cdot \frac{2}{6} = \frac{2}{7,776} = \frac{1}{3,888}.$$

For three or more dependent events the multiplication rule becomes more complicated, as is illustrated in Exercise 5.73 on page 137.

EXERCISES   5.55   If W is the event that a worker is well trained and Q is the event that he or she meets the production quota, express symbolically the probabilities that
      (a) a worker who is well trained will meet the production quota;
      (b) a worker who meets the production quota is not well trained;
      (c) a worker who is not well trained will not meet the production quota.

5.56   With reference to the preceding exercise, state in words what probabilities are expressed by
      (a) $P(W|Q)$;
      (b) $P(Q'|W)$;
      (c) $P(W'|Q')$.

5.57   A guidance department gives students various kinds of tests. If I is the event that a student scores high in intelligence, A is the event that a student rates high on a social adjustment scale, and N is the event that a student displays neurotic tendencies, express each of the following probabilities in symbolic form:
      (a) The probability that a student who scores high in intelligence will display neurotic tendencies.
      (b) The probability that a student who does not rate high on the social adjustment scale will not score high in intelligence.
      (c) The probability that a student who displays neurotic tendencies will neither score high in intelligence nor rate high on the social adjustment scale.
      (d) The probability that a student who scores high in intelligence and rates high on the social adjustment scale will not display any neurotic tendencies.

5.58   If E is the event that an applicant for a home mortgage is employed, G is the event that he has a good credit rating, and A is the event that the application is approved, state in words what probabilities are expressed by
      (a) $P(A|E)$;
      (b) $P(A|G)$;
      (c) $P(A'|E')$;
      (d) $P(A|E \cap G)$.

5.59 There are 60 applicants for a job in the news department of a television station. Some of them are college graduates and some are not, some of them have at least three years' experience and some have not, with the exact breakdown being

	College graduates	Not college graduates
At least three years' experience	12	6
Less than three years' experience	24	18

If the order in which the applicants are interviewed by the station manager is random, G is the event that the first applicant interviewed is a college graduate, and T is the event that the first applicant interviewed has at least three years' experience, determine each of the following probabilities directly from the entries and the row and column totals of the table:

(a) $P(G)$;
(b) $P(T')$;
(c) $P(G \cap T)$;
(d) $P(G' \cap T')$;
(e) $P(T|G)$;
(f) $P(G'|T')$.

5.60 Use the results of the preceding exercise to verify that

(a) $P(T|G) = \dfrac{P(G \cap T)}{P(G)}$;

(b) $P(G'|T') = \dfrac{P(G' \cap T')}{P(T')}$.

5.61 If the probabilities are 0.58, 0.25, and 0.19 that a tourist in London will visit Westminster Abbey, the British Museum, or both, find the probabilities that

(a) a tourist in London who visits Westminster Abbey will also visit the British Museum;

(b) a tourist in London who visits the British Museum will also visit Westminster Abbey.

5.62 An English professor figures that the probability is 0.75 that a term paper she receives will be well written. If the probability is 0.51 that such a term paper will be well written and also receive a good grade, what is the probability that a well-written term paper will receive a good grade?

5.63 The probability that a bus from Cleveland to Chicago will leave on time is 0.80, and the probability that it will leave on time and also arrive on time is 0.72.

(a) What is the conditional probability that if such a bus leaves on time it will also arrive on time?

(b) If the probability is 0.75 that such a bus will arrive on time, what is the conditional probability that if such a bus does not leave on time it will nevertheless arrive on time?

5.64 Given $P(A) = 0.50$, $P(B) = 0.30$, and $P(A \cap B) = 0.15$, verify that

(a) $P(A|B) = P(A)$;
(b) $P(A|B') = P(A)$;
(c) $P(B|A) = P(B)$;
(d) $P(B|A') = P(B)$.

5.65 If two cards are drawn from an ordinary deck of 52 playing cards, what are the probabilities that they will both be diamonds if the drawing is
   (a) with replacement;
   (b) without replacement?

5.66 Among 60 pieces of luggage loaded on a plane in San Francisco, 45 are destined for Seattle and 15 for Vancouver. If two of the pieces of luggage are sent to Portland by mistake and the "selection" is random, what are the probabilities that
   (a) both should have gone to Seattle;
   (b) both should have gone to Vancouver;
   (c) one should have gone to Seattle and one to Vancouver?

5.67 In a fifth-grade class of 18 boys and 12 girls, one pupil is chosen each week by lot to act as an assistant to the teacher. What is the probability that a girl will be chosen two weeks in a row if
   (a) the same pupil cannot serve two weeks in a row;
   (b) the restriction of part (a) is removed?

5.68 If A and B are independent events and $P(A) = 0.25$ and $P(B) = 0.60$, find
   (a) $P(A \cap B)$;   (c) $P(A \cup B)$;
   (b) $P(A|B)$;   (d) $P(A' \cap B')$.

5.69 For two rolls of a balanced die, find the probabilities of getting
   (a) two 5's;
   (b) first a 5 and then a number less than 5.

5.70 If the odds are 5 to 3 that event M will not occur, 2 to 1 that event N will occur, and 4 to 1 that they will not both occur, are the two events M and N independent?

5.71 If $P(A) = 0.80$, $P(C) = 0.35$, and $P(A \cap C) = 0.28$, are events A and C independent?

5.72 Find the probabilities of getting
   (a) eight heads in a row with a balanced coin;
   (b) no 3's in four rolls of a balanced die.

5.73 For three or more events which are not independent, the probability that they will all occur is obtained by multiplying the probability that one of the events will occur times the probability that a second of the events will occur given that the first event has occurred times the probability that a third of the events will occur given that the first two events have occurred, and so on. For instance, the probability of drawing without replacement three aces in a row from an ordinary deck of 52 playing cards is

$$\frac{4}{52} \cdot \frac{3}{51} \cdot \frac{2}{50} = \frac{1}{5,525}$$

   (a) A carton contains 12 shirts of which five have blemishes and the rest are good. What is the probability that if three of the shirts are randomly selected from the carton, they will all have blemishes?
   (b) If a person randomly picks four of the 15 gold coins a dealer has in stock, and six of the coins are counterfeits, what is the probability that the coins picked will all be counterfeits?
   (c) If five of a company's ten delivery trucks do not meet emission standards and three of them are chosen for inspection, what is the probability that none of the trucks chosen will meet emission standards?

(d) In a certain city, the probability that it will rain on a November day is 0.60, the probability that a rainy November day will be followed by another rainy day is 0.80, and the probability that a sunny November day will be followed by a rainy day is 0.30. What is the probability that it will rain, rain, not rain, and rain in this city on four consecutive November days?

(e) A department store which bills its charge-account customers once a month has found that if a customer pays promptly one month, the probability is 0.90 that he will also pay promptly the next month; however, if a customer does not pay promptly one month, the probability that he will pay promptly the next month is only 0.40.

(i) What is the probability that a customer who pays promptly one month will also pay promptly the next three months?

(ii) What is the probability that a customer who does not pay promptly one month will also not pay promptly the next two months and then make a prompt payment the month after that?

5.7
BAYES' THEOREM ★

Although the symbols $P(A|B)$ and $P(B|A)$ look very similar, there is a great difference between the probabilities which they represent. For instance, on page 130 we calculated the probability $P(G|N)$ that a name-brand dealer will provide good service under warranty, but what do we mean when we write $P(N|G)$? This is the probability that a tire dealer who provides good service under warranty is a name-brand dealer. To give another example, suppose that B represents the event that a person committed a recent burglary, and G represents the event that he or she is found guilty of the crime. Then, $P(G|B)$ is the probability that the person who committed the burglary will be found guilty of the crime, and $P(B|G)$ is the probability that the person who is found guilty of the burglary actually committed it. Thus, in both of these examples we turned things around—cause, so to speak, became effect and effect became cause.

Since there are many problems in statistics which involve such pairs of conditional probabilities, let us find a formula which expresses $P(B|A)$ in terms of $P(A|B)$ for any two events A and B. Equating the two expressions for $P(A \cap B)$ on page 133, we have

$$P(A) \cdot P(B|A) = P(B) \cdot P(A|B)$$

and, hence,

$$P(B|A) = \frac{P(B) \cdot P(A|B)}{P(A)}$$

after dividing the expressions on both sides of the equation by $P(A)$.

EXAMPLE   In a state where cars have to be tested for the emission of pollutants, 25 percent of all cars emit excessive amounts of pollutants. When tested, 99 percent of all cars that emit excessive amounts of pollutants will fail, but 17 percent of the cars that do not emit excessive amounts of pollutants will also fail. What is the probability that a car which fails the test actually emits excessive amounts of pollutants?

Solution   Letting A denote the event that a car fails the test and B the event that it emits excessive amounts of pollutants, we can translate the given percentages into probabilities and write $P(B) = 0.25$, $P(A|B) = 0.99$, and $P(A|B') = 0.17$.

Before we can calculate $P(B|A)$ by means of the formula given above, we will first have to determine $P(A)$, and to this end let us look at the tree diagram of Figure 5.12. Here A is reached either along the branch which passes through B or along the branch which passes through B', and the probabilities of this happening are, respectively, $(0.25)(0.99) = 0.2475$ and $(0.75)(0.17) = 0.1275$. Since the alternatives represented by the two branches are mutually exclusive, we find that $P(A) = 0.2475 + 0.1275 = 0.3750$, and substitution into the formula for $P(B|A)$ given above yields

$$P(B|A) = \frac{P(B) \cdot P(A|B)}{P(A)} = \frac{(0.25)(0.99)}{0.3750} = 0.66$$

This is the probability that a car which fails the test actually emits excessive amounts of pollutants.

With reference to the tree diagram of Figure 5.12 we can say that $P(B|A)$ is the probability that event A is reached via the upper branch of the tree, and we showed that its value is given by the ratio of the probability

FIGURE 5.12 *Tree diagram for emission testing example.*

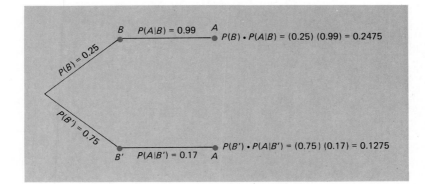

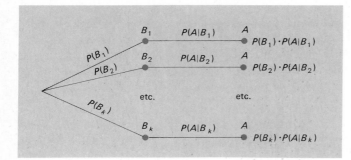

FIGURE 5.13 *Tree diagram for Bayes' theorem.*

associated with that branch to the sum of the probabilities associated with both branches of the tree. This argument can be generalized to the case where there are more than two possible "causes," namely, more than two branches leading to an event A. With reference to Figure 5.13 we can say that $P(B_i|A)$ is the probability that event A is reached via the ith branch of the tree (for $i = 1, 2, \ldots,$ or k), and it can be shown that its value is given by the ratio of the probability associated with the ith branch to the sum of the probabilities associated with all the branches leading to A. This result, which is named after the Methodist clergyman Thomas Bayes (1702–1761), is given by

Bayes' theorem

> *If B_1, B_2, ..., and B_k are mutually exclusive events of which one must occur, then*
>
> $$P(B_i|A) = \frac{P(B_i) \cdot P(A|B_i)}{P(B_1) \cdot P(A|B_1) + P(B_2) \cdot P(A|B_2) + \cdots + P(B_k) \cdot P(A|B_k)}$$
>
> *for $i = 1, 2, \ldots,$ or k.*

Note that the expression in the denominator actually equals $P(A)$; $P(B_1) \cdot P(A|B_1)$ is the probability of reaching A via the first branch, $P(B_2) \cdot P(A|B_2)$ is the probability of reaching A via the second branch, $\ldots, P(B_k) \cdot P(A|B_k)$ is the probability of reaching A via the kth branch, and the sum of all these probabilities equals $P(A)$. This rule for calculating $P(A)$ is often called the **Rule of Elimination** or the **Rule of Total Probability**.

EXAMPLE In a cannery, assembly lines I, II, and III account for 50, 30, and 20 percent of the total output. If 0.4 percent of the cans from assembly line I are improperly sealed, and the corresponding percentages for assembly lines II and III are 0.6 and 1.2 percent, what is the probability that an improperly sealed can (discovered at the final inspection of outgoing products) comes from assembly line I?

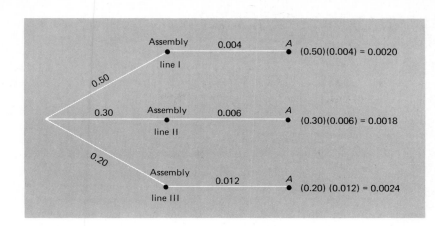

FIGURE 5.14 *Tree diagram for cannery example.*

Solution Letting A denote the event that a can is improperly sealed, and B_1, B_2, and B_3 denote the events that a can comes from assembly lines I, II, or III, we can translate the given percentages into probabilities and write $P(B_1) = 0.50$, $P(B_2) = 0.30$, $P(B_3) = 0.20$, $P(A|B_1) = 0.004$, $P(A|B_2) = 0.006$, and $P(A|B_3) = 0.012$. Then, picturing the situation as in Figure 5.14, we find that the probabilities associated with the three branches of the tree diagram are $(0.50)(0.004) = 0.0020$, $(0.30)(0.006) = 0.0018$, and $(0.20)(0.012) = 0.0024$. Thus, the probability that an improperly sealed can comes from assembly line I is

$$P(B_1|A) = \frac{0.0020}{0.0020 + 0.0018 + 0.0024} = 0.32$$

rounded to two decimals. Of course, if we had substituted directly into the formula for Bayes' theorem, all the calculations and the final result would have been the same.

As can be seen from the two examples of this section, Bayes' formula is a relatively simple mathematical rule. There can be no question about its validity, but criticism has frequently been raised about its applicability. This is because it involves a "backward" or "inverse" sort of reasoning, namely, reasoning from effect to cause. For instance, in the example on page 139 what brought about, or caused, a car's failing the test—the emission of excessive amounts of pollutants or an error in the testing procedure? Also, in the example immediately above, what produced, or caused, an improperly sealed can—assembly line I, assembly line II, or assembly line III?

★5.74 A hotel gets cars for its guests from three rental agencies, 20 percent from agency X, 40 percent from agency Y, and 40 percent from agency Z. If 14 percent of the cars from X, 4 percent from Y, and 8 percent from Z need tune-ups, what is the probability that a car needing a tune-up is delivered to one of the hotel's guests?

★5.75 In a T-maze, a rat is given food if it turns left and an electric shock if it turns right. On the first trial there is a fifty–fifty chance that a rat will turn either way; then, if it receives food on the first trial, the probability that it will turn left on the second trial is 0.68, and if it receives a shock on the first trial, the probability that it will turn left on the second trial is 0.84. What is the probability that a rat which turns left on the second trial will have turned left also on the first trial?

★5.76 At an electronics plant, it is known from past experience that the probability is 0.86 that a new worker who has attended the company's training program will meet his production quota, and that the corresponding probability is 0.35 for a new worker who has not attended the company's training program. If 80 percent of all new workers attend the training program, what is the probability that a new worker will meet his production quota?

★5.77 With reference to the preceding exercise, what is the probability that a new worker who meets the production quota will have attended the company's training program?

★5.78 Two firms V and W consider bidding on a road-building job which may or may not be awarded depending on the amounts of the bids. Firm V submits a bid and the probability is $\frac{3}{4}$ that it will get the job provided firm W does not bid. The odds are 3 to 1 that W will bid, and if it does, the probability that V will get the job is only $\frac{1}{3}$.
(a) What is the probability that V will get the job?
(b) If V gets the job, what is the probability that W did not bid?

★5.79 In a certain community, 8 percent of all adults over 50 have diabetes. If a doctor in this community correctly diagnoses 95 percent of all persons with diabetes as having the disease and incorrectly diagnoses 2 percent of all persons without diabetes as having the disease, what is the probability that an adult over 50 diagnosed by this doctor as having diabetes actually has the disease?

★5.80 An importer, expecting a shipment of ginger roots from Indonesia, knows from past experience that the odds are 4 to 1 against such a shipment getting lost. He also knows from past experience that half the shipments that do not get lost arrive within a month. If the shipment of ginger roots has not arrived within a month, what are the odds that it will still come?

★5.81 With reference to the assembly line example on page 140, find the probabilities that an improperly sealed can comes from
(a) assembly line II;
(b) assembly line III.

★5.82 A mail-order house employs three stock clerks, P, Q, and R, who pull items from shelves and assemble them for subsequent verification and packaging. P makes a mistake in an order (gets a wrong item or the wrong quantity) one time in a hundred, Q makes a mistake in an order five times in a hundred, and R makes a mistake in an order three times in a hundred. Of all the orders delivered for verification, P, Q, and R fill, respectively, 30, 40, and 30 percent. What is the probability that a mistake will be made in an order?

★5.83 With reference to the preceding exercise, if a mistake is found in a particular order, what is the probability that it was filled by Q?

★5.84 (From Miller, I., and Freund, J. E., *Probability and Statistics for Engineers*, *2nd ed*. Englewood Cliffs, N.J.: Prentice-Hall, Inc., 1977.) An explosion in an LNG storage tank in the process of being repaired could have occurred as the result of static electricity, malfunctioning electrical equipment, an open flame in contact with the liner, or purposeful action (industrial sabotage). Interviews with engineers who were analyzing the risks involved led to estimates that such an explosion would occur with probability 0.25 as a result of static electricity, 0.20 as a result of malfunctioning electric equipment, 0.40 as a result of an open flame, and 0.75 as a result of purposeful action. These interviews also yielded subjective estimates of the probabilities of the four causes of 0.30, 0.40, 0.15, and 0.15, respectively. Based on all this information, what is the most likely cause of the explosion?

★5.85 To get answers to sensitive questions, we sometimes use a method called the **randomized response technique.** Suppose, for instance, that we want to determine what percentage of the students at a large university smoke marijuana. We construct 20 flash cards, write "I smoke marijuana at least once a week" on 12 of the cards, where 12 is an arbitrary choice, and "I do not smoke marijuana at least once a week" on the others. Then, we let each student (in the sample interviewed) select one of the cards at random, and respond "yes" or "no" without divulging the question.
   (a) Establish a relationship between $P(Y)$, the probability that a student will give a "yes" response, and $P(M)$, the probability that a student randomly selected at that university smokes marijuana at least once a week.
   (b) If 106 of 250 students answered "yes" under these conditions, use the result of part (a) and $\frac{106}{250}$ as an estimate of $P(Y)$ to estimate $P(M)$.

5.8
CHECKLIST OF KEY TERMS
(with page references to their definitions)

5.9
REVIEW EXERCISES

★5.86 Convert each of the following probabilities to odds:
   (a) The probability of drawing a black Jack from an ordinary deck of 52 playing cards is $\frac{1}{26}$.
   (b) If a pollster randomly selects five of 24 households to be included in a survey, the probability is $\frac{5}{24}$ that any particular household will be included.
   (c) The probability of rolling 10 or less with a pair of balanced dice is $\frac{11}{12}$.

5.87 A small real estate office has four part-time salespersons. Using two coordinates so that (3, 1), for example, represents the event that three of the salespersons are at work and one of them is busy with a customer, and (2, 0) represents the event that two of the salespersons are at work but none of them is busy with a customer, draw a diagram similar to that of Figure 5.1, showing the 15 points of the corresponding sample space.

5.88 With reference to the preceding exercise, if each of the 15 points of the sample space has the probability $\frac{1}{15}$, find the probabilities that
   (a) all the salespersons that are at work are busy with customers;
   (b) at least three of the salespersons are at work;
   (c) at least three salespersons are busy with a customer;
   (d) none of the salespersons who are at work are busy with customers.

★5.89 There are two parties in a large city mayoral election. A is nominated by one party and the probability that he will be elected is $\frac{4}{5}$ provided that B is not nominated by the other party. The probability that B will be nominated is $\frac{3}{4}$ and the probability that A will be elected if B is nominated is $\frac{3}{5}$.
   (a) What is the probability that A will be elected?
   (b) If A is elected, what is the probability that B was not nominated?

5.90 If C and W are the events that a customer will order a cocktail or wine before dinner at a certain restaurant, $P(C) = 0.43$ and $P(W) = 0.21$, find the probabilities that a customer will
  (a) not order a cocktail before dinner;
  (b) order either a cocktail or wine before dinner;
  (c) order neither a cocktail nor wine before dinner.

5.91 If the probability is 0.26 that any one woman will name yellow or orange as her favorite color, what is the probability that four women, selected at random, will all name yellow or orange as their favorite color?

5.92 If Q is the event that a person is qualified for a job and G is the event that he or she will get the job, express in words what probabilities are represented by
  (a) $P(G|Q)$;   (c) $P(Q|G)$;
  (b) $P(G'|Q')$;   (d) $P(Q|G')$.

5.93 If $P(M) = 0.55$, $P(N) = 0.18$, and $P(M \cap N) = 0.099$, are the events M and N independent or dependent?

5.94 Discuss the following assertion: Since probabilities are measures of uncertainty, the probability we assign to a future event will always increase when we get more information.

★5.95 Convert each of the following odds to probabilities:
  (a) The odds are 19 to 5 that a given horse will not win the Kentucky Derby.
  (b) If five cards are drawn with replacement from an ordinary deck of 52 playing cards, the odds are 13 to 3 that at most three of them will be red.
  (c) If two persons are chosen at random from a group of ten men and twelve women, the odds are 40 to 37 that one man and one woman will be selected.

5.96 Suppose that the numbers 1, 2, 3, 4, 5, and 6 are used to denote that a committee of parents and teachers decides that a certain education program is terrible, poor, fair, good, very good, or excellent. If $L = \{2, 3, 4, 5\}$ and $R = \{4, 5, 6\}$, list the elements of the sample space comprising each of the following events, and also express the events in words:
  (a) L';   (c) $L \cap R$;
  (b) $L \cup R$;   (d) $L \cap R'$.

★5.97 If someone feels that 17 to 8 are fair odds that a paint job will be finished on time, what subjective probability does he assign to this event?

5.98 The probabilities are 0.15, 0.26, and 0.08 that a family driving through a Western city will spend the night at one of its hotels, at one of its motels, or at its campground. What is the probability that a family driving through this city will spend the night at one of these kinds of facilities?

5.99 The probabilities that a newspaper will receive 0, 1, 2, ..., 7, or at least 8 letters to the editor about an unpopular decision of the school board are 0.01, 0.02, 0.05, 0.14, 0.16, 0.20, 0.18, 0.15, and 0.09. What are the probabilities that the newspaper will receive
  (a) at most 4 letters to the editor about the school board decision;
  (b) at least 6;
  (c) from 3 to 5?

5.100 The probability that a new play will be well received in New York is 0.32, and the probability that it will be well received in New York as well as Chicago is 0.14.

What is the probability that a new play which was well received in New York will also be well received in Chicago?

5.101 If each point of the sample space of Figure 5.15 represents an outcome having the probability $\frac{1}{32}$, find

(a) $P(A)$;

(b) $P(B)$;

(c) $P(A \cap B)$;

(d) $P(A \cup B)$;

(e) $P(A' \cap B)$;

(f) $P(A' \cap B')$.

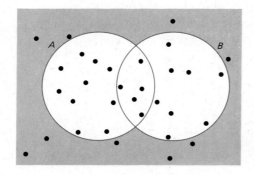

FIGURE 5.15 *Sample space for Exercise 5.101.*

5.102 As part of a promotional scheme in Arizona and New Mexico, a company distributing frozen foods will award a grand price of $100,000 to some person sending in his or her name on an entry blank, with the option of including a label from one of the company's products. A breakdown of the 225,000 entries received is shown in the following table:

	With label	*Without label*
Arizona	120,000	42,000
New Mexico	30,000	33,000

If the winner of the grand prize is chosen by lot, A represents the event that it will be won by an entry from Arizona, and L represents the event that it will be won by an entry which included a label, find each of the following probabilities:

(a) $P(A)$;

(b) $P(L)$;

(c) $P(A|L)$;

(d) $P(L|A)$;

(e) $P(A'|L')$;

(f) $P(L|A')$.

5.103 Suppose that in the preceding exercise the drawing is rigged so that by including a label each entry's probability of winning the grand prize is doubled. Recalculate the probabilities of parts (a) through (f).

★**5.104** The following illustrates how one's intuition can be misleading in connection with probabilities or odds: A box contains 100 beads, some red and some white.

One bead will be drawn, and you are asked to call beforehand whether it is going to be red or white. At what odds would you be willing to bet on this game if

    (a) you have no idea how many of the beads are red and how many are white?

    (b) you are told that 50 of the beads are red and 50 are white?

5.105 A library received 40 new books including 12 historical novels. If four of these books are selected at random, what is the probability that not one of them is a historical novel?

5.106 If A is the event that a university's football team is rated among the top twenty by AP and U is the event that it is rated among the top twenty by UPI, what events are represented by the four regions of the Venn diagram of Figure 5.16?

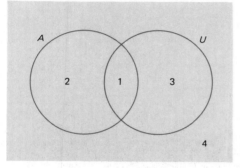

FIGURE 5.16 *Venn diagram for Exercise 5.106.*

★5.107 A movie producer feels that the odds are 8 to 1 that his new movie will not be rated G, 15 to 3 that it will not be rated PG, and 13 to 5 that it will not get either of these two ratings. Are the corresponding probabilities consistent?

5.108 The probability that George will get an M.A. degree is 0.40, and the probability that with an M.A. degree he will get a well-paying job is 0.85. What is the probability that he will get an M.A. degree and a well-paying job?

★5.109 During a time of war, a country uses lie detectors to uncover security risks. As is well known, lie detectors are not infallible, so let us suppose that the probabilities are 0.10 and 0.05 that they will fail to detect a security risk and incorrectly label a person a security risk. If 2 percent of the persons who are given the lie detector test are security risks, what is the probability that a person labeled a security risk by a lie detector actually is a security risk?

5.110 Explain why there must be a mistake in each of the following statements:

    (a) The probability that a new safety feature in cars will be able to prevent injuries is -0.02.

    (b) The probability that a student will get a B in a course is 0.11, but she is ten times as likely to get a C.

    (c) The probability that a teachers' conference will be well attended is 0.59, and the probability that it will not be well attended is 0.31.

5.111 If a student answers the 12 questions on a true–false test by flipping a balanced coin, what is the probability that he will answer all questions correctly?

5.112 If the probabilities are 0.20 that an item ordered by mail will arrive late, 0.12 that it will arrive in poor condition, and 0.05 that it will arrive late and in poor condition, what is the probability that the item will arrive late, in poor condition, or both?

5.113 In Figure 5.17, B is the event that a person traveling in Oregon will visit a friend in Bend, P is the event that she will visit a friend in Portland, and E is the event that she will visit a friend in Eugene. Explain in words what events are represented by the following regions or combinations of regions of the Venn diagram:
    (a) region 5;
    (b) regions 1 and 4 together;
    (c) regions 3 and 6 together;
    (d) regions 3, 5, and 6 together;
    (e) regions 5 and 8 together.

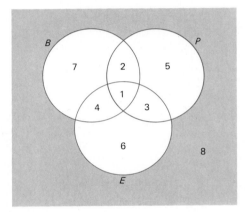

FIGURE 5.17 *Venn diagram for Exercise 5.113.*

★5.114 A retailer of automobile parts has four employees K, L, M, and N, who make mistakes in filling an order one time in 100, four times in 100, two times in 100, and six times in 100. Of all the orders filled, K, L, M, and N fill, respectively, 20, 40, 30, and 10 percent. If a mistake is found in a particular order, what are the probabilities that it was filled by K, L, M, or N?

5.115 Among a company's replacement parts for a given assembly, 20 percent are defective and the rest are good, 60 percent were bought from external sources and the rest were made by the company itself, and of those bought from external sources 80 percent are good and the rest are defective. What are the probabilities that a replacement part, randomly selected from this stock, is
    (a) company-made and good;
    (b) either defective or bought;
    (c) neither company-made nor good;
    (d) bought, given that it is defective?

★5.116 A horse breeder has entered one of his horses in a race in Florida and a race in Kentucky. If the odds are 3 to 1 that it will not win in Florida and 7 to 1 that it will not win in Kentucky, what are the odds that it will not win either race? Assume independence.

5.10
REFERENCES

More detailed, though still elementary, treatments of probability may be found in

FREUND, J. E., *Introduction to Probability*. Encino, Calif.: Dickenson Publishing Company, Inc., 1973.

GOLDBERG, S., *Probability—An Introduction*. Englewood Cliffs, N.J.: Prentice-Hall, Inc., 1960.

MOSTELLER, F., ROURKE, R. E. K., and THOMAS, G. B., *Probability with Statistical Applications, 2nd ed.* Reading, Mass.: Addison-Wesley Publishing Company, Inc., 1970.

SCHEAFFER, R. L. and MENDENHALL, W., *Introduction to Probability: Theory and Applications*. Boston: Duxbury Press, 1975.

and interesting information about probabilities and odds in

NEFT, D. S., COHEN, R. M., and DEUTCH, J. A., *The World Book of Odds*. New York: Grosset & Dunlap, Inc., 1978.

Subjective probability is discussed in

BOREL, E., *Elements of the Theory of Probability*. Englewood Cliffs, N.J.: Prentice-Hall, Inc., 1965.

KYBURG, H. E., JR., and SMOKLER, H. E., *Studies in Subjective Probability*. New York: John Wiley & Sons, Inc., 1964.

When decisions are made in the face of uncertainty, they are seldom based on probabilities alone. In most cases we must also know something about the consequences (profits, losses, penalties, or rewards) to which we are exposed. For instance, if we must decide whether to buy a new car, the knowledge that our old car will soon require repairs is not enough—to make an intelligent decision we must also know, among other things, the cost of the repairs and the trade-in value of our old car. To give another example, suppose that a building contractor has to decide whether to bid on a construction job which promises a profit of $120,000 with probability 0.20, or a loss of $27,000 (perhaps, due to bad estimates, strikes, or the late delivery of materials) with probability 0.80. Clearly, the probability that the contractor will make a profit is not very high, but on the other hand, the amount he stands to gain is much greater than the amount he stands to lose.

Both of these examples demonstrate the need for a method of combining probabilities and consequences, and this is why we introduce the concept of a mathematical expectation in Section 6.1. Then, in Sections 6.2 and 6.3, we give some examples which show how mathematical expectations are used in making decisions.

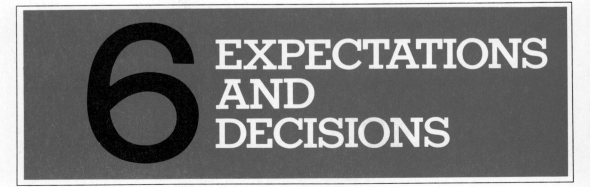

6 EXPECTATIONS AND DECISIONS

6.1
MATHEMATICAL EXPECTATION

If an insurance agent tells us that in the United States a 45-year-old woman can expect to live 33 more years, this does not mean that anyone really expects a 45-year-old woman to live until her 78th birthday and then die the next day. Similarly, if we read that a person living in the United States can expect to eat 13.2 pounds of cheese and 307.6 eggs a year, or that a child in the age group from 6 to 16 can expect to visit a dentist 1.9 times a year, it must be obvious that the word "expect" is not being used in its colloquial sense. A child cannot go to the dentist 1.9 times, and it would be surprising, indeed, if we found somebody who has actually eaten 13.2 pounds of cheese and 307.6 eggs in a given year. So far as 45-year-old women are concerned, some will live another 15 years, some will another 20 years, some will live another 36 years,..., and the life expectancy of "33 more years" will have to be interpreted as an average, or as we shall call it here, a **mathematical expectation.**

Originally, the concept of a mathematical expectation arose in connection with games of chance, and in its simplest form it is the product of the amount a player stands to win and the probability that he or she will win.

EXAMPLE What is our mathematical expectation if we stand to win $4 if and only if a balanced coin comes up heads?

Solution Assuming that the coin is randomly tossed, the probability of heads is $\frac{1}{2}$ and our mathematical expectation is $4 \cdot \frac{1}{2} = \$2$.

EXAMPLE What is our mathematical expectation if we buy one of 1,000 raffle tickets for a prize, a television set, worth $480?

Solution Since the probability that we will win the television set is $\frac{1}{1,000}$, our mathematical expectation is $480 \cdot \frac{1}{1,000} = \$0.48$ or 48 cents.

Thus, it would be foolish to spend more than 48 cents for the ticket, unless, of course, the proceeds of the raffle go to a worthy cause (or the difference can be credited to whatever pleasure a person may get from placing a bet).

In both of these examples there was a single prize, but in each case there were two possible payoffs—$4 or $0 in the first example and $480 or $0 in the other. Indeed, in the second example we can argue that 999 of the tickets will pay $0 and one of the tickets will pay $480 (or the equivalent in merchandise). Altogether, the 1,000 tickets will thus pay $480, or on the average 48 cents per ticket, and this is the mathematical expectation.

To generalize the concept of a mathematical expectation, let us consider the following modification of the raffle of the preceding example:

EXAMPLE What is our mathematical expectation if we buy one of 1,000 raffle tickets for a first prize of a television set worth $480, a second prize of a tape recorder worth $120, and a third prize of a radio worth $40?

Solution Now we can argue that 997 of the raffle tickets will not pay anything at all, one ticket will pay $480 (in merchandise), another will pay $120 (in merchandise), and a third will pay $40 (in merchandise); altogether, the 1,000 tickets will thus pay $480 + 120 + 40 = $640 (in merchandise), or on the average $0.64 per ticket. This is the mathematical expectation for each ticket. Looking at the problem in a different way, we could argue that if the raffle were repeated many times, we would lose 99.7 percent of the time (or with probability 0.997) and win each of the prizes 0.1 percent of the time (or with probability 0.001). On the average we would thus win

$$0(0.997) + 480(0.001) + 120(0.001) + 40(0.001) = \$0.64$$

which is the sum of the products obtained by multiplying each amount by the corresponding proportion or probability.

Generalizing from this example, let us now give the following definition:

Mathematical expectation

> If the probabilities of obtaining the amounts $a_1, a_2, \ldots,$ or a_k are $p_1, p_2, \ldots,$ and p_k, then the mathematical expectation is
>
> $$E = a_1 p_1 + a_2 p_2 + \cdots + a_k p_k$$

Each amount is multiplied by the corresponding probability, and the mathematical expectation, E, is given by the sum of all these products. In the $\sum$ notation, $E = \sum a \cdot p$.

In connection with this formula, it is important to keep in mind that the a's are positive when they represent profits, winnings, or gains (namely, amounts which we receive), and that they are negative when they represent losses, penalties, or deficits (namely, amounts which we have to pay).

EXAMPLE What is our mathematical expectation if we win $10 if a balanced coin comes up heads and lose $10 if it comes up tails?

Solution   The amounts are $a_1 = 10$ and $a_2 = -10$, the probabilities are $p_1 = \frac{1}{2}$ and $p_2 = \frac{1}{2}$, and the mathematical expectation is

$$E = 10 \cdot \tfrac{1}{2} + (-10) \cdot \tfrac{1}{2} = 0$$

A game like this, in which the mathematical expectation is zero and neither player is favored, is said to be **fair,** or **equitable.**

EXAMPLE   The probabilities are 0.22, 0.36, 0.28, and 0.14 that an investor will be able to sell a piece of property at a profit of \$2,500, at a profit of \$1,500, at a profit of \$500, or at a loss of \$500. What is the investor's expected profit?

Solution   Substituting $a_1 = 2,500$, $a_2 = 1,500$, $a_3 = 500$, $a_4 = -500$, $p_1 = 0.22$, $p_2 = 0.36$, $p_3 = 0.28$, and $p_4 = 0.14$ into the formula for E, we get

$$E = 2,500(0.22) + 1,500(0.36) + 500(0.28) + (-500)(0.14)$$

$$= \$1,160$$

Although we referred to the quantities $a_1, a_2, \ldots$, and a_k as "amounts," they need not be cash winnings, losses, penalties, or rewards. When we said on page 151 that a child in the age group from 6 to 16 goes to the dentist 1.9 times a year, we referred to a result which was obtained by multiplying $0, 1, 2, 3, 4, \ldots$, by the respective probabilities that a child in this age group will visit a dentist that many times a year.

EXAMPLE   If the probabilities are 0.06, 0.21, 0.24, 0.18, 0.14, 0.10, 0.04, 0.02, and 0.01 that an airline office at a certain airport will receive 0, 1, 2, 3, 4, 5, 6, 7, or 8 complaints per day about its luggage handling, how many such complaints can it expect per day?

Solution   The expected number is

$$E = 0(0.06) + 1(0.21) + 2(0.24) + 3(0.18) + 4(0.14)$$

$$+ 5(0.10) + 6(0.04) + 7(0.02) + 8(0.01)$$

$$= 2.75$$

In all of the examples in this section we were given the values of a and p (or the values of the a's and p's) and calculated E. Now let us consider an example in which we are given values of a and E to arrive at some result about p, and also an example in which we are given values of p and E to arrive at some result about a.

EXAMPLE   A recent college graduate must decide whether to accept a job paying $14,580 a year, or turn it down with the hope of getting another job paying $20,250. If she turns down the $14,580 job, how does she feel about her chances of getting the higher-paying job?

Solution   If she feels that the probability is p that she will get the higher-paying job, her mathematical expectation is $20{,}250p$. Since she feels that this expectation is preferable to the certainty of getting $14,580, we write

$$20{,}250p > 14{,}580$$

which yields $p > \dfrac{14{,}580}{20{,}250}$ and, hence $p > 0.72$. This assumes, of course, that salary is the only factor which affects her decision.

EXAMPLE   A friend says that he would "give his right arm" for our ticket to an NFL play-off game. To put this on a cash basis, we propose that he pay us $20 (the actual price of the ticket), but he will get the ticket only if he draws a jack, queen, king, or ace from an ordinary deck of 52 playing cards; otherwise, we keep the ticket and his $20. What is the ticket worth to him, if he feels that this arrangement is fair?

Solution   Since there are four jacks, four queens, four kings, and four aces, the probability that he will get the ticket is $\frac{16}{52}$, the probability that he will not get the ticket is $1 - \frac{16}{52} = \frac{36}{52}$, and the mathematical expectation associated with the gamble is

$$E = a \cdot \tfrac{16}{52} + 0 \cdot \tfrac{36}{52} = a \cdot \tfrac{16}{52}$$

where a is the amount, he feels, the ticket is worth. Putting this mathematical expectation equal to $20, which he considers a fair price to pay for taking the risk, we get $a \cdot \dfrac{16}{52} = 20$ and $a = \dfrac{52 \cdot 20}{16} = 65$. Thus, he feels that the ticket is worth $65.

EXERCISES

6.1   If a service club sells 4,000 raffle tickets for a cash prize of $800, what is the mathematical expectation of a person who buys one of the tickets?

6.2   A charitable organization raises funds by selling 2,000 raffle tickets for a $500 first prize and a $100 second prize. What is the mathematical expectation of a person who buys one of the tickets?

6.3   If someone gives us $10 each time we roll a 1 or a 2 with a balanced die, how much must we pay him each time we roll a 3, 4, 5, or 6 to make the game fair?

6.4   The winner of a tennis tournament gets $50,000 and the runner-up gets $25,000. What are the two finalists' mathematical expectations if
   (a) they are evenly matched;
   (b) their probabilities of winning are 0.60 and 0.40?

6.5   To introduce his new cars to the public, a dealer offers a first prize of $3,000 and a second prize of $500 to some lucky persons who come to his showroom and fill in entry cards. If 8,750 persons filled in cards and the winning cards are drawn at random, what is each entrant's mathematical expectation? Does this make it worthwhile to spend 50 cents on gasoline to drive to the dealer's showroom?

6.6   If the two league champions are evenly matched, the probabilities that a "best of seven" basketball play-off will take 4, 5, 6, or 7 games are $\frac{1}{8}$, $\frac{1}{4}$, $\frac{5}{16}$, and $\frac{5}{16}$. Under these conditions, how many games can we expect such a play-off to last?

6.7   A union wage negotiator feels that the probabilities are 0.40, 0.30, 0.20, and 0.10 that the union members will get a $1.50 an hour raise, a $1.00 raise, a 50-cent raise, or no raise at all. What is their expected raise?

6.8   An importer is offered a shipment of Dutch cheeses for $14,000, and the probabilities that he will be able to sell them for $18,000, $17,000, or $15,000 are 0.32, 0.55, and 0.13. What is the importer's expected gross profit?

6.9   A police chief knows that the probabilities of 0, 1, 2, 3, or 4 burglaries on any given day are 0.12, 0.25, 0.39, 0.18, and 0.06. How many burglaries can the police chief expect per day? It is assumed here that the probability of more than 4 burglaries is negligible.

6.10   The probabilities that a person who enters "The Department Store" will make 0, 1, 2, 3, 4, or 5 purchases are 0.11, 0.33, 0.31, 0.12, 0.09, and 0.04. How many purchases can a person entering this store be expected to make?

6.11   Defending a liability suit against a client, a lawyer must decide whether to charge a straight fee of $2,000 or a contingent fee of $10,000 which she will get only if her client wins the case. How does she feel about her client's chances, if she prefers the straight $2,000 fee?

6.12   A salesperson must choose between a straight salary of $21,600 and a salary of $18,000 plus a bonus of $7,200 if his sales exceed a certain quota. How does he assess his chances of exceeding the quota if he chooses the lower salary with the possibility of a bonus?

6.13   One contractor offers to do a road repair job for $45,000, while another contractor offers to do the job for $50,000 with a penalty of $12,500 if the job is not finished on time. If the person who lets out the contract for the job prefers the second offer, what does this tell us about her assessment of the probability that the second contractor will not finish the job on time?

6.14   Mr. Green has the choice of staying home and reading a good book or going to a party. If he goes to the party he might have a terrible time (to which he assigns a utility of 0), or he might have a wonderful time (to which he assigns a utility of 20 units). If he feels that the odds against his having a good time are 4 to 1 and he decides not to go, what can we say about the utility which he assigns to staying home and reading a good book?

6.15   Mr. Jones would like to beat Mr. Brown in an upcoming golf tournament, but his chances are nil unless he takes $500.00 worth of extra lessons, which (according to the pro at his club) will give him a fifty–fifty chance. If Mr. Jones assigns the utility U to his beating Mr. Brown and the utility $-\frac{1}{5}U$ to his losing to Mr. Brown, find U if Mr. Jones decides that it is just about worthwhile to spend the $500 on extra lessons.

6.2
DECISION MAKING★

When we are faced with uncertainties, mathematical expectations can often be used to great advantage in making decisions. In general, if we have to choose between two or more alternatives, it is considered "rational" to select the one with the "most promising" mathematical expectation: the one which maximizes expected profits, minimizes expected costs, maximizes expected tax advantages, minimizes expected losses, and so on.

EXAMPLE   A furniture manufacturer must decide whether to expand his plant capacity now or wait at least another year. His advisors tell him that if he expands now and economic conditions remain good, there will be a profit of $328,000 during the next fiscal year; if he expands now and there is a recession, there will be a loss (negative profit) of $80,000; if he waits at least another year and economic conditions remain good, there will be a profit of $160,000; and if he waits at least another year and there is a recession, there will be a small profit of $16,000. If the furniture manufacturer feels the probabilities for economic conditions remaining good or there being a recession are $\frac{1}{3}$ and $\frac{2}{3}$, would expanding his plant capacity now maximize his expected profit?

Solution   In problems like this, it usually helps to present the information in a table, such as the following:

	Expand now	Delay expansion
Economic conditions remain good	$328,000	$160,000
There is a recession	−$80,000	$16,000

As can be seen from this table, it will be advantageous to expand the plant capacity right away only if economic conditions remain good, and the furniture manufacturer's decision will, therefore, have to depend on the chances that this will be the case. Using the manufacturer's probabilities of $\frac{1}{3}$ and $\frac{2}{3}$ for economic conditions remaining good or there being a recession, we find that if he expands his plant capacity right way, the expected profit is

$$328,000 \cdot \tfrac{1}{3} + (-80,000) \cdot \tfrac{2}{3} = \$56,000$$

and if the expansion is delayed, the expected profit is

$$160{,}000 \cdot \tfrac{1}{3} + 16{,}000 \cdot \tfrac{2}{3} = \$64{,}000$$

Since the second of these two figures exceeds the first, it follows that delaying the expansion maximizes the furniture manufacturer's expected profit.

The way in which we have studied this problem is called a **Bayesian analysis.** In this kind of analysis, probabilities are assigned to the alternatives about which uncertainties exist (the **states of nature,** which in our example were economic conditions remaining good and a recession); then we choose whichever alternative promises the greatest expected profit or the smallest expected loss.

This approach to decision making has great intuitive appeal, but it is not without complications. If mathematical expectations are to be used for making decisions, it is essential that our appraisals of all relevant probabilities are close if not correct, and that we know the exact values of the payoffs associated with the various possibilities.

EXAMPLE With reference to the preceding example, suppose that an expert feels that the probabilities for economic conditions remaining good or there being a recession are 0.40 and 0.60. Based on these probabilities, what advice will maximize the furniture manufacturer's expected profit?

Solution The expert feels that if the furniture manufacturer expands his plant capacity right away, the expected profit is

$$328{,}000(0.40) + (-80{,}000)(0.60) = \$83{,}200$$

and if the expansion is delayed, the expected profit is

$$160{,}000(0.40) + 16{,}000(0.60) = \$73{,}600$$

Since the first of these two figures exceeds the second, the expert's advice to expand right away will, in his mind, maximize the expected profit.

EXAMPLE Suppose that the furniture manufacturer of our example is told by his accountant that the $328,000 figure is incorrect and that it should be $352,000. Based on his own appraisal of the probabilities for economic conditions remaining good or there being a recession, $\tfrac{1}{3}$ and $\tfrac{2}{3}$, will this affect his decision?

Solution Now, if he expands his plant capacity right away, the expected profit is

$$352{,}000 \cdot \tfrac{1}{3} + (-80{,}000) \cdot \tfrac{2}{3} = \$64{,}000$$

and if the expansion is delayed, the expected profit is

$$160{,}000 \cdot \tfrac{1}{3} + 16{,}000 \cdot \tfrac{2}{3} = \$64{,}000$$

Since the two figures are equal, we find that it does not matter in this case whether the furniture manufacturer expands his plant capacity now or delays the expansion.

6.3
STATISTICAL DECISION PROBLEMS★

Modern statistics, with its emphasis on inference, may be looked upon as the art, or science, of decision making under uncertainty. This approach to statistics, called **decision theory,** dates back only to the middle of this century and the publication of John von Neumann and Oscar Morgenstern's *Theory of Games and Economic Behavior* in 1944 and Abraham Wald's *Statistical Decision Functions* in 1950. Since the study of decision theory is quite complicated mathematically, we shall limit our discussion here to an example in which the method of the preceding section is applied to a problem that is of a statistical nature.

EXAMPLE A government agency has appointed five teams to study racial discrimination, and 1, 2, 5, 1, and 6 of the members of these teams favor liberal causes. The teams are randomly assigned to various cities, and a city manager hires a consultant to predict how many of the members of the team sent to his city will favor liberal causes. If the consultant is paid $100 plus a bonus of $200 which he receives only if his prediction is correct, what prediction maximizes the amount of money he can expect to get?

Solution If the consultant's prediction is 1, the mode of the five numbers, he will make $100 with probability $\tfrac{3}{5}$, $300 with probability $\tfrac{2}{5}$, and it follows that he can expect to make

$$100 \cdot \tfrac{3}{5} + 300 \cdot \tfrac{2}{5} = \$180$$

As can easily be verified, this is the best he can do—if his prediction is 2, 5, or 6, he can expect to make $140, and for any other prediction his expectation is $100. This illustrates the (perhaps obvious) fact that if one has to pick an exact value on the nose and there is no reward for being close, the best prediction is the mode.

To show how the consequences, penalties or rewards, dictate the choice of statistical methods of decision or prediction, let us consider the following variations of the above example:

EXAMPLE   With reference to the preceding example, suppose that the consultant is paid $300 minus an amount of money equal in dollars to 40 times the size of the error. What prediction maximizes the amount of money he can expect to get?

Solution   In this case it is the median which yields the best prediction. If the consultant's prediction is 2, the median of the five numbers, the size of the error will be 1, 0, 3, or 4, depending on whether 1, 2, 5, or 6 of the members of the team sent to the city will favor liberal causes. Correspondingly, he will get $260, $300, $180, or $140, and he can expect to make

$$260 \cdot \tfrac{2}{5} + 300 \cdot \tfrac{1}{5} + 180 \cdot \tfrac{1}{5} + 140 \cdot \tfrac{1}{5} = \$228$$

As can be verified, the consultant's expectation is less for any other number. For instance, if his prediction is 3, the mean of the five numbers, the size of the error will be 2, 1, 2, or 3 depending on whether 1, 2, 5, or 6 of the members of the team sent to the city will favor liberal causes. Correspondingly, he will get $220, $260, $220, or $180, and he can expect to make

$$220 \cdot \tfrac{2}{5} + 260 \cdot \tfrac{1}{5} + 220 \cdot \tfrac{1}{5} + 180 \cdot \tfrac{1}{5} = \$220$$

EXAMPLE   The mean comes into its own right when the penalty, the amount subtracted, increases more rapidly with the size of the error. Thus, suppose that the consultant of our example is paid $300 minus an amount of money equal in dollars to 20 times the square of the error. What prediction maximizes the amount of money he can expect to get?

Solution   If the consultant's prediction is 3, the mean of the five numbers, the size of the error will be 2, 1, 2, or 3 depending on whether 1, 2, 5, or 6 of the members of the team sent to the city will favor liberal causes. Correspondingly, he will get $220, $280, $220, or $120, and he can expect to make

$$220 \cdot \tfrac{2}{5} + 280 \cdot \tfrac{1}{5} + 220 \cdot \tfrac{1}{5} + 120 \cdot \tfrac{1}{5} = \$212$$

As can be verified, the consultant's expectation is less for any other prediction (see, for example, Exercise 6.27). This case is of special importance in statistics, as it ties in closely with the **method of least squares** which we shall study in Chapter 14. The idea of working with the squares of errors is justified on the grounds that in actual practice the seriousness of an error often increases very rapidly with the size of the error, more rapidly than the magnitude of the error itself.

The greatest difficulty in applying the methods of this chapter to realistic problems in statistics is that we seldom know the exact values of all the risks that are involved; that is, we seldom know the exact values of the "payoffs" corresponding to the various eventualities. For instance, if the FDA must decide whether or not to release a new drug for general use, how can it put a cash value on the damage that might be done by not waiting for a more thorough analysis of possible side effects, or on the lives that might be lost by not making the drug available to the public right away? Similarly, if a faculty committee must decide which of several applicants should be admitted to a medical school or, perhaps, receive a scholarship, how can they possibly foresee all the consequences that might be involved?

The fact that we seldom have adequate information about relevant probabilities also provides obstacles to finding suitable decision criteria; without them, is it "reasonable" to base decisions, say, on pessimism or optimism as in Exercises 6.22 and 6.23 on page 161? Questions like these are difficult to answer, but their analysis serves the important purpose of revealing the logic that underlies statistical thinking.

EXERCISES

★6.16 A truck driver has to deliver a load of building materials to one of two construction sites, which are 18 and 22 miles from the lumber yard, but he has misplaced the order telling him where the load should go. The two construction sites are 8 miles apart, and to complicate matters, the telephone at the lumberyard is out of order. If the driver feels that the probabilities are $\frac{1}{6}$ and $\frac{5}{6}$ that the load should go to the site which is 18 miles from the lumberyard or the one which is 22 miles from the lumberyard, where should he go first so as to minimize the expected distance he will have to drive?

★6.17 With reference to the preceding exercise, where should the driver go first so as to minimize the expected distance he will have to drive, if instead of $\frac{1}{6}$ and $\frac{5}{6}$ the two probabilities are
(a) $\frac{1}{3}$ and $\frac{2}{3}$;
(b) $\frac{1}{4}$ and $\frac{3}{4}$?

★6.18 The management of a mining company must decide whether to continue an operation at a certain location. If they continue and are successful, they will make a profit of $4,500,000; if they continue and are not successful, they will lose $2,700,000; if they do not continue but would have been successful if they had continued, they will lose $1,800,000 (for competitive reasons); and if they do not continue and would not have been successful if they had continued, they will make a profit of $450,000 (because funds allocated to the operation remain unspent). What decision would maximize the company's expected profit if it is felt that there is a fifty–fifty chance for success?

★6.19 With reference to the preceding exercise, show that it does not matter what they decide to do if it is felt that the probabilities for and against success are $\frac{1}{3}$ and $\frac{2}{3}$.

★6.20 With reference to the example on page 156, suppose that the $80,000 loss is in error and should be a $120,000 loss. What decision maximizes the furniture manufacturer's expected profit, if the probabilities are 0.40 and 0.60 that economic conditions will remain good or that there will be a recession?

★6.21 A retailer has shelf space for 4 highly perishable items which are destroyed at the end of the day if they are not sold. The unit cost of the item is $3.00, the selling price is $6.00, and the profit is thus $3.00 per item sold. How many items should the retailer stock so as to maximize his expected profit, if he knows that the probabilities of a demand for 0, 1, 2, 3, or 4 items are, respectively, 0.10, 0.30, 0.30, 0.20, and 0.10?

★6.22 In the absence of any information about relevant probabilities, a pessimist may well try to minimize the maximum loss or maximize the minimum profit, that is, use the **minimax** or **maximin criterion.**
   (a) With reference to the example on page 156, suppose that the furniture manufacturer has no information about the probabilities that economic conditions will remain good or that there will be a recession. What decision will maximize his minimum profit?
   (b) With reference to Exercise 6.16, suppose that the truck driver has no idea about the chances that the load of building materials should go to either site. Where should he go first so as to minimize the maximum distance he has to drive?

★6.23 In the absence of any information about relevant probabilities, an optimist may well try to minimize the minimum loss or maximize the maximum profit, that is, use the **minimin** or **maximax criterion.**
   (a) With reference to the example on page 156, suppose that the furniture manufacturer has no information about the probabilities that economic conditions will remain good or that there will be a recession. What decision will maximize his maximum profit?
   (b) With reference to Exercise 6.16, suppose that the truck driver has no idea about the chances that the load of building materials should go to either site. Where should he go first so as to minimize the minimum distance he has to drive?

★6.24 With reference to Exercise 6.18, suppose that the management of the mining company has no idea about the chances for success. What decision will
   (a) maximize the company's minimum profit;
   (b) maximize the company's maximum profit?

★6.25 With reference to the example in the text, suppose that the furniture manufacturer has the option of paying an infallible consultant $25,000 to find out for sure whether there will be a recession, before he decides whether or not to expand the capacity of his plant. This raises the question whether it is worthwhile to spend the $25,000. To answer it, let us take the furniture manufacturer's $\frac{1}{3}$ and $\frac{2}{3}$ probabilities that economic conditions will remain good or that there will be a recession. If he knew for sure whether or not there will be a recession, the right decision will yield a profit of $328,000 or a profit of $16,000. Since the corresponding probabilities are $\frac{1}{3}$ and $\frac{2}{3}$, the expected profit (with the help of the infallible expert) is

$$328,000 \cdot \tfrac{1}{3} + 16,000 \cdot \tfrac{2}{3} = \$120,000$$

This is called the **expected profit with perfect information.** On page 156 we showed that without the help of the infallible expert, the expected profit was either $56,000 or $64,000, so that there is an improvement of at least $120,000 − $64,000 = $56,000, which makes the $25,000 fee well worthwhile. It is customary to refer to the amount by which perfect information improves one's expectation, $56,000 in our example, as the **expected value of perfect information.**

  (a) With reference to Exercise 6.16 find the expected distance with perfect information and the expected value (in miles) of perfect information.
  (b) With reference to Exercise 6.18, find the expected profit with perfect information and the expected value of perfect information. Would it be worthwhile to spend $500,000 to find out for sure whether the operation will be a success, before deciding whether it should be continued?

★6.26 There are situations where the criteria we have discussed are outweighed by other considerations. For instance, with reference to the example on page 156, what may well be the furniture manufacturer's decision if
  (a) he knows that he will be bankrupt unless he makes a profit of at least $200,000 during the next fiscal year;
  (b) he knows that he will be bankrupt unless he shows a profit, no matter how small, during the next fiscal year?

★6.27 With reference to the example on page 159, where the consultant is paid $300 minus an amount of money equal in dollars to 20 times the square of the error, what can the consultant expect to get if
  (a) his prediction is 1;
  (b) his prediction is 2?

★6.28 The ages of the seven entries in an essay contest are 17, 17, 17, 18, 20, 21, and 23, and their chances of winning are all equal. If we want to predict the age of the winner and there is a reward for being right, but none for being close, what prediction maximizes the expected reward?

★6.29 With reference to the preceding exercise, what prediction maximizes the expected reward if
  (a) there is a penalty proportional to the size of the error;
  (b) there is a penalty proportional to the square of the error?

★6.30 Some of the used cars on a lot are priced at $895, some are priced at $1,395, some are priced at $1,795, and some are priced at $2,495. If we want to predict the price of the car which will be sold first, what prediction minimizes the maximum size of the error? What name did we give to this statistic in one of the exercises of Chapter 3?

★6.31 With reference to the retailer of Exercise 6.21, suppose that he has no idea about the potential demand for the item. How many of the items should he stock so as to minimize the maximum loss to which he may be exposed? Discuss the reasonableness of using the minimax criterion in a problem of this kind.

6.4
CHECKLIST OF KEY TERMS
(with page references to their definitions)

★ Bayesian analysis, 157
★ Decision theory, 158
   Equitable game, 153
★ Expected profit with perfect information, 162
★ Expected value of perfect information, 162
   Fair game, 153
   Mathematical expectation, 151
★ Maximax criterion, 161
★ Maximin criterion, 161
★ Minimax criterion, 161
★ Minimin criterion, 161
★ States of nature, 157

6.5
REVIEW EXERCISES

6.32 The probabilities that a person shopping at "The Bookstore" will buy 0, 1, 2, 3, or 4 books are 0.22, 0.54, 0.17, 0.06, and 0.01. How many books can a person shopping at this bookstore be expected to buy?

6.33 A grab bag contains 12 packages worth $1.30 apiece, 15 packages worth $0.90 apiece, and 25 packages worth $0.80 apiece. Is it worthwhile to pay $1.00 for the privilege of picking one of the packages at random?

★6.34 The credit manager of a mortgage company figures that if an applicant for a certain size home mortgage is a good risk and the company accepts him, the company's profit will be $4,110, and if he is a bad risk and the company accepts him, the company will lose $600. If the credit manager turns down the mortgage applicant, there will be no direct profit or loss either way. Which decision maximizes the mortgage company's expected profit if the credit manager feels that the probabilities are 0.10 and 0.90 that a certain loan applicant is a good risk or a bad risk?

★6.35 With reference to the preceding exercise, would the credit manager's decision be the same if he feels that the probabilities are 0.20 and 0.80 that the loan applicant is a good risk or a bad risk?

6.36 Mrs. Black feels that it is about a toss-up whether to accept $20 cash or to gamble on drawing a bead from an urn containing 15 white beads and 45 red beads, where she is to receive $2 if she draws a white bead or a bottle of fancy perfume if she draws a red bead. What value, or utility, does she attach to the bottle of perfume?

★6.37 The five kinds of cars considered by a police department average 41, 40, 40, 43, and 40 miles per gallon, and their chances of being chosen are all equal. If we want to predict the average miles per gallon of the kind of car they will choose and there is a reward for being right, but none for being close, what prediction maximizes the expected reward?

★6.38 With reference to the preceding exercise, what prediction maximizes the expected reward if a penalty proportional to the square of the error is subtracted from the reward?

6.39 An insurance company agrees to pay the promoter of a drag race $8,000 in case the event has to be canceled because of rain. If the company's actuary feels that a fair net premium for this risk is $1,280, what probability does she assign to the possibility that the race will have to be canceled because of rain?

★6.40 A contractor has to choose between two jobs. The first job promises a profit of $120,000 with a probability of $\frac{3}{4}$ or a loss of $30,000 (due to strikes and other delays) with a probability of $\frac{1}{4}$; the second job promises a profit of $180,000 with a probability of $\frac{1}{2}$ or a loss of $45,000 with a probability of $\frac{1}{2}$.

    (a) Which job should the contractor choose if he wants to maximize his expected profit?

    (b) Which job would the contractor probably choose if his business is in fairly bad shape and he will go broke unless he can make a profit of at least $150,000 on his next job?

6.41 The manufacturer of a new battery additive has to decide whether to sell his product for $1.00 a can, or for $1.25 with a "double-your-money-back-if-not-satisfied guarantee." How does he feel about the chances that a person will actually ask for double his money back if

    (a) he decides to sell the product for $1.00;

    (b) he decides to sell the product for $1.25 with the guarantee;

    (c) he cannot make up his mind?

★6.42 Ms. Cooper is planning to attend a convention in San Diego, and she must send in her room reservations immediately. The convention is so large that the activities are held partly in hotel A and partly in hotel B, and Ms. Cooper does not know whether the particular session she wants to attend will be held in hotel A or hotel B. She is planning to stay only one day, which would cost her $40.00 at hotel A and $36.40 at hotel B, but it will cost her an extra $6.00 for cab fare if she stays at the wrong hotel. Where should she make her reservation if she feels that the probability is 0.75 that the session she wants to attend will be held at hotel A and she wants to minimize her expected cost?

6.43 Two friends are betting on repeated flips of a balanced coin. One has $5 at the start and the other has $3, and after each flip the loser pays the winner $1. If p is the probability that the one who starts with $5 will win his friend's $3 before he loses his own $5, explain why $3p - 5(1 - p)$ should equal 0, and then solve the equation $3p - 5(1 - p) = 0$ for p. Generalize this result to the case where two players start with a dollars and b dollars, respectively.

6.6
REFERENCES

More detailed treatments of the subject matter of this chapter may be found in

BROSS, I. D. J., *Design for Decision*. New York: Macmillan Publishing Co., Inc., 1953.

JEFFREY, R. C., *The Logic of Decision*. New York: McGraw-Hill Book Company, 1965.

and in many textbooks on business statistics; for instance, in Chapter 7 of

FREUND, J. E., and WILLIAMS, F. J., *Elementary Business Statistics: The Modern Approach, 4th ed.* Englewood Cliffs, N.J.: Prentice-Hall, Inc., 1982.

In most statistical problems we are interested only in one aspect, or at most in a few aspects, of the outcomes of experiments. For instance, a student taking a true–false test may be interested only in the number of questions he answers correctly, since this determines his grade; a geologist may be interested only in the age of a rock sample and not in its hardness; and a sociologist may be interested only in the socioeconomic status of a person interviewed in a survey and not in her age or weight. Also, an agronomist may be interested in determining not only the yield per acre of a new variety of corn but also the temperature at which it will germinate; and an automotive engineer may be interested in the brightness and the durability of the headlights proposed for a new model car and also in their projected cost.

In these five examples, the student, the geologist, the sociologist, the agronomist, and the automotive engineer are all interested in numbers that are associated with the outcomes of situations involving an element of chance, or more specifically, in values of random variables.

In the study of random variables we are usually interested in the probabilities with which they take on the various values within their range, namely, in their probability distributions. The study of random variables and probability distributions in Sections 7.1 and 7.2 will be followed by a discussion of various special probability distributions in Sections 7.3 through 7.6, and descriptions of their most important properties in Sections 7.7, 7.8, and 7.9.

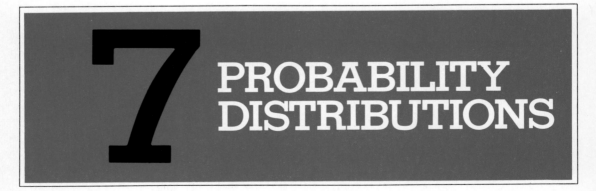

7 PROBABILITY DISTRIBUTIONS

7.1
RANDOM VARIABLES

To be more explicit about the concept of a random variable, let us consider Figure 7.1, which, like Figure 5.8, pictures the sample space for the example dealing with the two salespersons who hope to sell the two 1980 Chevettes. Note that we have added another number to each point—the number 0 to the point (0, 0); the number 1 to the points (1, 0) and (0, 1); and the number 2 to the points (2, 0), (1, 1), and (0, 2). In this way we have associated with each point of the sample space the total number of cars which, between them, the two salespersons will sell.

Since associating numbers with the points of a sample space is just a way of defining a function over the points of the sample space, random variables are really functions and not variables. Conceptually, though, most beginners find it easier to think of random variables simply as quantities which can take on different values depending on chance. For instance, the number of speeding tickets issued each day on the freeway between Indio and Blythe in California is a random variable, and so is the annual production of coffee in Brazil, the number of persons visiting Disneyland each week,

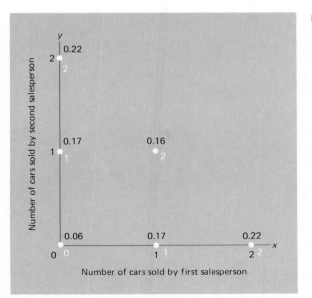

FIGURE 7.1 *Sample space with values of random variable.*

the wind velocity at Kennedy airport, the size of the audience at a baseball game, and the number of mistakes a person makes in typing a report.

Random variables are usually classified according to the number of values which they can assume. In this chapter we shall limit our discussion to **discrete random variables,** which can take on only a finite number of values, or a countable infinity of values (that is, as many values as there are whole numbers). Continuous random variables will be taken up in Chapter 8.

7.2
PROBABILITY DISTRIBUTIONS

The tables in the illustrations which follow serve to show what we mean by a **probability distribution.** With reference to Figure 7.1, if we add the probabilities associated with the respective points, we find that the random variable "the total number of 1980 Chevettes which, between them, the two sales-persons will sell" takes on the value 0 with probability 0.06, the value 1 with probability $0.17 + 0.17 = 0.34$, and the value 2 with probability $0.22 + 0.16 + 0.22 = 0.60$. All this is summarized in the following table:

Number of cars sold	Probability
0	0.06
1	0.34
2	0.60

As this table shows, a probability distribution is a correspondence which assigns probabilities to the values of a random variable. Another example of such a correspondence is given by the following table, which pertains to the number of points we roll with a balance die:

Number of points we roll with a die	Probability
1	$\frac{1}{6}$
2	$\frac{1}{6}$
3	$\frac{1}{6}$
4	$\frac{1}{6}$
5	$\frac{1}{6}$
6	$\frac{1}{6}$

Also, for four flips of a balanced coin there are the 16 equally likely possibilities, HHHH, HHHT, HHTH, HTHH, THHH, HHTT, HTHT,

THHT, HTTH, THTH, TTHH, HTTT, THTT, TTHT, TTTH, and TTTT, where H stands for heads and T for tails. Counting the number of heads in each case and using the formula $\frac{s}{n}$ for equally likely possibilities, we get the following probability distribution for the total number of heads:

Number of heads	Probability
0	$\frac{1}{16}$
1	$\frac{4}{16}$
2	$\frac{6}{16}$
3	$\frac{4}{16}$
4	$\frac{1}{16}$

Whenever possible, we try to express probability distributions by means of mathematical formulas which enable us to calculate directly the probabilities associated with the various values of a random variable. For instance, for the number of points we roll with a balanced die we can write

$$f(x) = \tfrac{1}{6} \qquad \text{for } x = 1, 2, 3, 4, 5, \text{ and } 6$$

where $f(1)$ represents the probability of rolling a 1, $f(2)$ represents the probability of rolling a 2, and so on, in the usual functional notation.[†]

To conclude this introduction to probability distributions, let us state the following two general rules which the values of all probability distributions must obey:

Since the values of a probability distribution are probabilities, they must be numbers on the interval from 0 to 1.

Since a random variable has to take on one of its values, the sum of all the values of a probability distribution must be equal to 1.

EXAMPLE   Check whether the following function can serve as the probability distribution of an appropriate random variable:

$$f(x) = \frac{x + 3}{15} \qquad \text{for } x = 1, 2, \text{ and } 3$$

Solution   Substituting $x = 1$, $x = 2$, and $x = 3$, we get $f(1) = \frac{4}{15}$, $f(2) = \frac{5}{15}$, and $f(3) = \frac{6}{15}$. Since none of these values is negative or greater than 1, and their sum is $\frac{4}{15} + \frac{5}{15} + \frac{6}{15} = 1$, the given function can serve as the probability distribution of a random variable.

[†] We shall write the probability that a random variable takes on the value x as $f(x)$, but we could just as well write it as $g(x)$, $h(x)$, $b(x)$, and so on.

7.3
THE BINOMIAL DISTRIBUTION

There are many applied problems in which we are interested in the probability that an event will occur x times out of n. For instance, we may be interested in the probability of getting 45 responses to 400 questionnaires sent out as part of a sociological study, the probability that 5 of 12 mice will survive for a given length of time after the injection of a cancer-inducing substance, the probability that 45 of 300 drivers stopped at a road block will be wearing their seat belts, or the probability that 66 of 200 television viewers (interviewed by a rating service) will recall what products were advertised on a given program. To borrow from the language of games of chance, we could say that in each of these examples we are interested in the probability of getting "*x successes in n trials*," or in other words, "*x* successes and $n - x$ failures in *n* attempts."

In the problems we shall study in this section, we shall always make the following assumptions:

The number of trials is fixed; the probability of a success is the same for each trial; and the trials are all independent (that is, what happens in any one trial does not affect the probability of a success in any other trial).

This means that the theory we shall develop will not apply, for example, if we are interested in the number of dresses a woman may try on before she buys one (where the number of trials is not fixed), if we check every hour whether traffic is congested at a certain intersection (where the probability of "success" is not constant), or if we are interested in the number of times that a person voted for the Republican candidate in the last five presidential elections (where the trials are not independent).

To solve problems which do meet the conditions listed in the preceding paragraph, we use a formula obtained in the following way: If p and $1 - p$ are the probabilities of a success and a failure on any given trial, then the probability of getting x successes and $n - x$ failures *in some specific order* is $p^x(1 - p)^{n-x}$; clearly, in this product of p's and $(1 - p)$'s there is one factor p for each success, one factor $1 - p$ for each failure, and the x factors p and $n - x$ factors $1 - p$ are all multiplied together by virtue of the generalized multiplication rule for more than two independent events. Since this probability applies to any point of the sample space which represents x successes and $n - x$ failures (in any specific order), we have only to count how many points of this kind there are, and then multiply $p^x(1 - p)^{n-x}$ by this number. Clearly, the number of ways in which we can select the x trials on which there is to be a success is $\binom{n}{x}$, the number of combinations of x objects selected from a set of n objects, and we have arrived at the following result:

Binomial distribution

> *The probability of getting x successes in n independent trials is*
>
> $$f(x) = \binom{n}{x} p^x (1 - p)^{n-x} \qquad \text{for } x = 0, 1, 2, \ldots, \text{ or } n$$
>
> *where p is the constant probability of a success for each trial.*

It is customary to say here that the number of successes in n trials is a random variable having the **binomial probability distribution,** or simply the **binomial distribution.** The binomial distribution is called by this name because for $x = 0, 1, 2, \ldots$, and n, the values of the probabilities are the successive terms of the binomial expansion of $[(1 - p) + p]^n$.

EXAMPLE   Write the formula for the binomial distribution of the number of times, in four flips, that a balanced coin comes up heads.

Solution   Substituting $n = 4$ and $p = \frac{1}{2}$ into the formula, we get

$$f(x) = \binom{4}{x}\left(\frac{1}{2}\right)^x \left(1 - \frac{1}{2}\right)^{4-x} = \binom{4}{x}\left(\frac{1}{2}\right)^4 = \frac{\binom{4}{x}}{16}$$

for $x = 0, 1, 2, 3$, and 4. For instance, for $x = 2$ we get

$$f(2) = \frac{\binom{4}{2}}{16} = \frac{6}{16}$$

and this agrees with the value given on page 169.

EXAMPLE   If the probability is 0.70 that any one registered voter (randomly selected from official rolls) will vote in a given election, what is the probability that two of five registered voters will vote in the election?

Solution   Substituting $x = 2$, $n = 5$, $p = 0.70$, and $\binom{5}{2} = 10$ into the formula, we get

$$f(2) = \binom{5}{2}(0.70)^2(1 - 0.70)^{5-2}$$

$$= 10(0.70)^2(0.30)^3$$

$$= 0.132$$

rounded to three decimals.

EXAMPLE   If the probability is 0.80 that a cleaning fluid will remove any one spot, what is the probability that it will remove six of eight spots?

Solution   Substituting $x = 6$, $n = 8$, $p = 0.80$, and $\binom{8}{6} = 28$ into the formula, we get

$$f(6) = \binom{8}{6}(0.80)^6(0.20)^{8-6}$$

$$= 28(0.80)^6(0.20)^2$$

$$= 0.294$$

rounded to three decimals.

In all these examples we took the values of the binomial coefficients, $\binom{4}{2} = 6$, $\binom{5}{2} = 10$, and $\binom{8}{6} = 28$, directly from Table X.

The following is an example in which we calculate all the probabilities of a binomial distribution:

EXAMPLE   The probability is 0.30 that a person returning from a trip to Europe will fail to declare at least one purchase on which there is duty. Find the probabilies that among six randomly selected persons returning from a trip to Europe there are 0, 1, 2, 3, 4, 5, or 6 who fail to declare at least one purchase on which there is duty.

Solution Substituting $n = 6$, $p = 0.30$, and $x = 0, 1, 2, 3, 4, 5,$ and 6 into the formula for the binomial distribution, we get

$$f(0) = \binom{6}{0}(0.30)^0(0.70)^6 = 0.118$$

$$f(1) = \binom{6}{1}(0.30)^1(0.70)^5 = 0.303$$

$$f(2) = \binom{6}{2}(0.30)^2(0.70)^4 = 0.324$$

$$f(3) = \binom{6}{3}(0.30)^3(0.70)^3 = 0.185$$

$$f(4) = \binom{6}{4}(0.30)^4(0.70)^2 = 0.060$$

$$f(5) = \binom{6}{5}(0.30)^5(0.70)^1 = 0.010$$

$$f(6) = \binom{6}{6}(0.30)^6(0.70)^0 = 0.001$$

where all the probabilities are rounded to three decimals. A histogram of this probability distribution is shown in Figure 7.2.

No doubt, the reader is not especially interested in tourists smuggling in goods from Europe, but let us stress the importance of the binomial distribution as a **mathematical model** by pointing out that the results of the preceding example apply also if the probability is 0.30 that the energy cell of a watch will last two years under normal usage, and we want to know the probabilities that, among six of these cells, 0, 1, 2, 3, 4, 5, or 6 will last two years under normal usage; if the probability is 0.30 that an embezzler will be caught and brought to trial, and we want to know the probabilities that, among six embezzlers, 0, 1, 2, 3, 4, 5, or 6 will be caught and brought to trial; if the probability is 0.30 that the head of a household owns some life insurance, and we want to know the probabilities that, among six heads of households, 0, 1, 2, 3, 4, 5, or 6 will own some life insurance; or if the probability is 0.30 that a person having a given disease will live for another ten years, and we want to know the probabilities that, among six persons having the disease, 0, 1, 2, 3, 4, 5, or 6 will live another ten years. The argument we have presented here is precisely like the one we used in Section 1.2, where we tried to impress upon the reader the generality of statistical techniques.

In actual practice, binomial probabilities are seldom obtained by direct substitution into the formula. Sometimes we use approximations such as those discussed later in this chapter and in Chapter 8, but more often we

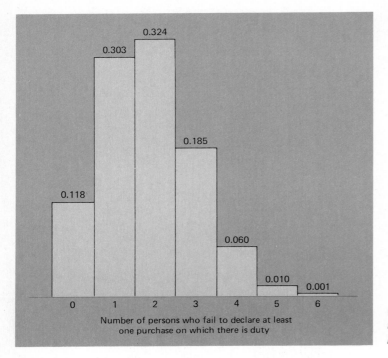

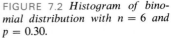

FIGURE 7.2 *Histogram of binomial distribution with n = 6 and p = 0.30.*

refer to special tables such as Table V at the end of this book or the more detailed tables listed among the references at the end of this chapter. Table V is limited to the binomial probabilities for $n = 2$ to $n = 20$, and $p = 0.05$, 0.1, 0.2, 0.3, 0.4, 0.5, 0.6, 0.7, 0.8, 0.9, and 0.95, all rounded to three decimals. Where values are omitted in the table, they are 0.0005 or less.

EXAMPLE   Suppose that the probability is 0.60 that a car stolen in a given city will be recovered. Use Table V to find the probabilities that
> (a) at least seven of ten cars stolen in this city will be recovered;
> (b) at most three of ten cars stolen in this city will be recovered.

Solution   (a) For $n = 10$ and $p = 0.60$ the entries in Table V corresponding to $x = 7$, 8, 9, and 10 are 0.215, 0.121, 0.040, and 0.006, and the probability that at least seven of ten cars will be recovered is

$$0.215 + 0.121 + 0.040 + 0.006 = 0.382$$

(b) For $n = 10$ and $p = 0.60$ the entries in Table V corresponding to $x = 1$, 2, and 3 are 0.002, 0.011, and 0.042; the probability that $x = 0$ is at most 0.0005 and it is not shown in Table V; so the probability that at most three of ten cars will be recovered is

$$0.002 + 0.011 + 0.042 = 0.055$$

EXAMPLE If the probability is 0.05 that any one person will dislike the taste of a new toothpaste, what is the probability that at least three of twenty randomly selected persons will dislike it?

Solution Adding the values in Table V for $n = 20$, $p = 0.05$, and $x = 3$, 4, and 5, we get

$$0.060 + 0.013 + 0.002 = 0.075$$

The values for $x = 6$ through 20 are all no greater than 0.0005, they are not shown in Table V, and we did not include them in the sum.

When we observe a value of a random variable having the binomial distribution—for instance, when we observe the number of heads in 25 flips of a coin, the number of seeds (in a package of 24 seeds) that germinate, the number of students (among 200 interviewed) who are opposed to a change in student activity fees, or the number of automobile accidents (among 20 investigated) that are due to drunk driving—we may say that we are **sampling a binomial population.**

EXERCISES

7.1 In each case determine whether the given values can be looked upon as the values of a probability distribution of a random variable which can take on only the values 1, 2, 3, and 4, and explain your answers:

    (a) $f(1) = 0.26, f(2) = 0.26, f(3) = 0.26, f(4) = 0.26$;

    (b) $f(1) = \frac{1}{9}, f(2) = \frac{2}{9}, f(3) = \frac{1}{3}, f(4) = \frac{1}{3}$;

    (c) $f(1) = 0.15, f(2) = 0.28, f(3) = 0.29, f(4) = 0.28$;

    (d) $f(1) = 0.33, f(2) = 0.37, f(3) = -0.03, f(4) = 0.33$.

7.2 Determine whether the following can be probability distributions (defined in each case only for the given values of x) and explain your answers:

    (a) $f(x) = \dfrac{1}{5}$ for $x = 0, 1, 2, 3, 4, 5$;

    (b) $f(x) = \dfrac{x + 1}{14}$ for $x = 1, 2, 3, 4$;

    (c) $f(x) = \dfrac{x - 2}{5}$ for $x = 1, 2, 3, 4, 5$;

    (d) $f(x) = \dfrac{x^2}{30}$ for $x = 0, 1, 2, 3, 4$.

7.3 In a given city, medical expenses are given as the reason for 75 percent of all personal bankruptcies. What is the probability that medical expenses will be given as the reason for two of the next four personal bankruptcies filed in that city?

7.4 A multiple-choice test consists of six questions and three answers to each question (of which only one is correct). If a student answers each question by rolling a balanced die and checking the first answer if he gets a 1 or a 2, the second answer if he gets a 3 or a 4, and the third answer if he gets a 5 or a 6, find the probabilities of getting

(a) exactly three correct answers;
(b) no correct answers;
(c) at least five correct answers.

7.5 If the probability is 0.15 that a set of tennis between two given professional players will go into a tie breaker, what is the probability that two of three sets between these two players will go into tie breakers?

7.6 If 40 percent of the mice used in an experiment will become very aggressive within one minute after having been administered an experimental drug, find the probability that exactly four of 12 mice which have been administered the drug become very aggressive within one minute, using

(a) the formula for the binomial distribution;
(b) Table V.

7.7 If it is true that 80 percent of all industrial accidents can be prevented by paying strict attention to safety regulations, find the probability that four of seven industrial accidents can thus be prevented, using

(a) the formula for the binomial distribution;
(b) Table V.

7.8 Suppose that a civil service examination is designed so that 70 percent of all persons with an IQ of 90 can pass it. Use Table V to find the probabilities that among 15 persons with an IQ of 90 who take the test

(a) at most six will pass;
(b) at least 12 will pass;
(c) from eight through 12 will pass.

7.9 A study shows that 50 percent of the families in a certain large metropolitan area have at least two cars. Find the probabilities that among 16 families randomly selected in this area

(a) nine have at least two cars;
(b) at most six have at least two cars;
(c) between eight and twelve, inclusive, have at least two cars.

7.10 An agricultural cooperative claims that 95 percent of the watermelons shipped out are ripe and ready to eat. Find the probabilities that among eighteen watermelons that are shipped out

(a) all eighteen are ripe and ready to eat;
(b) at least sixteen are ripe and ready to eat;
(c) at most fourteen are ripe and ready to eat.

7.11 A food distributor claims that 80 percent of her 6-ounce cans of mixed nuts contain at least three pecans. To check on this, a consumer testing service decides to examine eight of these 6-ounce cans of mixed nuts from a very large production lot, and reject the claim if fewer than six of them contain at least three pecans. Find the probabilities that the testing service will commit the error of

(a) rejecting the claim even though it is true;
(b) not rejecting the claim when in reality only 60 percent of the cans of mixed nuts contain at least three pecans;
(c) not rejecting the claim when in reality only 40 percent of the cans of mixed nuts contain at least three pecans.

7.12 A study shows that 70 percent of all patients coming to a certain medical clinic have to wait at least fifteen minutes to see their doctor. Find the probabilities that among ten patients coming to this clinic 0, 1, 2, 3, ..., or 10 have to wait at least fifteen minutes to see their doctor, and draw a histogram of this probability distribution.

★**7.13** In some situations where otherwise the binomial distribution applies, we are interested in the probabilities that the first success will occur on any given trial. For this to happen on the xth trial, it must be preceded by $x - 1$ failures for which the probability is $(1 - p)^{x-1}$, and it follows that the probability that the first success will occur on the xth trial is

Geometric distribution

$$f(x) = p(1 - p)^{x-1} \qquad for \ x = 1, 2, 3, 4, \ldots$$

This distribution is called the **geometric distribution** (because its successive values constitute a geometric progression) and it should be observed that there is a countable infinity of possibilities.[†] Using the formula, we find, for example, that for repeated rolls of a balanced die the probability that the first 6 will occur on the fifth roll is $\dfrac{1}{6}\left(\dfrac{5}{6}\right)^{5-1} = \dfrac{625}{7,776}$ or approximately 0.080.

  (a) When taping a television commercial, the probability that a certain actor will get his lines straight on any one take is 0.40. What is the probability that this actor will get his lines straight for the first time on the fourth take?

  (b) Suppose the probability is 0.25 that any given person will believe a rumor about the private life of a certain politician. What is the probability that the fifth person to hear the rumor will be the first one to believe it?

  (c) The probability is 0.70 that a child exposed to a certain contagious disease will catch it. What is the probability that the third child exposed to the disease will be the first one to catch it?

7.4
THE HYPERGEOMETRIC DISTRIBUTION ★

In Chapter 5 we spoke of sampling "with replacement" to illustrate the multiplication rule for independent events, and of sampling "without replacement" to illustrate the rule for dependent events. Since the binomial distribution applies only when the trials are all independent, it can be used

---

[†] As formulated in Chapter 5, the postulates of probability apply only when the sample space is finite. When the sample space is **countably infinite** (that is, when there are as many outcomes as there are whole numbers), as is the case here, the third postulate has to be modified so that for any sequence of mutually exclusive events $A_1, A_2, A_3, \ldots,$

$$P(A_1 \cup A_2 \cup A_3 \cup \cdots) = P(A_1) + P(A_2) + P(A_3) + \cdots$$

when we sample with replacement, but not when we sample without replacement. To introduce a probability distribution which applies when we sample without replacement, let us consider the following problem. A factory ships tape recorders in lots of 16, and when they arrive at their destination, an inspector randomly selects three from each lot. If they are all in good working condition, the whole lot is accepted; otherwise, the whole lot is inspected.

Clearly, this involves certain risks. A lot could be accepted even though 13 of the 16 tape recorders do not work, but more realistically, it may be of interest to know the probability that a lot will be accepted when, say, four of the 16 tape recorders are defective. This means that we must find the probability of three successes (tape recorders in good working condition) in three trials (among the three tape recorders inspected), and we might be tempted to argue that when 12 of the 16 tape recorders are in good working condition, the probability of a success is $\frac{12}{16} = 0.75$ and the desired probability is

$$f(3) = \binom{3}{3}(0.75)^3(1 - 0.75)^{3-3}$$

$$= 0.422$$

This result would be correct if sampling is with replacement and each tape recorder is replaced before the next one is selected. Otherwise, the assumption of independence is violated—the probability that the second tape recorder chosen is in good working condition is $\frac{11}{15}$ or $\frac{12}{15}$ depending on whether the first one is in good working condition or defective.

However, in realistic problems of sampling inspection we seldom, if ever, sample with replacement. To get the correct answer for our problem when sampling is without replacement, we might argue as follows. There are altogether $\binom{16}{3} = 560$ ways of selecting three of the 16 tape recorders, and they are all equiprobable since the selection is random. Among these, there are $\binom{12}{3} = 220$ ways of selecting three of the 12 tape recorders in good working condition, and it follows by the formula $\frac{s}{n}$ for equiprobable outcomes that the desired probability is $\frac{220}{560} = 0.393$.

To generalize the method used in this example, suppose that n objects are to be chosen from a set of a objects of one kind (successes) and b objects of another kind (failures), and that we are interested in the probability of getting "x successes and $n - x$ failures." Arguing as before, we can say that the x successes can be chosen in $\binom{a}{x}$ ways, the $n - x$ failures can be chosen in $\binom{b}{n-x}$ ways, and, hence, x successes and $n - x$ failures can be chosen in $\binom{a}{x} \cdot \binom{b}{n-x}$ ways. Also, n objects can be chosen from the whole set of $a + b$ objects in $\binom{a+b}{n}$ ways, and if we regard all these possibilities as equally likely, it follows that for sampling without replacement the probability of getting "x successes in n trials" is

Hypergeometric distribution

$$f(x) = \frac{\binom{a}{x} \cdot \binom{b}{n-x}}{\binom{a+b}{n}} \qquad for\ x = 0, 1, 2, \ldots, or\ n$$

This is the formula for the **hypergeometric distribution,** and it applies only when x does not exceed a and $n - x$ does not exceed b; clearly, we cannot very well get more successes, or failures, than there are in the whole set.

EXAMPLE   A mailroom clerk is supposed to send six of 15 packages to Europe by airmail, but he gets them all mixed up and randomly puts airmail postage on six of the packages. What is the probability that only three of the packages which are supposed to go by airmail will go by airmail?

Solution   Since $a = 6$, $b = 9$, $n = 6$, and $x = 3$, substitution into the formula for the hypergeometric distribution yields

$$f(3) = \frac{\binom{6}{3} \cdot \binom{9}{3}}{\binom{15}{6}} = \frac{20 \cdot 84}{5,005} = 0.336$$

EXAMPLE   Among a department store's 16 delivery trucks, five emit excessive amounts of pollutants. If eight of the trucks are randomly picked for inspection, what is the probability that this sample will include at least three of the trucks which emit excessive amounts of pollutants?

Solution   The probability we want to find is $f(3) + f(4) + f(5)$, where each term in this sum is to be calculated by means of the formula for the hypergeometric distribution with $a = 5$, $b = 11$, and $n = 8$. Substituting these values together with $x = 3, 4$, and 5, we get

$$f(3) = \frac{\binom{5}{3} \cdot \binom{11}{5}}{\binom{16}{8}} = \frac{10 \cdot 462}{12,870} = 0.359$$

$$f(4) = \frac{\binom{5}{4} \cdot \binom{11}{4}}{\binom{16}{8}} = \frac{5 \cdot 330}{12,870} = 0.128$$

$$f(5) = \frac{\binom{5}{5} \cdot \binom{11}{3}}{\binom{16}{8}} = \frac{1 \cdot 165}{12,870} = 0.013$$

and the probability that the sample will include at least three of the trucks which emit excessive amounts of pollutants is

$$0.359 + 0.128 + 0.013 = 0.500$$

In this section we introduced the hypergeometric distribution with an example in which we first erroneously used the binomial distribution. The error was not very large, though—we got 0.422 instead of 0.393—and in actual practice the binomial distribution is often used to approximate the hypergeometric distribution. It is generally agreed that this approximation is satisfactory if n constitutes less than 5 percent of $a + b$. The main advantages of the approximation are that the binomial distribution has been tabulated much more extensively than the hypergeometric distribution, and that it is generally easier to use.

7.5
THE POISSON DISTRIBUTION ⋆

If n is large and p is small, binomial probabilities are often approximated by means of the formula

Poisson distribution

$$f(x) = \frac{(np)^x \cdot e^{-np}}{x!} \qquad for \ x = 0, 1, 2, 3, \ldots$$

which is that for the **Poisson distribution.** The irrational number $e = 2.71828\ldots$ is the base of the system of natural logarithms, and values of e^{-np} may be obtained from Table XIII at the end of the book. A random variable having a Poisson distribution takes on the countably infinite set of values $x = 0, 1, 2, 3, \ldots$, but this poses no problems, since the probabilities usually become negligible after relatively few values of x.

EXAMPLE It is known that 2 percent of the books bound at a certain bindery have defective bindings. Use the Poisson approximation to the binomial distribution to find the probability that 5 of 400 books bound by this bindery will have defective bindings.

Solution Substituting $x = 5$ and $np = 400(0.02) = 8$ into the formula for the Poisson distribution and getting $e^{-8} = 0.00034$ from Table XIII, we obtain

$$f(5) = \frac{8^5 \cdot e^{-8}}{5!} = \frac{(32{,}768)(0.00034)}{120} = 0.093$$

Use of the formula for the binomial distribution would have been possible but very cumbersome.

EXAMPLE Records show that the probability is 0.00005 that a car will have a flat tire while driving through a certain tunnel. Use the Poisson approximation to the binomial distribution to find the probability that at least 2 of 10,000 cars passing through this tunnel will have flat tires.

Solution The answer is $1 - f(0) - f(1)$, where $f(0)$ and $f(1)$ are the values of the Poisson distribution with $np = 10{,}000(0.00005) = 0.5$ for $x = 0$ and $x = 1$. Getting $e^{-0.5} = 0.607$ from Table XIII, we find that

$$f(0) = \frac{(0.5)^0 (0.607)}{0!} = 0.607$$

$$f(1) = \frac{(0.5)^1 (0.607)}{1!} = 0.304$$

and, hence, that the desired probability is $1 - (0.607 + 0.304) = 0.089$.

The Poisson distribution also has many important applications which have no direct connection with the binomial distribution. In that case, np

is replaced by the parameter λ (Greek lowercase *lambda*) and we calculate the probability of getting x "successes" by means of the formula

Poisson distribution (with parameter λ)

$$f(x) = \frac{\lambda^x \cdot e^{-\lambda}}{x!} \qquad for \; x = 0, 1, 2, 3, \ldots$$

where the parameter λ is interpreted as the expected, or average, number of successes (see page 188).

This formula applies to many situations where we can expect a fixed number of "successes" per unit time (or for some other kind of unit), say, when a bank can expect to receive six bad checks per day, when 1.6 accidents can be expected per day at a busy intersection, when eight small pieces of meat can be expected in a frozen meat pie, when 5.6 imperfections can be expected per roll of cloth, when 0.14 complaint per passenger can be expected by an airline, and so on.

EXAMPLE   If a bank receives on the average $\lambda = 6$ bad checks per day, what is the probability that it will receive four bad checks on any given day?

Solution   Substituting $\lambda = 6$ and $x = 4$ into the formula, we get

$$f(4) = \frac{6^4 \cdot e^{-6}}{4!} = \frac{(1{,}296)(0.0025)}{24} = 0.135$$

EXAMPLE   If $\lambda = 5.6$ imperfections can be expected per roll of a certain kind of cloth, what is the probability that a roll will have three imperfections?

Solution   Substituting $\lambda = 5.6$ and $x = 3$ into the formula, we get

$$f(3) = \frac{(5.6)^3 \cdot e^{-5.6}}{3!} = \frac{(175.616)(0.0037)}{6} = 0.108$$

7.6
THE MULTINOMIAL DISTRIBUTION ★

An important generalization of the binomial distribution arises when there are more than two possible outcomes for each trial, the probabilities of the various outcomes remain the same for each trial, and the trials are all independent. This is the case, for example, when we repeatedly roll a die, where each trial has six possible outcomes; when students are asked whether they like a certain new recording, dislike it, or don't care; or when a U.S.

Department of Agriculture inspector grades beef as prime, choice, good, commercial, or utility.

If there are k possible outcomes for each trial and their probabilities are $p_1, p_2, \ldots,$ and p_k, it can be shown that the probability of x_1 outcomes of the first kind, x_2 outcomes of the second kind, $\ldots,$ and x_k outcomes of the kth kind in n trials is given by

Multinomial distribution

$$\frac{n!}{x_1! \cdot x_2! \cdot \ldots \cdot x_k!} \, p_1^{x_1} \cdot p_2^{x_2} \cdot \ldots \cdot p_k^{x_k}$$

This distribution is called the **multinomial distribution**.

EXAMPLE In a certain city, Channel 3 has 40 percent of the viewing audience on Friday nights, Channel 5 has 20 percent, Channel 10 has 10 percent, and Channel 12 has 30 percent. What is the probability that among seven television viewers randomly selected in that city on a Friday night, two will be watching Channel 3, one will be watching Channel 5, one will be watching Channel 10, and three will be watching Channel 12?

Solution Substituting $n = 7$, $x_1 = 2$, $x_2 = 1$, $x_3 = 1$, $x_4 = 3$, $p_1 = 0.40$, $p_2 = 0.20$, $p_3 = 0.10$, and $p_4 = 0.30$ into the formula, we get

$$\frac{7!}{2! \cdot 1! \cdot 1! \cdot 3!} (0.40)^2 (0.20)^1 (0.10)^1 (0.30)^3 = 0.036$$

EXERCISES

★7.14 Among the 15 applicants for a job, nine have college degrees. If two of the applicants are randomly chosen for interviews, find the probabilities that
  (a) neither has a college degree;
  (b) only one has a college degree;
  (c) both have college degrees.

★7.15 Among the 12 houses for sale in a development, nine have air conditioning. If four of the houses are randomly chosen for a full-page newspaper ad, what is the probability that three of them will have air conditioning?

★7.16 What is the probability that an IRS auditor will catch only two income tax returns with illegitimate deductions, if she randomly selects six returns from among 18 returns of which eight contain illegitimate deductions?

★7.17 To pass a quality control inspection, two batteries are chosen from each lot of 12 car batteries, and the lot is passed only if neither battery has any defects; otherwise, each of the batteries in the lot is checked. If the selection of the batteries is random, find the probabilities that a lot will
  (a) pass the inspection when one of the 12 batteries is defective;
  (b) fail the inspection when three of the batteries are defective;
  (c) fail the inspection when six of the batteries are defective.

★7.18 Among a person's 10 pairs of socks, four pairs need mending. If he randomly picks three pairs of these socks to take along on a trip, what are the probabilities that
  (a) none of the socks will need mending;
  (b) one pair will need mending;
  (c) two pairs will need mending;
  (d) all three pairs will need mending?

★7.19 Among the 200 employees of a company, 140 are union members and the others are nonunion. If four of the employees are chosen by lot to serve on a grievance committee, find the probability that two of them will be union members and the others nonunion, using
  (a) the formula for the hypergeometric distribution;
  (b) the binomial distribution as an approximation.

★7.20 A shipment of 100 burglar alarms contains four that are defective. If three of these burglar alarms are randomly selected and shipped to a customer, find the probability that she will get one bad unit using
  (a) the formula for the hypergeometric distribution;
  (b) the binomial approximation to the hypergeometric distribution.

★7.21 It is known from experience that 2 percent of the calls received by a switchboard are wrong numbers. Use the Poisson approximation to the binomial distribution to determine the probability that 3 of 200 calls received by the switchboard will be wrong numbers.

★7.22 If 0.8 percent of the fuses delivered to an arsenal are defective, use the Poisson approximation to the binomial distribution to determine the probability that in a random sample of 400 fuses, four will be defective.

★7.23 Records show that the probability is 0.0012 that a person will get food poisoning spending a day at a certain state fair. Use the Poisson approximation to the binomial distribution to find the probability that among 1,000 persons attending the fair, at most two will get food poisoning.

★7.24 If the number of complaints which a dry cleaning establishment receives per day is a random variable having the Poisson distribution with $\lambda = 3.3$, what is the probability that it will receive only two complaints on any given day?

★7.25 If the number of wild pigs seen on a two-hour Jeep trip in the Sonora desert is a random variable having the Poisson distribution with $\lambda = 0.8$, find the probabilities that on such a Jeep trip one will see
  (a) no wild pigs;
  (b) one wild pig;
  (c) two wild pigs;
  (d) more than two wild pigs.

★7.26 The number of minor injuries a football coach can expect during the course of a game is a random variable having the Poisson distribution with $\lambda = 4.4$. What is the probability that during the course of a game his team will have at most two minor injuries?

★7.27 The probabilities that a floodlight will last less than 100 hours, from 100 to 150 hours, or more than 150 hours are, respectively, 0.60, 0.30, and 0.10. What is the probability that among six such floodlights three will last less than 100 hours, two will last from 100 to 150 hours, and one will last more than 150 hours?

★7.28   It can easily be shown that the probabilities of getting two heads, a head and a tail, and two tails when flipping a pair of balanced coins are, respectively, $\frac{1}{4}$, $\frac{1}{2}$, and $\frac{1}{4}$. What is the probability of getting two heads once, a head and a tail twice, and two tails twice in five flips of a pair of balanced coins?

★7.29   The probabilities are 0.60, 0.20, 0.10, and 0.10 that a state income tax form will be filled out correctly, that it will contain only errors favoring the tax payer, that it will contain only errors favoring the government, and that it will contain both kinds of errors. What is the probability that among ten such tax forms (randomly selected for audit) seven will be filled out correctly, one will contain only errors favoring the tax payer, one will contain only errors favoring the government, and one will contain both kinds of errors?

7.7
THE MEAN
OF A PROBABILITY DISTRIBUTION

On page 153 we showed that an airline office at a certain airport can expect 2.75 complaints per day about its luggage handling, and we arrived at this result by adding the products obtained by multiplying 0, 1, 2, 3, ..., by the corresponding probabilities that it will receive 0, 1, 2, 3, ..., complaints about its luggage handling on any one day. If we apply the same argument to the illustrations of Section 7.2, we find that, between them, the two salespersons can be expected to sell

$$0(0.06) + 1(0.34) + 2(0.60) = 1.54$$

of the two 1980 Chevettes, the number of points we can expect in one roll of a balanced die is

$$1 \cdot \tfrac{1}{6} + 2 \cdot \tfrac{1}{6} + 3 \cdot \tfrac{1}{6} + 4 \cdot \tfrac{1}{6} + 5 \cdot \tfrac{1}{6} + 6 \cdot \tfrac{1}{6} = 3\tfrac{1}{2}$$

and the number of heads we can expect in four flips of a balanced coin is

$$0 \cdot \tfrac{1}{16} + 1 \cdot \tfrac{4}{16} + 2 \cdot \tfrac{6}{16} + 3 \cdot \tfrac{4}{16} + 4 \cdot \tfrac{1}{16} = 2$$

Of course, they cannot actually sell 1.54 cars and we cannot actually roll $3\tfrac{1}{2}$ with a die; like all mathematical expectations, these figures must be looked upon as averages.

In general, if a random variable takes on the values $x_1, x_2, x_3, \ldots,$ or x_k, with the probabilities $f(x_1), f(x_2), f(x_3), \ldots,$ and $f(x_k)$, its expected value (or its mathematical expectation) is given by

$$x_1 \cdot f(x_1) + x_2 \cdot f(x_2) + x_3 \cdot f(x_3) + \cdots + x_k \cdot f(x_k)$$

and it is customary to refer to this quantity as the **mean of the random variable,** or the **mean of its distribution.** Using the $\sum$ notation, we can write

Mean of probability distribution

$$\mu = \sum x \cdot f(x)$$

where the mean of the distribution of a random variable, like the mean of a population, is denoted by the Greek letter μ (*mu*). The notation is the same, for as we pointed out in connection with the binomial distribution, when we observe a value of a random variable, we may refer to its distribution as the population we are sampling. For instance, the histogram of Figure 7.2 on page 174 may be looked upon as the population we are sampling when we observe the number of successes in $n = 6$ trials with $p = 0.30$.

EXAMPLE With reference to the example on page 172, how many of six randomly selected persons returning from a trip to Europe can be expected not to declare at least one purchase on which there is duty?

Solution Substituting $x = 0, 1, 2, 3, 4, 5,$ and 6 and the probabilities on page 173 into the formula for μ, we get

$$\mu = 0(0.118) + 1(0.303) + 2(0.324) + 3(0.185) + 4(0.060)$$
$$+ 5(0.010) + 6(0.001)$$
$$= 1.802$$

When a random variable can take on many different values, the calculation of μ may become laborious. For instance, if we want to know how many among 800 customers entering a store can be expected to make a purchase, and the probability that any one of them will make a purchase is 0.40, we might first calculate the 801 probabilities corresponding to $0, 1, 2, \ldots,$ or 800 of them making a purchase. However, if we think for a moment, we might argue that in the long run 40 percent of the customers will make a purchase, 40 percent of 800 is 320, and, hence, we can expect that 320 of the 800 customers will make a purchase. Similarly, if a balanced coin is flipped 1,000 times, we can argue that in the long run heads will come up 50 percent of the time, and, hence, that we can expect $1,000(0.50) = 500$ heads. These two values are, indeed, correct; both problems deal with binomial distributions, and it can be shown that in general

Mean of binomial distribution

$$\mu = n \cdot p$$

for the mean of a binomial distribution. In words, the mean of a binomial distribution is simply the product of the number of trials and the probability of success on an individual trial.

EXAMPLE   With reference to the example on page 172, find the mean of the probability distribution of the number of persons, among six returning from a trip to Europe, who fail to declare at least one purchase on which there is duty.

Solution   Since we are dealing with a binomial distribution with $n = 6$ and $p = 0.30$, $\mu = 6(0.30) = 1.80$. The small difference of 0.002 between this exact value and the value obtained before is due to rounding the probabilities on page 173 to three decimals.

EXAMPLE   Find the mean of the probability distribution of the number of heads obtained in four flips of a balanced coin.

Solution   For a binomial distribution with $n = 4$ and $p = \frac{1}{2}$, we get $\mu = 4 \cdot \frac{1}{2} = 2$, and this agrees with the result obtained on page 185.

It is important to remember that the formula $\mu = n \cdot p$ applies only to binomial distributions. There are other formulas for other distributions; for instance, for the hypergeometric distribution the formula for the mean is

Mean of hypergeometric distribution

$$\mu = \frac{n \cdot a}{a + b}$$

EXAMPLE   Among 12 school buses, five have worn brakes. If six of the school buses are randomly picked for inspection, how many of them can be expected to have worn brakes?

Solution   We have here a hypergeometric distribution with $a = 5$, $b = 7$, and $n = 6$, so that substitution into the above formula yields

$$\mu = \frac{6 \cdot 5}{5 + 7} = 2.5$$

This should not come as a surprise; half of the school buses are selected and half of the ones with worn brakes are expected to be included in the sample.

Also, the mean of the Poisson distribution is simply $\mu = \lambda$. Formal proofs of all these special formulas may be found in any textbook on mathematical statistics.

7.8
THE STANDARD DEVIATION
OF A PROBABILITY DISTRIBUTION

In Chapter 3 we saw that there are many situations in which we must describe, in addition to the mean or some other measure of location, the variability (spread, or dispersion) of a set of data. As we indicated in that chapter, the most widely used statistical measures of variation are the variance and its square root, the standard deviation, which both measure variability by averaging the squared deviations from the mean. For probability distributions, we measure variability in almost the same way, but instead of averaging the squared deviations from the mean, we calculate their expected value. If x is a value of some random variable whose probability distribution has the mean μ, the deviation from the mean is $x - \mu$ and we define the **variance of the probability distribution** as the expected value of the squared deviation from the mean, namely, as

Variance of probability distribution

$$\sigma^2 = \sum (x - \mu)^2 \cdot f(x)$$

where the summation extends over all values taken on by the random variable. As in the preceding section, and for the same reason, we denote descriptions of probability distributions with the same symbols as descriptions of populations. The square root of the variance defines the **standard deviation of a probability distribution**, and we write

Standard deviation of probability distribution

$$\sigma = \sqrt{\sum (x - \mu)^2 \cdot f(x)}$$

EXAMPLE   Use the probabilities obtained in the example on page 173 to determine the standard deviation of the probability distribution of the number of persons, among six returning from a trip to Europe, who fail to declare at least one purchase on which there is duty.

Solution   Here $\mu = 6(0.30) = 1.80$ and we arrange the calculations as follows:

Number of persons	Probability	Deviation from mean	Squared deviation from mean	$(x - \mu)^2 f(x)$
0	0.118	−1.8	3.24	0.38232
1	0.303	−0.8	0.64	0.19392
2	0.324	0.2	0.04	0.01296
3	0.185	1.2	1.44	0.26640
4	0.060	2.2	4.84	0.29040
5	0.010	3.2	10.24	0.10240
6	0.001	4.2	17.64	0.01764

$$\sigma^2 = 1.26604$$

The values in the column on the right were obtained by multiplying each squared deviation from the mean by its probability, and their sum is the variance of the distribution. Also, the standard deviation is $\sigma = \sqrt{1.26604} = 1.13$.

The calculations were easy in this example because the deviations from the mean were small numbers given to one decimal. If the deviations from the mean are large numbers, or if they are given to several decimals, it is usually worthwhile to simplify the calculations by using the following computing formula:

Computing formula for the variance of a probability distribution

$$\sigma^2 = \sum x^2 \cdot f(x) - \left[\sum x \cdot f(x)\right]^2$$

The advantage of this formula is that we do not have to work with the deviations from the mean. Instead, we subtract $\mu^2 = \left[\sum x \cdot f(x)\right]^2$ from the sum of the products obtained by multiplying the square of each value of the random variable by the corresponding probability.

EXAMPLE   On page 153 we showed that if the probabilities are 0.06, 0.21, 0.24, 0.18, 0.14, 0.10, 0.04, 0.02, and 0.01 that an airline office at a certain airport will receive 0, 1, 2, 3, 4, 5, 6, 7, or 8 complaints per day about its luggage handling, the mean of this probability distribution is $\sum x \cdot f(x) = 2.75$. Find the variance and the standard deviation of this probability distribution.

Solution First we calculate

$$\sum x^2 \cdot f(x) = 0^2(0.06) + 1^2(0.21) + 2^2(0.24) + 3^2(0.18)$$
$$+ 4^2(0.14) + 5^2(0.10) + 6^2(0.04)$$
$$+ 7^2(0.02) + 8^2(0.01)$$
$$= 10.59$$

Then, substituting this value and $\sum x \cdot f(x) = 2.75$ into the computing formula, we get

$$\sigma^2 = 10.59 - (2.75)^2 = 3.03$$

and, hence, $\sigma = \sqrt{3.03} = 1.74$.

As in the case of the mean, the calculation of the variance or the standard deviation can generally be simplified when we deal with special kinds of probability distributions. For instance, for the binomial distribution we have the formula

***Standard deviation
of binomial
distribution***

$$\sigma = \sqrt{np(1 - p)}$$

EXAMPLE Use this formula to verify the result obtained on page 189 for the example dealing with the number of persons, among six returning from a trip to Europe, who fail to declare at least one purchase on which there is duty.

Solution For the binomial distribution with $n = 6$ and $p = 0.30$, the formula yields

$$\sigma^2 = 6(0.30)(0.70) = 1.26$$

and this exact value differs from the results we obtained before by the small rounding error $1.26604 - 1.26 = 0.00604$.

EXAMPLE Find the variance of the probability distribution of the number of heads obtained in four flips of a balanced coin.

Solution The variance of the binomial distribution with $n = 4$ and $p = \frac{1}{2}$ is

$$\sigma^2 = 4 \cdot \tfrac{1}{2} \cdot \tfrac{1}{2} = 1$$

There also exist formulas for the standard deviation of other special distributions, and they may be found in more advanced texts.

7.9
CHEBYSHEV'S THEOREM

Intuitively speaking, the variance and the standard deviation of a probability distribution measure its spread or its dispersion: When σ is small, the probability is high that we will get a value close to the mean, and when σ is large, we are more likely to get a value far away from the mean. This important idea is expressed rigorously in Chebyshev's theorem, which we introduced in Section 3.8 as it pertains to frequency distributions. For probability distributions, the theorem can be stated as follows:

Chebyshev's theorem

> *The probability that a random variable will take on a value within k standard deviations of the mean is at least*
>
> $$1 - \frac{1}{k^2}$$

For instance, the probability of getting a value within two standard deviations of the mean (a value between $\mu - 2\sigma$ and $\mu + 2\sigma$) is at least $1 - \frac{1}{2^2} = \frac{3}{4}$, and the probability of getting a value within five standard deviations of the mean (a value between $\mu - 5\sigma$ and $\mu + 5\sigma$) is at least $1 - \frac{1}{5^2} = \frac{24}{25}$. When Chebyshev's theorem is used to illustrate the relationship between the standard deviation of a probability distribution and its spread or dispersion, the constant k is usually chosen arbitrarily—it can be any positive number, but the theorem becomes trivial when k is 1 or less.

EXAMPLE The number of customers who visit a Buick dealer's showroom on a Saturday morning is a random variable with $\mu = 18$ and $\sigma = 3.5$. What does Chebyshev's theorem with, say, $k = 3$ tell us about the number of customers who will visit the dealer's showroom on a Saturday morning?

Solution Since $\mu - 3\sigma = 18 - 3(3.5) = 7.5$ and $\mu + 3\sigma = 18 + 3(3.5) = 28.5$, we can assert with a probability of at least $1 - \frac{1}{3^2} = \frac{8}{9}$, or approximately 0.89, that there will be anywhere from 8 to 28 customers.

EXAMPLE What does Chebyshev's theorem with, say, $k = 6$ tell us about the number of heads we may get in 400 flips of a balanced coin?

Solution For the binomial distribution with $n = 400$ and $p = \frac{1}{2}$, the mean and the standard deviation are

$$\mu = np = 400 \cdot \tfrac{1}{2} = 200$$

and

$$\sigma = \sqrt{np(1 - p)} = \sqrt{400 \cdot \tfrac{1}{2} \cdot \tfrac{1}{2}} = 10$$

Thus, $\mu - 6\sigma = 200 - 6(10) = 140$ and $\mu + 6\sigma = 200 + 6(10) = 260$, and we can assert with a probability of at least $1 - \dfrac{1}{6^2} = \dfrac{35}{36}$, or approximately 0.97, that we will get between 140 and 260 heads.

If, in this example, we convert the numbers of heads into proportions, we can assert with a probability of at least $\frac{35}{36}$ that the proportion of heads we get in 400 flips of a balanced coin will lie between $\frac{140}{400} = 0.35$ and $\frac{260}{400} = 0.65$. To continue this argument, the reader will be asked to show in Exercise 7.50 on page 195 that the probability is at least $\frac{35}{36}$ that for 10,000 flips of a balanced coin the proportion of heads will lie between 0.47 and 0.53, and that for 1,000,000 flips of a balanced coin it will lie between 0.497 and 0.503. This provides support for the Law of Large Numbers, which we mentioned in Section 4.4 in connection with the frequency interpretation of probability.

In actual practice, Chebyshev's theorem is rarely used, since the probability "at least $1 - \dfrac{1}{k^2}$" is in many cases unnecessarily small. For instance, in the preceding example we showed that the probability of getting a value within six standard deviations of the mean is at least $\frac{35}{36}$, or about 0.97, but the actual probability that this will happen for a random variable having the binomial distribution with $n = 400$ and $p = \frac{1}{2}$ is about 0.999999998. "At least 0.97" is not wrong, but it may well not tell us enough.

EXERCISES

7.30 Suppose that the probabilities are 0.4, 0.3, 0.2, and 0.1 that among three recently married couples 0, 1, 2, or 3 will be divorced within two years. Use the formulas which define μ and σ^2 to find
(a) the mean of this probability distribution;
(b) the variance of this probability distribution.

7.31 Use the computing formula to rework part (b) of the preceding exercise.

7.32 The following table gives the probabilities that a probation officer will receive 0, 1, 2, 3, 4, 5, or 6 reports of probation violations on any given day:

Number of violations	0	1	2	3	4	5	6
Probability	0.15	0.22	0.31	0.18	0.09	0.04	0.01

Use the formulas which define μ and σ to find
   (a) the mean of this probability distribution;
   (b) the standard deviation of this probability distribution.

7.33 Use the computing formula to rework part (b) of the preceding exercise.

7.34 The probabilities that a building inspector will observe 0, 1, 2, 3, 4, or 5 violations of the building code in a home built in a large development are, respectively, 0.41, 0.22, 0.17, 0.13, 0.05, and 0.02. Find the mean and the standard deviation of this probability distribution.

7.35 As can easily be verified by means of the formula for the binomial distribution (or by listing all 32 possibilities), the probabilities of getting 0, 1, 2, 3, 4, or 5 heads in five flips of a balanced coin are $\frac{1}{32}$, $\frac{5}{32}$, $\frac{10}{32}$, $\frac{10}{32}$, $\frac{5}{32}$, and $\frac{1}{32}$. Find the mean of this probability distribution using
   (a) the formula which defines μ;
   (b) the special formula for the mean of a binomial distribution.

7.36 With reference to the preceding exercise, find the variance of the probability distribution using
   (a) the formula which defines σ^2;
   (b) the computing formula for σ^2;
   (c) the special formula for the variance of a binomial distribution.

7.37 The probabilities of rolling a 2, 3, 4, 5, 6, 7, 8, 9, 10, 11, or 12 with a pair of balanced dice are $\frac{1}{36}$, $\frac{2}{36}$, $\frac{3}{36}$, $\frac{4}{36}$, $\frac{5}{36}$, $\frac{6}{36}$, $\frac{5}{36}$, $\frac{4}{36}$, $\frac{3}{36}$, $\frac{2}{36}$, and $\frac{1}{36}$. Find the mean and the standard deviation of this probability distribution.

7.38 A study shows that 70 percent of all first-class letters between two cities are delivered within 48 hours. Find the mean and the variance of the distribution of the number of first-class letters between the two cities, among twelve randomly selected, which are delivered within 48 hours, using
   (a) Table V, the formula which defines μ, and the computing formula for σ^2;
   (b) the special formulas for the mean and the variance of a binomial distribution.

7.39 If 95 percent of certain radial tires last at least 30,000 miles, find the mean and the standard deviation of the distribution of the number of these tires, among 20 selected at random, that last at least 30,000 miles, using
   (a) Table V, the formula which defines μ, and the computing formula for σ;
   (b) the special formulas for the mean and the standard deviation of a binomial distribution.

7.40 Find the mean and the standard deviation of the distribution of each of the following random variables (having binomial distributions):
   (a) The number of heads obtained in 676 flips of a balanced coin.
   (b) The number of 3's obtained in 720 rolls of a balanced die.
   (c) The number of persons (among 600 invited) who will attend the opening of a new branch bank, when the probability is 0.30 that any one of them will attend.
   (d) The number of defectives in a sample of 600 parts made by a machine, when the probability is 0.04 that any one of the parts is defective.
   (e) The number of students (among 800 interviewed) who do not like the food served at the university cafeteria, when the probability is 0.65 that any one of them does not like the food.

★7.41 On page 180 we showed that if five of a department store's 16 delivery trucks emit excessive amounts of pollutants and eight of them are randomly picked for inspection, the probabilities are 0.359, 0.128, and 0.013 that 3, 4, or 5 of the trucks will emit excessive amounts of pollutants.
(a) Show that the probabilities are 0.013, 0.128, and 0.359 that 0, 1, or 2 of the eight trucks will emit excessive amounts of pollutants.
(b) Use the formula which defines μ to calculate the mean of the distribution of the number of the department store's trucks, among eight inspected, which emit excessive amounts of pollutants.
(c) Use the special formula for the mean of a hypergeometric distribution to rework part (b), and compare the two results.

★7.42 Among ten faculty members considered for promotions there are six men and four women.
(a) If two of them are chosen at random, find the probabilities that 0, 1, or 2 women will be included.
(b) Use the probabilities obtained in part (a) to calculate the mean and the standard deviation of this probability distribution.
(c) Use the special formula for the mean of a hypergeometric distribution to verify the value obtained for μ in part (b).

★7.43 The probabilities that there will be 0, 1, 2, 3, 4, or 5 fires caused by lightning during a summer storm are, respectively, 0.449, 0.360, 0.144, 0.038, 0.008, and 0.001. Calculate the mean of this Poisson distribution with $\lambda = 0.8$, and use the result to verify the special formula $\mu = \lambda$ mentioned on page 188.

★7.44 Calculate the variance of the probability distribution of the preceding exercise, and use the result to verify the formula $\sigma^2 = \lambda$ for the variance of a Poisson distribution with the parameter λ.

★7.45 If the number of gamma rays emitted per second by a certain radioactive substance is a random variable having the Poisson distribution with $\lambda = 2.5$, the probabilities that it will emit 0, 1, 2, 3, 4, 5, 6, 7, 8, or 9 gamma rays in any one second are, respectively, 0.082, 0.205, 0.256, 0.214, 0.134, 0.067, 0.028, 0.010, 0.003, and 0.001.
(a) Calculate the mean and the standard deviation of this probability distribution.
(b) Use the results of part (a) to verify the formulas $\mu = \lambda$ and $\sigma = \sqrt{\lambda}$ for the mean and the standard deviation of a Poisson distribution with the parameter λ.

7.46 If a student answers the 144 questions of a true–false test by flipping a balanced coin—heads is "true" and tails is "false"—what does Chebyshev's theorem with $k = 4$ tell us about the number of correct answers he will get?

7.47 The daily number of customers served lunch on a weekday by a certain restaurant is a random variable with $\mu = 142$ and $\sigma = 12$. According to Chebyshev's theorem, with what probability can we assert that between 82 and 202 customers will be served lunch by the restaurant on any given weekday?

7.48 The number of marriage licenses issued in a certain city during the month of June averages $\mu = 134$ with a standard deviation of $\sigma = 7.5$.

(a) What does Chebyshev's theorem with $k = 9$ tell us about the number of marriage licenses issued there during any particular month of June?

(b) According to Chebyshev's theorem, with what probability can we assert that between 74 and 194 marriage licenses will be issued there during any particular month of June?

7.49 The annual number of rainy days in a certain city is a random variable with $\mu = 126$ and $\sigma = 9$.

(a) What does Chebyshev's theorem with $k = 12$ tell us about the number of days it will rain in the given city in any one year?

(b) According to Chebyshev's theorem, with what probability can we assert that it will rain in the given city between 96 and 156 days in any one year?

7.50 Use Chebyshev's theorem to show that the probability is at least $\frac{35}{36}$ that

(a) in 10,000 flips of a balanced coin there will be between 4,700 and 5,300 heads, and hence the proportion of heads will be between 0.47 and 0.53;

(b) in 1,000,000 flips of a balanced coin there will be between 497,000 and 503,000 heads, and hence the proportion of heads will be between 0.497 and 0.503.

7.10
TECHNICAL NOTE (Simulation) ★

In recent years, simulation techniques have been applied to many problems in the various sciences, and if the processes being simulated involve an element of chance, we refer to these techniques as **Monte Carlo methods**. Such methods have been used, for instance, to study traffic flow on proposed freeways, to conduct war "games," to study the spread of epidemics or human behavior during times of natural disaster (say, a flood or an earthquake), and to study the scattering of neutrons or the collisions of photons with electrons. Also, in business research, such methods are used to solve inventory problems, production scheduling, or the effects of advertising campaigns, and to study many other situations involving overall planning and organization. In most cases, this will eliminate the cost of building and operating expensive equipment, and in some instances it can be used when direct experimentation is impossible.

Although Monte Carlo methods are sometimes based on actual gambling devices, it is usually expedient to use published tables of **random numbers** (or **random digits**). Such tables consist of many pages on which the digits 0, 1, 2, ..., and 9 are set down in a "random" fashion, much as they would appear if they were generated one at a time by a chance or gambling device giving each digit the same probability of $\frac{1}{10}$. Some early tables of random numbers were copied from pages of census data or from tables of 20-place logarithms, but they were found to be deficient in various ways.

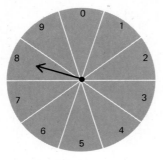

FIGURE 7.3 *Spinner for generating random digits.*

Nowadays, such tables are made with the use of electronic computers, but it would be possible to generate a table with a carefully constructed spinner like that of Figure 7.3.

To illustrate the use of a table of random numbers, let us show how we can play "heads or tails" without actually flipping a coin. Letting 0, 2, 4, 6, and 8 represent heads and 1, 3, 5, 7, and 9 represent tails, we might arbitrarily choose the fourth column of the first page of Table XI (that is, page 517), start at the top and go down the page. Thus, we get 3, 9, 8, 1, 5, 1, 6, 3, 2, 4, . . . , and we interpret this as tail, tail, head, tail, tail, tail, head, tail, head, head, ...

Repeated flips of any number of coins, say, three coins, can be simulated in the same way. If we arbitrarily choose the first three columns of the second page of Table XI (that is, page 518), starting with the sixteenth row and going down the page, we read out the three-digit random numbers 659, 900, 972, 219, 411, 237, 599, 826, 838, 619, . . . , and, counting the number of even digits in each case, we interpret this as 1, 2, 1, 1, 1, 1, 0, 3, 2, 1, . . . , heads. If we did not want to count even digits in each case, we could make use of the fact that the probabilities of getting 0, 1, 2, or 3 heads are $\frac{1}{8}$, $\frac{3}{8}$, $\frac{3}{8}$, and $\frac{1}{8}$, and use the following scheme:

Number of heads	Probability	Random digits
0	$\frac{1}{8}$	0
1	$\frac{3}{8}$	1, 2, 3
2	$\frac{3}{8}$	4, 5, 6
3	$\frac{1}{8}$	7

Ignoring the digits 8 and 9 wherever they may occur, we would thus interpret the random digits 8, 1, 3, 0, 7, 4, 3, 6, 9, 4, 8, 3, 5, 8, 0, . . . , in the twelfth row of page 517, going from left to right, as 1, 1, 0, 3, 2, 1, 2, 2, 1, 2, 0, . . . , heads.

Of the two methods used in the preceding example, the first has the disadvantage that we have to count how many even digits there are in each case; the second has the disadvantage that some of the digits have to be ignored. To avoid such waste of effort and time, we could have used the following scheme:

Number of heads x	Probability of x heads	Probability of x or less heads	Random numbers
0	0.125	0.125	000–124
1	0.375	0.500	125–499
2	0.375	0.875	500–874
3	0.125	1.000	875–999

Here we used three-digit random numbers because the probabilities are given to three decimals, and we allocated 125 (or one-eighth) of the 1,000 random numbers from 000 to 999 to 0 heads, 375 (or three-eighths) to 1 head, 375 (or three-eighths) to 2 heads, and 125 (or one-eighth) to 3 heads. The column of cumulative probabilities was added to facilitate the assignment of the random numbers. Observe that in each case the last random digit is one less than the number formed by the three decimal digits of the corresponding cumulative probability. With this scheme, if we arbitrarily use the 16th, 17th, and 18th columns of page 518, starting with the sixth row and going down the page, we get 974, 611, 345, 664, 041, 203, 531, 421, 031, 925, ... , and we interpret this as 3, 2, 1, 2, 0, 1, 2, 1, 0, 3, ..., heads.

The method we have illustrated here with reference to a game of chance can be used to simulate observations of any random variable with a given probability distribution.

EXAMPLE   Suppose that the probabilities are 0.008, 0.037, 0.089, 0.146, 0.179, 0.175, 0.143, 0.100, 0.062, 0.034, 0.016, 0.007, 0.003, and 0.001 that 0, 1, 2, 3, ..., or 13 cars will arrive at a tollbooth of a bridge in any one minute during the early afternoon.

(a) Distribute the three-digit random numbers from 000 to 999 among the fourteen values of this random variable, so that they can be used to simulate the arrival of cars at the tollbooth.

(b) Use the 6th, 7th, and 8th column of page 517, starting with the 21st row and going down the page, to simulate the arrival of cars at the tollbooth during 20 one-minute intervals during the early afternoon.

Solution   **(a)** Calculating the cumulative probabilities and following the suggestion on page 197, we arrive at the following scheme:

Number of cars	Probability	Cumulative probability	Random numbers
0	0.008	0.008	000–007
1	0.037	0.045	008–044
2	0.089	0.134	045–133
3	0.146	0.280	134–279
4	0.179	0.459	280–458
5	0.175	0.634	459–633
6	0.143	0.777	634–776
7	0.100	0.877	777–876
8	0.062	0.939	877–938
9	0.034	0.973	939–972
10	0.016	0.989	973–988
11	0.007	0.996	989–995
12	0.003	0.999	996–998
13	0.001	1.000	999

(b) Following the instructions, we get the random numbers 836, 712, 524, 762, 325, 081, 960, 594, 473, 370, 305, 178, 523, 184, 368, 864, 676, 975, 553, and 618, and this means that 7, 6, 5, 6, 4, 2, 9, 5, 5, 4, 4, 3, 5, 3, 4, 7, 6, 10, 5, and 5 cars arrived at the tollbooth in 20 one-minute intervals during the early afternoon.

EXERCISES

★7.51   Letting any five digits represent "head" and the other five digits "tail," use random numbers to simulate 100 flips of a balanced coin.

★7.52   Using the digits 1, 2, 3, 4, 5, and 6 to represent the corresponding faces of a die (and omitting 0, 7, 8, and 9), simulate 120 rolls of a balanced die.

★7.53   The probabilities that a real estate broker will sell 0, 1, 2, 3, 4, 5, or 6 houses in a week are 0.14, 0.27, 0.27, 0.18, 0.09, 0.04, and 0.01.
(a) Distribute the two-digit random numbers from 00 through 99 among these seven possibilities so that the corresponding random numbers can be used to simulate the number of houses the real estate broker sells in a week.
(b) Use the results of part (a) to simulate the real estate broker's weekly sales during 25 consecutive weeks.

★7.54   Suppose that the probabilities are 0.41, 0.37, 0.16, 0.05, and 0.01 that there will be 0, 1, 2, 3, or 4 UFO sightings in a certain region on any one day.
(a) Distribute the two-digit random numbers from 00 through 99 among the five values of this random variable, so that the corresponding random numbers can be used to simulate the sighting of UFO's in the given region.
(b) Use the result of part (a) to simulate the sighting of UFO's in the given region on 30 days.

★7.55 Suppose the probabilities are 0.2466, 0.3452, 0.2417, 0.1128, 0.0395, 0.0111, 0.0026, and 0.0005 that there will be 0, 1, 2, 3, 4, 5, 6, or 7 polluting spills in the Great Lakes on any one day.

(a) Distribute the four-digit random numbers from 0000 through 9999 among the eight values of this random variable, so that the corresponding random numbers can be used to simulate polluting spills in the Great Lakes.

(b) Use the results of part (a) to simulate the number of polluting spills in the Great Lakes on 40 consecutive days.

★7.56 Depending on the availability of parts, a company can manufacture 3, 4, 5, or 6 units of a certain item per week with corresponding probabilities of 0.10, 0.40, 0.30, and 0.20. The probabilities that there will be a weekly demand for 0, 1, 2, 3, ..., or 8 units are 0.05, 0.10, 0.30, 0.30, 0.10, 0.05, 0.05, 0.04, and 0.01. If a unit is sold during the week that it is made, it will yield a profit of $100; this profit is reduced by $20 for each week that a unit has to be stored. Use random numbers to simulate the operations of this company for 50 consecutive weeks and estimate their expected weekly profit.

7.11
CHECKLIST OF KEY TERMS
(with page references to their definitions)

Binomial distribution, 171
Binomial population, 175
Chebyshev's theorem, 191
Countably infinite sample space, 177
Discrete random variable, 168
★ Geometric distribution, 177
★ Hypergeometric distribution, 179
Mean of probability distribution, 186
★ Multinomial distribution, 183
★ Poisson distribution, 180
Probability distribution, 168
★ Random numbers, 195
Random variable, 166
★ Simulation, 195
Standard deviation of probability distribution, 188
Variance of probability distribution, 188

7.12
REVIEW EXERCISES

7.57 If 12 percent of all medical students want to specialize in surgery, what are the probabilities that in a random sample of three medical students

(a) one wants to specialize in surgery;

(b) at least one wants to specialize in surgery?

★7.58 A panel of 300 persons chosen for jury duty includes 30 under 25 years of age. Since the jury of 12 persons chosen from this panel to judge a narcotics violation does not include anyone under 25 years of age, the youthful defendant's attorney complains that this jury is not really representative. Indeed, he argues, if the selection were random, the probability of having one of the 12 jurors under 25 years of age should be *many times* the probability of having none of them under 25 years of age. Actually, what is the ratio of these two probabilities?

7.59 In each case determine whether the given values can be looked upon as the values of a probability distribution of a random variable which can take on the values 1, 2, 3, 4, and 5, and explain your answers:
(a) $f(1) = 0.18, f(2) = 0.20, f(3) = 0.22, f(4) = 0.20, f(5) = 0.18$;
(b) $f(1) = 0.05, f(2) = 0.05, f(3) = 0.05, f(4) = -0.05, f(5) = 0.90$;
(c) $f(1) = 0.07, f(2) = 0.23, f(3) = 0.50, f(4) = 0.19, f(5) = 0.01$.

7.60 The probabilities are 0.22, 0.34, 0.25, 0.13, 0.05, and 0.01 that 0, 1, 2, 3, 4, or 5 of a doctor's patients will come down with the flu during the first week of January. Find the mean of this probability distribution.

7.61 Find the standard deviation of the probability distribution of the preceding exercise.

★7.62 If the probability that a burglar will get caught on any given "job" is 0.20, what is the probability that he will get caught for the first time on his fifth "job?"

★7.63 Among 25 workers on a picket line, 14 are men and 11 are women. If a television news reporter randomly picks four of them to be shown on camera, what are the probabilities that this will include
(a) two men and two women;
(b) one man and three women?

★7.64 Suppose that the probabilities are 0.46, 0.27, 0.15, 0.08, and 0.04 that it will take an accountant 1, 2, 3, 4, or 5 hours to complete a certain income tax form.
(a) Distribute the two-digit random numbers from 00 to 99 among the five values of this random variable, so that they can be used to simulate the amount of time it takes the accountant to complete such forms.
(b) Use the results of part (a) to simulate the amount of time it takes the accountant to complete 25 such forms.

★7.65 A certain supermarket carries four grades of ground beef, and the probabilities that a shopper will buy the poorest, third best, second best, or best grade are, respectively, 0.10, 0.20, 0.40, and 0.30. Find the probability that among 12 (randomly chosen) shoppers buying ground beef at this supermarket, one will choose the poorest kind, two will choose the third best kind, six will buy the second best kind, and three will buy the best kind.

7.66 Find the mean of the binomial distribution with $n = 8$ and $p = 0.50$, using
(a) the formula which defines μ;
(b) the special formula for the mean of a binomial distribution.

★7.67 If 5 percent of all rats carry a certain disease, what is the probability that three of 100 rats, constituting a random sample, will carry the disease?

7.68 During the month of August, the daily number of persons visiting a certain tourist attraction is a random variable with $\mu = 1,200$ and $\sigma = 80$.
(a) What does Chebyshev's theorem with $k = 7$ tell us about the number of persons who will visit the tourist attraction on an August day?

(b) According to Chebyshev's theorem, with what probability can we assert that between 1,000 and 1,400 persons will visit the tourist attraction on an August day?

7.69   An experiment consists of rolling a pair of balanced dice, one green and one red.

(a) Letting (2, 4) represent the outcome where the green die comes up 2 and the red die comes up 4, letting (5, 1) represent the outcome where the green die comes up 5 and the red die comes up 1, and so forth, draw a diagram similar to that of Figure 7.1 which shows the 36 possible outcomes.

(b) On the diagram obtained in part (a), write next to each point the corresponding total rolled with the pair of dice, and, assuming that each point has the probability $\frac{1}{36}$, construct the probability distribution for the total number of points rolled with a pair of "honest" dice; that is, construct a table showing the probabilities of rolling a total of 2, 3, 4, ..., 11, or 12.

(c) Verify that the equation of the probability distribution of part (b) can be written as

$$f(x) = \frac{6 - |x - 7|}{36} \qquad \text{for } x = 2, 3, \ldots, 11, \text{ or } 12$$

where the absolute value $|x - 7|$ equals $x - 7$ or $7 - x$, whichever is positive or zero.

★7.70   The size of an animal population is sometimes estimated by the **capture–recapture method**. In this method, n_1 of the animals are captured, marked, and released. Later, n_2 of the animals are captured, x of them are found to be marked, and all this information is used to estimate N, the size of the population. If $n_1 = 3$ rare owls are captured, marked, and released, and later $n_2 = 4$ such owls are captured and $x = 1$ of them is found to be marked, for what value of N is the probability of getting $x = 1$ a maximum? (*Hint:* Try $N = 9, 10, 11, 12, 13,$ and 14.)

★7.71   If we have to determine all the values of a binomial distribution, it is sometimes helpful to calculate $f(0)$ using the formula for the binomial distribution, and then calculate the other values, one after the other, using the formula

$$\frac{f(x + 1)}{f(x)} = \frac{n - x}{x + 1} \cdot \frac{p}{1 - p}$$

(a) Verify this formula by substituting for $f(x)$ and $f(x + 1)$ the corresponding expressions given by the formula for the binomial distribution.

(b) Use this method to find all the values of the binomial distribution with $n = 6$ and $p = \frac{1}{4}$, writing them as common fractions with the denominator 4,096.

★7.72   Among 600 plants exposed to excessive radiation, 90 show abnormal growth. If a scientist collects the seed of three of the plants chosen at random, find the probability that he will get the seed from one plant with abnormal growth and two plants with normal growth by using

(a) the hypergeometric distribution;

(b) the binomial distribution as an approximation.

7.73 The probabilities that 0, 1, 2, 3, 4, 5, or 6 fires will break out in a national forest on a July morning are 0.120, 0.207, 0.358, 0.162, 0.096, 0.043, and 0.014. Find the mean and the variance of this probability distribution.

★7.74 If four of 18 mediaeval gold coins are counterfeits and two of them are randomly selected to be sold at auction, what are the probabilities that
(a) neither coin is a counterfeit;
(b) one of the coins is a counterfeit;
(c) both coins are counterfeits?

★7.75 With reference to the preceding exercise, find the mean of the probability distribution of the number of counterfeits, using
(a) the formula which defines μ;
(b) the special formula for the mean of a hypergeometric distribution.

7.76 Determine whether the following can be probability distributions (defined in each case only for the given values of x) and explain your answers:
(a) $f(x) = \dfrac{x}{10}$ for $x = 0, 1, 2, 3, 4$;

(b) $f(x) = \dfrac{x-1}{2}$ for $x = 0, 1, 2, 3$;

(c) $f(x) = \dfrac{1}{5}\binom{2}{x}$ for $x = 0, 1,$ and 2.

7.77 A study shows that 90 percent of the families in a certain metropolitan area have a color television set. Find the probabilities that among 18 families randomly selected in this area
(a) 14 have a color television set;
(b) at least 15 have a color television set;
(c) at most 13 have a color television set.

★7.78 The number of blossoms on a rare plant is a random variable having the Poisson distribution with $\lambda = 2.4$. What are the probabilities that such a plant will have
(a) no blossoms;
(b) two blossoms;
(c) four blossoms?

7.79 Find the mean and the variance of the binomial distribution with $n = 324$ and $p = 0.50$.

7.80 According to Chebyshev's theorem, with what probability can we assert that in 40,000 flips of a balanced coin the proportion of heads will be between 0.45 and 0.55?

7.13
REFERENCES

More detailed tables of binomial probabilities may be found in
ROMIG, H. G., *50–100 Binomial Tables.* New York: John Wiley & Sons, Inc., 1953.
Tables of the Binomial Probability Distribution, National Bureau of Standards Applied Mathematics Series No. 6. Washington, D.C.: U.S. Government Printing Office, 1950.

and a detailed table of Poisson probabilities is given in

MOLINA, E. C., *Poisson's Exponential Binomial Limit*. Princeton, N.J.: D. Van Nostrand Company, Inc., 1947.

Among the many published tables of random numbers, one of the most widely used is

RAND CORPORATION, *A Million Random Digits with 100,000 Normal Deviates*. New York: Macmillan Publishing Co., Inc., third printing 1966.

$\mathbf{C}$ontinuous sample spaces and continuous random variables arise when we deal with quantities that are measured on a continuous scale—for instance, when we measure the speed of a car, the amount of alcohol in a person's blood, the net weight of a package of frozen food, or the amount of tar in a cigarette. Although there exist continuums of possibilities in situations like these, in practice we always round measurements to the nearest whole unit or to a few decimals. Thus, we shall be interested in probabilities associated with intervals or regions of a sample space, and not in probabilities associated with individual points, For instance, we may want to know the probability that at a given time a car is moving between 50 and 55 miles per hour (not at exactly $16\pi = 50.26548246\ldots$ miles per hour), or that a package of frozen food weighs more than 5.95 ounces (not exactly $\sqrt{35.9} = 5.99166087\ldots$ ounces).

In this chapter we shall learn how to determine, and work with, probabilities relating to continuous sample spaces and continuous random variables. The concept of a continuous distribution will be introduced in Section 8.1, followed by that of a normal distribution in Section 8.2. Various applications of the normal distribution will be discussed in Sections 8.3 through 8.6.

8 THE NORMAL DISTRIBUTION

8.1
CONTINUOUS DISTRIBUTIONS

In histograms, the frequencies, percentages, or proportions associated with the various classes are given by the heights of the rectangles, or by their areas if the class intervals are all equal; as we saw in Chapter 7, this is true also for histograms representing probability distributions. In the continuous case, we also represent probabilities by means of areas, as is illustrated in Figure 8.1, but instead of areas of rectangles we consider areas under continuous curves. The histogram on the left in Figure 8.1 represents the probability distribution of a random variable which takes on only the values 0, 1, 2, . . . , 9, and 10, and the probability that it will take on the value 3, for example, is given by the area of the white rectangle. The diagram on the right in Figure 8.1 pertains to a continuous random variable which can take on any value on the interval from 0 to 10, and the probability that it will take on a value on the interval from 2.5 to 3.5 is given by the area of the white region under the curve. Similarly, the area of the dark region under the curve gives the probability that the continuous random variable will take on a value greater than or equal to 8.

Continuous curves such as the one shown on the right in Figure 8.1 are the graphs of functions called **probability densities,** or informally, **continuous distributions.** A probability density is characterized by the fact that

The area under the curve between any two values *a* and *b* (see Figure 8.2) gives the probability that a random variable having the continuous distribution will take on a value on the interval from *a* to *b*.

It follows from this that the values of a probability density should not be negative, and that the total area under the curve (representing the certainty that a random variable must take on one of its values) is always equal to 1.

For instance, if we approximate a family income distribution with a smooth curve as in Figure 8.3, we can determine what proportion of the incomes falls into any given interval (or the probability that the income of a family, chosen at random, will fall into the interval) by looking at the corresponding area under the curve. By comparing the area of the white region on the right in Figure 8.3 with the total area under the curve (representing 100 percent), we can judge by eye that roughly 10 to 12 percent of the families

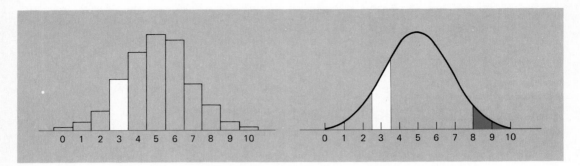

FIGURE 8.1 *Histogram of a probability distribution and graph of a continuous distribution.*

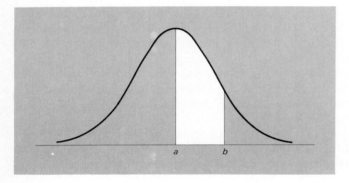

FIGURE 8.2 *Continuous distribution.*

FIGURE 8.3 *Curve approximating family income distribution.*

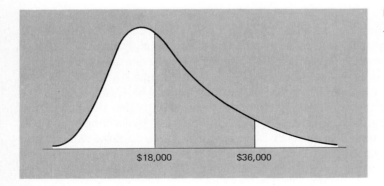

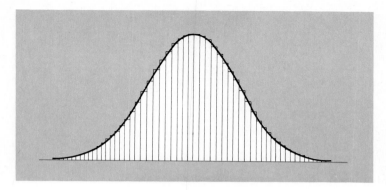

FIGURE 8.4 *Continuous distribution approximated by histogram of probability distribution.*

have incomes of $36,000 or more. Similarly, it can be seen that about 40 to 45 percent of the families have incomes of $18,000 or less.

Statistical descriptions of continuous distributions are as important as descriptions of probability distributions or distributions of observed data, but most of them, including the mean and the standard deviation, cannot be defined without the use of calculus. Nevertheless, we can always picture a continuous distribution as being approximated by a histogram of a probability distribution (see Figure 8.4) whose mean and standard deviation can be calculated. Then, if we choose histograms with narrower and narrower classes, the means and the standard deviations of the corresponding probability distributions will approach the mean and the standard deviation of the continuous distribution. Actually, the mean and the standard deviation of a continuous distribution measure the same properties as the mean and the standard deviation of a probability distribution—the expected value of a random variable having the given distribution, and the expected value of its squared deviations from the mean. More intuitively, the mean μ of a continuous distribution is a measure of its "center" or "middle," and the standard deviation σ of a continuous distribution is a measure of its "dispersion" or "spread."

8.2
THE NORMAL DISTRIBUTION

Among the many continuous distributions used in statistics, the **normal distribution** is by far the most important. Its study dates back to eighteenth-century investigations into the nature of experimental errors. It was observed that discrepancies among repeated measurements of the same physical quantity displayed a surprising degree of regularity; their patterns (distribution), it was found, could be closely approximated by a certain kind of continuous distribution curve, referred to as the "normal curve of errors"

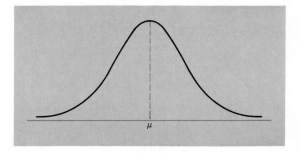

FIGURE 8.5 *Normal distribution.*

and attributed to the laws of chance. The mathematical properties of this kind of continuous distribution curve and its theoretical basis were first investigated by Pierre Laplace (1749–1827), Abraham de Moivre (1667–1745), and Karl Friedrich Gauss (1777–1855).

The graph of a normal distribution is a bell-shaped curve that extends indefinitely in both directions. Although this may not be apparent from a small drawing like that of Figure 8.5, the curve comes closer and closer to the horizontal axis without ever reaching it, no matter how far we go in either direction away from the mean. Fortunately, it is seldom necessary to extend the tails of a normal distribution very far because the area under that part of the curve lying more than four or five standard deviations away from the mean is for most practical purposes negligible. (This can be seen from the values given at the bottom of Table I on page 497.)

An important feature of the normal distribution is that its mathematical equation is such that we can determine the area under the curve between any two points on the horizontal scale if we know its mean and its standard deviation; in other words, there is one and only one normal distribution with a given mean μ and a given standard deviation σ.

Since the equation of the normal distribution depends on μ and σ, we get different curves and, hence, different areas for different values of μ and σ. For instance, Figure 8.6 shows the superimposed graphs of two normal

FIGURE 8.6 *Two normal distributions.*

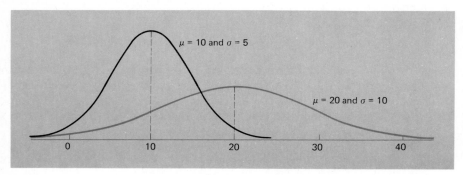

distributions, one with $\mu = 10$ and $\sigma = 5$ and the other with $\mu = 20$ and $\sigma = 10$. As can be seen, the area under the curve, say, between 12 and 15, is not the same for the two distributions.

In practice, we find areas under the graphs of normal distributions, or simply areas under normal curves, in special tables, such as Table I at the end of the book. As it is physically impossible, and also unnecessary, to construct separate tables of normal-curve areas for all conceivable pairs of values of μ and σ, we tabulate these areas only for the normal distribution with $\mu = 0$ and $\sigma = 1$, called the **standard normal distribution.** Then, we obtain areas under any normal curve by performing the change of scale (see Figure 8.7) which converts the units of measurement from the original scale, or x-scale, into **standard units, standard scores,** or z-**scores,** by means of the formula

Standard units

$$z = \frac{x - \mu}{\sigma}$$

In this new scale, the z-scale, a value of z simply tells us how many standard deviations the corresponding value of x lies above or below the mean of its distribution.

The entries in Table I are the areas under the standard normal curve between the mean $z = 0$ and $z = 0.00, 0.01, 0.02, \ldots, 3.08$, and 3.09, and also $z = 4.00$, $z = 5.00$, and $z = 6.00$. In other words, the entries in Table I are areas under the standard normal curve like that of the white region in Figure 8.8.

Table I has no entries corresponding to negative values of z, for these are not needed by virtue of the symmetry of any normal curve about its mean.

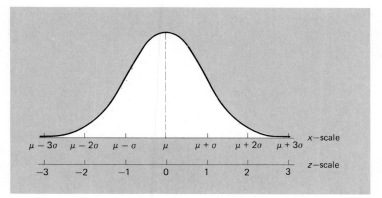

FIGURE 8.7 *Change of scale to standard units.*

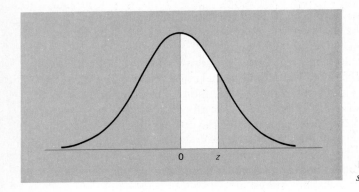

FIGURE 8.8 *Tabulated areas under the standard normal curve.*

EXAMPLE   Find the area under the standard normal curve between $z = -1.20$ and $z = 0$.

Solution   As can be seen from Figure 8.9, the area under the curve between $z = -1.20$ and $z = 0$ equals the area under the curve between $z = 0$ and $z = 1.20$. So, we look up the entry for $z = 1.20$ and get 0.3849.

Questions concerning areas under normal distributions arise in various ways, and the ability to find any desired area quickly can be a big help. Although the table gives only areas between $z = 0$ and selected positive values of z, we often have to find areas to the left or to the right of given positive or negative values of z, or areas between two given values of z. This is easy, provided we remember exactly what areas are represented by the entries in Table I, and also that the standard normal distribution is symmetrical about $z = 0$, so that the area under the curve to the left of $z = 0$ and that to the right of $z = 0$ are both equal to 0.5000.

FIGURE 8.9 *Area under normal curve.*

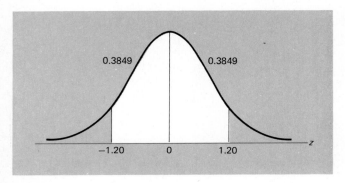

EXAMPLE Find the area under the standard normal curve which lies
(a) to the left of $z = 0.94$;
(b) to the right of $z = -0.65$;
(c) to the right of $z = 1.76$;
(d) to the left of $z = -0.85$;
(e) between $z = 0.87$ and $z = 1.28$;
(f) between $z = -0.34$ and $z = 0.62$.

Solution For each part see Figure 8.10. (a) The area to the left of $z = 0.94$ is
0.5000 plus the entry in Table I corresponding to $z = 0.94$, or
$0.5000 + 0.3264 = 0.8264$; (b) the area to the right of $z = -0.65$
is 0.5000 plus the entry in Table I corresponding to $z = 0.65$, or
$0.5000 + 0.2422 = 0.7422$; (c) the area to the right of $z = 1.76$ is
0.5000 minus the entry in Table I corresponding to $z = 1.76$, or
$0.5000 - 0.4608 = 0.0392$; (d) the area to the left of $z = -0.85$ is
0.5000 minus the entry in Table I corresponding to $z = 0.85$, or
$0.5000 - 0.3023 = 0.1977$; (e) the area between $z = 0.87$ and

FIGURE 8.10 *Areas under normal curves.*

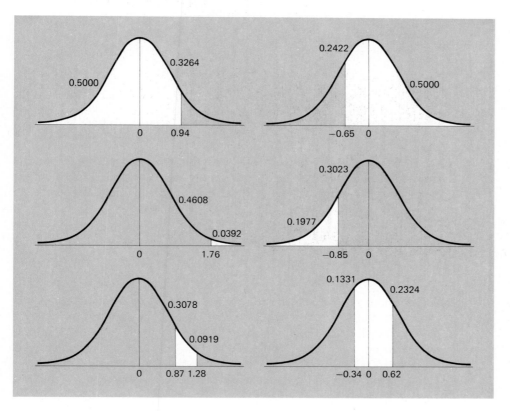

$z = 1.28$ is the difference between the entries in Table I corresponding to $z = 0.87$ and $z = 1.28$, or $0.3997 - 0.3078 = 0.0919$; **(f)** the area between $z = -0.34$ and $z = 0.62$ is the sum of the entries in Table I corresponding to $z = 0.34$ and $z = 0.62$, or $0.1331 + 0.2324 = 0.3655$.

EXAMPLE   With reference to Figure 8.6, find the area under the curve between 12 and 15 for the normal distributions with

(a) $\mu = 10$ and $\sigma = 5$;
(b) $\mu = 20$ and $\sigma = 10$.

Solution   **(a)** The values of z corresponding to $x = 12$ and $x = 15$ are

$$z = \frac{12 - 10}{5} = 0.40 \quad \text{and} \quad z = \frac{15 - 10}{5} = 1.00$$

the corresponding entries in Table I are 0.1554 and 0.3413, and the area under the curve between 12 and 15 (the area of the white region of the upper diagram of Figure 8.11) is $0.3413 - 0.1554 = 0.1859$; **(b)** the values of z corresponding to $x = 12$ and $x = 15$ are

$$z = \frac{12 - 20}{10} = -0.80 \quad \text{and} \quad z = \frac{15 - 20}{10} = -0.50$$

the corresponding entries in Table I are 0.2881 and 0.1915, and the area under the curve between 12 and 15 (the area of the white region of the lower diagram of Figure 8.11) is $0.2881 - 0.1915 = 0.0966$.

There are also problems in which we are given areas under normal curves and asked to find the corresponding values of z. The results of the example which follows will be used extensively in subsequent chapters.

EXAMPLE   If z_α denotes the value of z for which the area under the standard normal curve to its right is equal to α (Greek lowercase *alpha*), find

(a) $z_{0.01}$;
(b) $z_{0.05}$.

Solution   **(a)** It can be seen from Figure 8.12 that $z_{0.01}$ corresponds to an entry of $0.5000 - 0.0100 = 0.4900$ in Table I; since the nearest entry is 0.4901 corresponding to $z = 2.33$, we find that $z_{0.01} = 2.33$; **(b)** also from Figure 8.12, we see that $z_{0.05}$ corresponds to an entry of $0.5000 - 0.0500 = 0.4500$ in Table I; since the two nearest entries are 0.4495 and 0.4505 corresponding to $z = 1.64$ and $z = 1.65$, we find that $z_{0.05} = 1.645$.

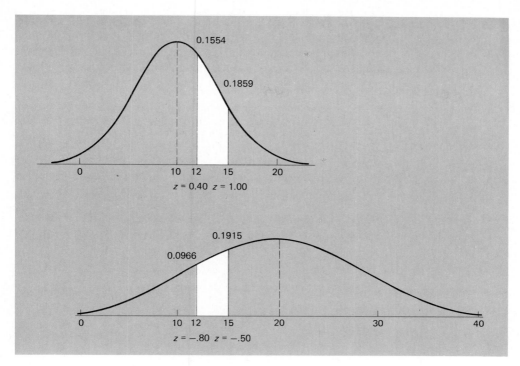

FIGURE 8.11 *Areas under normal curves.*

Table I also enables us to verify the remark on page 59 that for frequency distributions having the general shape of the cross section of a bell, about 68 percent of the values will lie within one standard deviation of the mean, about 95 percent will lie within two standard deviations of the mean, and about 99.7 percent will lie within three standard deviations of the mean. These percentages apply to frequency distributions having the general shape of a normal distribution, and the reader will be asked to verify them

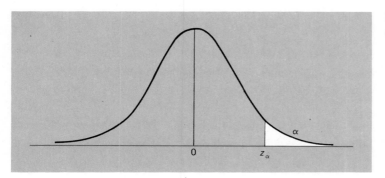

FIGURE 8.12 *Diagram for determination of z_α.*

in parts (a), (b), and (c) of Exercise 8.6 on page 215. The results of parts (d) and (e) of that exercise show that, although the "tails" extend indefinitely in both directions, the area under a standard normal curve beyond $z = 4$ or $z = 5$ is negligible.

EXERCISES

8.1 Suppose that a continuous random variable takes on values on the interval from 2 to 10 and that the graph of its distribution, called a **uniform density,** is given by the horizontal line of Figure 8.13.

(a) What probability is represented by the white region of the diagram and what is its value?

(b) What is the probability that the random variable will take on a value less than 7? Is this probability the same as the probability that the random variable will take on a value less than or equal to 7?

(c) What is the probability that the random variable will take on a value between 2.7 and 8.8?

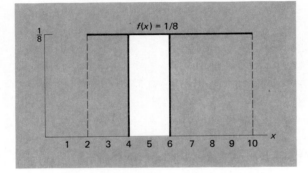

FIGURE 8.13 *Uniform density.*

8.2 Find the area under the standard normal curve which lies
(a) between $z = 0$ and $z = 0.87$
(b) between $z = -1.66$ and $z = 0$;
(c) to the right of $z = 0.48$;
(d) to the right of $z = -0.27$;
(e) to the left of $z = 1.30$;
(f) to the left of $z = -0.79$.

8.3 Find the area under the standard normal curve which lies
(a) between $z = 0.55$ and $z = 1.12$;
(b) between $z = -1.05$ and $z = -1.75$;
(c) between $z = -1.95$ and $z = 0.44$.

8.4 Find the area under the standard normal distribution which lies
(a) between $z = -0.72$ and $z = 0.75$;
(b) to the right of $z = -2.20$;
(c) to the left of $z = -0.15$;
(d) between $z = 2.15$ and $z = 2.35$;
(e) to the right of $z = 1.40$;
(f) between $z = -0.36$ and $z = -0.24$;
(g) to the left of $z = 0.93$.

8.5 Find z if
(a) the normal-curve area between 0 and z is 0.4726;
(b) the normal-curve area to the left of z is 0.9868;
(c) the normal-curve area to the right of z is 0.7704;
(d) the normal-curve area to the left of z is 0.3085;
(e) the normal-curve area to the right of z is 0.1314;
(f) the normal-curve area between $-z$ and z is 0.8502;
(g) the normal-curve area between $-z$ and z is 0.9700.

8.6 Find the normal-curve area between $-z$ and z if
(a) $z = 1$;
(b) $z = 2$;
(c) $z = 3$;
(d) $z = 4$;
(e) $z = 5$.

8.7 Verify that
(a) $z_{0.005} = 2.575$;
(b) $z_{0.025} = 1.96$.

8.8 A random variable has a normal distribution with the mean $\mu = 80.0$ and the standard deviation $\sigma = 4.8$. What are the probabilities that this random variable will take on a value
(a) less than 87.2;
(b) greater than 76.4;
(c) between 81.2 and 86.0;
(d) between 71.6 and 88.4?

8.9 A normal distribution has the mean $\mu = 62.4$. Find its standard deviation if 20 percent of the area under the curve lies to the right of 79.2.

8.10 A random variable has a normal distribution with $\sigma = 10$. If the probability that the random variable will take on a value less than 82.5 is 0.8212, what is the probability that it will take on a value greater than 58.3?

★8.11 Another continuous distribution, called the **exponential distribution,** has many important applications. If a random variable has an exponential distribution with the mean μ, the probability that it will take on a value between 0 and any given positive value x is $1 - e^{-x/\mu}$ (see Figure 8.14). Here e is the constant which

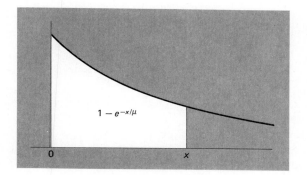

FIGURE 8.14 *Exponential distribution.*

$1 - e^{-x/\mu}$

appears also in the formula for the Poisson distribution, and values of $e^{-x/\mu}$ can be obtained directly from Table XIII.

(a) Find the probabilities that a random variable having an exponential distribution with $\mu = 10$ takes on a value less than 3, a value between 4 and 6, and a value greater than 15.

(b) The lifetime of a certain electronic component is a random variable which has an exponential distribution with a mean of 2,000 hours. What is the probability that such a component will last at most 1,800 hours? What is the probability that such a component will last anywhere from 4,000 hours to 5,000 hours?

(c) According to medical research, the time between successive reports of a rare tropical disease is a random variable having an exponential distribution with a mean of 120 days. What is the probability that the time between successive reports of the disease will exceed 48 days?

8.3
A CHECK FOR "NORMALITY" ★

There are various ways in which we can test whether an observed frequency distribution has roughly the shape of a normal distribution. The one we shall discuss here is largely subjective, but it has the decided advantage that it is very easy to perform.

To illustrate this technique, let us refer again to the sulfur oxides emission data which we used as an example in Chapters 2 and 3. First we convert the cumulative frequencies in the table on page 18 into cumulative percentages by dividing each of them by 80, the total frequency, and multiplying by 100. This yields

Tons of sulfur oxides	Cumulative percentage
Less than 4.95	0.00
Less than 8.95	3.75
Less than 12.95	16.25
Less than 16.95	33.75
Less than 20.95	65.00
Less than 24.95	86.25
Less than 28.95	97.50
Less than 32.95	100.00

where we showed the class boundaries instead of the class limits (4.95, for instance, instead of 5.0).

Before we plot this cumulative percentage distribution on the special graph paper shown in Figure 8.15, let us briefly examine its scales. As can be seen from Figure 8.15, the cumulative percentage scale is already printed

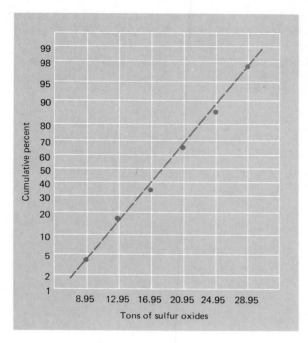

FIGURE 8.15 *Normal probability paper.*

on the graph paper in the special way which makes it suitable for our purpose. The other scale consists of equal subdivisions. This kind of graph paper is called **normal probability paper,** or **arithmetic probability paper,** and it can be obtained in the bookstores of most colleges and universities.

Now, the criterion for deciding whether an observed frequency distribution has roughly the shape of a normal distribution may be stated as follows:

> If we plot the cumulative "less than" percentages which correspond to the class boundaries of the distribution on normal probability paper and the points follow the general pattern of a straight line, we consider this as positive evidence that the distribution has roughly the shape of a normal distribution.

Returning to our example and plotting the cumulative percentages as in Figure 8.15, we find that the points are all close to the dashed line, and hence conclude that the distribution of the sulfur oxides emission data has roughly the shape of a normal distribution. Note that in Figure 8.15 we did not plot the cumulative percentages corresponding to 4.95 and 32.95; as we have pointed out earlier, we never quite reach 0 or 100 percent of the area under a normal curve, no matter how far away from the mean we go in either direction.

It must be understood that the method described here is only a crude way of deciding whether an observed distribution follows the general pattern of a normal curve; indeed, only large and obvious departures from a straight line constitute real evidence that a distribution does not follow the pattern of a normal curve. A more rigorous way of checking for "normality" is shown in Exercise 12.63 on page 359.

EXERCISES ★8.12 Use normal probability paper to check whether the distribution of Exercise 2.21 on page 28 has roughly the shape of a normal distribution.

★8.13 The following is the distribution of the times required by 200 persons to complete a certain job application:

Time (minutes)	Number of men
24 or less	15
25–29	50
30–34	75
35–39	40
40–44	15
45 or over	5

Use normal probability paper to check whether this distribution has roughly the shape of a normal distribution.

★8.14 The following is the distribution of the numbers of inquiries a realty firm received about 500 pieces of property:

Number of inquiries	Frequency
3– 6	55
7–10	227
11–14	170
15–18	42
19–22	6

Use normal probability paper to check whether this distribution has roughly the shape of a normal distribution.

★8.15 Normal probability paper can be used to obtain crude estimates of the mean and the standard deviation of a distribution which has roughly the shape of a normal curve. To estimate the mean, we have only to observe that since the normal distribution is symmetrical about the mean, the area under the curve to the left of the mean is 0.5000. Hence, if we check the 50 percent mark on the vertical scale and go horizontally to the line we fit to the points (for instance, the dashed line of Figure 8.15), then the corresponding value on the horizontal scale provides an estimate of the mean of the distribution. To estimate the standard deviation, we observe that the areas under the curve to the left of $z = -1$ and $z = +1$ are roughly 0.16 and 0.84. Hence, if we check 16 percent and 84 percent on the vertical scale, we can judge by the straight line we have fitted to the points what values on the horizontal scale correspond to $z = -1$ and $z = +1$; their difference divided by 2 provides an estimate of the standard deviation of the distribution.

(a) Use this method to estimate the mean and the standard deviation of the distribution of the sulfur oxides emission data from Figure 8.15. Compare the results with the values, $\bar{x} = 18.85$ and $s = 5.55$, previously obtained in the text.

(b) Use this method to estimate the mean and the standard deviation of the distribution of Exercise 8.13.

8.4
APPLICATIONS
OF THE NORMAL DISTRIBUTION

Let us now consider some applied problems in which we shall assume that the distribution of the data, or the distribution of the random variable under consideration, can be approximated closely with a normal curve.

EXAMPLE If the amount of cosmic radiation to which a person is exposed while flying by jet across the United States is a random variable having a normal distribution with $\mu = 4.35$ mrem and $\sigma = 0.59$ mrem, find the probabilities that a person on such a flight will be exposed to

(a) more than 5.00 mrem of cosmic radiation;

(b) anywhere from 3.00 to 4.00 mrem of cosmic radiation.

Solution **(a)** This probability is given by the area of the white region of the upper diagram of Figure 8.16, namely, the area under the curve to the right of

$$z = \frac{5.00 - 4.35}{0.59} = 1.10$$

Since the entry in Table I corresponding to $z = 1.10$ is 0.3643, we find that the probability is $0.5000 - 0.3643 = 0.1357$, or approximately 0.14, that a person will be exposed to more than 5.00 mrem of cosmic radiation on such a flight. **(b)** This probability is given

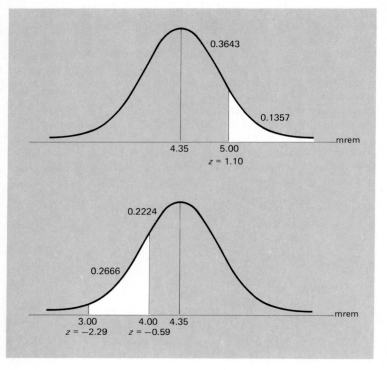

FIGURE 8.16 *Diagrams for cosmic radiation example.*

by the area of the white region of the lower diagram of Figure 8.16, namely, the area under the curve between

$$z = \frac{3.00 - 4.35}{0.59} = -2.29 \quad \text{and} \quad z = \frac{4.00 - 4.35}{0.59} = -0.59$$

Since the entries in Table I corresponding to $z = 2.29$ and $z = 0.59$ are, respectively, 0.4890 and 0.2224, we find that the probability is $0.4890 - 0.2224 = 0.2666$, or approximately 0.27, that a person will be exposed to anywhere from 3.00 to 4.00 mrem of cosmic radiation on such a flight.

EXAMPLE The actual amount of instant coffee which a filling machine puts into "6-ounce" jars varies from jar to jar, and it may be looked upon as a random variable having a normal distribution with a standard deviation of 0.04 ounce. If only 2 percent of the jars are to contain less than 6 ounces of coffee, what must be the mean fill of these jars?

Solution We are given $\sigma = 0.04$, $x = 6.00$, a normal-curve area (that of the white region of Figure 8.17), and we are asked to find μ. Since the

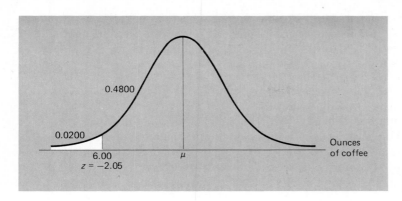

FIGURE 8.17 *Diagram for instant coffee filling example.*

value of z for which the entry in Table I comes closest to $0.5000 - 0.0200 = 0.4800$ is 2.05, we have

$$-2.05 = \frac{6.00 - \mu}{0.04}$$

and, solving for μ, we get

$$6.00 - \mu = -2.05(0.04) = -0.082$$

and then

$$\mu = 6.00 + 0.082 = 6.082 \text{ ounces}$$

or 6.08 ounces rounded to the nearest hundredth of an ounce. This "giveaway" may not be satisfactory so far as the coffee processor is concerned, and he may well want to reduce the variability of the filling machine (see Exercise 8.21 on page 226).

Although the normal distribution is a continuous distribution which applies to continuous random variables, it is often used to approximate distributions of **discrete random variables,** which can take on only a finite number of values or as many values as there are positive integers. Quite often, this yields satisfactory results, provided that we make the **continuity correction** illustrated in the next example.

EXAMPLE In a study of aggressive behavior, male white mice, returned to the group in which they live after four weeks of isolation, averaged 18.6 fights in the first five minutes with a standard deviation of 3.3 fights. If it can be assumed that the distribution of this random variable (the number of fights into which such a mouse gets under the stated conditions) can be approximated closely with a normal distribution, what is the probability that such a mouse will get into at least 15 fights in the first five minutes?

Solution   The answer is given by the area of the white region of Figure 8.18; the area to the right of 14.5, not 15. The reason for this is that the number of fights in which such a mouse gets involved is a whole number. Hence, if we want to approximate the distribution of this random variable with a normal curve, we must "spread" its values over a continuous scale, and we do this by representing each whole number k by the interval from $k - \frac{1}{2}$ to $k + \frac{1}{2}$. For instance, 5 is represented by the interval from 4.5 to 5.5, 10 is represented by the interval from 9.5 to 10.5, 20 is represented by the interval from 19.5 to 20.5, and the probability of 15 or more is given by the area under the curve to the right of 14.5. Accordingly, we get

$$ z = \frac{14.5 - 18.6}{3.3} = -1.24 $$

and it follows from Table I that the area of the white region of Figure 8.18—the probability that such a mouse will get into at least 15 fights in the first five minutes—is $0.5000 + 0.3925 = 0.8925$, or approximately 0.89.

All the examples of this section dealt with random variables having normal distributions, or distributions which can be approximated closely with normal curves. When we observe a value (or values) of a random variable having a normal distribution, we may say that we are sampling a **normal population**; this is consistent with the terminology introduced at the end of Section 7.3.

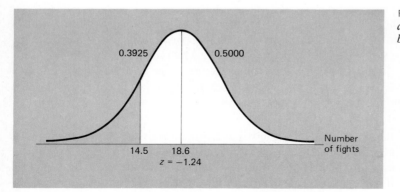

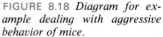

FIGURE 8.18 *Diagram for example dealing with aggressive behavior of mice.*

8.5
THE NORMAL APPROXIMATION
TO THE BINOMIAL DISTRIBUTION

The normal distribution is sometimes introduced as a continuous distribution which provides a very close approximation to the binomial distribution when n, the number of trials, is very large and p, the probability of a success on an individual trial, is close to $\frac{1}{2}$. Figure 8.19 shows the histograms of binomial distributions having $p = \frac{1}{2}$ and $n = 2, 5, 10,$ and 25, and it can be seen that with increasing n these distributions approach the symmetrical bell-shaped pattern of the normal distribution. In fact, a normal curve with the mean $\mu = np$ and the standard deviation $\sigma = \sqrt{np(1 - p)}$ can often be used to approximate a binomial distribution even when n is fairly small and p differs from $\frac{1}{2}$, but is not too close to either 0 or 1. A good rule of thumb is to use this approximation only when np and $n(1 - p)$ are both greater than 5.

The following examples illustrate the normal approximation to the binomial distribution.

EXAMPLE Find the probability of getting 6 heads and 10 tails in 16 flips of a balanced coin, using

      (a) the formula for the binomial distribution;

      (b) the normal approximation to the binomial distribution with $n = 16$ and $p = \frac{1}{2}$.

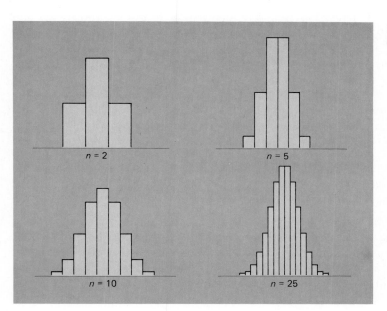

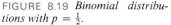
FIGURE 8.19 *Binomial distributions with $p = \frac{1}{2}$.*

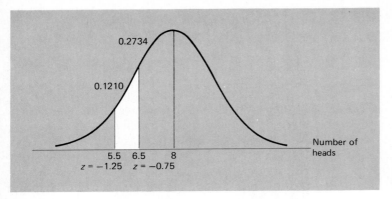

FIGURE 8.20 *Normal-curve approximation to binomial distribution.*

Solution (a) Substituting $n = 16$, $x = 6$, and $p = \frac{1}{2}$ into the formula for the binomial distribution, we get

$$f(6) = \binom{16}{6}\left(\frac{1}{2}\right)^{6}\left(1 - \frac{1}{2}\right)^{10} = 8{,}008\left(\frac{1}{2}\right)^{16} = \frac{8{,}008}{65{,}536}$$

or 0.1222 rounded to four decimals. **(b)** To find the normal-curve approximation to this probability, we use the continuity correction and represent 6 heads by the interval from 5.5 to 6.5 (see Figure 8.20). Since $\mu = 16 \cdot \frac{1}{2} = 8$ and $\sigma = \sqrt{16 \cdot \frac{1}{2} \cdot \frac{1}{2}} = 2$, we have in standard units $z = \dfrac{5.5 - 8}{2} = -1.25$ for $x = 5.5$ and $z = \dfrac{6.5 - 8}{2} = -0.75$ for $x = 6.5$. The corresponding entries in Table I are 0.3944 and 0.2734, and the approximate probability of getting 6 heads and 10 tails in 16 flips of a balanced coin is $0.3944 - 0.2734 = 0.1210$. This differs by only 0.0012 from the result of part (a).

The normal-curve approximation to the binomial distribution is particularly useful in problems where we would otherwise have to use the formula for the binomial distribution repeatedly to obtain the values of many different terms.

EXAMPLE What is the probability that at least 70 of 100 mosquitos will be killed by a new insect spray, if the probability is 0.75 that any one of them will be killed by the spray?

Solution If we tried to find the answer by using the formula for the binomial distribution, we would have to find the sum of the probabilities corresponding to 70, 71, 72,..., 99, and 100 successes. This would obviously involve a tremendous amount of work, but using the normal-curve approximation, we need only find the area of the white region of Figure 8.21, the area to the right of 69.5. Here we

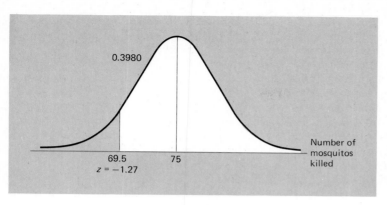

FIGURE 8.21 *Normal-curve approximation to binomial distribution.*

are again using the continuity correction according to which 70 is represented by the interval from 69.5 to 70.5, 71 is represented by the interval from 70.5 to 71.5, and so on.

Since $\mu = 100(0.75) = 75$ and $\sigma = \sqrt{100(0.75)(0.25)} = 4.33$, we find that in standard units 69.5 becomes

$$z = \frac{69.5 - 75}{4.33} = -1.27$$

and that the probability is $0.3980 + 0.5000 = 0.8980$. The actual value of the probability, looked up in the tables by Romig listed among the references at the end of Chapter 7, is 0.8962. So, the error of the approximation is only 0.0018.

EXERCISES

8.16 In an experiment to determine the amount of time required to assemble an "easy to assemble" toy, the assembly time was found to be a random variable having approximately a normal distribution with $\mu = 12.9$ minutes and $\sigma = 2.0$ minutes. What are the probabilities that this kind of toy can be assembled in
   (a) less than 11.5 minutes;
   (b) anywhere from 12.0 to 13.8 minutes?

8.17 Suppose that during periods of transcendental meditation the reduction of a person's oxygen consumption may be looked upon as a random variable having a normal distribution with $\mu = 38.6$ cc per minute and $\sigma = 4.3$ cc per minute. Find the probabilities that during a period of transcendental meditation a person's oxygen consumption will be reduced by
   (a) at least 45.5 cc per minute;
   (b) at most 36.0 cc per minute;
   (c) anywhere from 32.0 to 42.0 cc per minute;
   (d) anywhere from 40.0 to 45.0 cc per minute.

8.18 The burning time of an experimental rocket is a random variable which has a normal distribution with $\mu = 4.36$ seconds and $\sigma = 0.04$ second. What are the probabilities that this kind of rocket will burn for
  (a) less than 4.25 seconds;
  (b) more than 4.40 seconds;
  (c) 4.30 to 4.42 seconds?

8.19 The sardines processed by a cannery have a mean length of 4.54 inches with a standard deviation of 0.25 inch. If the distribution of the lengths of the sardines can be approximated closely with a normal distribution, what percentage of the sardines are
  (a) shorter than 4.00 inches;
  (b) between 4.40 and 4.60 inches long?

8.20 With reference to the preceding exercise, above which length are the longest 10 percent of the sardines?

8.21 With reference to the filling-machine example on page 220, show that 97.7 percent of the jars will contain at least 6 ounces of coffee if the machine is adjusted so that $\mu = 6.02$ ounces and $\sigma = 0.01$ ounce.

8.22 In a photographic process, the developing time of prints may be looked upon as a random variable having a normal distribution with a mean of 15.28 seconds and a standard deviation of 0.12 second. Find the probabilities that it will take
  (a) at least 15.50 seconds to develop one of the prints;
  (b) at most 15.00 seconds to develop one of the prints;
  (c) from 15.10 to 15.40 seconds to develop one of the prints;
  (d) from 15.05 to 15.15 seconds to develop one of the prints.

8.23 With reference to the preceding exercise, below which figure are the lowest 5 percent of the developing times?

8.24 In a very large class in European history, the final examination grades have a mean of 71.6 and a standard deviation of 12.6. If it is reasonable to approximate the distribution of these grades with a normal distribution, what percentage of the grades should exceed 79?

8.25 If the yearly number of major earthquakes, the world over, is a random variable whose distribution can be closely approximated with a normal distribution having $\mu = 20.8$ and $\sigma = 4.5$, find the probabilities that there will be
  (a) exactly 18 major earthquakes in any given year;
  (b) at least 22 major earthquakes in any given year;
  (c) from 20 to 25 major earthquakes, inclusive, in any given year.

8.26 A taxicab driver knows from experience that the number of fares he will pick up in an evening is a random variable with $\mu = 23.7$ and the standard deviation $\sigma = 4.2$. Assuming that the distribution of this random variable can be approximated with a normal distribution, find the probabilities that in an evening the driver will pick up
  (a) exactly 20 fares;
  (b) at least 18 fares;
  (c) at most 25 fares;
  (d) from 15 to 21 fares.

8.27 Use the normal-curve approximation to find the probability of getting 7 heads and 7 tails in 14 flips of a balanced coin, and compare the result with the exact value (rounded to three decimals) in Table V.

8.28 A student answers each of the 48 questions on a multiple-choice test, each with four possible answers, by randomly drawing a card from an ordinary deck of 52 playing cards and checking the first, second, third, or fourth answer depending on whether the card drawn is a spade, heart, diamond, or club. Use the normal-curve approximation to find the probabilities that the student will get

    (a) exactly 15 correct answers;

    (b) at least 15 correct answers.

Compare these results with the corresponding exact values, which are 0.0767 and 0.1999 (rounded to four decimals) according to the National Bureau of Standards Tables listed among the references at the end of Chapter 7.

8.29 If 20 percent of the loan applications received by a bank are refused, what is the probability that among 225 loan applications at most 40 will be refused?

8.30 A television network claims that its Tuesday night movie regularly has 34 percent of the total viewing audience. If this claim is correct, what is the probability that among 400 Tuesday night television viewers at least 150 will be watching the network's movie?

8.31 If 70 percent of all persons flying across the Atlantic Ocean feel the effect of the time difference for at least 24 hours, what is the probability that among 150 persons flying across the Atlantic Ocean, at least 100 will feel the effect of the time difference for at least 24 hours?

8.32 If 22 percent of all patients with high blood pressure have bad side effects from a certain kind of medicine, what is the probability that among 120 patients with high blood pressure who are treated with this medicine more than 30 will have bad side effects?

8.33 If 62 percent of all clouds seeded with silver iodide show spectacular growth, what is the probability that among 40 clouds seeded with silver iodide at most 20 will show spectacular growth?

8.34 To avoid accusations of sexism or worse, the author of a mathematics text decides by the flip of a balanced coin whether to use "he" or "she" whenever the occasion arises in exercises and examples. If he runs into this problem 80 times while revising one of his books, what is the probability that he will use "she" at least 48 times?

8.35 To illustrate the law of large numbers which we mentioned in Section 4.4. find the probabilities that the proportion of heads will be anywhere from 0.49 to 0.51 when a balanced coin is flipped

    (a) 100 times;

    (b) 1,000 times;

    (c) 10,000 times.

8.6
TECHNICAL NOTE (Simulation) ★

There are many ways in which we can simulate observations of continuous random variables, but the theory on which they are based is beyond the scope of this text. Limiting ourselves to random variables having normal distributions, we shall illustrate here only the use of published tables of **random normal numbers,** also called **random normal deviates.** Such tables consist of many pages on which numbers (rounded to three decimals in Table XII at the end of the book) are set down in a random fashion, much as they would appear if they were generated one at a time by a chance or gambling device which "produces" values of a random variable having the standard normal distribution.

To simulate values of a random variable having a normal distribution with a given mean μ and a given standard deviation σ, we look up values of z in a table of random normal numbers, and then change them into values of x, the random variable under consideration, by using the formula

Changing from standard units

$$x = \mu + \sigma z$$

This follows directly from the formula $z = \dfrac{x - \mu}{\sigma}$ for converting to standard units. As in the use of ordinary random numbers, the choice of the page and the place from which to start should, in practice, be left to chance.

EXAMPLE   In a certain city, the time it takes the police to respond to emergencies is a random variable having approximately a normal distribution with $\mu = 7.3$ minutes and $\sigma = 0.6$ minute. Simulate the time it takes the police in this city to respond to five emergencies, using the eighth column of the table on page 522, starting with the sixth row and going down the page.

Solution   The values we get from the table are -0.745, 0.655, -1.115, 0.027, and -2.520, and the corresponding response times are $7.3 + 0.6(-0.745) = 6.9$ minutes, $7.3 + 0.6(0.655) = 7.7$ minutes, $7.3 + 0.6(-1.115) = 6.6$ minutes, $7.3 + 0.6(0.027) = 7.3$ minutes, and $7.3 + 0.6(-2.520) = 5.8$ minutes.

EXERCISES   ★8.36   The amount of time it takes a person to learn how to operate a certain machine is a random variable having a normal distribution with $\mu = 5.6$ hours and $\sigma = 1.2$ hours. Simulate the amount of time it takes eight persons to learn how to operate the machine.

★8.37   The distribution of the grades which college-bound high school seniors get on a certain standardized test can be approximated closely by a normal distribution with $\mu = 54.3$ and $\sigma = 6.2$. Rounding the results to the nearest whole numbers, simulate the grades which 12 college-bound high school seniors get on this test.

★8.38   Suppose that the increase in the pulse rate of a person performing a certain task is a random variable whose distribution can be closely approximated by a normal distribution with $\mu = 28.40$ and $\sigma = 4.17$. Simulate the increase in the pulse rate of 20 persons performing this task. Round the results to the nearest whole numbers.

★8.39 The distribution of the weights of the grapefruits shipped by a large orchard can be closely approximated by a normal distribution with $\mu = 19.6$ ounces and $\sigma = 2.2$ ounces. Simulate the weight of 24 grapefruits shipped by this orchard.

8.7
CHECKLIST OF KEY TERMS
(with page references to their definitions)

8.8
REVIEW EXERCISES

8.40 Suppose that the amount of time which tourists spend in a famous museum is a random variable with $\mu = 43.4$ minutes and $\sigma = 6.8$ minutes. Assuming that the distribution of this random variable can be approximated closely with a normal curve, find the probabilities that a tourist will spend
  (a) at most 36.0 minutes in the museum;
  (b) from 40.0 to 50.0 minutes in the museum.

8.41 Use the normal-curve approximation to find the probability of getting 12 heads in 20 flips of a balanced coin, and compare the result with the value given in Table V.

★8.42 If the time it takes to fill in a new income tax form is a random variable having a normal distribution with $\mu = 32.6$ minutes and $\sigma = 3.5$ minutes, use Table XII to simulate the time it takes 12 persons to fill in the new income tax form.

8.43 Find the area under the standard normal curve which lies
 (a) between $z = 0$ and $z = 1.83$;
 (b) to the left of $z = 2.50$;
 (c) to the right of $z = -0.64$;
 (d) to the right of $z = 1.24$;
 (e) to the left of $z = -0.71$.

8.44 A random variable has a normal distribution with $\mu = 102.4$ and $\sigma = 3.6$. What are the probabilities that this random variable will take on a value
 (a) less than 107.8;
 (b) greater than 99.7;
 (c) between 106.9 and 110.5;
 (d) between 96.1 and 104.2?

8.45 Find the values of
 (a) $z_{0.02}$;
 (b) $z_{0.10}$.

★8.46 The following is the distribution of the miles per gallon obtained with 200 cars:

Miles per gallon	Frequency
18.0–19.9	2
20.0–21.9	10
22.0–23.9	19
24.0–25.9	39
26.0–27.9	62
28.0–29.9	37
30.0–31.9	20
32.0–33.9	8
34.0–35.9	3

Use normal probability paper to check whether this distribution has roughly the shape of a normal curve.

8.47 If 85 percent of all scorpion stings cause extensive discomfort, what is the probability that among 100 scorpion stings more than 90 will cause extensive discomfort?

8.48 Suppose that a random variable takes on values on the interval from 0 to 4 and that the graph of its distribution, called a **triangular density**, is given by the white line of Figure 8.22. Verify that the total area under the curve is equal to 1 and find the probabilities that the random variable will take on a value
 (a) less than 1;
 (b) greater than 2;
 (c) between 2.5 and 3.5.

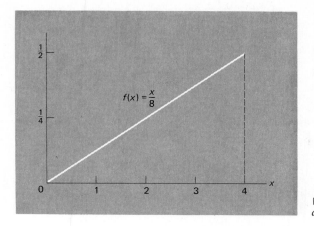

FIGURE 8.22 *Triangular density.*

8.49 The head of the complaint department of a department store knows from experience that the number of complaints she receives per day is a random variable with $\mu = 48.4$ and $\sigma = 5.5$. Assuming that the distribution of the number of complaints has roughly the shape of a normal distribution, find the probabilities that in one day she will receive
    (a) more than 55 complaints;
    (b) at least 55 complaints;
    (c) anywhere from 45 to 55 complaints.

8.50 A random variable has a normal distribution with $\sigma = 4.0$. If the probability is 0.9713 that this random variable will take on a value less than 77.6, what is the probability that it will take on a value between 65.0 and 68.0?

8.51 The average time required to perform job A is 78.5 minutes with a standard deviation of 16.2 minutes, and the average time required to perform job B is 103.2 minutes with a standard deviation of 11.3 minutes. Assuming normal distributions, what proportion of the time will job A take longer than the average job B, and what proportion of the time will job B take less time than the average job A?

8.52 If 75 percent of the persons shopping at a certain shopping mall live within ten miles of the mall, what is the probability that among 160 persons shopping there fewer than 115 live within ten miles?

8.53 Find the area under the standard normal curve which lies
    (a) between $z = 0$ and $z = -1.11$;
    (b) between $z = -0.63$ and $z = 0.63$;
    (c) between $z = 0.40$ and $z = 0.55$;
    (d) between $z = -1.18$ and $z = -0.68$;
    (e) between $z = -1.22$ and $z = 1.82$.

8.54 The weekly number of muggings reported in a certain precinct is a random variable with $\mu = 25.3$ and $\sigma = 2.5$. Assuming that the distribution of this random variable can be approximated closely with a normal distribution, find the probabilities that in any one week there will be
    (a) exactly 27 muggings;
    (b) at least 27 muggings.

8.55 Find z if
    (a) the normal-curve area between 0 and z is 0.2019;
    (b) the normal-curve area to the right of z is 0.8810;
    (c) the normal-curve area to the right of z is 0.0336;
    (d) the normal-curve area between $-z$ and z is 0.2662.

8.9
REFERENCES

More detailed tables of normal-curve areas may be found in many handbooks of statistical tables; for instance, in

    PEARSON, E. S., and HARTLEY, H. O., *Biometrika Tables for Statisticians,*
        3rd ed. Cambridge: Cambridge University Press, 1966.

Extensive tables of random normal numbers are given in the RAND Corporation tables listed on page 203.

The purpose of most statistical investigations is to make sound generalizations on the basis of samples about the populations from which the samples came. Note the word "sound," because the question of when and under what conditions samples permit such generalizations is not easily answered. For instance, if we want to estimate the average amount of money a person spends on a vacation, would we take as our sample the amounts spent by deluxe-class passengers on a four-week cruise; or would we attempt to estimate, or predict, the wholesale price of all farm products on the basis of the price of fresh asparagus alone? Obviously not, but just what vacationers and which farm products we should include in our samples, and how many of them, is neither intuitively clear nor self-evident.

In most of the methods we shall study in this book, it will be assumed that we are dealing with a particular kind of sample called a random sample. This attention to random samples, which we shall discuss in Section 9.1, is due to their permitting valid, or logical, generalizations. As we shall see, however, random sampling is not always feasible, or even desirable, and some other sampling procedures will be mentioned briefly in Sections 9.2 through 9.5. Then, in Sections 9.6 through 9.9, we shall see how statistics (that is, quantities determined from samples) can be expected to vary from sample to sample.

9 SAMPLING AND SAMPLING DISTRIBUTIONS

9.1
RANDOM SAMPLING

In the beginning of Chapter 3 we distinguished between populations and samples, stating that a population consists of all conceivably or hypothetically possible observations (instances, or occurrences) of a given phenomenon, while a sample is simply part of a population. For the work which follows, let us now distinguish between two kinds of populations—**finite populations** and **infinite populations**.

A finite population is one which consists of a finite number, or fixed number, of elements (items, objects, measurements, or observations). Examples of finite populations are the net weights of the 10,000 cans of paint in a production lot, the SAT scores of all the freshmen admitted to a certain university in 1981, and the daily high temperatures recorded at a weather station during the years 1975–1982.

In contrast to finite populations, a population is said to be infinite if there is, hypothetically at least, no limit to the number of elements it can contain. The population which consists of the results of all hypothetically possible rolls of a pair of dice is an infinite population, and so is the population which consists of all conceivably possible measurements of the weight of a piece of rock.

To introduce the idea of **random sampling from a finite population,** let us first ask how many different samples of size n can be drawn from a finite population of size N. Referring to the rule for combinations on page 90, we find that, with a change of letters, the answer is $\binom{N}{n}$.

EXAMPLE How many different samples of size n can be drawn from a finite population of size N if
(a) $n = 2$ and $N = 12$;
(b) $n = 3$ and $N = 100$?

Solution (a) $\binom{12}{2} = \dfrac{12 \cdot 11}{2!} = 66;$ (b) $\binom{100}{3} = \dfrac{100 \cdot 99 \cdot 98}{3!} = 161{,}700$

Based on the result that there are $\binom{N}{n}$ different samples, let us now give the following definition of a **simple random sample** (or more briefly, a

random sample) from a finite population:

> A sample of size n from a finite population of size N is random if it is chosen in such a way that each of the $\binom{N}{n}$ possible samples has the same probability, $\dfrac{1}{\binom{N}{n}}$, of being selected.

For instance, if a finite population consists of the $N = 5$ elements a, b, c, d, and e (which might be the incomes of five persons, the weights of five guinea pigs, or the prices of five commodities), there are $\binom{5}{3} = 10$ possible samples of size $n = 3$ consisting, respectively, of the elements abc, abd, abe, acd, ace, ade, bcd, bce, bde, and cde. If we choose one of these samples in such a way that each sample has the probability $\frac{1}{10}$ of being selected, we call this sample a random sample.

What remains to be seen is how we draw random samples in actual practice. In a simple case like the one described immediately above, we could write each of the $\binom{N}{n}$ possible samples on a slip of paper, put the slips of paper in a hat, shuffle them thoroughly, and then draw one without looking. Such a procedure is obviously impractical, if not impossible, in more realistically complex problems of sampling; we mention it here only to make the point that the selection of a random sample must depend entirely on chance.

Fortunately, we can take a random sample without actually resorting to the tedious process of listing all possible samples. We can list instead the N individual elements of a finite population, and then take a random sample by choosing the elements to be included in the sample one at a time without replacement, making sure that in each of the successive drawings each of the remaining elements of the population has the same chance of being selected. This leads to the same probability, $\dfrac{1}{\binom{N}{n}}$, for each possible sample. For instance, to take a random sample of size $n = 12$ from the population which consists of the amounts of sales tax collected by a city's 247 drugstores in December 1981 we could write each of the 247 figures on a slip of paper, mix them up thoroughly in a bag, a box, or a hat, and then draw (without looking) twelve slips one after the other without replacement.

Even this relatively easy procedure can be simplified in actual practice; usually, the simplest way to take a random sample from a finite population is to refer to a table of random numbers. As we pointed out on page 195, published tables of random numbers (such as the one from which Table XI of this book is excerpted) consist of pages on which the digits $0, 1, 2, \ldots,$

and 9 are set down in much the same fashion as they might appear if they had been generated by a chance or gambling device giving each digit the same probability of $\frac{1}{10}$ of appearing at any given place in the table.

To illustrate the use of random numbers in random sampling, let us refer again to the problem of sampling the amounts of sales tax collected by a city's 247 drugstores in December 1981. Numbering the stores 001, 002, 003, ..., 246, and 247 (say, in the order in which the stores are listed in the telephone directory), we arbitrarily pick a starting place in the table and then move in any direction, reading out three-digit numbers. For instance, if we arbitrarily use the 26th, 27th, and 28th columns of the table on page 518, starting with the sixth row and going down the page, we get

046   230   079   022   119   150   056   064   193   232   040   146

where we ignored the tabled numbers greater than 247; had any number recurred, we would have ignored it too. The numbers we got here are numbers assigned to the drugstores—the corresponding sales tax figures constitute the desired random sample.

When lists are available and items are, or can readily be, numbered, it is easy to draw random samples from finite populations with the aid of random number tables. Unfortunately, however, it is often impossible to proceed in the way we have just described. For instance, if we want to use a sample to estimate the mean outside diameter of thousands of ball bearings packed in a large crate, or if we want to estimate the mean height of the trees in a forest, it would be impossible to number the ball bearings or the trees, choose random numbers, and then locate and measure the corresponding ball bearings or trees. In these and in many similar situations, all we can do is proceed according to the dictionary definition of the word "random," namely, "haphazardly, without aim or purpose." That is, we must not select or reject any element of a population because of its seeming typicalness or lack of it, nor must we favor or ignore any part of a population because of its accessibility or lack of it, and so forth. With some reservations, such samples can often be treated as if they were, in fact, random samples.

Until now we have discussed only random sampling from finite populations. The concept of a random sample from an infinite population is more difficult to define, but a few simple examples will help to explain the basic characteristics of such a sample. For instance, we consider the results of 15 tosses of a coin as a sample from a hypothetically infinite population, since there is no limit to the number of times the coin could be tossed. If the probability of heads is the same for each toss and the tosses are all independent, we say that the sample is random. Also we would be sampling from an infinite population if we sample with replacement from a finite population, and our sample would be random if in each draw all elements of the population have the same chance of being selected and successive draws are independent.

In general, we require that the selection of each item in a random sample from an infinite population must be controlled by the same probabilities and that successive selections must be independent.

EXERCISES

9.1 How many different samples of size $n = 2$ can be selected from a finite population of size
(a) $N = 6$;   (b) $N = 10$;   (c) $N = 25$?

9.2 How many different samples of size $n = 3$ can be selected from a finite population of size
(a) $N = 10$;   (b) $N = 25$;   (c) $N = 50$?

9.3 What is the probability of each possible sample if
(a) a random sample of size 4 is to be drawn from a finite population of size 12;
(b) a random sample of size 5 is to be drawn from a finite population of size 22?

9.4 With reference to the example on page 235, where we listed all possible random samples of size $n = 3$ from the finite population which consists of the elements a, b, c, d, and e, what is the probability that any specific element, say, the element c, will be contained in the sample?

9.5 List the $\binom{6}{2} = 15$ possible samples of size $n = 2$ that can be drawn from the finite population whose elements are denoted a, b, c, d, e, and f.

9.6 With reference to the preceding exercise, what is the probability that a random sample of size $n = 2$ from the given finite population will include the element denoted by the letter f?

9.7 List all possible choices of four of the following six airlines: TWA, American, Western, Continental, PSA, and Delta. If a person randomly selects four of these airlines to study their safety records, find
(a) the probability of each possible sample;
(b) the probability that Western will be included in the sample.

9.8 A bacteriologist wants to double-check a sample of $n = 8$ of the 754 blood specimens analyzed by a medical laboratory in a given month. If he numbers the specimens 001, 002, ... , 753, and 754, which ones (by number) will he select if he chooses them by using the sixth, seventh, and eighth columns of the table on page 520, going down the page starting with the sixteenth row?

9.9 A county assessor wants to reassess a random sample of 15 of 8,019 one-family homes. If she numbers them 0001, 0002, ... , 8,018, and 8,019, which ones (by numbers) will she select if she chooses them by using the 11th, 12th, 13th, and 14th columns of the table on page 517, starting with the fourth row and going down the page.

9.10 A sociologist wants to include ten of the 83 counties in Michigan in a survey. If he numbers these counties 01, 02, ... , 82, and 83, which ones (by number) will he include in the survey if he selects them by using the 21st and 22nd columns of the table on page 518, starting with the tenth row and going down the page.

9.11 The employees of a company have badges numbered serially from 1 through 615. Use the third, fourth, and fifth columns of the table on page 519, starting with the eighth row and going down the page, to select (by number) a random sample of eight of the company's employees to serve on a grievance committee.

9.12 On page 235 we said that a random sample can be drawn from a finite population by choosing the elements to be included in the sample one at a time, making sure that in each of the successive drawings each of the remaining elements of the population has the same chance of being selected. To verify that this will give the correct probability to each sample, let us refer to the example on page 235, where we dealt with random samples of size 3 drawn from the finite population which consists of the elements a, b, c, d, and e. To find the probability of drawing any particular sample (say, b, c, and e), we can argue that the probability of getting one of these three letters on the first draw is $\frac{3}{5}$, the probability of then getting one of the remaining two letters on the second draw is $\frac{2}{4}$, and the probability of then getting the third letter on the third draw is $\frac{1}{3}$. Multiplying these three probabilities, we find that the probability of getting the particular sample is $\frac{3}{5} \cdot \frac{2}{4} \cdot \frac{1}{3} = \frac{1}{10}$, and this agrees with the value obtained on page 235.

(a) Use the same kind of argument to verify that for each possible random sample of size 3, drawn one at a time from a finite population of size 100, the probability is $1 \left/ \binom{100}{3} \right. = \dfrac{1}{161,700}$.

(b) Use the same kind of argument to verify in general that for each possible random sample of size n, drawn one at a time from a finite population of size N, the probability is $1 \left/ \binom{N}{n} \right.$.

9.13 Making use of the fact that among the $\binom{N}{n}$ samples of size n which can be drawn from a finite population of size N there are $\binom{N-1}{n-1}$ which contain a specific element, show that the probability that any specific element of the population will be contained in a random sample of size n is $\dfrac{n}{N}$.

9.2
SAMPLE DESIGNS ★

So far we discussed only random samples, and we did not even consider the possibility that under certain conditions there may be samples which are better (say, easier to obtain, cheaper, or more informative) than random samples, and we did not go into any details about the question of what might be done when random sampling is impossible. Indeed, there are many other ways of selecting a sample from a population, and there is an extensive literature devoted to the subject of designing sampling procedures.

In statistics, a **sample design** is a definite plan, completely determined before any data are actually collected, for obtaining a sample from a given population. Thus, the plan to take a simple random sample of 12 of a city's

247 drugstores by using a table of random numbers in a prescribed way constitutes a sample design. In what follows, we shall discuss briefly some of the most important kinds of sample designs.

9.3
SYSTEMATIC SAMPLING ⋆

In some instances, the most practical way of sampling is to select, say, every 20th name on a list, every 12th house on one side of a street, every 50th piece coming off an assembly line, and so on. Sampling of this sort is called **systematic sampling,** and an element of randomness is usually introduced into this kind of sampling by using random numbers to pick the unit with which to start. Although a systematic sample may not be a random sample in accordance with the definition, it is often reasonable to treat systematic samples as if they were random samples; indeed, in some instances, systematic samples actually provide an improvement over simple random samples inasmuch as the samples are spread more evenly over the entire populations.

The real danger in systematic sampling lies in the possible presence of hidden periodicities. For instance, if we inspect every 40th piece made by a particular machine, the results would be very misleading if, because of a regularly recurring failure, every 10th piece produced by the machine is blemished. Also, a systematic sample might yield biased results if we interview the residents of every 10th house along a certain route and it so happens that each house thus chosen is a corner house on a double lot.

9.4
STRATIFIED SAMPLING ⋆

When we know something about the makeup of a population (that is, if we know something about its composition) and this is of relevance to our investigation, we may be able to improve on random sampling by **stratification.** This is a procedure which consists of stratifying (or dividing) the population into a number of non-overlapping subpopulations, or **strata,** and then taking a sample from each stratum. If the items selected from each stratum constitute a simple random sample, the entire procedure (first stratification and then simple random sampling) is called **stratified (simple) random sampling.**

Suppose, for instance, that we want to estimate the mean weight of four persons on the basis of a sample of size 2; the (unknown) weights of the four persons are 115, 135, 185, and 205 pounds, so that μ, the mean weight we want to estimate, is 160 pounds. If we take an ordinary random sample of size 2 from this population, the $\binom{4}{2} = 6$ possible samples are 115 and 135, 115 and 185, 115 and 205, 135 and 185, 135 and 205, and 185 and 205, and the corresponding means are 125, 150, 160, 160, 170, and 195. Observe that since each of these samples has the probability $\frac{1}{6}$, there is a probability of $\frac{1}{3}$ that the mean of such a sample will differ from $\mu = 160$ by 35 and, hence, provide a poor estimate of the mean weight of the four persons.

Now suppose that we know that the two persons who weigh 185 and 205 pounds are men, and the two persons who weigh 115 and 135 pounds are women. Stratifying our sample (by sex) and randomly choosing one of the two men and one of the two women, we find that there are only the four stratified samples 115 and 185, 115 and 205, 135 and 185, and 135 and 205. The means of these samples are 150, 160, 160, and 170, and since none of the means differs from $\mu = 160$ by more than 10, our chances of getting a good (close) estimate of the mean weight of the four persons are greatly improved.

Essentially, the goal of stratification is to form strata in such a way that there is some relationship between being in a particular stratum and the answer sought in the statistical study, and that within the separate strata there is as much homogeneity (uniformity) as possible. In our example there is such a connection between sex and weight and there is much less variability in weight within each of the two groups than there is within the entire population.

In the above example, we used **proportional allocation,** which means that the sizes of the samples from the different strata are proportional to the sizes of the strata. In general, if we divide a population of size N into k strata of size $N_1, N_2, \ldots,$ and N_k, and take a sample of size n_1 from the first stratum, a sample of size n_2 from the second stratum, ..., and a sample of size n_k from the kth stratum, we say that the allocation is proportional if

$$\frac{n_1}{N_1} = \frac{n_2}{N_2} = \cdots = \frac{n_k}{N_k}$$

or if these ratios are as nearly equal as possible. In the example dealing with the weights we have $N_1 = 2, N_2 = 2, n_1 = 1,$ and $n_2 = 1,$ so that

$$\frac{n_1}{N_1} = \frac{n_2}{N_2} = \frac{1}{2}$$

and the allocation is, indeed, proportional.

It can easily be shown (see Exercise 9.18 on page 243) that allocation is proportional if

Sample sizes for proportional allocation

$$n_i = \frac{N_i}{N} \cdot n \qquad \text{for } i = 1, 2, \ldots, \text{and } k$$

where n is the total size of the sample, that is, $n = n_1 + n_2 + \cdots + n_k$. When necessary, we use the integers closest to the values given by this formula.

EXAMPLE A stratified sample of size $n = 60$ is to be taken from a population of size $N = 4,000$, which consists of three strata of size $N_1 = 2,000$, $N_2 = 1,200$, and $N_3 = 800$. If the allocation is to be proportional, how large a sample must be taken from each stratum?

Solution Substituting into the formula, we get

$$n_1 = \frac{2,000}{4,000} \cdot 60 = 30$$

$$n_2 = \frac{1,200}{4,000} \cdot 60 = 18$$

and

$$n_3 = \frac{800}{4,000} \cdot 60 = 12$$

The preceding example illustrates proportional allocation, but there exist other ways of allocating portions of a sample to the different strata; one of these, called **optimum allocation,** is described in Exercise 9.22 on page 244. Stratification is not restricted to a single variable of classification, or characteristic, and populations are often stratified according to several characteristics. For instance, in a system-wide survey designed to determine the attitude of its students, say, toward a new tuition plan, a state college system with 17 colleges might stratify the students with respect to class standing, sex, major, and college. So, part of the sample would be allocated to junior women majoring in engineering in college A, part to senior men majoring in English in college L, and so on. Up to a point, stratification like this, called **cross stratification,** will usually increase the precision (reliability) of estimates and other generalizations, and it is widely used, particularly in opinion sampling and market surveys.

In stratified sampling, the cost of taking random samples from the individual strata is often so expensive that interviewers are simply given quotas to be filled from the different strata, with few (if any) restrictions on how they are to be filled. For instance, in determining voters' attitudes toward increased medical coverage for elderly persons, an interviewer working a certain area might be told to interview 6 male self-employed homeowners under 30 years of age, 10 female wage earners in the 45–60 age bracket who live in apartments, 3 retired males over 60 who live in trailers, and so on, with the actual selection of the individuals being left to the interviewer's discretion. This is called **quota sampling,** and it is a convenient, relatively inexpensive, and sometimes necessary procedure, but as it is often executed, the resulting samples do not have the essential features of random samples. In the absence of any controls on their choice, interviewers naturally tend to select individuals who are most readily available—persons who work in the same building, shop in the same store, or perhaps reside in the same general area. Quota samples are thus essentially **judgment samples,** and inferences based on such samples generally do not lend themselves to any sort of formal statistical evaluation.

9.5
CLUSTER SAMPLING ⋆

To illustrate another important kind of sampling, suppose that a large foundation wants to study the changing patterns of family expenditures in the San Diego area. In attempting to complete schedules for 1,200 families, the foundation finds that simple random sampling is practically impossible, since suitable lists are not available and the cost of contacting families scattered over a wide area (with possibly two or three callbacks for the not-at-homes) is very high. One way in which a sample can be taken in this situation is to divide the total area of interest into a number of smaller, non-overlapping areas, say, city blocks. A number of these blocks are then randomly selected, with the ultimate sample consisting of all (or samples of) the families residing in these blocks.

In the kind of sampling described in the above example, called **cluster sampling,** the total population is divided into a number of relatively small subdivisions, which are themselves clusters of still smaller units, and then some of these subdivisions, or clusters, are randomly selected for inclusion in the overall sample. If the clusters are geographic subdivisions, as in our example, this kind of sampling is also called **area sampling.** To give another example of cluster sampling, suppose that the Dean of Students of a university wants to know how fraternity men at the school feel about a certain new regulation. He can take a cluster sample by interviewing some or all of the members of several randomly selected fraternities.

Although estimates based on cluster samples are usually not as reliable as estimates based on simple random samples of the same size (see Exercise 9.35 on page 257), they are usually more reliable per unit cost. Referring again to the survey of family expenditures in the San Diego area, it is easy to see that it may well be possible to take a cluster sample several times the size of a simple random sample for the same cost. It is much cheaper to visit and interview families living close together in clusters than families selected at random over a wide area.

In practice, several of the methods of sampling we have discussed may well be used in the same study. For instance, if government statisticians want to study the attitude of American elementary school teachers toward certain federal programs, they might first stratify the country by states or some other geographic subdivisions. To take a sample from each stratum, they might then use cluster sampling, subdividing each stratum into a number of smaller geographic subdivisions (say, school districts), and finally they might use simple random sampling or systematic sampling to select a sample of elementary school teachers within each cluster.

EXERCISES

★9.14 The following are the numbers of commercial FM radio stations in operation in 1978 in the 50 states (listed in alphabetic order): 69, 6, 32, 53, 194, 43, 23, 8, 111, 86, 6, 16, 129, 95, 71, 40, 86, 61, 28, 37, 38, 109, 66, 68, 77, 21, 31, 14, 17, 34, 29, 117, 85, 12, 130, 50, 33, 125, 7, 55, 18, 74, 185, 19, 12, 73, 51, 32, 91, and 11. List the ten possible systematic samples of size five that can be taken from this list by starting with one of the first ten numbers and then taking each tenth number on the list.

★9.15 The following are the numbers of restaurants in twenty cities: 18, 23, 12, 10, 20, 11, 7, 11, 116, 34, 9, 8, 19, 25, 9, 83, 28, 9, 13, and 15.
   (a) List the five possible systematic samples of size 4 that can be taken from this list by starting with one of the first five numbers and then taking each fifth number on the list.
   (b) Calculate the mean of each of the five samples obtained in part (a) and verify that their mean equals the average (mean) number of restaurants in the given twenty cities.

★9.16 To generalize the example on page 239, suppose that we want to estimate the mean weight of six persons, whose (unknown) weights are 115, 125, 135, 185, 195, and 205 pounds.
   (a) List all possible random samples of size 2 which can be taken from this population, calculate their means, and determine the probability that the mean of such a sample will differ by more than 5 from $\mu = 160$, the mean weight of the six persons.
   (b) Supposing that the first three weights are those of women and the other three are those of men, list all possible stratified samples of size 2 which can be taken by randomly choosing one of the three women and one of the three men, calculate their means, and determine the probability that the mean of such a sample will differ by more than 5 from $\mu = 160$, the mean weight of the six persons.

★9.17 Among the 300 persons empaneled for jury duty, 150 are whites, 100 are blacks, and 50 are Chicanos. In how many ways can we choose a stratified 2 percent sample of these 300 persons,
   (a) if one-third of the sample is to be allocated to each group;
   (b) if the allocation is to be proportional?

★9.18 If the sample sizes for the individual strata are calculated according to the formula on page 240, verify that
   (a) the allocation is, indeed, proportional;
   (b) the sum of the sample sizes is actually equal to n.

★9.19 A stratified sample of size $n = 80$ is to be taken from a population of size $N = 2,000$, which consists of four strata for which $N_1 = 500$, $N_2 = 1,200$, $N_3 = 200$, and $N_4 = 100$. If we use proportional allocation, how large a sample must we take from each stratum?

★9.20 A stratified sample of size $n = 400$ is to be taken from a population of size $N = 50,000$, which consists of five strata for which $N_1 = 20,000$, $N_2 = 15,000$, $N_3 = 5,000$, $N_4 = 8,000$, and $N_5 = 2,000$. If the allocation is to be proportional, how large a sample must be taken from each stratum?

★9.21 With reference to Exercise 9.16, list all possible cluster samples of size 2 which can be taken by randomly choosing either two of the three women or two of the three men, calculate their means, and determine the probability that the mean of such a sample will differ by more than 5 from $\mu = 160$, the mean weight of the six persons. If we compare this probability with those obtained in parts (a) and (b) of Exercise 9.16, what does this tell us about the relative merits of simple random sampling, stratified sampling, and cluster sampling in the given situation?

★9.22 In stratified sampling with proportional allocation, the importance of differences in stratum size is accounted for by letting the larger strata contribute relatively more items to the sample. However, strata differ not only in size but also in variability, and it would seem reasonable to take larger samples from the more variable strata and smaller samples from the less variable strata. If we let σ_1, $\sigma_2, \ldots$, and σ_k denote the standard deviations of the k strata, we can account for both, differences in stratum size and differences in stratum variability, by requiring that

$$\frac{n_1}{N_1 \sigma_1} = \frac{n_2}{N_2 \sigma_2} = \cdots = \frac{n_k}{N_k \sigma_k}$$

or that these ratios be as nearly equal as possible. This is called **optimum allocation**, and it can be shown that it leads to the formula

Sample sizes for optimum allocation

$$n_i = \frac{n \cdot N_i \sigma_i}{N_1 \sigma_1 + N_2 \sigma_2 + \cdots + N_k \sigma_k} \qquad \textit{for } i = 1, 2, \ldots, \textit{and } k$$

For instance, if we wanted to use optimum allocation in the example on page 241 and the standard deviations of the three strata are $\sigma_1 = 8$, $\sigma_2 = 15$, and $\sigma_3 = 32$, substitution of these values together with $n = 60$, $N_1 = 2,000$, $N_2 = 1,200$, and $N_3 = 800$ into the above formula would yield

$$n_1 = \frac{60 \cdot 2,000 \cdot 8}{2,000 \cdot 8 + 1,200 \cdot 15 + 800 \cdot 32}$$

which equals 16 rounded to the nearest integer, and $n_2 = 18$ and $n_3 = 26$.

(a) A population is divided into two strata so that $N_1 = 10,000$, $N_2 = 30,000$, $\sigma_1 = 45$, and $\sigma_2 = 60$. How should a sample of size $n = 100$ be allocated to the two strata if we use optimum allocation?

(b) A population is divided into three strata so that $N_1 = 5,000$, $N_2 = 2,000$, $N_3 = 3,000$, $\sigma_1 = 15$, $\sigma_2 = 18$, and $\sigma_3 = 5$. How should a sample of size $n = 84$ be allocated to the three strata if we use optimum allocation?

★9.23 If we want to estimate the mean of a population on the basis of a stratified sample, we first calculate the means of the k samples, $\bar{x}_1, \bar{x}_2, \ldots$, and $\bar{x}_k$, and then we combine them using the formula

$$\bar{x}_w = \frac{N_1 \bar{x}_1 + N_2 \bar{x}_2 + \cdots + N_k \bar{x}_k}{N_1 + N_2 + \cdots + N_k}$$

This is a weighted mean of the individual $\bar{x}$'s, and the weights are the sizes of the strata.

For instance, the records of a casualty insurance company show that among 3,800 claims filed against the company over a period of time, 2,600 were minor claims (under \$200), while the other 1,200 were major claims (\$200 or more). To estimate the average size of these claims, the company takes a 1 percent sample, proportionally allocated to the two strata, with the following results (rounded to the nearest dollar):

Minor claims: 42, 115, 63, 78, 45, 148, 195, 66, 18, 73, 55, 89, 170, 41, 92, 103, 22, 138, 49, 62, 88, 113, 29, 71, 58, 83

Major claims: 246, 355, 872, 649, 253, 338, 491, 860, 755, 502, 488, 311

(a) Find the means of these two samples and then determine their weighted mean, using as weights the sizes of the two strata, as an estimate of the mean size of the 3,800 claims.

(b) Verify that the result of part (a) equals the ordinary mean of the 38 claims; namely, that proportional allocation is **self-weighting**.

★9.24 Verify symbolically that for stratified sampling with proportional allocation the weighted mean given by the formula of the preceding exercise equals the ordinary mean of all the sample values obtained for all the strata. (*Hint*: Make use of the fact that $n_i \cdot \bar{x}_i$ equals the sum of the values obtained for the ith stratum.)

9.6
SAMPLING DISTRIBUTIONS

Let us now introduce the concept of the **sampling distribution** of a statistic, probably the most basic concept of statistical inference. As we shall see, this concept is closely related to the idea of chance variation, or chance fluctuations, which we mentioned earlier to emphasize the need for measuring the variability of data. In this chapter we shall concentrate mainly on the sample mean and its sampling distribution, but in some of the exercises at the end of this section and in later chapters we shall consider also the sampling distributions of other statistics.

There are two ways of approaching the study of sampling distributions. One, based on appropriate mathematical theory, leads to what is called a **theoretical sampling distribution**; the other, based (in reality or by simulation) on repeated samples from the same population, leads to what is called an **experimental sampling distribution.** The latter will prove to be very useful in our study because it provides experimental verification of theorems which cannot be derived formally at the level of this book.

To present the idea of a theoretical sampling distribution, let us construct the one for the mean of random samples of size $n = 2$ from the finite population of size $N = 5$, whose elements are the numbers 3, 5, 7, 9, and 11. The mean of this population is

$$\mu = \frac{3 + 5 + 7 + 9 + 11}{5} = 7$$

and its standard deviation is

$$\sigma = \sqrt{\frac{(3 - 7)^2 + (5 - 7)^2 + (7 - 7)^2 + (9 - 7)^2 + (11 - 7)^2}{5}}$$

$$= \sqrt{8}$$

Now, there are $\binom{5}{2} = 10$ random samples of size $n = 2$ which can be drawn from this population. They are

3 and 5,	3 and 7,	3 and 9,	3 and 11,	5 and 7
5 and 9,	5 and 11,	7 and 9,	7 and 11,	9 and 11

and their means are 4, 5, 6, 7, 6, 7, 8, 8, 9, and 10. Since each sample has the probability $\frac{1}{10}$, we thus obtain the following theoretical sampling distribution of the mean for random samples of size $n = 2$ from the given population:

$\bar{x}$	Probability
4	$\frac{1}{10}$
5	$\frac{1}{10}$
6	$\frac{2}{10}$
7	$\frac{2}{10}$
8	$\frac{2}{10}$
9	$\frac{1}{10}$
10	$\frac{1}{10}$

A histogram of this distribution is shown in Figure 9.1.

An examination of this sampling distribution reveals some pertinent information relative to the problem of estimating the mean of the given population on the basis of a random sample of size 2. For instance, we see that corresponding to $\bar{x} = 6$, 7, or 8, the probability is $\frac{6}{10}$ that a sample mean will not differ from the population mean by more than 1, and that corresponding to $\bar{x} = 5$, 6, 7, 8, or 9, the probability is $\frac{8}{10}$ that a sample mean will not differ from the population mean by more than 2.

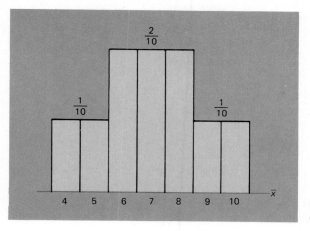

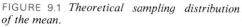

FIGURE 9.1 *Theoretical sampling distribution of the mean.*

Further useful information about this theoretical sampling distribution of the mean can be obtained by calculating its mean $\mu_{\bar{x}}$ and its standard deviation $\sigma_{\bar{x}}$. (The subscript $\bar{x}$ is used here to distinguish these parameters from those of the original population.) Following the definitions of the mean and the variance of a probability distribution on pages 186 and 188, we get

$$\mu_{\bar{x}} = 4 \cdot \frac{1}{10} + 5 \cdot \frac{1}{10} + 6 \cdot \frac{2}{10} + 7 \cdot \frac{2}{10} + 8 \cdot \frac{2}{10} + 9 \cdot \frac{1}{10} + 10 \cdot \frac{1}{10}$$

$$= 7$$

and

$$\sigma_{\bar{x}}^2 = (4 - 7)^2 \cdot \frac{1}{10} + (5 - 7)^2 \cdot \frac{1}{10} + (6 - 7)^2 \cdot \frac{2}{10} + (7 - 7)^2 \cdot \frac{2}{10}$$

$$+ (8 - 7)^2 \cdot \frac{2}{10} + (9 - 7)^2 \cdot \frac{1}{10} + (10 - 7)^2 \cdot \frac{1}{10}$$

$$= 3$$

so that $\sigma_{\bar{x}} = \sqrt{3}$. Observe that $\mu_{\bar{x}}$, the mean of the sampling distribution of $\bar{x}$, equals μ, the mean of the population, and that $\sigma_{\bar{x}}$, the standard deviation of the sampling distribution of $\bar{x}$, is smaller than σ, the standard deviation of the population. These relationships are of fundamental importance, and we shall return to them in the next section.

Let us now turn to the problem of constructing an experimental sampling distribution of the mean, hoping thereby to gain further insight into the chance fluctuations of sample means. Suppose that, in connection with the purchase of new equipment, a tow truck operator wants to determine how many service calls he can expect to receive on a weekday afternoon.

Suppose, furthermore, that in a sample of five weekday afternoons he received 19, 15, 11, 12, and 21 service calls. The mean of this sample is

$$\bar{x} = \frac{19 + 15 + 11 + 12 + 21}{5} = 15.6$$

and in the absence of any other information he may use this figure as an estimate of μ, the number of service calls he can expect to receive on a weekday afternoon. It stands to reason, however, that if the sample had consisted of the number of service calls he received on five other weekdays, the mean would probably not have been 15.6. Indeed, if the operator of the tow trucks took several samples, each consisting of the number of service calls he received on five weekdays, he might get such discrepant estimates of μ as 13.6, 18.4, 12.2, and 17.0.

To show how the means of such samples might vary purely as the result of chance, we assumed that the number of service calls the tow truck operator receives on a weekday afternoon is a random variable having the Poisson distribution (see Section 7.5) with the mean $\mu = 16$, and we simulated the drawing of 50 random samples of size $n = 5$ with the use of random numbers. The results are shown in the following table:

Sample	Number of service calls	Sample	Number of service calls
1	16, 15, 14, 12, 18	26	10, 18, 19, 13, 20
2	20, 18, 16, 19, 14	27	14, 19, 16, 13, 21
3	21, 18, 17, 26, 25	28	18, 14, 23, 23, 14
4	16, 10, 9, 19, 15	29	17, 16, 11, 17, 11
5	17, 16, 20, 17, 7	30	16, 13, 10, 14, 20
6	20, 8, 17, 16, 13	31	13, 17, 22, 19, 18
7	22, 21, 16, 15, 13	32	15, 13, 16, 14, 21
8	11, 23, 12, 20, 14	33	16, 15, 8, 12, 23
9	17, 22, 21, 16, 20	34	16, 15, 11, 20, 13
10	18, 13, 15, 11, 12	35	17, 17, 16, 21, 14
11	15, 11, 14, 14, 18	36	18, 20, 14, 26, 18
12	22, 15, 13, 19, 11	37	13, 16, 17, 11, 6
13	20, 17, 11, 19, 15	38	17, 19, 15, 19, 16
14	15, 16, 16, 15, 17	39	11, 13, 18, 23, 18
15	15, 16, 17, 17, 16	40	12, 25, 21, 18, 8
16	13, 15, 15, 13, 18	41	21, 12, 14, 17, 16
17	12, 11, 19, 17, 16	42	19, 10, 15, 16, 18
18	15, 26, 19, 20, 15	43	20, 10, 15, 15, 19
19	15, 11, 21, 8, 17	44	16, 10, 26, 14, 20
20	12, 21, 10, 15, 16	45	18, 12, 13, 19, 9
21	10, 11, 9, 11, 11	46	15, 14, 21, 17, 11
22	12, 12, 24, 11, 5	47	12, 13, 13, 14, 12
23	16, 18, 14, 9, 11	48	18, 8, 21, 14, 15
24	17, 9, 18, 16, 9	49	20, 17, 16, 18, 19
25	11, 17, 19, 20, 17	50	19, 17, 16, 13, 15

Next we calculated the mean number of service calls for each of the 50 samples, getting

$$
\begin{array}{cccccccccc}
15.0 & 17.4 & 21.4 & 13.8 & 15.4 & 14.8 & 17.4 & 16.0 & 19.2 & 13.8 \\
14.4 & 16.0 & 16.4 & 15.8 & 16.2 & 14.8 & 15.0 & 19.0 & 14.4 & 14.8 \\
10.4 & 12.8 & 13.6 & 13.8 & 16.8 & 16.0 & 16.6 & 18.4 & 14.4 & 14.6 \\
17.8 & 15.8 & 14.8 & 15.0 & 17.0 & 19.2 & 12.6 & 17.2 & 16.6 & 16.8 \\
16.0 & 15.6 & 15.8 & 17.2 & 14.2 & 15.6 & 12.8 & 15.2 & 18.0 & 16.0 \\
\end{array}
$$

and finally we grouped these means into a table having the classes 9.5–10.5, 10.5–11.5,..., and 20.5–21.5. (There is no risk of ambiguity here, since division of an integer by 5 always leaves a remainder of 0, 2, 4, 6, or 8 tenths.) The resulting distribution is shown as a histogram in Figure 9.2.

Now, inspection of this experimental sampling distribution of $\bar{x}$ tells us a great deal about the way in which the means of random samples drawn from the given population tend to scatter among themselves due to chance. For instance, we find that the smallest mean is 10.4 and the largest is 21.4; furthermore, 31 out of 50 (or 62 percent) of the means are between 14.5 and 17.5, and 48 out of 50 (or 96 percent) of the means are between 12.5 and 19.5. Since the mean of the Poisson population from which the samples came is known to be $\mu = 16$, we find that 62 percent of the sample means are "off" (differ from the population mean) by less than 1.5, and that 96 percent of the sample means are "off" by less than 3.5.

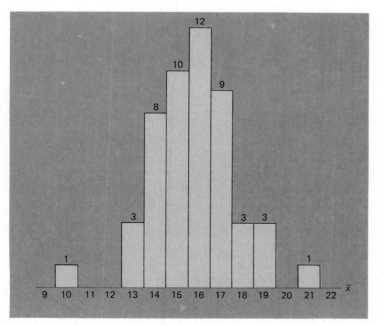

FIGURE 9.2 *Experimental sampling distribution of the mean.*

The two examples of this section served to introduce the concept of a sampling distribution, but they do not reflect what we do in actual practice. They are really only learning devices, for in actual practice we can rarely enumerate all possible samples, and we base inferences on one sample, not on fifty.

9.7
THE STANDARD ERROR OF THE MEAN

In most practical situations we can determine how close a sample mean might be to the mean of the population from which it came, by referring to two theorems, one given below and the other on page 253, which express essential facts about sampling distributions of the mean. The first of these theorems expresses formally what we discovered in connection with the example on page 248 — the mean of the sampling distribution of $\bar{x}$ equals the mean of the population sampled, and the standard deviation of the sampling distribution of $\bar{x}$ is smaller than the standard deviation of the population. It may be phrased as follows:

For random samples of size n taken from a population having the mean μ and the standard deviation σ, the theoretical sampling distribution of $\bar{x}$ has the mean $\mu_{\bar{x}} = \mu$ and the standard deviation

Standard error of the mean

$$\sigma_{\bar{x}} = \frac{\sigma}{\sqrt{n}} \quad or \quad \sigma_{\bar{x}} = \frac{\sigma}{\sqrt{n}} \cdot \sqrt{\frac{N - n}{N - 1}}$$

depending on whether the population is infinite or finite of size N.

It is customary to refer to $\sigma_{\bar{x}}$, the standard deviation of the sampling distribution of the mean, as the **standard error of the mean.** Its role in statistics is fundamental, since it measures the extent to which sample means can be expected to fluctuate, or vary, due to chance. Clearly, some knowledge of this variability is essential in determining how well an $\bar{x}$ estimates the mean of the population from which the sample came. As intuition suggests correctly, the smaller $\sigma_{\bar{x}}$ is (the less the $\bar{x}$'s are spread out) the better are our chances that the estimate will be close. What determines the size of $\sigma_{\bar{x}}$, and hence the goodness of an estimate, can be seen from the formulas above. Both formulas show that the standard error of the mean increases as the variability of the population increases (in fact, it is directly proportional to σ), and that it decreases as the number of items in the sample increases. With respect to the latter, note that both formulas yield $\sigma_{\bar{x}} = \sigma$ for $n = 1$, and that $\sigma_{\bar{x}} = 0$ only for $n = N$; in other words, $\sigma_{\bar{x}}$ takes on values between 0 and σ, and is 0 only when the sample includes the entire population.

EXAMPLE When we sample from an infinite population, what happens to the standard error of the mean (and, hence, to the size of the error we are exposed to when we use $\bar{x}$ as an estimate of μ) if the sample size is increased from $n = 50$ to $n = 200$?

Solution The ratio of the two standard errors is

$$\frac{\dfrac{\sigma}{\sqrt{200}}}{\dfrac{\sigma}{\sqrt{50}}} = \frac{\sqrt{50}}{\sqrt{200}} = \sqrt{\frac{50}{200}} = \sqrt{\frac{1}{4}} = \frac{1}{2}$$

so that the standard error is only half of what it was before.

The factor $\sqrt{\dfrac{N - n}{N - 1}}$ in the second formula for $\sigma_{\bar{x}}$ is called the **finite population correction factor,** for without it the two formulas for $\sigma_{\bar{x}}$ (for infinite and finite populations) are the same. In practice, it is omitted unless the sample constitutes at least 5 percent of the population, for otherwise it is so close to 1 that it has little effect on the value of $\sigma_{\bar{x}}$.

EXAMPLE Find the value of the finite population correction factor for $n = 100$ and $N = 10,000$.

Solution Substituting $n = 100$ and $N = 10,000$, we get

$$\sqrt{\frac{N - n}{N - 1}} = \sqrt{\frac{10,000 - 100}{10,000 - 1}} = 0.995$$

This is so close to 1 that the correction factor would ordinarily be omitted in practice.

To get a feeling for the two formulas for $\sigma_{\bar{x}}$, let us return to the two illustrations of the preceding section.

EXAMPLE With reference to the illustration on page 247, verify that the second of the two formulas on page 250 also yields $\sigma_{\bar{x}} = \sqrt{3}$.

Solution Substituting $n = 2$, $N = 5$, and $\sigma = \sqrt{8}$ into this formula, we get

$$\sigma_{\bar{x}} = \frac{\sqrt{8}}{\sqrt{2}} \cdot \sqrt{\frac{5 - 2}{5 - 1}} = \frac{\sqrt{8}}{\sqrt{2}} \cdot \sqrt{\frac{3}{4}} = \sqrt{\frac{8}{2} \cdot \frac{3}{4}} = \sqrt{3}$$

EXAMPLE With reference to the illustration on page 248, use the mean and the standard deviation of the 50 sample means as estimates of $\mu_{\bar{x}}$ and $\sigma_{\bar{x}}$, and compare them with the corresponding values expected in accordance with the theorem for random samples of size $n = 5$ from a population having the Poisson distribution with the mean $\mu = 16$ and, hence, the standard deviation $\sigma = \sqrt{16} = 4$ (see Exercise 7.45).

Solution The mean and the standard deviation of the 50 means are 15.75 and 1.94, and the corresponding values expected in accordance with the theorem are

$$\mu_{\bar{x}} = 16 \quad \text{and} \quad \sigma_{\bar{x}} = \frac{4}{\sqrt{5}} = 1.79$$

Since the experimental values, 15.75 and 1.94, are fairly close to the theoretical values, 16 and 1.79, this small-scale sampling experiment provides evidence which tends to support the theory.

9.8
THE CENTRAL LIMIT THEOREM

When we use a sample mean to estimate the mean of a population, we usually attach a probability to a measure of the error of our estimate. For instance, if we use Chebyshev's theorem, we can assert with a probability of at least $1 - \frac{1}{k^2}$ that the mean of a random sample will differ from the mean of the population from which it came by less than $k \cdot \sigma_{\bar{x}}$. In other words, when we use the mean of a random sample to estimate the mean of a population, we can assert with a probability of at least $1 - \frac{1}{k^2}$ that our error will be less than $k \cdot \sigma_{\bar{x}}$.

EXAMPLE Based on Chebyshev's theorem with, say, $k = 2$, what can we say about the possible size of our error, if we intend to use the mean of a random sample of size $n = 64$ to estimate the mean of an infinite population with $\sigma = 20$?

Solution Substituting $\sigma = 20$ and $n = 64$ into the first of the two formulas for the standard error of the mean, we get

$$\sigma_{\bar{x}} = \frac{20}{\sqrt{64}} = 2.5$$

and it follows that we can assert with a probability of at least $1 - \frac{1}{2^2} = 0.75$ that the error will be less than $k \cdot \sigma_{\bar{x}} = 2(2.5) = 5$.

This shows that we can make probability statements about errors of estimates without having to go through the tedious (if not impossible) process of constructing the corresponding theoretical sampling distribution.

Chebyshev's theorem applies to any distribution, and it may always be used as in the preceding example, but there exists another basic theorem of statistics which enables us in a great many instances to make much stronger probability statements. This theorem, which is the second of the two theorems referred to on page 250, is called the **central limit theorem** and it may be stated as follows:

Central limit
theorem

> *If n (the sample size) is large, the theoretical sampling distribution of the mean can be approximated closely with a normal distribution.*

This theorem is of fundamental importance in statistics, because it justifies the use of normal-curve methods in a wide range of problems; it applies to infinite populations, and also to finite populations when n, though large, constitutes but a small portion of the population. It is difficult to say precisely how large n must be so that the central limit theorem applies, but unless the population distribution has a very unusual shape, $n = 30$ is usually regarded as sufficiently large. When the population we are sampling has, itself, roughly the shape of a normal curve, the sampling distribution of the mean can be approximated with a normal distribution regardless of the size of n.

To illustrate the importance of the central limit theorem, let us re-examine the preceding example and also the experimental sampling distribution of Section 9.6.

EXAMPLE   Based on the central limit theorem, what is the probability that the error will be less than 5 when we use the mean of a random sample of size $n = 64$ to estimate the mean of a population with $\sigma = 20$?

Solution   The probability is given by the area of the white region under the curve in Figure 9.3; that is, by the normal-curve area between

$$z = \frac{-5}{20/\sqrt{64}} = -2 \text{ and } z = \frac{5}{20/\sqrt{64}} = 2. \text{ The entry in Table I cor-}$$

responding to $z = 2$ is 0.4772, so the probability asked for is 0.4772 + 0.4772 = 0.9544. Compared to Chebyshev's theorem, according to which the probability is "at least 0.75," we can now make the much stronger statement that the probability is 0.9544 that the mean of a random sample of size $n = 64$ from the given population will differ from the mean of the population by less than 5.

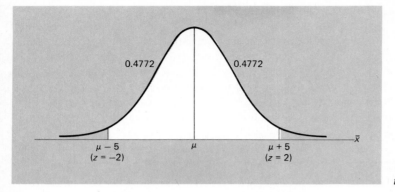

FIGURE 9.3 *Sampling distribution of the mean.*

EXAMPLE   When we constructed the experimental sampling distribution in Section 9.6, we took 50 random samples of size $n = 5$ from a population with $\sigma = 4$ and found that 31 of their means (or 62 percent) differed from the mean of the population by less than 1.5. Based on the central limit theorem, what is the probability that the difference between a sample mean and the mean of the population will be less than 1.5, when we take a random sample of size $n = 5$ from a population with $\sigma = 4$?

Solution   The probability is given by the area of the white region under the curve in Figure 9.4; that is, by the normal-curve area between

$$z = \frac{-1.5}{4/\sqrt{5}} = -0.84 \qquad \text{and} \qquad z = \frac{1.5}{4/\sqrt{5}} = 0.84$$

Since the entry in Table I corresponding to $z = 0.84$ is 0.2995, the probability is $0.2995 + 0.2995 = 0.5990$, or about 0.60. This is remarkably close to the 0.62 probability (62 percent) which we got in the sampling experiment of Section 9.6.

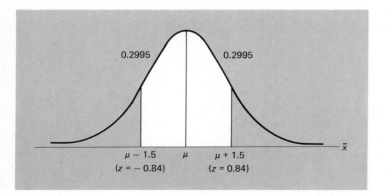

FIGURE 9.4 *Sampling distribution of the mean.*

9.9
SOME FURTHER CONSIDERATIONS

The main goal of this chapter has been to introduce the concept of a sampling distribution, and the one which we chose for this purpose was the sampling distribution of the mean. Observe, however, that instead of the mean we could have studied the median or some other statistic and investigated its chance fluctuations. So far as the corresponding theory is concerned, this would, of course, have required different formulas for the standard errors, namely, different formulas for the standard deviations of the respective sampling distributions. For instance, for infinite populations, the **standard error of the median** is approximately $\sigma_{\tilde{x}} = 1.25 \cdot \dfrac{\sigma}{\sqrt{n}}$, where n is the size of the sample and σ is the population standard deviation. Note that comparison of the two formulas $\sigma_{\bar{x}} = \dfrac{\sigma}{\sqrt{n}}$ and $\sigma_{\tilde{x}} = 1.25 \cdot \dfrac{\sigma}{\sqrt{n}}$ reflects the fact that the mean is generally more reliable than the median.

EXAMPLE How large a random sample do we have to take so that its mean is as reliable an estimate of the mean of a symmetrical infinite population as the median of a random sample of size $n = 200$? (For symmetrical populations, the mean of the sampling distributions of $\bar{x}$ and $\tilde{x}$ are both equal to the population mean μ.)

Solution Equating the two standard errors, we get

$$\frac{\sigma}{\sqrt{n}} = 1.25 \cdot \frac{\sigma}{\sqrt{200}}$$

which, solved for n, yields 128.

Also, a point worth repeating is that the two examples of Section 9.6 were meant to be teaching aids, designed to convey the idea of a sampling distribution. They do not reflect what we do in actual practice, where we ordinarily base an inference on one sample and not 50, and we do not enumerate all possible samples. In Chapter 10 and in subsequent chapters we shall go further into the problem of translating theory concerning sampling distributions into methods of evaluating the merits and disadvantages of statistical procedures.

Another fact worth noting concerns the $\sqrt{n}$ appearing in the denominator of the formula for the standard error of the mean. As n becomes larger and larger and we gain more and more information, our generalizations should be subject to smaller errors, and in general, our results (inferences) should be more reliable and more precise. However, the $\sqrt{n}$ in the formulas for $\sigma_{\bar{x}}$ illustrates the fact that gains in precision or reliability are not proportional to increases in the size of the sample. That is, doubling the size of the sample does not double the reliability of $\bar{x}$ as an estimate of the mean of a population, and so on. As is apparent from the formula $\sigma_{\bar{x}} = \dfrac{\sigma}{\sqrt{n}}$ for samples from infinite populations, we must take four times as large a sample to cut the standard error in half, and nine times as large a sample to triple the reliability; namely, to divide the standard error by 3. This clearly suggests that it seldom pays to take excessively large samples.

EXERCISES

9.25 Random samples of size 2 are drawn from the finite population which consists of the numbers 5, 6, 7, 8, 9, and 10.
  (a) Show that the mean of this population is $\mu = 7.5$ and that its standard deviation is $\sigma = \sqrt{\frac{35}{12}}$.
  (b) List the 15 possible random samples of size 2 that can be drawn from this finite population and calculate their means.
  (c) Using the results of part (b) and assigning each possible sample the probability $\frac{1}{15}$, construct the sampling distribution of the mean for random samples of size 2 from the given finite population.
  (d) Calculate the mean and the standard deviation of the probability distribution obtained in part (c) and verify the results with the use of the theorem on page 250.

9.26 Repeat parts (b), (c), and (d) of the preceding exercise for random samples of size $n = 3$ from the given finite population.

9.27 The finite population of Exercise 9.25 can be converted into an infinite population if we sample with replacement, that is, if we take a random sample of size 2 by first drawing one value and replacing it before drawing the second.
  (a) List the 36 possible ordered samples of size 2 that can be drawn with replacement from the given population and calculate their means. (By "ordered" we mean that the samples 7 and 10 and 10 and 7, for example, are considered different samples.)
  (b) Using the results of part (a) and assigning each possible sample the probability $\frac{1}{36}$, construct the sampling distribution of the mean for random samples of size 2 drawn with replacement from the given population.
  (c) Calculate the mean and the standard deviation of the probability distribution obtained in part (b), and verify the results with the use of the theorem on page 250.

9.28 Convert the 50 samples on page 248 into 25 samples of size $n = 10$ by combining samples 1 and 26, samples 2 and 27, ..., and samples 25 and 50. Calculate the mean of each of these samples of size 10 and determine their mean and their standard deviation. Compare this mean and this standard deviation with the corresponding values expected in accordance with the theorem on page 250.

9.29 When we sample from an infinite population, what happens to the standard error of the mean (and, hence, to the size of the error we are exposed to when we use $\bar{x}$ to estimate μ) if the sample size is
(a) increased from 60 to 240;
(b) increased from 200 to 450;
(c) increased from 25 to 225;
(d) decreased from 640 to 40?

9.30 What is the value of the finite population correction factor when
(a) $n = 5$ and $N = 200$;
(b) $n = 10$ and $N = 300$;
(c) $n = 100$ and $N = 5{,}000$?

9.31 Show that if the mean of a random sample of size n is used to estimate the mean of an infinite population with the standard deviation σ, there is a fifty–fifty chance that the error (that is, the difference between $\bar{x}$ and μ) will be less than

$0.6745 \cdot \dfrac{\sigma}{\sqrt{n}}$. It has been the custom to refer to this quantity as the **probable error**

of the mean; nowadays, this term is used mainly in military applications.
(a) If a random sample of size $n = 64$ is taken from an infinite population with $\sigma = 20.6$, what is the size of the probable error of the mean?
(b) If a random sample of size $n = 100$ is drawn from a very large population (consisting of the fines paid for speeding violations in a certain county in 1981) which has the standard deviation $\sigma = \$8.25$, determine the probable error of the mean and explain its significance.

9.32 With reference to Exercise 9.15 on page 243, assign each of the systematic samples the probability $\frac{1}{5}$, determine the standard deviation of the corresponding sampling distribution of the mean, and compare it with $\sigma_{\bar{x}} = 12.24$, the standard error of the mean for random samples of size $n = 4$ from the given finite population.

9.33 In the example on page 239, we compared stratified samples from a population which consists of four weights with ordinary random samples of the same size.
(a) Assigning each of the random samples on page 240 the probability $\frac{1}{6}$, show that the mean and the standard deviation of this sampling distribution of the mean are $\mu_{\bar{x}} = 160$ and $\sigma_{\bar{x}} = 21.0$.
(b) Assigning each of the stratified samples on page 240 the probability $\frac{1}{4}$, show that the mean and the standard deviation of this sampling distribution of the mean are $\mu_{\bar{x}} = 160$ and $\sigma_{\bar{x}} = 7.1$.

★9.34 If $\bar{x}$ is the mean of a stratified random sample of size n obtained by proportional allocation from a finite population of size N, which consists of k strata of size $N_1, N_2, \ldots$, and N_k, then

$$\sigma_{\bar{x}}^2 = \sum_{i=1}^{k} \frac{(N - n)N_i^2}{nN^2(N_i - 1)} \cdot \sigma_i^2$$

where $\sigma_1^2, \sigma_2^2, \ldots$, and σ_k^2 are the corresponding variances for the individual strata. Use this formula to verify the result of part (b) of the preceding exercise.

9.35 In the example on page 239, the two cluster samples, which consist of choosing and weighing either the two women or the two men, have the means 125 and 195. Assigning the probability $\frac{1}{2}$ to each of these means, calculate the standard deviation of this sampling distribution of the mean, and compare it with the corresponding values for simple random sampling and stratified sampling obtained in parts (a) and (b) of Exercise 9.33.

9.36 The mean of a random sample of size $n = 36$ is used to estimate the mean of an infinite population having the standard deviation $\sigma = 9$. What can we assert about the probability that the error will be less than 4.5, if we use
   (a) Chebyshev's theorem;
   (b) the central limit theorem?

9.37 The mean of a random sample of size $n = 25$ is used to estimate the mean of a very large population (consisting of the attention spans of persons over 65), which has a standard deviation of $\sigma = 2.4$ minutes. What can we assert about the probability that the error will be less than 1.2 minutes, if we use
   (a) Chebyshev's theorem;
   (b) the central limit theorem?

9.38 The mean of a random sample of size $n = 144$ is used to estimate the mean of a very large population (consisting of the weights of certain animals), which has a standard deviation of 3.6 ounces. If we use the central limit theorem, what can we assert about the probability that the error will be
   (a) less than 0.50 ounce;
   (b) less than 1.20 ounces;
   (c) greater than 0.20 ounce;
   (d) greater than 0.40 ounce?

9.39 When we constructed the experimental sampling distribution in Section 9.6, we took 50 random samples of size $n = 5$ from a population with $\sigma = 4$ and found that 48 of the sample means (or 96 percent) differed from the mean of the population by less than 3.5. Based on the central limit theorem, what is the probability that the difference between the sample mean and the population mean will be less than 3.5, when we take a random sample of size $n = 5$ from a population with $\sigma = 4$? Compare this figure with the 0.96 proportion obtained in the sampling experiment.

9.40 If measurements of the specific gravity of a metal can be looked upon as a sample from a normal population having the standard deviation $\sigma = 0.025$, what is the probability that the mean of a random sample of size $n = 16$ will be "off" by at most 0.01?

9.41 If the distribution of the weights of all men traveling by air between Dallas and El Paso has a mean of 163 pounds and a standard deviation of 18 pounds, what is the probability that the combined weight of 36 men traveling on such a flight is more than 6,012 pounds?

9.42 Find the medians of the 50 samples on page 248 and group them as we grouped the corresponding means.
   (a) Construct a histogram of this **experimental sampling distribution of the median** and compare it with that of Figure 9.2.
   (b) Calculate the standard deviation of the 50 medians, or of their distribution, and compare the result with the value expected in accordance with

   the formula $\sigma_{\bar{x}} = 1.25 \cdot \dfrac{\sigma}{\sqrt{n}}$.

9.43 Show that the mean of a random sample of size $n = 256$ is as reliable an estimate of the mean of a symmetrical infinite population as the median of a random sample of size $n = 400$.

9.44 How large a random sample do we have to take so that its median is as reliable an estimate of the mean of a symmetrical infinite population as the mean of a random sample of size $n = 144$?

★**9.45** Calculating the standard deviations of the 50 samples on page 248, we obtain the following **experimental sampling distribution of** s:

Sample standard deviation	Frequency
0.5–1.5	4
1.5–2.5	6
2.5–3.5	12
3.5–4.5	16
4.5–5.5	8
5.5–6.5	2
6.5–7.5	2

where there are no ambiguities since none of the values is exactly 1.5, 2.5, ..., or 6.5.

(a) Calculate the mean and the standard deviation of this sampling distribution.

(b) For large samples, the formula $\dfrac{\sigma}{\sqrt{2n}}$ is sometimes used for the **standard error of a sample standard deviation.** Substituting $\sigma = 4$ and $n = 5$ into this formula, calculate the value of this standard error for the given example and compare it with the value of the standard deviation obtained for the experimental sampling distribution in part (a).

9.10
TECHNICAL NOTE (Simulation) ★

The experimental sampling distribution of Section 9.6 was obtained by simulating values of a random variable having the Poisson distribution with the parameter $\lambda = 16$. The technique we used is described in Section 7.10, and it was performed with the scheme at the top of page 260.

The probabilities were copied from a table of Poisson probabilities, and the three-digit random numbers were assigned to the values of the random variable by working with the cumulative probabilities, as suggested in the example on page 197. In our experiment, we randomly chose a page from a table of random numbers, three adjacent columns and a place from which to start, and read off three-digit random numbers, five for each of the 50 samples.

Number of service calls	Probability	Random numbers
5	0.001	000
6	0.003	001–003
7	0.006	004–009
8	0.012	010–021
9	0.021	022–042
10	0.034	043–076
11	0.050	077–126
12	0.066	127–192
13	0.082	193–274
14	0.093	275–367
15	0.099	368–466
16	0.099	467–565
17	0.093	566–658
18	0.083	659–741
19	0.070	742–811
20	0.056	812–867
21	0.043	868–910
22	0.031	911–941
23	0.022	942–963
24	0.014	964–977
25	0.009	978–986
26	0.006	987–992
27	0.003	993–995
28	0.002	996–997
29	0.001	998
30	0.001	999

Simulations like this are very useful in the study of sampling distributions; particularly, when the theory is complicated and no tables (such as the one for the normal distribution) are available.

EXERCISES

★9.46 Use the above scheme and Table XI to simulate 25 random samples of size $n = 4$ from the given Poisson distribution with $\mu = 16$ and $\sigma = 4$, calculate the mean and the standard deviation of their means, and compare the results with those expected in accordance with the theorem on page 250.

★9.47 Use the scheme of the example on page 198 and Table XI to simulate 20 random samples, each consisting of the numbers of cars that arrive at the tollbooth of the bridge in $n = 6$ one-minute intervals, and calculate their means. Also calculate the standard deviation of these means and compare it with the value expected in accordance with the theorem on page 250, given that the population standard deviation is $\sigma = 2.2$.

★9.48 Use Table XII to simulate 40 random samples of size $n = 3$ from a standard normal population and calculate their means. Also calculate the mean and the standard deviation of these means and compare them with the values expected in accordance with the theorem on page 250.

★9.49 Determine the medians of the 40 samples of the preceding exercise, calculate their standard deviation, and compare it with the value expected in accordance with the standard error formula given in Section 9.9.

★9.50 For random samples of size n from a normal population, the **sampling distribution** of s^2 has the mean σ^2 and the standard deviation $\sigma^2\sqrt{\dfrac{2}{n-1}}$. To verify this theory experimentally, use Table XII to simulate twenty random samples of size $n = 8$ from a standard normal population, calculate s^2 for each of the twenty samples, and then determine the mean and the standard deviation of these twenty values of s^2.

★9.51 Based on fairly advanced theory, it can be shown that for random samples of size $n = 3$ from a normal population, the mean of the sampling distribution of the sample range is approximately 1.69σ where σ is the population standard deviation. Thus, for a random sample of size $n = 3$ we can use the sample range divided by 1.69 as an estimate of σ; for other sample sizes we divide by different constants, as is explained in Exercise 11.9 on page 314. Verify this theory experimentally by determining the ranges of the 40 random samples of Exercise 9.48, calculate their mean, and compare it with $1.69\sigma = 1.69 \cdot 1 = 1.69$.

9.11
CHECKLIST OF KEY TERMS
(with page references to their definitions)

★ Area sampling, 242
Central limit theorem, 253
★ Cluster sampling, 242
★ Cross stratification, 241
Experimental sampling distribution, 245
Finite population, 234
Finite population correction factor, 251
Infinite population, 234
★ Judgment sample, 242
★ Optimum allocation, 241, 244
Probable error of the mean, 257
★ Proportional allocation, 240
★ Quota sampling, 241
Random sample, 235
★ Sample design, 238
Sampling distribution, 245
Simple random sample, 234
Standard error of the mean, 250
Standard error of the median, 255
★ Stratified sampling, 239
★ Systematic sampling, 239
Theoretical sampling distribution, 245

9.12
REVIEW EXERCISES

9.52 The mean of a random sample of size $n = 36$ is used to estimate the mean of a population having a normal distribution with $\sigma = 18$. With what probability can we assert that the error will be less than 9, if we use
   (a) Chebyshev's theorem;
   (b) the central limit theorem?

★**9.53** Among 80 persons interviewed for jobs by a government agency, 40 are whites, 20 are blacks, 10 are Orientals, and 10 are Chicanos. In how many ways can a 10 percent stratified sample be chosen from among the persons interviewed, if
   (a) one-fourth of the sample is to be allocated to each group;
   (b) the allocation is proportional?

9.54 Suppose that a customs official wants to check 24 of 855 shipments listed on a ship's manifest. Which ones (by numbers) will she check, if she numbers the shipments from 001 through 855 and chooses a random sample by means of random numbers, using the last three columns of the table on page 517, starting with the 11th row and going down the page?

9.55 If random samples of size $n = 3$ are chosen from a finite population of size $N = 80$, what is the probability of each possible sample?

9.56 Random samples of size $n = 2$ are drawn from the population which consists of the numbers 1, 3, 5, and 7.
   (a) Verify that the mean of this finite population is $\mu = 4$ and that its standard deviation is $\sigma = \sqrt{5}$.
   (b) List the six possible samples of size $n = 2$ that can be drawn from this finite population, calculate their means, and, assigning each of these values the probability $\frac{1}{6}$, construct the theoretical sampling distribution of the mean for random samples of size $n = 2$ from the given population.
   (c) Calculate the mean and the standard deviation of the sampling distribution obtained in part (b), and verify the results using the theorem on page 250.

9.57 Suppose that in the preceding exercise sampling is with replacement, so that the population is (hypothetically, at least) infinite; that is, there is no limit to the number of observations we could make.
   (a) List the 16 possible samples of size $n = 2$ that can thus be drawn from the given population (counting 3 and 5, for example, and 5 and 3 as different samples), calculate their means, and assigning each of these values the probability $\frac{1}{16}$, construct the theoretical sampling distribution of the mean for random samples of size $n = 2$ drawn with replacement from the given population.
   (b) Calculate the mean and the standard deviation of the sampling distribution obtained in part (a), and verify the results using part (a) of the preceding exercise and the theorem on page 250.

★**9.58** A stratified sample of size $n = 168$ is to be taken from a population of size $N = 10,000$, which consists of three strata for which $N_1 = 5,000$, $N_2 = 2,000$, $N_3 = $

3,000, $\sigma_1 = 15$, $\sigma_2 = 18$, and $\sigma_3 = 5$. If we use optimum allocation, how large a sample will we take from each stratum?

★9.59 With reference to Exercise 9.16 show that
   (a) $\sigma_{\bar{x}} = 22.7$ for the means of the random samples of part (a) of that exercise;
   (b) $\sigma_{\bar{x}} = 5.8$ for the means of the stratified samples of part (b) of that exercise.

★9.60 Use the formula of Exercise 9.34 to verify part (b) of the preceding exercise.

9.61 What is the value of the finite population correction factor when
   (a) $n = 30$ and $N = 120$;
   (b) $n = 20$ and $N = 500$?

9.62 When we sample from an infinite population, what happens to the standard error of the mean if the sample size is
   (a) increased from 20 to 500;
   (b) decreased from 490 to 40?

★9.63 A stratified sample of size $n = 90$ is to be taken from a group of 2,000 persons including 600 who are not junior college or college graduates, 800 who are only junior college graduates, 400 who are college graduates but hold no advanced degrees, and 200 who hold advanced degrees. What part of the sample should be allocated to each of these strata, if the allocation is to be proportional?

9.64 How many different samples of size $n = 4$ can be selected from a finite population of size
   (a) $N = 18$;
   (b) $N = 30$;
   (c) $N = 100$?

9.65 What is the probability of each possible sample if a random sample of size $n = 5$ is to be drawn from a finite population of size $N = 24$?

9.66 The mean of a random sample of size $n = 81$ is used to estimate the mean annual growth of certain plants. Assuming that $\sigma = 3.6$ mm for such data, use the central limit theorem to find the probabilities that this estimate will be off either way by
   (a) less than 1.0 mm;
   (b) less than 0.5 mm.

9.13
REFERENCES

Derivations of the formulas for the standard error of the mean and more general formulations (and proofs) of the central limit theorem may be found in most textbooks on mathematical statistics; for instance, in

FREUND, J. E., and WALPOLE, R. E., *Mathematical Statistics*, 3rd ed. Englewood Cliffs, N.J.: Prentice-Hall, Inc., 1980.

Details on sampling are given in

SCHEAFFER, R. L., MENDENHALL, W., and OTT, L., *Elementary Survey Sampling*. Boston: Duxbury Press, 1979.
WILLIAMS, W. H., *A Sampler on Sampling*. New York: John Wiley & Sons, Inc., 1978.

Traditionally, problems of statistical inference have been divided into problems of estimation, in which we estimate various unknown parameters (statistical descriptions) of populations, tests of hypotheses, in which we either accept or reject specific assertions about populations or their parameters, and problems of prediction, in which we predict future values of random variables. In this chapter, and in those that immediately follow, we shall concentrate on problems of estimation and tests of hypotheses; methods of prediction will be taken up in Chapters 14 and 15.

After a general introduction to problems of estimation in Section 10.1, Sections 10.2 through 10.4 are devoted to the estimation of means. Then, after a general introduction to tests of hypotheses in Sections 10.5 through 10.8, Sections 10.9 and 10.10 are devoted to tests of hypotheses concerning means, and Sections 10.11 and 10.12 are devoted to tests of hypotheses concerning the means of two populations.

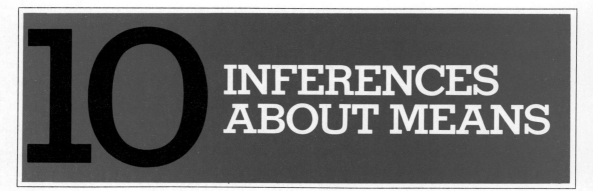

10 INFERENCES ABOUT MEANS

10.1
PROBLEMS OF ESTIMATION

Problems of estimation arise everywhere, in science, in business, and in everyday life. In science, a psychologist may want to determine the average time that it takes an adult person to react to a visual stimulus, an engineer may need to know how much variability there is in the strength of a new alloy, and a biologist may want to determine what percentage of certain insects are born physically defective. In business, a retailer may want to determine the average income of all families living within a mile of a proposed new store, a union official may have to know how much variation there is in the time it takes members to get to work, and a banker may want to find out what proportion of all installment loan payments she can expect to be late. Finally, in everyday life, we may want to determine how long it takes on the average to mend a pair of sox, we may be interested in the variation we can expect in a child's performance in school, and we may want to find out what percentage of all one-car accidents are due to driver fatigue.

In each of these three sets of examples, the first deals with the estimation of a mean, the second with the estimation of a measure of variation, and the third with the estimation of a percentage or proportion. Since the statistical treatment of such problems of inference differs, we shall devote this chapter to inferences about means, and then take up inferences about variability and inferences about proportions in Chapters 11 and 12.

10.2
THE ESTIMATION OF MEANS

To illustrate some of the problems we face in the estimation of a mean, let us refer to the example used in Chapter 2 to illustrate the construction of a frequency distribution. In that example, we dealt with data on the sulfur oxides emission of an industrial plant, and as we saw later, their mean is $\bar{x} = 18.85$ tons. Now, if the data constitute a random sample, we can use this figure as an estimate of μ, the plant's "true" average daily emission of sulfur oxides.

An estimate like this is called a **point estimate,** since it consists of a single number, or a single point on the real number scale. Although this is the most common way in which estimates are expressed, it leaves room for many questions. For instance, it does not tell us on how much information the estimate is based, and it does not tell us anything about the possible size of the error. And of course, we must expect an error. This should be clear from the discussion of sampling distributions in Chapter 9, where we saw that the reliability of the mean as an estimate of μ depends on two things—the size of the sample and the size of the population standard deviation σ. Thus, we might supplement the estimate, $\bar{x} = 18.85$ tons, with the information that it is the mean of a random sample of size $n = 80$, whose standard deviation is $s = 5.55$ tons. Although this does not tell us the actual value of σ, the sample standard deviations can serve as an estimate of this quantity.

Scientific reports often present sample means in this way, together with the values of n and s, but this does not supply the reader with a coherent picture unless he has had some formal training in statistics. To make the supplementary information meaningful also to the layman, let us refer to the two theorems of Chapter 9 about the sampling distribution of the mean. According to the second of the two theorems, the central limit theorem, the sampling distribution of the mean can, for large samples, be approximated closely with a normal curve. Hence, we can assert with probability $1 - \alpha$ that a sample mean $\bar{x}$ will differ from the population mean μ by at most $z_{\alpha/2}$ standard errors of the mean. As defined on page 212, $z_{\alpha/2}$ is the z value for which the area to its right under the standard normal curve is $\alpha/2$ (see Figure 10.1). Since the standard error of the mean is $\sigma_{\bar{x}} = \dfrac{\sigma}{\sqrt{n}}$ for samples from infinite populations, we can thus assert with probability $1 - \alpha$ that $\bar{x}$ will differ from μ by at most $z_{\alpha/2} \cdot \dfrac{\sigma}{\sqrt{n}}$. In other words,

When we use $\bar{x}$ as an estimate of μ, the probability is $1 - \alpha$ that this estimate will be "off" either way by at most

Maximum error of estimate

$$E = z_{\alpha/2} \cdot \frac{\sigma}{\sqrt{n}}$$

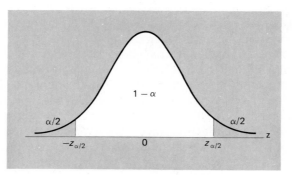

FIGURE 10.1 *Standard normal distribution.*

This result applies when n is large ($n \geq 30$) and the population is infinite, or large enough so that the finite population correction factor can be ignored. The two most widely used values for $1 - \alpha$ are 0.95 and 0.99, and from Table I (see also Exercise 8.7 on page 215) we find that the corresponding values of $z_{\alpha/2}$ are $z_{0.025} = 1.96$ and $z_{0.005} = 2.575$.

EXAMPLE An efficiency expert intends to use the mean of a random sample of size $n = 150$ to estimate the average mechanical aptitude (as measured by a certain test) of assembly-line workers in a large industry. If, based on experience, the efficiency expert can assume that $\sigma = 6.2$ for such data, what can he assert with probability 0.99 about the maximum size of his error?

Solution Substituting $n = 150$, $\sigma = 6.2$, and $z_{0.005} = 2.575$ into the formula for E, we get

$$E = 2.575 \cdot \frac{6.2}{\sqrt{150}} = 1.30$$

Thus, the efficiency expert can assert with probability 0.99 that his error will be at most 1.30.

Suppose now that the efficiency expert of our example actually collects his data and gets $\bar{x} = 69.5$. Can he still assert with probability 0.99 that the error is at most 1.30? After all, $\bar{x} = 69.5$ differs from the true average by at most 1.30 or it does not, and he does not know which. Well, he can, but it must be understood that the 0.99 probability applies to the method which he used to determine the maximum error (getting the sample data and using the formula for E) and not directly to the parameter he is trying to estimate. To make this distinction, it has become the custom to use the word "**confidence**" here instead of "probability." In general, we make probability statements about future values of random variables (say, the potential error of an estimate) and confidence statements once the data have been obtained. Accordingly, we would say in our example that the efficiency expert can be 99 percent confident that the error of his estimate, $\bar{x} = 69.5$, is at most 1.30.

Use of the formula for E involves one complication. To be able to judge the size of the error we might make when we use $\bar{x}$ as an estimate of μ, we must know the value of the population standard deviation σ. Since this is not the case in most practical situations, we have no choice but to replace σ with an estimate, usually the sample standard deviation s. In general, this is considered to be reasonable provided that the sample size is 30 or more.

EXAMPLE With reference to the air pollution example on page 66, what can we assert with 95 percent confidence about the maximum error, if we use $\bar{x} = 18.85$ tons as an estimate of the plant's true average daily emission of sulfur oxides?

Solution   Substituting $n = 80$, $z_{0.025} = 1.96$, and $s = 5.55$ for σ into the formula for E, we find that we can assert with 95 percent confidence that the error is at most

$$E = 1.96 \cdot \frac{5.55}{\sqrt{80}} = 1.22 \text{ tons}$$

Of course, the error is at most 1.22 or it is not, but if we had to bet, 95 to 5 (or 19 to 1) would be fair odds that it is at most 1.22.

The formula for E can also be used to determine the sample size that is needed to attain a desired degree of precision. Suppose that we want to use the mean of a large random sample to estimate the mean of a population, and we want to be able to assert with $(1 - \alpha)100$ percent confidence that the error of this estimate is at most some prescribed quantity E. As before, we write

$$E = z_{\alpha/2} \cdot \frac{\sigma}{\sqrt{n}}$$

and upon solving this equation for n we get

Sample size
$$n = \left[\frac{z_{\alpha/2} \cdot \sigma}{E} \right]^2$$

EXAMPLE   The dean of a college wants to use the mean of a random sample to estimate the average amount of time students take to get from one class to the next, and she wants to be able to assert with 95 percent confidence that the error is at most 0.25 minute. If it can be presumed from experience that $\sigma = 1.50$ minutes, how large a sample will she have to take?

Solution   Substituting $z_{0.025} = 1.96$, $E = 0.25$, and $\sigma = 1.50$ into the formula for n, we get

$$n = \left[\frac{1.96 \cdot 1.50}{0.25} \right]^2 = 139$$

rounded up to the nearest integer. Thus, a random sample of size $n = 139$ is required for the estimate.

As can be seen from the formula and also from the example, this method has the shortcoming that it can be used only when we know (at least approximately) the value of the population standard deviation. For this reason, we sometimes begin with a small sample, and then use the sample standard deviation as an estimate of σ to determine whether more data are required.

Let us now introduce a different way of presenting the information provided by a sample mean and the assessment of its potential error. In what follows, we shall make use of the fact that for large random samples from infinite populations, the sampling distribution of the mean is approximately normal with the mean μ and the standard deviation $\sigma_{\bar{x}} = \dfrac{\sigma}{\sqrt{n}}$, so that

$$z = \frac{\bar{x} - \mu}{\sigma/\sqrt{n}}$$

is a value of a random variable having approximately the standard normal distribution. Since the probability is $1 - \alpha$ that a random variable having the standard normal distribution will take on a value between $-z_{\alpha/2}$ and $z_{\alpha/2}$ (see Figure 10.1), or that

$$-z_{\alpha/2} < z < z_{\alpha/2}$$

we can substitute the above expression for z into this inequality and get

$$-z_{\alpha/2} < \frac{\bar{x} - \mu}{\sigma/\sqrt{n}} < z_{\alpha/2}$$

Now, if we apply some simple algebra, we can rewrite this inequality as

Large-sample confidence interval for μ

$$\bar{x} - z_{\alpha/2} \cdot \frac{\sigma}{\sqrt{n}} < \mu < \bar{x} + z_{\alpha/2} \cdot \frac{\sigma}{\sqrt{n}}$$

and we can assert with probability $1 - \alpha$ that it will be satisfied for any given sample. In other words, we can assert with $(1 - \alpha)100$ percent confidence that the interval from $\bar{x} - z_{\alpha/2} \cdot \dfrac{\sigma}{\sqrt{n}}$ to $\bar{x} + z_{\alpha/2} \cdot \dfrac{\sigma}{\sqrt{n}}$, determined on the basis of a large random sample, contains the population mean we are trying to estimate. When σ is unknown and n is at least 30, we replace σ by the sample standard deviation s.

An interval like this is called a **confidence interval,** its endpoints are called **confidence limits,** and $(1 - \alpha)100$ percent is called the **degree of confidence.** As before, the values most often used for the degree of confidence are 95 percent and 99 percent, and the corresponding values of $z_{\alpha/2}$ are 1.96 and 2.575. In contrast to point estimates, estimates given in the form of a confidence interval are called **interval estimates.**

EXAMPLE   In the air pollution example on page 66 we had $n = 80$, $\bar{x} = 18.85$, and $s = 5.55$. Construct a 95 percent large-sample confidence interval for the plant's true average daily emission of sulfur oxides.

Solution   Substituting into the confidence interval formula, we get

$$18.85 - 1.96 \cdot \frac{5.55}{\sqrt{80}} < \mu < 18.85 + 1.96 \cdot \frac{5.55}{\sqrt{80}}$$

$$17.63 < \mu < 20.07$$

Of course, the interval from 17.63 to 20.07 contains the plant's true average daily emission of sulfur oxides or it does not, but we are 95 percent confident that it does. If we wanted to bet, 95 to 5 (or 19 to 1) would be fair odds that it does.

Had we asked for a 99 percent confidence interval in the preceding example, we would have obtained $17.25 < \mu < 20.45$, and it should be observed that this interval is wider than the 95 percent interval. This illustrates the important fact that "the surer we want to be, the less we have to be sure of." In other words, if we increase the degree of confidence, the confidence interval becomes wider and thus tells us less about the quantity we want to estimate.

10.3
THE ESTIMATION OF MEANS
(Small Samples)

In the preceding section we assumed that the sample is large enough, $n \geq 30$, to treat the sampling distribution of the mean as if it were a normal distribution, and that (when necessary) σ can be replaced by s in the formula for the standard error of the mean. To develop corresponding theory that applies also to small samples, we must now assume that the population we are sampling has roughly the shape of a normal distribution. We can then base our methods on the statistic

$$t = \frac{\bar{x} - \mu}{s/\sqrt{n}}$$

whose sampling distribution is called the *t* **distribution**. More specifically, it is called the **Student-*t* distribution** or **Student's *t* distribution**, as it was first investigated by W. S. Gosset, who published his writings under the pen name "Student." As is shown in Figure 10.2, the shape of this distribution is

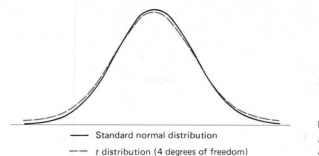

—— Standard normal distribution	FIGURE 10.2 *Standard*
—— *t* distribution (4 degrees of freedom)	*normal distribution and t distribution.*

very much like that of a normal distribution, and it is symmetrical with zero mean. The exact shape of the t distribution depends on the quantity $n - 1$, the sample size less one, called the **number of degrees of freedom.**[†]

For the standard normal distribution, we defined $z_{\alpha/2}$ in such a way that the area under the curve to its right equals $\alpha/2$, and, hence, the area under the curve between $-z_{\alpha/2}$ and $z_{\alpha/2}$ equals $1 - \alpha$. As is shown in Figure 10.3, the corresponding values for the t distribution are $-t_{\alpha/2}$ and $t_{\alpha/2}$. Since these values depend on $n - 1$, the number of degrees of freedom, they must be looked up in a special table, such as Table II at the end of this book. This table contains among others the values of $t_{0.025}$ and $t_{0.005}$ for 1 through 29 degrees of freedom, and it can be seen that $t_{0.025}$ and $t_{0.005}$ approach the corresponding values for the standard normal distribution as the number of degrees of freedom becomes large.

Since the t distribution, like the standard normal distribution, is symmetrical about its mean $\mu = 0$ (see Figure 10.3), we can now duplicate the argument on page 269 and thus arrive at the following **$(1 - \alpha)100$ percent small-sample confidence interval** for μ:

Small-sample confidence interval for μ

$$\bar{x} - t_{\alpha/2} \cdot \frac{s}{\sqrt{n}} < \mu < \bar{x} + t_{\alpha/2} \cdot \frac{s}{\sqrt{n}}$$

The only difference between this confidence interval formula and the large-sample formula (with s substituted for σ) is that $t_{\alpha/2}$ takes the place of $z_{\alpha/2}$.

[†] It is difficult to explain at this time why one should want to assign a special name to $n - 1$. However, we shall see later in this chapter that there are other applications of the t distribution, where the number of degrees of freedom is defined in a different way. The reason for the term "degrees of freedom" lies in the fact that if we know $n - 1$ of the deviations from the mean, then the nth is automatically determined (see Exercise 3.71). Since the sample standard deviation measures variation in terms of the squared deviations from the mean, we can thus say that this estimate of σ is based on $n - 1$ independent quantities or that we have $n - 1$ degrees of freedom.

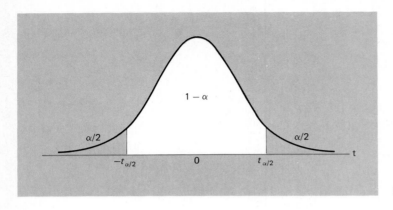

FIGURE 10.3 *t distribution.*

EXAMPLE To test the durability of a new paint for white centerlines, a highway department painted test strips across heavily traveled roads in eight different locations, and electronic counters showed that they deteriorated after being crossed by (to the nearest hundred) 142,600, 167,800, 136,500, 108,300, 126,400, 133,700, 162,000, and 149,400 cars. Construct a 95 percent confidence interval for the average number of crossings this paint can withstand before it deteriorates.

Solution The mean and the standard deviation of these values are $\bar{x} = 140,800$ and $s = 19,200$ (to the nearest hundred), and since $t_{0.025}$ for $8 - 1 = 7$ degrees of freedom equals 2.365, substitution into the formula yields

$$140,800 - 2.365 \cdot \frac{19,200}{\sqrt{8}} < \mu < 140,800 + 2.365 \cdot \frac{19,200}{\sqrt{8}}$$

$$124,700 < \mu < 156,900$$

for the 95 percent confidence interval.

The method which we used earlier to determine the maximum error we risk with $(1 - \alpha)100$ percent confidence when we use a sample mean to estimate the mean of a population can easily be adapted to small samples, provided that the population we are sampling has roughly the shape of a normal distribution. All we have to do is substitute s for σ and $t_{\alpha/2}$ for $z_{\alpha/2}$ in the formula for the maximum error E on page 266.

EXAMPLE While performing a certain task under simulated weightlessness, the pulse rate of twelve astronauts increased on the average by

27.33 beats per minute with a standard deviation of 4.28 beats per minute. If we use $\bar{x} = 27.33$ as an estimate of the true average increase of the pulse rate of astronauts performing the given task, what can we assert with 99 percent confidence about the maximum size of our error?

Solution   Substituting $s = 4.28$, $n = 12$, and $t_{0.05} = 3.106$ (the entry in Table II for 11 degrees of freedom) into this new formula for E, we get

$$ E = t_{\alpha/2} \cdot \frac{s}{\sqrt{n}} = 3.106 \cdot \frac{4.28}{\sqrt{12}} = 3.84 $$

Thus, if we use the mean $\bar{x} = 27.33$ beats per minute as an estimate of the true average increase of the pulse rate of astronauts performing the given task, we can be 99 percent confident that the error of this estimate is at most 3.84 beats per minute.

10.4
THE ESTIMATION OF MEANS
(A Bayesian Method) ⋆

In recent years there has been mounting interest in methods of inference in which parameters (for example, the population mean μ or the population standard deviation σ) are looked upon as random variables having **prior distributions** which reflect how a person feels about the different values that a parameter can take on. Such prior considerations are then combined with direct sample evidence to obtain **posterior distributions** of the parameters, on which subsequent inferences are based. Since the method used to combine the prior considerations with the direct sample evidence is based on a generalization of Bayes' theorem of Section 5.7, we refer to such inferences as **Bayesian.**

In this section we shall present a Bayesian method of estimating the mean of a population. As we said, our prior feelings about the possible values of μ are expressed in the form of a prior distribution, and like any distribution, this kind of distribution has a mean and a standard deviation. We shall designate these values μ_0 and σ_0 and call them the **prior mean** and the **prior standard deviation.**

If we are sampling a population having the mean μ (which we want to estimate) and the standard deviation σ, if the sample is large enough to apply the central limit theorem (or if the population is normal), and if the prior distribution of μ has roughly the shape of a normal distribution, it can be shown that the posterior distribution of μ is also a normal distribution with the mean

Posterior mean

$$\mu_1 = \frac{\dfrac{n}{\sigma^2} \cdot \bar{x} + \dfrac{1}{\sigma_0^2} \cdot \mu_0}{\dfrac{n}{\sigma^2} + \dfrac{1}{\sigma_0^2}}$$

and the standard deviation σ_1 given by the formula

Posterior standard deviation

$$\frac{1}{\sigma_1^2} = \frac{n}{\sigma^2} + \frac{1}{\sigma_0^2}$$

Since μ_1, the posterior mean, may be used as an estimate of the mean of the population, let us examine some of its most important features. We note first that μ_1 is a weighted mean of $\bar{x}$ and μ_0, and that the weights are $\dfrac{n}{\sigma^2}$ and $\dfrac{1}{\sigma_0^2}$, the reciprocals of the variances of the distribution of $\bar{x}$ and the prior distribution of μ. We see also that when no direct information is available and $n = 0$, the weight assigned $\bar{x}$ is 0, the formula reduces to $\mu_1 = \mu_0$, and the estimate is based entirely on the prior distribution. However, as more and more direct evidence becomes available (that is, as n becomes larger and larger), the weight shifts more and more toward the direct sample evidence, the sample mean $\bar{x}$. Finally, we see that when the prior feelings about the possible values of μ are vague (that is, when σ_0 is relatively large), the estimate will be based to a greater extent on $\bar{x}$. However, when there is a great deal of variability in the population which yields the direct sample evidence (that is, when σ is relatively large), the estimate will be based to a greater extent on μ_0.

EXAMPLE   An investor who is planning to open a new travel agency feels most strongly that he should net on the average $\mu_0 = \$3,600$ per month; also, the subjective prior distribution which he attaches to the various possible values of μ has roughly the shape of a normal distribution with the standard deviation $\sigma_0 = \$130$. If during nine months the operation of the travel agency nets \$3,810, \$3,690, \$3,350, \$3,400, \$3,320, \$3,250, \$3,430, \$3,600, and \$3,670, what is the posterior probability that the travel agency will net on the average between \$3,500 and \$3,600 per month?

Solution The mean and the standard deviation of the sample data are $\bar{x} = 3,502$ and $s = 195$. Using this sample standard deviation to estimate the unknown σ, and substituting $n = 9$, $\bar{x} = 3,502$, $s = 195$, $\mu_0 = 3,600$, and $\sigma_0 = 130$ into the formulas for the posterior mean and the posterior standard deviation, we get

$$\mu_1 = \frac{\dfrac{9}{195^2} \cdot 3,502 + \dfrac{1}{130^2} \cdot 3,600}{\dfrac{9}{195^2} + \dfrac{1}{130^2}} = \$3,522$$

and

$$\frac{1}{\sigma_1^2} = \frac{9}{195^2} + \frac{1}{130^2} \qquad \text{and} \qquad \sigma_1 = 58.1$$

Having found the mean and the standard deviation of the posterior distribution of μ pictured in Figure 10.4, we must now determine the area of the white region under the curve, the area under the standard normal distribution between $z = \dfrac{3,500 - 3,522}{58.1} = -0.38$ and $z = \dfrac{3,600 - 3,522}{58.1} = 1.34$. The entries corresponding to 0.38 and 1.34 in Table I are 0.1480 and 0.4099, so we get $0.1480 + 0.4099 = 0.5579$, or approximately 0.56, for the posterior probability that the travel agency will net on the average between \$3,500 and \$3,600 per month.

FIGURE 10.4 *Posterior distribution of μ.*

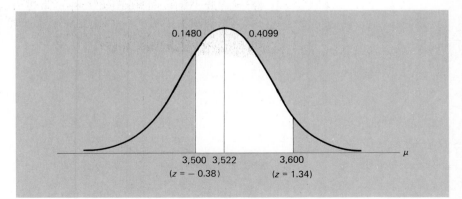

This brief introduction to **Bayesian inference** serves to bring out the following points: (1) In Bayesian statistics the parameter about which an inference is made is looked upon as a random variable having a distribution of its own, and (2) this kind of inference permits the use of direct as well as collateral information. To clarify the second point, let us add that in the travel-agency example the subjective prior distribution of the investor may have been based on a subjective evaluation of various factors (business conditions in general, for instance, and indirect information about other travel agencies), and this evaluation was combined with the figures which were actually observed for the nine months.

EXERCISES

10.1 To estimate the average time required for certain repairs, an efficiency expert timed a random sample of 40 automobile mechanics in the performance of this task, getting a mean of 13.62 minutes and a standard deviation of 1.45 minutes. With 95 percent confidence, what can he say about the maximum error in estimating the true average time required for the repairs as 13.62 minutes?

10.2 With reference to the preceding exercise, construct a 95 percent confidence interval for the true average time it takes an automobile mechanic to perform the repairs.

10.3 A study of the annual growth of certain cacti showed that 64 of them, selected at random in a desert region, grew on the average 52.8 mm with a standard deviation of 4.5 mm. Construct a 95 percent confidence interval for the true average annual growth of the given kind of cactus.

10.4 With reference to the preceding exercise, what can we assert with 99 percent confidence about the maximum size of the error, if we use $\bar{x} = 52.8$ mm as an estimate of the true average annual growth of the given kind of cactus?

10.5 A study conducted by an airline showed that 120 of its passengers disembarking at Kennedy airport, a random sample, had to wait on the average 9.45 minutes with a standard deviation of 1.84 minutes to get their luggage.
   (a) What can be said with 99 percent confidence about the maximum size of the error, if the airline uses $\bar{x} = 9.45$ minutes as an estimate of the true average time it takes its passengers to get their luggage when disembarking at Kennedy airport?
   (b) Construct a 95 percent confidence interval for the true average time it takes one of the airline's passengers to get his or her luggage when disembarking at Kennedy airport.

10.6 In an arithmetic achievement test, a random sample of 150 sixth graders from a very large school district has a mean score of 78.1 with a standard deviation of 9.5.
   (a) Construct a 99 percent confidence interval for the mean score which all the sixth graders in the school district would get if they took the test.
   (b) With 98 percent confidence, what can we say about the maximum error, if we use $\bar{x} = 78.1$ as an estimate of the true mean score which all the sixth graders in the school district would get if they took the test?

10.7 A sample survey conducted in a large city in 1981 showed that 200 families spent on the average $105.76 per week on food with a standard deviation of $10.17.
  (a) Construct a 98 percent confidence interval for the actual average weekly food expenditures of families in the population sampled.
  (b) What can we say with 90 percent confidence about the maximum error, if we use $\bar{x} = \$105.76$ as an estimate of the actual average weekly food expenditures of families in the population sampled?

10.8 A power company takes a random sample from its very extensive files and finds that the amounts owed on 180 delinquent accounts have a mean of $27.35 and a standard deviation of $6.31. If it uses $\bar{x} = \$27.35$ as an estimate of the average size of all its delinquent accounts, with what confidence can it assert that this estimate is off by at most $0.50?

10.9 A random sample of 60 cans of pear halves has a mean weight of 16.1 ounces and a standard deviation of 0.3 ounce. If $\bar{x} = 16.1$ ounces is used as an estimate of the mean weight of all the cans of pear halves in the large lot from which the sample came, with what confidence can we assert that the error of this estimate is at most 0.1 ounce?

10.10 If a sample constitutes an appreciable portion of a finite population (say, 5 percent or more), the various formulas of Section 10.2 must be modified by applying the finite population correction factor. For instance, the formula for E on page 266 becomes

$$E = z_{\alpha/2} \cdot \frac{\sigma}{\sqrt{n}} \cdot \sqrt{\frac{N - n}{N - 1}}$$

  (a) A sample of 50 scores on the admission test for a law school is randomly drawn from the scores of the 400 persons who applied to the law school in 1981. If the sample mean and standard deviation are $\bar{x} = 563$ and $s = 87$, what can we assert with 95 percent confidence about the maximum size of the error, if $\bar{x} = 563$ is used as an estimate of the average score of all the applicants?
  (b) A random sample of 40 drums of a chemical, drawn from among 200 such drums whose weights are assumed to have the standard deviation $\sigma = 12.2$ pounds, has a mean weight of 240.8 pounds. If we estimate the mean weight of all 200 drums as 240.8 pounds, what can we assert with 99 percent confidence about the maximum error?
  (c) Rework part (b) of Exercise 10.6, given that there are 900 sixth graders in the school district.

10.11 Use the finite population correction factor, $\sqrt{\dfrac{N - n}{N - 1}}$, to modify the large-sample confidence interval formula on page 269, and thus make it applicable to problems in which a sample constitutes a substantial portion of a large finite population.
  (a) With reference to part (a) of the preceding exercise, construct a 99 percent confidence interval for the average score of all the persons who applied to the law school in 1981.
  (b) Rework part (a) of Exercise 10.6, given that there are 900 sixth graders in the school district.

★10.12 A stratified random sample of size $n = 330$ is allocated proportionally to the four strata of a finite population for which $N_1 = 2,000$, $N_2 = 6,000$, $N_3 = 10,000$, $N_4 = 4,000$, $\sigma_1 = 20$, $\sigma_2 = 25$, $\sigma_3 = 15$, and $\sigma_4 = 30$ cm.

    (a) If the mean of this sample is used to estimate the mean of the population, what can we assert with 95 percent confidence about the maximum error?

    (b) If the mean of the sample is 112.54 cm, construct a 99 percent confidence interval for the mean of the population.

(*Hint*: Use the standard error formula of Exercise 9.34 on page 257.)

10.13 In a study of television viewing habits, it is desired to estimate the average number of hours that teenagers spend watching per week. If it is reasonable to assume that $\sigma = 3.2$ hours, how large a sample is needed so that it will be possible to assert with 95 percent confidence that the sample mean is off by at most 20 minutes?

10.14 It is desired to estimate the mean number of hours of continuous use until a certain kind of computer will first require repairs. If it can be assumed that $\sigma = 48$ hours, how large a sample is needed so that it can be asserted with probability 0.99 that the sample mean will be off by at most 10 hours?

10.15 Suppose that we want to estimate the mean score of persons over 50 on a current events test, and we want to be able to assert with 95 percent confidence that the error of our estimate, the mean of a random sample, is at most 2.5. How large a sample will we need, if it can be assumed that $\sigma = 7.3$?

10.16 A major truck stop has kept extensive records on various transactions with its customers. If a random sample of 18 of these records show average sales of 63.84 gallons of diesel fuel with a standard deviation of 2.75 gallons, construct a 99 percent confidence interval for the mean of the population sampled.

10.17 With reference to the preceding exercise, what can we assert with 95 percent confidence about the maximum error if we use $\bar{x} = 63.84$ gallons as an estimate of the truck stop's average sales of diesel fuel?

10.18 Nine bearings made by a certain process have a mean diameter of 0.505 cm and a standard deviation of 0.004 cm. With 95 percent confidence, what can be said about the maximum error if $\bar{x} = 0.505$ cm is used as an estimate of the average diameter of bearings made by the process?

10.19 With reference to the preceding exercise, construct a 98 percent confidence interval for the average diameter of bearings made by the process.

10.20 In an air pollution study, an experiment station obtained a mean of 2.26 micrograms of suspended benzene-soluble organic matter per cubic meter with a standard deviation of 0.56 from a random sample of $n = 8$ different specimens.

    (a) If the mean of this sample is used to estimate the corresponding true mean, what can the station assert with 99 percent confidence about the maximum error?

    (b) Construct a 98 percent confidence interval for the true mean of the population sampled.

10.21 In six test runs it took 13, 14, 12, 16, 12, and 11 minutes to assemble a certain mechanical device. Construct a 95 percent confidence interval for the actual mean time it takes to assemble the device.

10.22 Five containers of a commercial solvent (randomly selected from a large production lot) weigh 19.5, 19.3, 20.0, 19.0, and 19.7 pounds.

    (a) If the mean of this sample is used to estimate the mean of the population sampled, what can we assert with 95 percent confidence about the maximum error?

    (b) Construct a 99 percent confidence interval for the mean weight of all the containers of the solvent from which the sample came.

10.23 A dentist finds in a routine check that six prison inmates require 2, 3, 6, 0, 4, and 3 fillings.

    (a) If he uses the mean of this sample to estimate the true mean of the population sampled, what can he say with 99 percent confidence about the maximum error?

    (b) Construct a 90 percent confidence interval for the average number of fillings required by the inmates of this very large prison.

★10.24 An actuary feels that the prior distribution of the average annual losses for a certain kind of liability coverage is a normal distribution with the mean $\mu_0 = \$103.50$ and the standard deviation $\sigma_0 = \$4.30$. She also knows that for any one policy the losses vary from year to year with the standard deviation $\sigma = \$20.85$. If the losses of a particular policy like this average $185.26 over a period of ten years, find the posterior mean as a Bayesian estimate of its true average annual losses.

★10.25 With reference to the preceding exercise, what is the posterior probability that the policy's true average annual losses are between $120.00 and $130.00?

★10.26 A college professor is making up a final examination in history which is to be given to a large group of students. His feelings about the average grade they should get is expressed subjectively by a normal distribution which has the mean $\mu_0 = 65.2$ and the standard deviation $\sigma_0 = 1.5$.

    (a) What prior probability does the professor assign to the event that the average grade of all the students will lie on the interval from 63.0 to 68.0?

    (b) If the examination is tried on a random sample of 40 of the students and their grades have the mean $\bar{x} = 72.9$ and the standard deviation $s = 7.4$, calculate the posterior mean as a point estimate of the average grade which all the students will get on the test.

    (c) Use the information given in part (b) to calculate the posterior probability which the professor should assign to the event that the average grade of all the students will lie on the interval from 63.0 to 68.0.

10.5
TESTS OF HYPOTHESES

In each of the problems mentioned in Section 10.1, somebody was interested in determining the true value of a quantity, so they were all problems of estimation. They would have concerned **tests of hypotheses,** however, if the psychologist had wanted to decide whether the average time it takes an adult person to react to the stimulus is really 0.44 second, if the engineer had wanted to determine whether the variability of the strength of the new alloy is less than that of a known substance, if the biologist had wanted to check another biologist's claim that 2.3 percent of the given insects are born physically defective, if the retailer had wanted to find out whether the actual average family income of all families living within a mile of the proposed new store is $22,500, . . . , and if a D.P.S. engineer had wanted to investigate the claim that 14.5 percent of all one-car accidents are due to driver fatigue. Now it must be decided in each case whether to accept or reject a hypothesis, namely, whether to accept or reject an assertion or claim about the parameter of a population.

10.6
TWO KINDS OF ERRORS

At this point it may seem to make little difference whether we state the problems as in Section 10.1 or as in Section 10.5, but it will soon become apparent that a number of considerations arise in connection with tests of hypotheses that are not present in problems of estimation.

To illustrate the nature of the problems we face when testing a statistical hypothesis, suppose that the members of an airport's planning commission are considering the possibility of redesigning the parking facilities. In connection with this, they first want to check the claim (made by the operator of the parking facilities) that on the average cars remain in the short-term parking area for 42.5 minutes. So, they instruct a member of their staff to take a random sample of 50 ticket stubs showing time of arrival and time of departure, with the intention of accepting the claim if the mean of the sample falls anywhere from 40.5 to 44.5 minutes. Otherwise, they will reject the claim, and in either case they will take whatever action is thus called for in their plans.

This provides a clear-cut criterion for accepting or rejecting the claim, but unfortunately it is not infallible. Since the decision is based on a sample, there is the possibility that the sample mean may be less than 40.5 minutes or greater than 44.5 minutes even though the true mean is $\mu = 42.5$ minutes, and there is also the possibility that the sample mean may fall on the interval from 40.5 minutes to 44.5 minutes even though the true mean is, say, $\mu = 45.5$ minutes. Thus, before adopting the criterion, it would seem wise to investigate the chances that it may lead to a wrong decision.

Assuming that it is known from similar studies that $\sigma = 7.6$ minutes for this kind of data, let us first investigate the possibility that the sample mean may be less than 40.5 minutes or greater than 44.5 minutes even though the true mean is $\mu = 42.5$ minutes. The probability that this will happen purely due to chance is given by the sum of the areas of the white regions of Figure 10.5, and it can easily be determined by approximating the sampling distribution of the mean with a normal distribution. Assuming that the

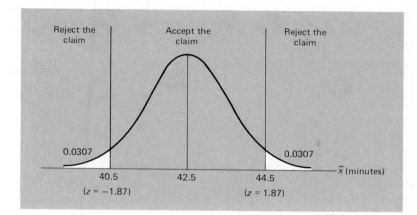

FIGURE 10.5 *Test criterion.*

population sampled is large enough to be treated as infinite, we have $\sigma_{\bar{x}} = \dfrac{\sigma}{\sqrt{n}} = \dfrac{7.6}{\sqrt{50}} = 1.07$, and the dividing lines of the criterion, in standard units, are

$$z = \frac{40.5 - 42.5}{1.07} = -1.87 \quad \text{and} \quad z = \frac{44.5 - 42.5}{1.07} = 1.87$$

It follows from Table I that the area in each "tail" of the sampling distribution of Figure 10.5 is $0.5000 - 0.4693 = 0.0307$, and hence that the probability of getting a value in either tail of the sampling distribution (and erroneously rejecting the hypothesis $\mu = 42.5$ minutes) is $0.0307 + 0.0307 = 0.0614$, or approximately 0.06. Whether this is an acceptable risk is for the members of the planning commission to decide; it would have to depend on the consequences of their making such an error.

Let us now consider the other possibility, where the test fails to detect that μ is not equal to 42.5 minutes. Suppose again, for the sake of argument, that the true mean is $\mu = 45.5$ minutes, so that the probability of getting a sample mean on the interval from 40.5 minutes to 44.5 minutes (and, hence, erroneously accepting the claim that $\mu = 42.5$ minutes) is given by the area of the white region of Figure 10.6. The mean of the sampling distribution is now 45.5, its standard deviation is as before $\sigma_{\bar{x}} = \dfrac{7.6}{\sqrt{50}} = 1.07$, and the dividing lines of the criterion, in standard units, are

$$z = \frac{40.5 - 45.5}{1.07} = -4.67 \quad \text{and} \quad z = \frac{44.5 - 45.5}{1.07} = -0.93$$

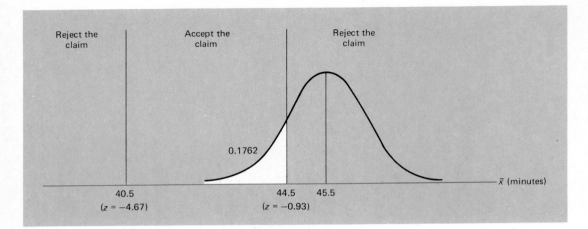

FIGURE 10.6 *Test criterion.*

Since the area under the curve to the left of $z = -4.67$ is negligible, it follows from Table I that the area of the white region of Figure 10.6 is $0.5000 - 0.3238 = 0.1762$, or approximately 0.18. Again, it is up to the members of the planning commission to decide whether this represents an acceptable risk.

The situation described in this example is typical of testing a statistical hypothesis, and it may be summarized in the following table, where we refer to the hypothesis being tested as hypothesis H:

	Accept H	*Reject H*
H is true	Correct decision	Type I error
H is false	Type II error	Correct decision

If hypothesis H is true and accepted or false and rejected, the decision is in either case correct. If hypothesis H is true but rejected, it is rejected in error, and if hypothesis H is false but accepted, it is accepted in error. The first of these errors is called a **Type I error** and the probability of committing it is designated by the Greek letter α (*alpha*); the second is called a **Type II error** and the probability of committing it is designated by the Greek letter β (*beta*). Thus, in our example we showed that for the given test criterion $\alpha = 0.06$, and $\beta = 0.18$ when $\mu = 45.5$ minutes.

The scheme outlined above is reminiscent of what we did in Section 6.2. Analogous to the decision which the furniture manufacturer had to make in the example on page 156, we now have to decide whether to accept or reject hypothesis H. It is difficult to carry this analogy much further, though, for in actual practice we can seldom associate cash payoffs with the various possibilities, as we did in that example.

In calculating the probability of a Type II error in our example, we arbitrarily chose the alternative value $\mu = 45.5$ minutes. However, in this problem, as in most others, there are infinitely many other alternatives, and for each one of them there is a positive probability β of erroneously accepting the hypothesis H. So, in practice we choose some key alternative values and calculate the corresponding probabilities β of committing a Type II error, or we sidestep the issue by proceeding as in Section 10.8. If we do calculate β for various alternative values of μ and plot these probabilities as in Figure 10.7, we obtain a curve which is called the **operating characteristic curve,** or simply the **OC-curve,** of the test criterion. Since the probability of a Type II error is the probability of accepting the hypothesis H when it is false, we "completed the picture" in Figure 10.7 by plotting at $\mu = 42.5$ minutes the probability of accepting the hypothesis H when it is true, namely, $1 - \alpha = 1 - 0.06 = 0.94$. Accordingly, the vertical scale is labeled "Probability of accepting H."

Examination of the curve of Figure 10.7 shows that the probability of accepting the hypothesis H is greatest when it is true, and that it is still high for small departures from $\mu = 42.5$ minutes. However, for larger and larger departures from $\mu = 42.5$ minutes in either direction, the probabilities of

FIGURE 10.7 *Operating characteristic curve.*

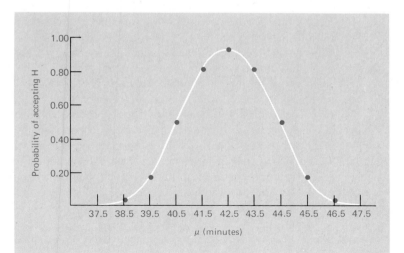

failing to detect them (and erroneously accepting the hypothesis *H*) become smaller and smaller. In Exercise 10.32 the reader will be asked to verify some of the probabilities plotted in Figure 10.7.

If we had plotted the probabilities of rejecting *H* instead of those of accepting *H*, we would have obtained the graph of the **power function** of the test criterion rather than its operating characteristic curve. The concept of an OC-curve is used more widely in applications, especially in industrial applications, while the concept of a power function is used more widely in matters that are of theoretical interest. A detailed study of operating characteristic curves and power functions would go considerably beyond the scope of this text, and the purpose of our example was mainly to show how statistical methods can be used to measure and control the risks to which one is exposed when testing hypotheses. Of course, the methods we have discussed here are not limited to the particular problem concerning the average length of time that cars are parked in a given area—hypothesis *H* could have been the hypothesis that the average age of divorced women at the time of divorce is 35.6 years, the hypothesis that an antibiotic is 83 percent effective, the hypothesis that a computer-assisted method of instruction will on the average raise students' scores on a standard achievement test by 8.4 points, and so forth.

10.7
NULL HYPOTHESES

In the example of the preceding section, we had less trouble with Type I errors than with Type II errors, because we formulated the hypothesis *H* as a **simple hypothesis** about the parameter μ; that is, we formulated the hypothesis *H* so that μ took on a single value and the probability of a Type I error could be calculated. Had we formulated instead a **composite hypothesis** about the parameter μ, say, $\mu \neq 42.5$ minutes, $\mu < 42.5$ minutes, or $\mu > 42.5$ minutes, where in each case μ can take on more than one possible value, we could not have calculated the probability of a Type I error without specifying how much μ differs from, is less than, or is greater than 42.5 minutes.[†]

To be able to calculate, or specify, the probability of a Type I error, it is customary to formulate hypotheses to be tested as simple hypotheses. We do this even when it requires that we hypothesize the opposite of what we hope to prove. For instance, if we want to show that one method of teaching computer programming is more effective than another, we hypothesize that the two methods are equally effective. Similarly, if we want to

---

[†] Note that we are applying the terms "simple hypothesis" and "composite hypothesis" to hypotheses about specific parameters; some statisticians use the term "simple hypothesis" only when the hypothesis completely specifies the population.

show that one method of irrigating the soil is more expensive than another, we hypothesize that the two methods are equally expensive; and if we want to show that a new copper-bearing steel has a higher yield strength than ordinary steel, we hypothesize that the two yield strengths are the same. Since we hypothesize that there is no difference in the effectiveness of the two teaching methods, no difference in the cost of the two methods of irrigation, and no difference in the yield strength of the two kinds of steel, we call hypotheses like these **null hypotheses** and denote them by H_0. Nowadays, the term "null hypothesis" is used for any hypothesis set up primarily to see whether it can be rejected, and the idea of setting up a null hypothesis is common even in nonstatistical thinking. It is precisely what we do in criminal proceedings, where an accused is presumed to be innocent until his guilt has been established beyond a reasonable doubt. The presumption of innocence is a null hypothesis.

10.8
SIGNIFICANCE TESTS

In the example of Section 10.6, the probabilities of Type II errors had to be calculated for various alternative values of μ. The example which follows shows how Type II errors can sometimes be sidestepped altogether.

Experience shows that in a given city licensed drivers average 1.4 traffic tickets per year with a standard deviation of 0.6. To confirm her suspicion that licensed drivers over 65 average more than 1.4 traffic tickets per year, a social scientist checks the 1981 records of 40 randomly selected licensed drivers over 65 in the given city, and bases her decision on the following criterion:

> Reject the null hypothesis $\mu = 1.4$ (and accept the alternative hypothesis $\mu > 1.4$) if the 40 licensed drivers over 65 average more than 1.6 traffic tickets per year; otherwise, reserve judgment (perhaps, pending further checks).

If judgment is reserved as in this criterion, there is no possibility of a Type II error; no matter what happens, the null hypothesis is never accepted.

The procedure we have just outlined is referred to as a **significance test.** If the difference between what we expect under the null hypothesis and what we observe in a sample is too large to be reasonably attributed to chance, we reject the null hypothesis. If the difference between what we expect and what we observe is so small that it may well be attributed to chance, we say that the results are **not statistically significant.** We then accept the null hypothesis or reserve judgment, depending on whether a definite decision one way or the other is required.

With reference to the criterion above, the probability that the sample mean will exceed 1.6 when $\mu = 1.4$ is given by the area of the white region of Figure 10.8, and it equals 0.0174, as the reader will be asked to verify in Exercise 10.33. Since this probability is very small, it would seem reasonable to claim that the difference between $\mu = 1.4$ and a value of $\bar{x}$ greater than 1.6 is too large to be attributed to chance, and hence that the null hypothesis must be wrong. If the sample mean does not exceed 1.6, the social scientist's suspicion is not confirmed. Observe that we did not say that her suspicion is wrong, or unjustified, when the sample mean does not exceed 1.6—we merely said that it is not confirmed.

Returning to the airport parking example of Section 10.6, we could convert the criterion on page 280 into that of a significance test by writing

> **Reject the hypothesis μ = 42.5 minutes (and accept the alternative μ ≠ 42.5 minutes) if the mean of the 50 sample values is less than 40.5 minutes or greater than 44.5 minutes; reserve judgment if the mean falls anywhere from 40.5 to 44.5 minutes.**

So far as the rejection of the null hypothesis is concerned, the criterion has remained unchanged and the probability of a Type I error is still 0.06. However, so far as its acceptance is concerned, we are now playing it safe by reserving judgment.

Reserving judgment in a significance test is similar to what happens in court proceedings where the prosecution does not have sufficient evidence to get a conviction, but where it would be going too far to say that the

FIGURE 10.8 *Test criterion.*

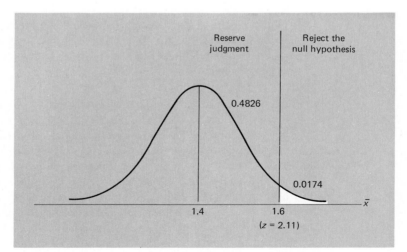

defendant definitely did not commit the crime. In general, whether one can afford the luxury of reserving judgment in any given situation depends entirely on the nature of the situation. If a decision must be reached one way or the other, there is no way of avoiding the risk of committing a Type II error.

Since the general problem of testing hypotheses and constructing statistical decision criteria may seem confusing, at least to the beginner, it will help to proceed systematically as outlined in the following five steps:

1. **We formulate a simple null hypothesis and an appropriate alternative hypothesis which is to be accepted when the null hypothesis must be rejected.**[†]

In the airport parking example the null hypothesis is $\mu = 42.5$ minutes, and the alternative hypothesis is $\mu \neq 42.5$ minutes (because the planning commission wants protection against the possibilities that 42.5 minutes may be too high or too low). We refer to this kind of alternative as a **two-sided alternative.** In the traffic ticket example the null hypothesis is $\mu = 1.4$, and the alternative hypothesis is $\mu > 1.4$ (because the social scientist hopes to be able to show that licensed drivers over 65 average more than 1.4 tickets per year). This is called a **one-sided alternative.** We can also write a one-sided alternative with the inequality going the other way. For instance, if we hope to be able to show that the average time required to do a certain job is less than 15 minutes, we would test the null hypothesis $\mu = 15$ minutes against the alternative hypothesis $\mu < 15$ minutes.

As in the three examples of the preceding paragraph, alternative hypotheses usually specify that the population mean (or whatever other parameter may be of concern) is not equal to, greater than, or less than the value assumed under the null hypothesis. For any given problem, the choice of an appropriate alternative depends mostly on what we hope to be able to show, or better, perhaps, where we want to put the burden of proof.

EXAMPLE An appliance manufacturer is considering the purchase of a new machine for stamping out sheet metal parts. If μ_0 is the average number of good parts stamped out per hour by his old machine and μ is the corresponding average for the new machine, the manufacturer wants to test the null hypothesis $\mu = \mu_0$ against a suitable alternative. What should the alternative be if

(a) he does not want to buy the new machine unless it is more productive than the old one;

(b) he wants to buy the new machine (which has some other nice features) unless it is less productive than the old one.

[†] See also discussion on page 288.

Solution **(a)** The manufacturer should use the alternative hypothesis $\mu > \mu_0$ and purchase the new machine only if the null hypothesis can be rejected; **(b)** the manufacturer should use the alternative hypothesis $\mu < \mu_0$ and purchase the new machine unless the null hypothesis is rejected.

Having formulated a null hypothesis and an alternative hypothesis, we then proceed with the following step:

2. **We specify the probability of a Type I error; if possible, desired, or necessary, we may also specify the probabilities of Type II errors for particular alternatives.**

The probability of a Type I error is also called the **level of significance,** and it is usually set at $\alpha = 0.05$ or $\alpha = 0.01$. Testing a null hypothesis at the level of significance $\alpha = 0.05$ simply means that we are fixing the probability of rejecting the null hypothesis even though it is true at 0.05. The decision to use $\alpha = 0.05$, $\alpha = 0.01$, or some other value, depends mostly on the consequences of committing a Type I error in the given situation. Observe, however, that we cannot make the probability of a Type I error too small, because this will have the tendency to make the probabilities of serious Type II errors too large, or in a significance test make it virtually impossible to get significant results. In practice, the choice of α depends on the risks one can take; in the exercises in this book, the level of significance will always be specified.

Step 2 cannot be performed unless the null hypothesis is a simple hypothesis, but this is not as restrictive as it may seem. To illustrate, let us investigate briefly what might be done if the social scientist of our example wants to allow for the possibility that licensed drivers over 65 in the given city may average fewer than 1.4 traffic tickets per year, and hence test the null hypothesis $\mu \leq 1.4$ against the alternative hypothesis $\mu > 1.4$. Now, the probability of a Type I error cannot be calculated since we do not know which value of $\mu \leq 1.4$ to use, but observe that if μ is less than 1.4, the normal curve of Figure 10.8 is shifted to the left, and the area under the curve to the right of 1.6 becomes less than 0.0174. Thus, if the null hypothesis is $\mu \leq 1.4$, we can say that the probability of a Type I error is at most 0.0174, and we write $\alpha \leq 0.0174$. In general, if the null hypothesis is of the form $\mu \leq \mu_0$ or $\mu \geq \mu_0$, we can specify only the maximum probability of a Type I error, and by performing the test as if the null hypothesis were $\mu = \mu_0$, we protect ourselves against the worst possibility (see also Exercise 10.80 on page 308).

After the null hypothesis, the alternative hypothesis, and the level of significance have been specified, the remaining steps are

3. **Based on the sampling distribution of an appropriate statistic, we construct a criterion for testing the null hypothesis against the given alternative.**

4. We calculate from the data the value of the statistic on which the decision is to be based.
5. We decide whether to reject the null hypothesis, whether to accept it, or whether to reserve judgment.

In the airport parking example we studied the criterion using the normal curve approximation to the sampling distribution of the mean; in general, step 3 depends not only on the statistic on which we want to base the decision and on its sampling distribution, but also on the alternative hypothesis we happen to choose. In the airport parking example we used a **two-sided criterion (two-sided test** or **two-tailed test)** with the two-sided alternative $\mu \neq 42.5$ minutes, rejecting the null hypothesis for small or for large values of the sample mean; in the traffic ticket example we used a **one-sided criterion (one-sided test** or **one-tailed test)** with the one-sided alternative $\mu > 1.4$, rejecting the null hypothesis only for large values of the sample mean. In general, a test is called two-sided or two-tailed if the null hypothesis is rejected for values of the **test statistic** falling into either tail of its sampling distribution, and it is called one-sided or one-tailed if the null hypothesis is rejected only for values of the test statistic falling into one specified tail of its sampling distribution.

As part of step 3 we must also specify whether the alternative to rejecting the null hypothesis is to accept it or to reserve judgment. This, as we have said, depends on whether we must make a decision one way or the other, or whether the circumstances permit that we delay a decision pending further study. In exercises and examples, the phrase "whether or not" will sometimes be used to indicate that a decision must be reached one way or the other.

In connection with step 5, let us point out that we often accept null hypotheses with the tacit hope that we are not exposed to overly high risks of committing serious Type II errors. Of course, if it is necessary we can calculate enough probabilities of Type II errors to get an overall picture from the operating characteristic curve of the test criterion.

Before we discuss various special tests about means in the remainder of this chapter, let us point out that the concepts we have introduced here apply equally well to hypotheses concerning proportions, standard deviations, the randomness of samples, relationships among several variables, and so on.

EXERCISES   10.27 Suppose that a psychological testing service is asked to check whether an executive is emotionally fit to assume the presidency of a large corporation. What type of error would it commit if it erroneously rejects the null hypothesis that the executive is fit for the job? What type of error would it commit if it erroneously accepts the null hypothesis that the executive is fit for the job?

10.28 Suppose we want to test the hypothesis that an antipollution device for cars is effective. Explain under what conditions we would commit a Type I error and under what conditions we would commit a Type II error.

10.29 Whether an error is a Type I error or a Type II error depends on how we formulate the hypothesis we want to test. To illustrate this, rephrase the hypothesis of Exercise 10.28 so that the Type I error becomes a Type II error, and vice versa.

10.30 A professor of education is concerned with the effectiveness of a method of computer-assisted instruction.
    (a) What hypothesis is she testing if she would commit a Type I error by erroneously concluding that the method is effective?
    (b) What hypothesis is she testing if she would commit a Type II error by erroneously concluding that the method is effective?

10.31 Suppose that for a given population with $\sigma = 8.4$ inches we want to test the null hypothesis $\mu = 80.0$ inches against the alternative hypothesis $\mu < 80.0$ inches on the basis of a random sample of size $n = 100$.
    (a) If the null hypothesis is rejected when $\bar{x} < 78.0$ inches and otherwise it is accepted, what is the probability of a Type I error?
    (b) If the null hypothesis is $\mu \geq 80.0$ inches instead of $\mu = 80.0$ inches and the criterion is the same as in part (a), what can we say about the probability of a Type I error?

10.32 With reference to the operating characteristic curve of Figure 10.7, verify that the probabilities of Type II errors are
    (a) 0.82 when $\mu = 41.5$ minutes or $\mu = 43.5$ minutes;
    (b) 0.50 when $\mu = 40.5$ minutes or $\mu = 44.5$ minutes;
    (c) 0.18 when $\mu = 39.5$ minutes or $\mu = 45.5$ minutes;
    (d) 0.03 when $\mu = 38.5$ minutes or $\mu = 46.5$ minutes.

10.33 Verify for the traffic ticket example on page 285 that the probability of a Type I error is 0.0174.

10.34 Suppose that in the airport parking example on page 280 the criterion is changed so that the null hypothesis $\mu = 42.5$ minutes is rejected if the sample mean is less than 41.0 minutes or greater than 44.0 minutes; otherwise, the null hypothesis is accepted.
    (a) What is the probability of a Type I error?
    (b) Find the probabilities of Type II errors when $\mu = 38.5, 39.5, 40.5, 41.5, 43.5, 44.5, 45.5,$ or 46.5 minutes, and plot the operating characteristic curve.

10.35 Suppose that in the airport parking example on page 280 the sample size is increased from 50 to 60, while the criterion remains as stated on page 280.
    (a) What is the probability of a Type I error?
    (b) Find the probabilities of Type II errors when $\mu = 38.5, 39.5, 40.5, 41.5, 43.5, 44.5, 45.5,$ or 46.5 minutes, and plot the operating characteristic curve.

10.36 To reduce the probability of a Type I error in the airport parking example on page 280, the criterion is modified so that the null hypothesis $\mu = 42.5$ minutes is rejected if the sample mean is less than 40.0 minutes or greater than 45.0 minutes; otherwise, the null hypothesis is accepted.
    (a) Show that this reduces the probability of a Type I error from 0.06 to 0.02.

(b) Show that for $\mu = 45.5$ the probability of a Type II error is increased from 0.18 to 0.32.

(c) Show that for $\mu = 46.5$ the probability of a Type II error is increased from 0.03 to 0.08.

10.37 The average drying time of a manufacturer's paint is 20 minutes. Investigating the effectiveness of a modification in the chemical composition of his paint, the manufacturer wants to test the null hypothesis $\mu = 20$ minutes against a suitable alternative, where μ is the average drying time of the modified paint.

(a) What alternative hypothesis should the manufacturer use if he does not want to make the modification in the chemical composition of the paint unless it is definitely superior with respect to drying time?

(b) What alternative hypothesis should the manufacturer use if the new process is actually cheaper and he wants to make the modification unless it increases the drying time of the paint?

10.38 With reference to the preceding exercise, what simple null hypothesis and what alternative hypothesis should the manufacturer use if he does not want to make the modification unless it decreases the drying time by at least five minutes?

10.39 A city police department is considering replacing the tires on its cars with radial tires. If μ_1 is the average number of miles the old tires last and μ_2 is the average number of miles the new tires will last, the null hypothesis to be tested is $\mu_1 = \mu_2$.

(a) What alternative hypothesis should the department use if it does not want to buy the radial tires unless they are definitely proved to give better mileage? In other words, the burden of proof is put on the radial tires and the old tires are to be kept unless the null hypothesis can be rejected.

(b) What alternative hypothesis should the department use if it is anxious to get the new tires (which have some other good features) unless they actually give poorer mileage than the old tires? Note that now the burden of proof is on the old tires, which will be kept only if the null hypothesis can be rejected.

(c) What alternative hypothesis should the department use so that the rejection of the null hypothesis can lead either to keeping the old tires or to buying the new ones?

10.9
TESTS CONCERNING MEANS

Having used tests concerning means to illustrate the basic principles of hypothesis testing, let us now see how we proceed in actual practice. Suppose, for instance, that an oceanographer wants to test, on the basis of the mean of a random sample of size $n = 35$ and at the 0.05 level of significance, whether the average depth of the ocean in a certain area is 72.4 fathoms, as has been recorded. From information gathered in similar studies, she can expect that the variability of her measurements is given by $\sigma = 2.1$ fathoms.

Following the outline of the preceding section, the oceanographer begins with steps 1 and 2 by writing

1. *Null hypothesis*: $\mu = 72.4$ fathoms
   Alternative hypothesis: $\mu \neq 72.4$ fathoms
2. *Level of significance*: $\alpha = 0.05$

The alternative hypothesis is two-sided because the oceanographer will want to reject the null hypothesis if her sample mean is significantly less than or significantly greater than 72.4 fathoms.

Next, in step 3 we shall depart somewhat from the procedure used in our earlier examples. In both the airport parking example and the traffic ticket example, we stated the test criterion in terms of values of $\bar{x}$; now we shall base it on the statistic

Statistic for test concerning mean

$$z = \frac{\bar{x} - \mu_0}{\sigma / \sqrt{n}}$$

where μ_0 is the value of the mean assumed under the null hypothesis. The reason for working with standard units, or z-values, is that it enables us to formulate criteria which are applicable to a great variety of problems, not just one.

If we approximate the sampling distribution of the mean, as before, with a normal distribution, we can now use the test criteria shown in Figure 10.9; depending on the choice of the alternative hypothesis, the dividing lines, or **critical values,** of the criteria are $-z_\alpha$ or z_α for the one-sided alternatives, and they are $-z_{\alpha/2}$ and $z_{\alpha/2}$ for the two-sided alternative. As before, z_α and $z_{\alpha/2}$ are such that the area to their right under the standard normal curve is α and $\alpha/2$. Symbolically, we can formulate these criteria as follows:

Alternative hypothesis	Reject the null hypothesis if	Accept the null hypothesis or reserve judgment if
$\mu < \mu_0$	$z < -z_\alpha$	$z \geq -z_\alpha$
$\mu > \mu_0$	$z > z_\alpha$	$z \leq z_\alpha$
$\mu \neq \mu_0$	$or \begin{array}{l} z < -z_{\alpha/2} \\ z > z_{\alpha/2} \end{array}$	$-z_{\alpha/2} \leq z \leq z_{\alpha/2}$

If $\alpha = 0.05$, the dividing lines of the criteria are -1.645 or 1.645 for the one-sided alternatives, and -1.96 and 1.96 for the two-sided alternative; if $\alpha = 0.01$, the dividing lines of the criteria are -2.33 or 2.33 for the one-

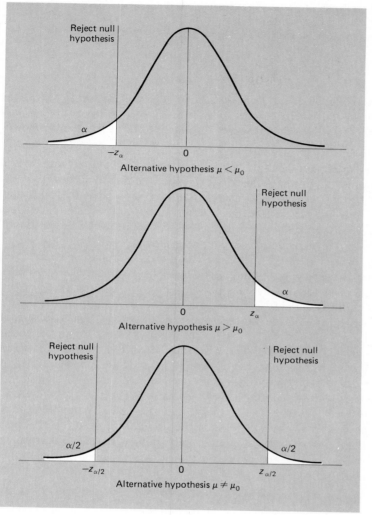

Reject null
hypothesis

α

$-z_\alpha$ 0

Alternative hypothesis $\mu < \mu_0$

Reject null
hypothesis

α

0 z_α

Alternative hypothesis $\mu > \mu_0$

Reject null
hypothesis

Reject null
hypothesis

$\alpha/2$ $\alpha/2$

$-z_{\alpha/2}$ 0 $z_{\alpha/2}$

Alternative hypothesis $\mu \neq \mu_0$

FIGURE 10.9 *Test criteria.*

sided alternatives, and -2.575 and 2.575 for the two-sided alternative. All these values come directly from Table I.

Returning now to the oceanographer who is concerned with the average depth of the ocean in a certain area, suppose that the mean of her 35 measurements is 73.2 fathoms. So, she continues by writing

3. *Criterion*: Reject the null hypothesis if $z < -1.96$ or $z > 1.96$, where

$$z = \frac{\bar{x} - \mu_0}{\sigma/\sqrt{n}}$$

and otherwise accept it.

4. *Calculations*:

$$z = \frac{73.2 - 72.4}{2.1/\sqrt{35}} = 2.25$$

5. *Decision*: Since $z = 2.25$ exceeds 1.96, the null hypothesis must be rejected; to put it another way, the difference between $\bar{x} = 73.2$ fathoms and $\mu = 72.4$ fathoms is too large to be attributed to chance.

If the oceanographer had used the 0.01 level of significance in this example, she would not have been able to reject the null hypothesis because $z = 2.25$ falls between -2.575 and 2.575. This illustrates the important point that the level of significance should always be specified before a significance test is actually performed. This will spare us the temptation of later choosing a level of significance which happens to suit our purpose.

In problems like this, some research workers accompany the calculated values of z with corresponding **tail probabilities,** or *p*-**values,** namely, with the probabilities of getting a difference between $\bar{x}$ and μ_0 greater than or equal to that actually observed. For instance, in the oceanographer example, the tail probability is given by the total area under the standard normal curve to the left of $z = -2.25$ and to the right of $z = 2.25$, and it equals $2(0.5000 - 0.4878) = 0.0244$. This value falls between 0.01 and 0.05, which agrees with our earlier results, but observe that giving a tail probability does not relieve us of the responsibility of specifying the level of significance before the test is actually performed.

The test we have described in this section is essentially an approximate **large-sample test**; it is exact only when the population we are sampling is normal and σ is known. In most practical situations where σ is unknown, we must make the further approximation of substituting for it the sample standard deviation s.

EXAMPLE   A trucking firm suspects the claim that the average lifetime of certain tires is at least 28,000 miles. To check the claim, the firm puts 40 of these tires on its trucks and gets a mean lifetime of 27,563 miles with a standard deviation of 1,348 miles. What can it conclude if the probability of a Type I error is to be at most 0.01?

Solution
1. *Null hypothesis*: $\mu \geq 28,000$ miles
   Alternative hypothesis: $\mu < 28,000$ miles
2. *Level of significance*: $\alpha \leq 0.01$
3. *Criterion*: Since the probability of a Type I error is greatest when $\mu = 28,000$ miles, we proceed as if we were testing the null hypothesis $\mu = 28,000$ miles against the alternative

hypothesis $\mu < 28{,}000$ miles at the level of significance $\alpha = 0.01$. Thus, the null hypothesis must be rejected if $z < -2.33$, where

$$z = \frac{\bar{x} - \mu_0}{\sigma/\sqrt{n}}$$

with σ replaced by s and $\mu_0 = 28{,}000$; otherwise, accept the null hypothesis or reserve judgment.

4. *Calculations*:

$$z = \frac{27{,}563 - 28{,}000}{1{,}348/\sqrt{40}} = -2.05$$

5. *Decision*: Since $z = -2.05$ is not less than -2.33, the null hypothesis cannot be rejected; in other words, the trucking firm's suspicion that $\mu < 28{,}000$ miles is not confirmed by its data.

10.10
TESTS CONCERNING MEANS
(Small Samples)

When we do not know the value of the population standard deviation and the sample is small, $n < 30$, we shall assume, as on page 270, that the population we are sampling has roughly the shape of a normal distribution, and base our decision on the statistic

Statistic for small-sample test concerning mean

$$t = \frac{\bar{x} - \mu_0}{s/\sqrt{n}}$$

whose sampling distribution is the t distribution with $n - 1$ degrees of freedom. The criteria for this **one-sample t test** concerning means based on the above t statistic are those of Figure 10.9 and the table on page 292 with z replaced by t and z_α and $z_{\alpha/2}$ replaced by t_α and $t_{\alpha/2}$. As we explained on page 271, t_α and $t_{\alpha/2}$ are values for which the area to their right under the t distribution is equal to α and $\alpha/2$. All the dividing lines of such tests may be read from Table II, with the number of degrees of freedom equal to $n - 1$.

EXAMPLE   Suppose that we want to test, on the basis of a random sample of size $n = 5$, whether or not the fat content of a certain kind of ice cream exceeds 12 percent. What can we conclude about the null hypothesis $\mu = 12$ percent at the 0.01 level of significance, if the sample has the mean $\bar{x} = 12.7$ percent and the standard deviation $s = 0.38$ percent?

Solution
1. *Null hypothesis*: $\mu = 12$ percent
   Alternative hypothesis: $\mu > 12$ percent
2. *Level of significance*: $\alpha = 0.01$
3. *Criterion*: Reject the null hypothesis if $t > 3.747$, the value of $t_{0.01}$ for $5 - 1 = 4$ degrees of freedom, where

$$t = \frac{\bar{x} - \mu_0}{s/\sqrt{n}}$$

and otherwise accept it.
4. *Calculations*:

$$t = \frac{12.7 - 12}{0.38/\sqrt{5}} = 4.12$$

5. *Decision*: Since $t = 4.12$ exceeds 3.747, the null hypothesis must be rejected; in other words, the fat content of the given kind of ice cream exceeds 12 percent. The exact tail probability, or *p*-value, cannot be determined from Table II, but it is 0.0073.

10.11
DIFFERENCES BETWEEN MEANS

There are many statistical problems in which we must decide whether an observed difference between two sample means can be attributed to chance. For instance, we may want to know whether there is really a difference in the average gasoline consumption of two kinds of cars, if sample data show that one kind averaged 24.6 miles per gallon while, under the same conditions, the other kind averaged 25.7 miles per gallon. Similarly, we may want to decide on the basis of samples whether men can perform a certain task faster than women, whether one kind of ceramic insulator is more brittle than another, whether the average diet in one country is more nutritious than that in another country, and so on.

The method we shall use to test whether an observed difference between two sample means can be attributed to chance, or whether it is statistically significant, is based on the following theory: If $\bar{x}_1$ and $\bar{x}_2$ are the means of two independent random samples, then the sampling distribution of the statistic $\bar{x}_1 - \bar{x}_2$ has the mean $\mu_1 - \mu_2$ and the standard deviation

$$\sqrt{\frac{\sigma_1^2}{n_1} + \frac{\sigma_2^2}{n_2}}$$

where μ_1, μ_2, σ_1, and σ_2 are the means and the standard deviations of the two populations sampled. It is customary to refer to the standard deviation of this sampling distribution as the **standard error of the difference between two means.**

By "independent" samples we mean that the selection of one sample is in no way affected by the selection of the other. Thus, the theory does not apply to "before and after" kinds of comparisons, nor does it apply, say, if we want to compare the IQ's of husbands and wives. A special method for comparing the means of dependent samples is explained in Exercise 10.64 on page 303.

If we limit ourselves to large samples (neither n_1 nor n_2 should be less than 30), we can base the test of the null hypothesis $\mu_1 = \mu_2$ on the statistic

Statistic for large-sample test concerning difference between two means

$$z = \frac{\bar{x}_1 - \bar{x}_2}{\sqrt{\dfrac{\sigma_1^2}{n_1} + \dfrac{\sigma_2^2}{n_2}}}$$

which has approximately the standard normal distribution. Indeed, this statistic is a z-value, for we calculate it by subtracting from $\bar{x}_1 - \bar{x}_2$ the mean of its sampling distribution, which under the null hypothesis is $\mu_1 - \mu_2 = 0$, and then dividing by the standard error of the difference between two means. Depending on whether the alternative hypothesis is $\mu_1 < \mu_2$, $\mu_1 > \mu_2$, or $\mu_1 \neq \mu_2$, the criteria we use for the test are again those shown in Figure 10.9 and also in the table on page 292, with $\mu_1 - \mu_2$ substituted for μ and 0 substituted for μ_0.

The test we have described here is essentially an approximate large-sample test; it is exact only when both of the populations we are sampling are normal. In most practical situations where σ_1 and σ_2 are unknown, we must make the further approximation of substituting for them the sample standard deviations s_1 and s_2.

EXAMPLE   In a study designed to test whether or not there is a difference between the average heights of adult females born in two different countries, random samples yielded the following results:

$$n_1 = 120 \qquad \bar{x}_1 = 62.7 \qquad s_1 = 2.50$$
$$n_2 = 150 \qquad \bar{x}_2 = 61.8 \qquad s_2 = 2.62$$

where the measurements are in inches. Use the 0.05 level of significance to test the null hypothesis that the corresponding population means are equal against the alternative hypothesis that they are not equal.

Solution

1. *Null hypothesis*: $\mu_1 = \mu_2$
   Alternative hypothesis: $\mu_1 \neq \mu_2$
2. *Level of significance*: $\alpha = 0.05$
3. *Criterion*: Reject the null hypothesis if $z < -1.96$ or $z > 1.96$, where

$$z = \frac{\bar{x}_1 - \bar{x}_2}{\sqrt{\dfrac{\sigma_1^2}{n_1} + \dfrac{\sigma_2^2}{n_2}}}$$

with s_1 and s_2 substituted for σ_1 and σ_2; otherwise, accept it.

4. *Calculations*:

$$z = \frac{62.7 - 61.8}{\sqrt{\dfrac{(2.50)^2}{120} + \dfrac{(2.62)^2}{150}}} = 2.88$$

5. *Decision*: Since $z = 2.88$ exceeds 1.96, the null hypothesis must be rejected; in other words, the sample data show that there is a difference between the true average heights of adult females in the two given countries.

10.12
DIFFERENCES BETWEEN MEANS
(Small Samples)

As in Section 10.10, a small-sample test of the significance of the difference between two means may be based on an appropriate t statistic. For this test we must assume that we have independent random samples and that the populations we are sampling have roughly the shape of normal distributions with equal standard deviations. Then, we can base the test of the null hypothesis $\mu_1 = \mu_2$ on the statistic

Statistic for small-sample test concerning difference between two means

$$t = \frac{\bar{x}_1 - \bar{x}_2}{\sqrt{\dfrac{(n_1 - 1)s_1^2 + (n_2 - 1)s_2^2}{n_1 + n_2 - 2} \cdot \left(\dfrac{1}{n_1} + \dfrac{1}{n_2}\right)}}$$

whose sampling distribution is the t distribution with $n_1 + n_2 - 2$ degrees of freedom. Depending on whether the alternative hypothesis is $\mu_1 < \mu_2$, $\mu_1 > \mu_2$, or $\mu_1 \neq \mu_2$, the criteria we use for the test, called the **two-sample** t **test,** are again those shown in Figure 10.9 and also in the table on page 292, with t substituted throughout for z, $\mu_1 - \mu_2$ substituted for μ, and 0 substituted for μ_0.

EXAMPLE   The following random samples are measurements of the heat-producing capacity (in millions of calories per ton) of specimens of coal from two mines:

$$Mine\ 1: \quad 8,400,\ 8,230,\ 8,380,\ 7,860,\ 7,930$$

$$Mine\ 2: \quad 7,510,\ 7,690,\ 7,720,\ 8,070,\ 7,660$$

Use the 0.05 level of significance to test whether the difference between the means of these two samples is significant.

Solution
1. *Null hypothesis*: $\mu_1 = \mu_2$
   Alternative hypothesis: $\mu_1 \neq \mu_2$
2. *Level of significance*: $\alpha = 0.05$
3. *Criterion*: Reject the null hypothesis if $t < -2.306$ or $t > 2.306$, where t is given by the above formula and 2.306 is the value of $t_{0.025}$ for $5 + 5 - 2 = 8$ degrees of freedom; otherwise state that the difference between the two sample means is not significant.
4. *Calculations*: The means and the variances of the two samples are $\bar{x}_1 = 8,160$, $\bar{x}_2 = 7,730$, $s_1^2 = 63,450$, and $s_2^2 = 42,650$, so that

$$t = \frac{8,160 - 7,730}{\sqrt{\dfrac{4(63,450) + 4(42,650)}{5 + 5 - 2} \cdot \left(\dfrac{1}{5} + \dfrac{1}{5}\right)}}$$

$$= 2.95$$

5. *Decision*: Since $t = 2.95$ exceeds 2.306, the null hypothesis must be rejected; in other words, we conclude that the average heat-producing capacity of the coal from the two mines is not the same.

EXERCISES

10.40 A law student, who wants to check a professor's claim that convicted embezzlers spend on the average 12.3 months in jail, takes a random sample of 35 such cases from court files. Using his results, $\bar{x} = 11.5$ months and $s = 3.8$ months, and the 0.05 level of significance, should the student accept the null hypothesis $\mu = 12.3$ months or the alternative hypothesis $\mu \neq 12.3$ months?

10.41 According to the norms established for a reading comprehension test, eighth graders should average 83.2 with a standard deviation of 8.6. If 45 randomly selected eighth graders from a certain school district averaged 86.7, test the null hypothesis $\mu = 83.2$ against the alternative hypothesis $\mu > 83.2$ at the 0.01 level of significance, and thus decide whether to accept or reject the district superintendent's claim that her eighth graders are above average.

10.42 The security department of a factory wants to know whether or not the true average time required by the night watchman to walk his round is 25 minutes. If, in a random sample of 32 rounds, the night watchman averaged 25.8 minutes with a standard deviation of 1.5 minutes, determine at the 0.01 level of significance whether this is sufficient evidence to reject the null hypothesis $\mu = 25$ minutes.

10.43 Tests performed with a random sample of 40 diesel engines produced by a large manufacturer showed that they have a mean thermal efficiency of 31.8 percent with a standard deviation of 2.2 percent. Based on this information and at the 0.05 level of significance, should the persons performing the tests accept the null hypothesis $\mu = 32.3$ or the alternative hypothesis that the true average thermal efficiency of the manufacturer's diesel engines is less than 32.3 percent?

10.44 In a study of new sources of food, it is reported that a pound of a certain kind of fish yields on the average 2.41 ounces of FPC (fish-protein concentrate), which is used to enrich various food products (including flour). Yes or no, is this figure supported by a study in which 30 samples of this kind of fish yielded on the average 2.44 ounces of FPC (per pound of fish) with a standard deviation of 0.07 ounce, if we use
    (a) the level of significance $\alpha = 0.05$;
    (b) the level of significance $\alpha = 0.01$?

10.45 With reference to the tire-mileage example on page 294, find the tail probability corresponding to the value obtained for z.

10.46 An ambulance service claims that on the average it takes no more than 8.9 minutes to reach its destination in emergency calls. To check this claim, the agency which licenses ambulance services has the ambulance service timed on 50 emergency calls, getting a mean of 9.3 minutes and a standard deviation of 1.8 minutes. If the probability of a Type I error is to be at most 0.05, does this constitute evidence against the ambulance service's claim?

10.47 If the null hypothesis $\mu = \mu_0$ is to be tested against the simple alternative hypothesis $\mu = \mu_A$, and the probabilities of Type I and Type II errors are to have the given values α and β, we must take a random sample of size

$$n = \frac{\sigma^2(z_\alpha + z_\beta)^2}{(\mu_A - \mu_0)^2}$$

where σ^2 is the variance of the population. Since this formula is based on normal-curve theory, the sample must be large or the population sampled must have roughly the shape of a normal distribution.
    Suppose, for instance, that for a population with $\sigma = 6$ we want to test the null hypothesis $\mu = 200$ pounds against the alternative hypothesis $\mu = 198$ pounds. If α and β are both to be 0.05, we find that

$$n = \frac{6^2(1.645 + 1.645)^2}{(198 - 200)^2} = 97.4$$

or 98 rounded up to the nearest integer. Thus, a random sample of size $n = 98$ will expose us to the specified risks.

    (a) Suppose that we want to test the null hypothesis $\mu = \$250$ against the alternative hypothesis $\mu = \$255$ for a population whose standard deviation is $\sigma = \$12$. How large a random sample must we take, if α and β are both to be 0.01?

    (b) Suppose that we want to test the null hypothesis $\mu = 64$ inches against the alternative hypothesis $\mu = 61$ inches for a population whose standard deviation is $\sigma = 7.2$ inches. How large a random sample must we take, if α is to be 0.05 and β is to be 0.01? Also, for what values of $\bar{x}$ will the null hypothesis be rejected?

10.48 A random sample of 12 graduates of a secretarial school averaged 73.8 words per minute with a standard deviation of 7.9 words per minute on a typing test. At the 0.01 level of significance, should we accept or reject an employer's claim that the school's graduates average less than 75.0 words per minute?

10.49 A soft-drink vending machine is set to dispense 6.0 ounces per cup. If the machine is tested nine times, yielding a mean cup fill of 6.2 ounces with a standard deviation of 0.15 ounce, is this evidence at the 0.05 level of significance that the machine is on the average overfilling cups?

10.50 In an experiment with a new tranquilizer, the pulse rate of 16 patients was determined before they were given the tranquilizer and again five minutes later, and their pulse rate was found to be reduced on the average by 6.8 beats with a standard deviation of 1.9. Using the 0.05 level of significance, what can we conclude about the claim that this tranquilizer will reduce the pulse rate on the average by 7.5 beats in five minutes?

10.51 A random sample from a company's very extensive files shows that orders for a certain piece of machinery were filled, respectively, in 12, 10, 17, 14, 13, 18, 11, and 9 days. Use the 0.01 level of significance to test the claim that on the average such orders are filled in 9.5 days. Choose the alternative hypothesis in such a way that rejection of the null hypothesis $\mu = 9.5$ days implies that it takes longer than that.

10.52 The yield of alfalfa from six test plots is 1.4, 1.8, 1.1, 1.9, 2.2, and 1.2 tons per acre. Test at the 0.05 level of significance whether this supports the contention that the average yield for this kind of alfalfa is 1.5 tons per acre.

10.53 A reading teacher wants to determine whether a certain student has an average reading speed of $\mu = 750$ words per minute or whether her average reading speed is less than 750 words per minute. What can the teacher conclude at the 0.05 level of significance if in six minutes the student reads, respectively, 730, 750, 740, 700, 700, and 760 words?

10.54 Five measurements of the tar content of a certain kind of cigarette yielded 14.5, 14.2, 14.4, 14.3, and 14.6 mg/cig (milligrams per cigarette). Show that the difference between the mean of this sample, $\bar{x} = 14.4$, and the average tar content claimed by the cigarette manufacturer, $\mu = 14.0$, is significant at the 0.05 level of significance.

10.55 Suppose that in the preceding exercise the first measurement is recorded incorrectly as 16.0 instead of 14.5. Show that now the difference between the mean of the sample, $\bar{x} = 14.7$, and the average tar content claimed by the cigarette manufacturer, $\mu = 14.0$, is not significant at $\alpha = 0.05$. Explain the apparent paradox that even though the difference between $\bar{x}$ and μ has increased, it is no longer significant.

10.56 A sample study was made of the number of business lunches that executives claim as deductible expenses per month. If 40 executives in the insurance industry averaged 8.7 such deductions with a standard deviation of 1.9 in a given month, while 50 bank executives averaged 7.6 with a standard deviation of 2.1, test at the 0.05 level of significance whether the difference between these two sample means is significant.

10.57 An investigation of two kinds of photocopying equipment showed that 60 failures of the first kind of equipment took on the average 84.2 minutes to repair with a standard deviation of 19.4 minutes, while 60 failures of the second kind of equipment took on the average 91.6 minutes to repair with a standard deviation of 18.8 minutes. Test at the 0.01 level of significance whether the difference between these two sample means is significant.

10.58 Suppose that we want to investigate whether or not men and women earn comparable wages in a certain industry. If sample data show that 60 men earn on the average $\bar{x}_1 = \$282.50$ per week with a standard deviation of $s_1 = \$15.60$, while 60 women earn on the average $\bar{x}_2 = \$266.10$ per week with a standard deviation of $s_2 = \$18.20$, test the null hypothesis $\mu_1 = \mu_2$ against the alternative hypothesis $\mu_1 > \mu_2$ at the 0.01 level of significance.

10.59 Six guinea pigs injected with 0.5 mg of a medication took on the average 15.4 seconds to fall asleep with a standard deviation of 2.2 seconds, while six other guinea pigs injected with 1.5 mg of the medication took on the average 11.2 seconds to fall asleep with a standard deviation of 2.6 seconds. Use the 0.05 level of significance to test whether or not the increase in dosage from 0.5 mg to 1.5 mg reduces the average amount of time it takes a guinea pig to fall asleep.

10.60 Twelve measurements each of the hydrogen content (in percent number of atoms) of gases collected from the eruption of two volcanos yielded $\bar{x}_1 = 41.2$, $\bar{x}_2 = 45.8$, $s_1 = 5.2$, and $s_2 = 6.7$. Decide, at the 0.05 level of significance, whether to accept or reject the null hypothesis that there is no difference (with regard to hydrogen content) in the composition of the gases in the two eruptions.

10.61 The following are the numbers of sales which a random sample of nine salesmen of industrial chemicals in California and a random sample of six salesmen of industrial chemicals in Oregon made over a fixed period of time:

California: 41, 47, 62, 39, 56, 64, 37, 61, 52

Oregon: 34, 63, 45, 55, 24, 43

Use the 0.01 level of significance to test whether the difference between the means of these two samples is significant.

10.62 As part of an industrial training program, some trainees are instructed by method A, which is straight teaching-machine instruction, and some are instructed by method B, which also involves the personal attention of an instructor. If random samples of size 10 are taken from large groups of trainees instructed by each of these two methods, and the scores which they obtained in an appropriate achievement test are

$$Method\ A:\quad 71,\quad 75,\quad 65,\quad 69,\quad 73,\quad 66,\quad 68,\quad 71,\quad 74,\quad 68$$

$$Method\ B:\quad 72,\quad 77,\quad 84,\quad 78,\quad 69,\quad 70,\quad 77,\quad 73,\quad 65,\quad 75$$

test whether to accept the null hypothesis $\mu_B \le \mu_A$ or the alternative hypothesis $\mu_B > \mu_A$ so that the probability of a Type I error does not exceed 0.05. (Here μ_A and μ_B are the true means of the two populations of scores sampled.)

10.63   In some problems we are interested in testing whether the difference between the means of two populations is equal to, less than, or greater than a given constant. So, we test the null hypothesis $\mu_1 - \mu_2 = \delta\ (delta)$, where δ is the given constant, against an appropriate alternative hypothesis. To perform this kind of test, we substitute $\bar{x}_1 - \bar{x}_2 - \delta$ for $\bar{x}_1 - \bar{x}_2$ in the numerator of the z-statistic on page 297, or the t-statistic on page 298, and otherwise proceed in the same way as before.

With reference to the illustration of Section 10.12, suppose that we want to test whether the heat-producing capacity of the coal from the first mine exceeds that of the coal from the second mine by more than 100 (million calories per ton). Thus, we shall want to test the null hypothesis $\mu_1 - \mu_2 = 100$ against the alternative hypothesis $\mu_1 - \mu_2 > 100$, and the test statistic becomes

$$t = \frac{8,160 - 7,730 - 100}{\sqrt{\dfrac{253,800 + 170,600}{5 + 5 - 2} \cdot \left(\dfrac{1}{5} + \dfrac{1}{5}\right)}} = 2.27$$

Since this exceeds 1.860, the value of $t_{0.05}$ for $5 + 5 - 2 = 8$ degrees of freedom, it follows that the null hypothesis must be rejected. We conclude that the heat-producing capacity of the coal from the first mine exceeds that of the coal from the second mine by more than 100 (million calories per ton).

(a) Sample surveys conducted in a large county in 1950 and again in 1970 showed that in 1950 the average height of 400 ten-year-old boys was 53.2 inches with a standard deviation of 2.4 inches, while in 1970 the average height of 500 ten-year-old boys was 54.5 inches with a standard deviation of 2.5 inches. Use the 0.05 level of significance to test whether the true average increase in height is 0.5 inch or whether it is greater than that.

(b) To test the claim that the resistance of electric wire can be reduced by more than 0.050 ohm by alloying, 25 values obtained for alloyed wire yielded $\bar{x}_2 = 0.083$ ohm and $s_2 = 0.003$ ohm, and 25 values obtained for standard wire yielded $\bar{x}_1 = 0.136$ ohm and $s_1 = 0.002$ ohm. Use the 0.05 level of significance to test whether the true average reduction is 0.050 ohm or whether it is more than that.

10.64   If we want to study the effectiveness of a new diet on the basis of weights "before and after," or if we want to study whatever differences there may be between the IQ's of husbands and wives, the methods of Sections 10.11 and 10.12 cannot be used. The samples are not independent; in fact, in each case the data are *paired*. To handle data of this kind, we work with the (signed) differences of the paired data and test whether these differences may be looked upon as a sample from a population for which $\mu = 0$. If the sample is small, we use the t test; otherwise, we use the large-sample test of Section 10.9. Apply this technique to determine the effectiveness of an industrial safety program on the basis of the following data (collected over a period of one year) on the average weekly loss of man-hours due to accidents in twelve plants "before and after" the program was put into operation: 37 and 28, 72 and 59, 26 and 24, 125 and 120, 45 and 46, 54 and 43, 13 and 15, 79 and 75, 12 and 18, 34 and 29, 39 and 35, and 26 and 24. Use the 0.05 level of significance to decide whether or not the safety program is effective.

10.65 The following data were obtained in an experiment designed to check whether there is a systematic difference in the weights (in grams) obtained with two different scales:

	Scale I	Scale II
Rock specimen 1	12.13	12.17
Rock specimen 2	17.56	17.61
Rock specimen 3	9.33	9.35
Rock specimen 4	11.40	11.42
Rock specimen 5	28.62	28.61
Rock specimen 6	10.25	10.27
Rock specimen 7	23.37	23.42
Rock specimen 8	16.27	16.26
Rock specimen 9	12.40	12.45
Rock specimen 10	24.78	24.75

Use the method of the preceding exercise and the 0.01 level of significance to test whether the difference between the means of the weights obtained with the two scales is significant.

10.13
CHECKLIST OF KEY TERMS
(with page references to their definitions)

10.14
REVIEW EXERCISES

10.66 A random sample of 40 cans of pineapple slices has a mean weight of 15.85 ounces and a standard deviation of 0.23 ounce. If this mean is used as an estimate of the mean weight of all the cans of pineapple slices in the large production lot from which the 40 cans came, with what confidence can we assert that this estimate is "off" by at most 0.06 ounce?

10.67 A random sample of 54 shirts worn by soldiers in a tropical climate has an average useful life of 63.9 washings with a standard deviation of 4.5. Under moderate weather conditions, such shirts are known to have an average useful life of

81.6 washings. At the 0.01 level of significance, can we conclude that their use in a tropical climate reduces the average useful life of such shirts?

10.68 Suppose we want to test the hypothesis that solar heating unit A is more efficient than solar heating unit B. Explain under what conditions we would be committing a Type I error and under what conditions we would be committing a Type II error.

10.69 In a French restaurant, the chef receives 26, 21, 14, 22, 18, and 20 orders for coq au vin on six different nights. Construct a 95 percent confidence interval for the number of orders for coq au vin the chef can expect per night.

10.70 A general achievement test is standardized so that eighth graders should average 79.4 with a standard deviation of 4.8. If the superintendent of a large school district hopes to show that eighth graders in her district average better than that, and she has the test given to 32 eighth graders in her district (presumably a random sample), by how much must their average exceed 79.4 to make the difference significant at the 0.01 level of significance?

10.71 The technique of Exercise 10.63 can also be used to construct confidence intervals for the difference $\delta = \mu_1 - \mu_2$ between two population means. In the large-sample case, we modify the z-statistic on page 297 as indicated in that exercise, substitute it for the middle term of $-z_{\alpha/2} < z < z_{\alpha/2}$, and manipulate this inequality algebraically so that the middle term is δ; in the small-sample case we proceed similarly with the t-statistic on page 298.

    (a) With reference to the example on page 297, construct a 99 percent confidence interval for the difference between the true average heights of adult females born in the two countries.

    (b) With reference to the example on page 299, construct a 98 percent confidence interval for the difference between the true average heat-producing capacities of coal from the two mines.

10.72 It is desired to test the null hypothesis $\mu = 0$ against the alternative hypothesis $\mu > 0$ on the basis of the mean of a random sample of size $n = 9$ from a normal population with $\sigma = 5$. If the probability of a Type I error is to be $\alpha = 0.05$,

    (a) verify that the null hypothesis must be rejected when $\bar{x} > 2.74$;

    (b) calculate β for $\mu = 2.50, 5.00$, and 7.50, and draw a rough sketch of the operating characteristic curve.

★10.73 A sales manager's feelings about the average monthly demand for a certain product may be described by means of a normal distribution with $\mu_0 = 4{,}800$ units and $\sigma_0 = 260$ units.

    (a) What probability does she, thus, assign to the true average monthly demand being somewhere on the interval from 4,500 to 5,000 units?

    (b) If data for ten months show an average demand of 4,702 units with a standard deviation of 380 units, find the mean and the standard deviation of the posterior distribution.

    (c) How would the probability of part (a) be modified in the light of the information given in part (b)?

10.74 An efficiency expert wants to determine the average time it takes a person to buy a week's groceries at a supermarket. If preliminary studies show that it is reasonable to let $\sigma = 2.6$ minutes, how large a sample will be needed to be able to assert with 95 percent confidence that the mean of the sample is "off" by at most 0.2 minute?

10.75 In a study of the effectiveness of certain exercises in weight reduction, a group of 16 persons engaged in these exercises for one month and showed the following results:

Weight before	Weight after	Weight before	Weight after
211	198	172	166
180	173	155	154
171	172	185	181
214	209	167	164
182	179	203	201
194	192	181	175
160	161	245	233
182	182	146	142

Use the method of Exercise 10.64 and the 0.05 level of significance to test whether or not the exercises are effective in reducing weight.

10.76 In a study of the relationship between family size and intelligence, 40 "only children" had an average IQ of 101.5 with a standard deviation of 6.7 and 50 "firstborns" in two-child families had an average IQ of 105.9 with a standard deviation of 5.8. Use the 0.05 level of significance to test whether the difference between these two means is significant.

10.77 Five containers of a commercial solvent randomly selected from a large production lot weigh 19.7, 19.5, 20.2, 19.2, and 19.9 pounds.

(a) What can we assert with 95 percent confidence about the maximum size of our error in estimating the mean of the population sampled as 19.7 pounds?

(b) Construct a 99 percent confidence interval for the mean weight of all the containers in the production lot.

10.78 To compare some of the features of the confidence interval formula on page 269 and the test described on page 292, suppose that a random sample of size n is taken from a normal population with the mean μ and the standard deviation σ. What relationship is there between the set of values covered by the $(1 - \alpha)100$ percent confidence interval for μ and the set of values of μ_0 for which the null hypothesis $\mu = \mu_0$ cannot be rejected at the level of significance α against the two-sided alternative $\mu \neq \mu_0$?

10.79 To compare two kinds of bumper guards, six of each kind were mounted on certain imported compact cars. Then each car was run into a concrete wall at 5 miles per hour, and the following are the costs of the repairs (in dollars):

Bumper guard 1: 107, 128, 123, 135, 102, 119

Bumper guard 2: 123, 114, 105, 101, 110, 131

Use the 0.05 level of significance to test whether the difference between the mean repair costs is significant.

10.80 To check whether the true average wing span of a certain kind of insect is anywhere from 18.2 mm to 18.6 mm, a biologist decides to take a random sample of size $n = 60$ with the intention of rejecting the null hypothesis $18.2 \leq \mu \leq 18.6$ if the sample mean is less than 18.0 mm or greater than 18.8 mm; otherwise, he will accept it. If it is known from similar studies that $\sigma = 0.8$ mm for this kind of data, what is the maximum probability of a Type I error?

10.81 A random sample of size $n = 50$ is taken from a population which has the standard deviation $\sigma = 5.5$ grams. If we use the mean of this sample to estimate the mean of the population, with what confidence can we assert that the error is at most 1.2 grams?

10.82 During the investigation of an alleged unfair trade practice, the Federal Trade Commission takes a random sample of 49 "3-ounce" candy bars from a large shipment. If the mean and the standard deviation of the sample are 2.94 ounces and 0.12 ounce, show that, at the 0.01 level of significance, the commission has grounds upon which to proceed against the manufacturer on the unfair practice of shortweight selling.

10.83 It is desired to estimate the average number of days of continuous use until a new refrigerator of a certain kind will first require repairs. If it can be assumed that $\sigma = 232$ days, how large a sample is needed so that it will be possible to assert with 95 percent confidence that the sample mean is "off" by at most 30 days?

10.84 To compare freshmen's knowledge of history, samples of 50 freshmen from each of two universities were given a special test. If those from the first university have an average score of 67.4 with a standard deviation of 5.0, while those from the second university have an average score of 62.8 with a standard deviation of 4.6, test at the 0.05 level of significance whether the difference between the two sample means is significant.

10.85 Measurements of the amount of chloroform (micrograms per liter) in 36 specimens of the drinking water of a city yielded the following results: $\bar{x} = 34.8$ and $s = 4.9$. Use this information to construct a 99 percent confidence interval for the average amount of chloroform in the drinking water of this city.

10.86 In ten trials, a car ran for 28, 27, 21, 26, 29, 26, 29, 28, 29, and 27 miles with a gallon of a certain kind of gasoline. Does this support or refute the claim that with this kind of gasoline the car averages at least 28 miles per gallon, if the probability of a Type I error is to be at most 0.05?

10.87 Suppose that a law firm has one secretary whom it suspects of making more mistakes than the average of all its secretaries.
   (a) If the law firm decides that it will let the secretary go, provided this suspicion is confirmed on the basis of observations made on the secretary's performance, what hypothesis and alternative should the law firm set up?
   (b) If the law firm decides to let the secretary go unless he can prove himself significantly better than the average of all its secretaries, what hypothesis and alternative should the law firm set up?

10.15
REFERENCES

An informal introduction to interval estimation is given under the heading of "How to be precise though vague." in

MORONEY, M. J., *Facts from Figures*. London: Penguin Books, 1956.

A discussion of the theoretical foundation of the t distribution as well as other mathematical details omitted in this book may be found in most textbooks on mathematical statistics. Some of the theory of Bayesian estimation pertaining to Section 10.4 is discussed in Chapter 10 of the book by Freund and Walpole listed on page 263.

In the preceding chapter we learned how to judge the size of the error in estimating a population mean, how to construct confidence intervals for means, and how to perform tests of hypotheses about the means of one and of two populations. As we shall see in this and in subsequent chapters, very similar methods apply to inferences about other population parameters.

In this chapter we shall concentrate on population standard deviations, or population variances, which are not only important in their own right, but which must sometimes be estimated before inferences about other parameters can be made. This is the case, for example, when we make inferences about population means and must know or estimate the value of σ.

Section 11.1 is devoted to the estimation of σ and σ^2, and Sections 11.2 and 11.3 deal with tests of hypotheses about these parameters.

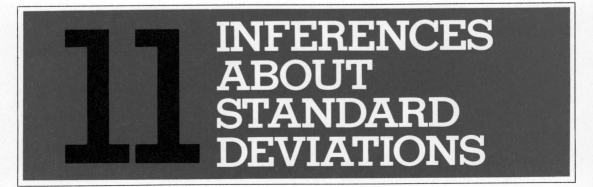

11 INFERENCES ABOUT STANDARD DEVIATIONS

11.1
THE ESTIMATION OF σ

There are various ways in which we can estimate the standard deviation of a population (see, for example, Exercises 8.15 and 11.9). Among them, the sample standard deviation is by far most popular, and to show how it is used to make inferences about σ, let us begin with a confidence interval for σ based on s, which requires that the population we are sampling has roughly the shape of a normal distribution. In that case, the statistic

Chi-square statistic

$$\chi^2 = \frac{(n-1)s^2}{\sigma^2}$$

called "chi-square," has as its sampling distribution an important continuous distribution called the **chi-square distribution**. (χ is the Greek lowercase letter *chi*.) The mean of this distribution is $n - 1$ and, as with the t distribution, we call this quantity the number of degrees of freedom, or simply the **degrees of freedom**. An example of a chi-square distribution is shown in Figure 11.1; unlike the normal and t distributions, its domain is restricted to the non-negative real numbers.

Analogous to z_α and t_α, we now define χ^2_α as the value for which the area to its right under the chi-square distribution is equal to α; like t_α, this

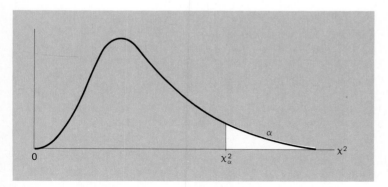

FIGURE 11.1 *Chi-square distribution.*

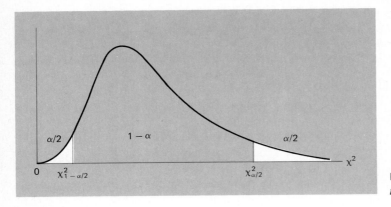

FIGURE 11.2 *Chi-square distribution.*

value depends on the number of degrees of freedom. Thus, $\chi^2_{\alpha/2}$ is such that the area to its right under the curve is $\alpha/2$, while $\chi^2_{1-\alpha/2}$ is such that the area to its left under the curve is $\alpha/2$ (see Figure 11.2). We make this distinction because the chi-square distribution, unlike the normal and t distributions, is not symmetrical. Among others, values of $\chi^2_{0.995}$, $\chi^2_{0.975}$, $\chi^2_{0.025}$, and $\chi^2_{0.005}$ are given in Table III at the end of the book for $1, 2, 3, \ldots$, and 30 degrees of freedom.

From Figure 11.2, we find that we can assert with probability $1 - \alpha$ that a random variable having the chi-square distribution will take on a value between $\chi^2_{1-\alpha/2}$ and $\chi^2_{\alpha/2}$. Applying this result to the χ^2 statistic given above, we can assert with probability $1 - \alpha$ that

$$\chi^2_{1-\alpha/2} < \frac{(n-1)s^2}{\sigma^2} < \chi^2_{\alpha/2}$$

Now, if we proceed as on page 269, we can rewrite this inequality as

Confidence interval for σ^2

$$\frac{(n-1)s^2}{\chi^2_{\alpha/2}} < \sigma^2 < \frac{(n-1)s^2}{\chi^2_{1-\alpha/2}}$$

which is a **$(1 - \alpha)100$ percent confidence interval for σ^2**, the population variance; if we take the square root of each term, we get a $(1 - \alpha)100$ percent confidence interval for σ, the population standard deviation.

EXAMPLE A random sample of $n = 5$ specimens of a certain kind of ice cream has a mean fat content of $\bar{x} = 12.7$ percent and a standard deviation of $s = 0.38$ percent. Construct a 95 percent confidence interval for σ, the standard deviation of the population sampled.

Solution Since we have $5 - 1 = 4$ degrees of freedom, we find from Table III that $\chi^2_{0.975} = 0.484$ and $\chi^2_{0.025} = 11.143$. Thus, substitution into the formula above yields

$$\sqrt{\frac{4(0.38)^2}{11.143}} < \sigma < \sqrt{\frac{4(0.38)^2}{0.484}}$$

and

$$0.23 \text{ percent} < \sigma < 1.09 \text{ percent}$$

The kind of confidence interval we have described here is often referred to as a **small-sample confidence interval,** because it is used mainly when n is small, less than 30, and, of course, only when we can assume that the population sampled has roughly the shape of a normal distribution. Otherwise, we make use of the theory that for large samples, $n \geq 30$, the sampling distribution of s can be approximated with a normal distribution having the mean σ and the standard deviation $\dfrac{\sigma}{\sqrt{2n}}$ (see Exercise 9.45 on page 259). So, we can assert with probability $1 - \alpha$ that

$$-z_{\alpha/2} < \frac{s - \sigma}{\dfrac{\sigma}{\sqrt{2n}}} < z_{\alpha/2}$$

and simple algebra leads to the following large-sample confidence interval for the population standard deviation σ:

Large-sample confidence interval for σ

$$\frac{s}{1 + \dfrac{z_{\alpha/2}}{\sqrt{2n}}} < \sigma < \frac{s}{1 - \dfrac{z_{\alpha/2}}{\sqrt{2n}}}$$

EXAMPLE With reference to the example on page 66, where we showed that $s = 5.55$ tons for a large industrial plant's emission of sulfur oxides on $n = 80$ days, construct a 95 percent confidence interval for the standard deviation of the population sampled.

Solution Substituting $n = 80$, $s = 5.55$, and $z_{\alpha/2} = 1.96$ into the above confidence interval formula, we get

$$\frac{5.55}{1 + \dfrac{1.96}{\sqrt{160}}} < \sigma < \frac{5.55}{1 - \dfrac{1.96}{\sqrt{160}}}$$

and

$$4.80 < \sigma < 6.57$$

This means that we are 95 percent confident that the interval from 4.80 tons to 6.57 tons contains σ, the true standard deviation of the plant's daily emission of sulfur oxides.

EXERCISES

11.1 With reference to Exercise 10.18 on page 278, construct a 95 percent confidence interval for σ, the standard deviation of all the bearings made by the given process.

11.2 With reference to Exercise 10.20 on page 278, construct a 99 percent confidence interval for the true standard deviation of the amount of benzene-soluble organic matter in the air at the experiment station.

11.3 With reference to Exercise 10.49 on page 301, construct a 98 percent confidence interval for the true variance of the vending machine's fill per cup.

11.4 With reference to Exercise 10.23 on page 279, construct a 95 percent confidence interval for σ, the standard deviation of the number of fillings required by all the inmates of the very large prison.

11.5 With reference to Exercise 10.51 on page 301, construct a 99 percent confidence interval for σ^2, the true variance of the time it takes to fill orders for the given kind of machinery.

11.6 With reference to Exercise 10.1 on page 276, construct a 95 percent confidence interval for the standard deviation of the time it takes to perform such repairs.

11.7 With reference to Exercise 10.3 on page 276, construct a 99 percent confidence interval for σ, the true standard deviation of the given cacti's annual growth.

11.8 With reference to Exercise 10.44 on page 300, construct a 90 percent confidence interval for the true standard deviation of the FPC yield of the given kind of fish.

11.9 When we deal with very small samples, good estimates of the population standard deviation can often be obtained on the basis of the sample range (the largest sample value minus the smallest). Such quick estimates of σ are given by the sample range divided by the divisor d, which depends on the size of the sample; for samples from populations having roughly the shape of a normal distribution, its values are shown in the following table for $n = 2, 3, \ldots$, and 12:

n	2	3	4	5	6	7	8	9	10	11	12
d	1.13	1.69	2.06	2.33	2.53	2.70	2.85	2.97	3.08	3.17	3.26

For instance, if a psychiatrist saw 11, 9, 14, 8, 7, 11, 15, and 10 patients on eight days, we find that $s = 2.77$, $r = 15 - 7 = 8$, and $d = 2.85$ for $n = 8$. Thus, $\dfrac{8}{2.85} = 2.81$ as well as 2.77 are estimates of the true standard deviation of the number of patients the psychiatrist sees per day. As can be seen, the difference between the two estimates is small.

(a) With reference to Exercise 10.21 on page 278, use this method to estimate the true standard deviation of the time it takes to assemble the mechanical device, and compare the result with the sample standard deviation s.

(b) With reference to Exercise 10.54 on page 301, use this method to estimate the standard deviation of the tar content of the given kind of cigarette, and compare the result with the sample standard deviation s.

11.2
TESTS CONCERNING STANDARD DEVIATIONS

In this section we shall consider the problem of testing the null hypothesis that a population standard deviation equals a specified constant σ_0, or that a population variance equals σ_0^2. This kind of test is required whenever we want to test the uniformity of a product, process, or operation. For instance, if we want to test whether a certain kind of glass is sufficiently homogeneous for making delicate optical equipment, whether the intelligence of a group of students is sufficiently uniform so that they can be taught in one class, whether a lack of uniformity in certain workers' performance may call for stricter supervision, and so forth.

The test of the null hypothesis $\sigma = \sigma_0$, the hypothesis that a population standard deviation equals a specified constant, is based on the same assumptions, the same statistic, and the same sampling theory as the small-sample confidence interval for σ. Assuming that our sample is random and comes from a population having roughly the shape of a normal distribution, we base our decision on the statistic

Statistic for test concerning standard deviation

$$\chi^2 = \frac{(n - 1)s^2}{\sigma_0^2}$$

where n and s^2 are the sample size and the sample variance, and σ_0 is the value of the population standard deviation assumed under the null hypothesis. If the null hypothesis is true, the sampling distribution of this statistic is the chi-square distribution with $n - 1$ degrees of freedom; hence, the criteria for testing the null hypothesis $\sigma = \sigma_0$ against the alternative hypothesis $\sigma < \sigma_0$, $\sigma > \sigma_0$, or $\sigma \neq \sigma_0$ are as shown in Figure 11.3. For the one-sided alternative $\sigma < \sigma_0$, we reject the null hypothesis for values of χ^2 falling into the left-hand tail of its sampling distribution; for the one-sided alternative $\sigma > \sigma_0$, we reject the null hypothesis for values of χ^2 falling into the right-hand tail of its sampling distribution; and for the two-sided alternative $\sigma \neq \sigma_0$, we reject the null hypothesis for values of χ^2 falling into either tail of its sampling distribution. The quantities χ_α^2, $\chi_{\alpha/2}^2$, and $\chi_{1-\alpha/2}^2$, for $n = 1, 2, 3, \ldots,$ or 30 degrees of freedom, are given in Table III.

FIGURE 11.3 *Test criteria.*

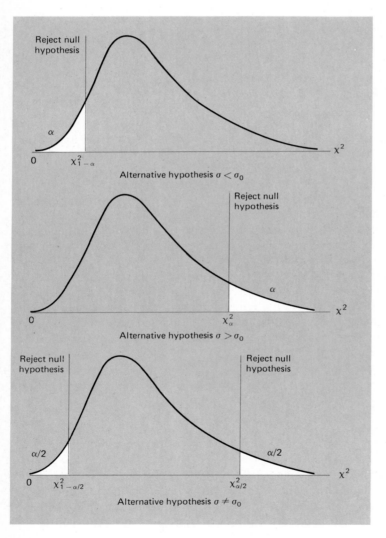

EXAMPLE   An automotive engineer, interested in certain safety features, must know whether or not the standard deviation of the time it takes drivers to react in an emergency situation is less than 0.010 second. Use the 0.05 level of significance to test the null hypothesis $\sigma = 0.010$ against the alternative hypothesis $\sigma < 0.010$ on the basis of a random sample of size $n = 15$ for which $s = 0.006$ second.

Solution
1. *Null hypothesis*: $\sigma = 0.010$
   Alternative hypothesis: $\sigma < 0.010$
2. *Level of significance*: $\alpha = 0.05$
3. *Criterion*: Reject the null hypothesis if $\chi^2 < 6.571$, the value of $\chi^2_{0.95}$ for $15 - 1 = 14$ degrees of freedom, where

$$\chi^2 = \frac{(n-1)s^2}{\sigma_0^2}$$

   and otherwise accept it.
4. *Calculations*:

$$\chi^2 = \frac{14(0.006)^2}{(0.010)^2} = 5.04$$

5. *Decision*: Since $\chi^2 = 5.04$ is less than 6.571, the null hypothesis must be rejected; in other words, the standard deviation of the time it takes drivers to react in emergency situations is less than 0.010 second.

When n is large, $n \geq 30$, we can base tests of the null hypothesis $\sigma = \sigma_0$ on the same theory as the large-sample confidence intervals of the preceding section. That is, we use the statistic

Statistic for large-sample test concerning standard deviation

$$z = \frac{s - \sigma_0}{\sigma_0/\sqrt{2n}}$$

whose sampling distribution is approximately the standard normal distribution, and the criteria of Figure 10.9.

EXAMPLE   The specifications for the mass production of certain springs require, among other things, that the standard deviation of their compressed lengths should not exceed 0.040 cm. If a random sample of size $n = 35$ from a certain production lot yields $s = 0.053$ and the probability of a Type I error is not to exceed 0.01, does this constitute evidence for the null hypothesis $\sigma \leq 0.040$ or for the alternative hypothesis $\sigma > 0.040$?

Solution   **1.** *Null hypothesis*: $\sigma \le 0.040$
Alternative hypothesis: $\sigma > 0.040$

2. *Level of significance*: $\alpha \le 0.01$

3. *Criterion*: Since the probability of a Type I error is greatest when $\sigma = 0.040$, we proceed as if we were testing the null hypothesis $\sigma = 0.040$ against the given alternative hypothesis at the level of significance $\alpha = 0.01$. Thus, the null hypothesis must be rejected if $z > 2.33$, where

$$z = \frac{s - \sigma_0}{\sigma_0 / \sqrt{2n}}$$

Otherwise, it must be accepted.

4. *Calculations*:

$$z = \frac{0.053 - 0.040}{0.040 / \sqrt{70}} = 2.72$$

5. *Decision*: Since $z = 2.72$ exceeds 2.33, the null hypothesis must be rejected; in other words, the springs in the production lot sampled do not meet the specifications.

11.3
TESTS CONCERNING TWO STANDARD DEVIATIONS

In this section we shall discuss a test concerning the equality of the standard deviations, or variances, of two populations. This test is often used in connection with the two-sample t test of Section 10.12, which requires that the variances of the two populations must be equal. For instance, in the example on page 299 dealing with the heat-producing capacity of the coal from two mines, we had $s_1^2 = 63,450$ and $s_2^2 = 42,650$, and despite what may seem to be a large difference, we assumed that the population variances were equal. We could not discuss the rationale of this assumption at that time and did not show the work in the text, but we actually tested—and were unable to reject—the null hypothesis that the two populations have equal variances, before we performed the t test for the significance of the difference between the two sample means.

Given independent random samples of size n_1 and n_2 from two populations, we usually base tests of the equality of the population standard deviations (or variances) on the ratios s_1^2/s_2^2 or s_2^2/s_1^2, where s_1^2 and s_2^2 are the two sample variances. If we assume that the two populations sampled have roughly the shape of normal distributions, it can be shown that the sampling distribution of such a ratio, appropriately called a **variance ratio**, is a continuous

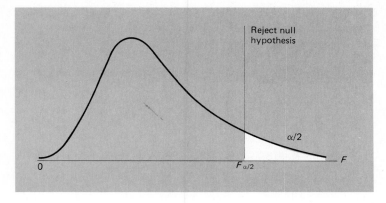

FIGURE 11.4 *F distribution*.

distribution called the *F* **distribution.** This distribution depends on the two parameters $n_1 - 1$ and $n_2 - 1$, the degrees of freedom in the sample estimates, s_1^2 and s_2^2, of the unknown population variances. One difficulty with this distribution is that most tables give only values of $F_{0.05}$ (defined in the same way as $z_{0.05}$, $t_{0.05}$, and $\chi_{0.05}^2$) and $F_{0.01}$, so we can work only with the right-hand tail of the distribution. For this reason, we base our decision about the equality of two population standard deviations, σ_1 and σ_2 (or variances σ_1^2 and σ_2^2) on the statistic

Statistic for test concerning the equality of two standard deviations

$$F = \frac{s_1^2}{s_2^2} \quad or \quad \frac{s_2^2}{s_1^2} \quad whichever\ is\ larger$$

With this statistic we reject the null hypothesis $\sigma_1 = \sigma_2$ at the level of significance α, and accept the alternative hypothesis $\sigma_1 \neq \sigma_2$, when the observed value of F exceeds $F_{\alpha/2}$ (see Figure 11.4). By using the right-hand tail area of $\alpha/2$ instead of α, we compensate for the fact that we always use the larger of the two variance ratios. The necessary values of $F_{\alpha/2}$ for $\alpha = 0.10$ or 0.02, that is, $F_{0.05}$ and $F_{0.01}$, are given in Table IV at the end of the book. The number of degrees of freedom for the numerator is $n_1 - 1$ or $n_2 - 1$, depending on whether we are using the ratio s_1^2/s_2^2 or the ratio s_2^2/s_1^2; correspondingly, the number of degrees of freedom for the denominator is $n_2 - 1$ or $n_1 - 1$.

EXAMPLE   In the example about the coal from two mines on page 299, we had $s_1^2 = 63,450$ and $s_2^2 = 42,650$ for two random samples of size $n = 5$. Use the 0.02 level of significance to test whether there is any real evidence that the standard deviations of the two populations are not equal.

Solution 1. *Null hypothesis*: $\sigma_1 = \sigma_2$
           Alternative hypothesis: $\sigma_1 \neq \sigma_2$
       2. *Level of significance*: $\alpha = 0.02$
       3. *Criterion*: Reject the null hypothesis if $F > 16.0$, the value of $F_{0.01}$ for $5 - 1 = 4$ and $5 - 1 = 4$ degrees of freedom, where

$$F = \frac{s_1^2}{s_2^2} \quad \text{or} \quad \frac{s_2^2}{s_1^2}$$

whichever is larger; otherwise, accept it.
       4. *Calculations*:

$$F = \frac{s_1^2}{s_2^2} = \frac{63{,}450}{42{,}650} = 1.49$$

       5. *Decision*: Since $F = 1.49$ does not exceed 16.0, the null hypothesis $\sigma_1 = \sigma_2$ cannot be rejected.

Since the test described here is very sensitive to departures from the underlying assumptions, it must be used with some caution. This is why, in actual practice, it is sometimes replaced by the method suggested at the end of Section 16.4.

EXERCISES

11.10 In a random sample, the amounts of time which 18 women took to complete the written test for their driver's licenses had a standard deviation of 2.1 minutes. Test the null hypothesis $\sigma = 2.5$ minutes against the alternative $\sigma \neq 2.5$ minutes at the 0.05 level of significance.

11.11 If 10 determinations of the specific heat of iron had a standard deviation of 0.0086, test the null hypothesis that $\sigma = 0.0100$ for such determinations. Use the alternative hypothesis $\sigma < 0.0100$ and the 0.05 level of significance.

11.12 With reference to Exercise 10.52 on page 301, test the null hypothesis $\sigma = 0.1$ ton per acre against the alternative hypothesis $\sigma > 0.1$ ton per acre at the 0.01 level of significance.

11.13 With reference to Exercise 10.53 on page 301, test the null hypothesis $\sigma = 30$ words per minute against the alternative hypothesis $\sigma \neq 30$ words per minute at the 0.05 level of significance.

11.14 Past data indicate that the variance of measurements made on sheet metal stampings by experienced inspectors is 0.18 square inch. If a new inspector measures 100 stampings with a variance of 0.25 square inch, test at the 0.05 level of significance whether the inspector is making satisfactory measurements (against the alternative that their variability is excessive).

11.15 With reference to Exercise 10.40 on page 299, test at the 0.01 level of significance whether $\sigma = 4.2$ months for the average time that convicted embezzlers spend in jail.

11.16 With reference to Exercise 10.42 on page 300, test at the 0.05 level of significance whether $\sigma = 2.0$ minutes or $\sigma < 2.0$ minutes.

11.17 Two different lighting techniques are compared by measuring the intensity of light at selected locations in areas lighted by the two methods. If 12 measurements of the first technique have a standard deviation of 2.6 foot-candles and 16 measurements of the second technique have a standard deviation of 4.4 foot-candles, test the null hypothesis $\sigma_1 = \sigma_2$ against the alternative hypothesis $\sigma_1 \neq \sigma_2$ at the 0.10 level of significance.

11.18 With reference to Exercise 10.60 on page 302, test at the 0.02 level of significance whether $\sigma_1 = \sigma_2$ or $\sigma_1 \neq \sigma_2$.

11.19 With reference to Exercise 10.61 on page 302, test at the 0.02 level of significance whether $\sigma_1 = \sigma_2$ or $\sigma_1 \neq \sigma_2$.

11.4
CHECKLIST OF KEY TERMS
(with page references to their definitions)

Chi-square distribution, 311
Chi-square statistic, 311
Degrees of freedom, 311
F distribution, 319
Variance ratio, 318

11.5
REVIEW EXERCISES

11.20 In a random sample of the scores on 12 rounds of golf played on her home course, a golf professional averaged 71.6 with a standard deviation of 1.48. Use the 0.01 level of significance to test the null hypothesis $\sigma = 1.10$ against the alternative hypothesis that her game is actually less consistent.

11.21 The refractive indices of 16 pieces of glass (randomly selected from a large shipment purchased by an optical firm) have $s = 0.011$. Construct a 95 percent confidence interval for the true standard deviation of the population sampled.

11.22 Independent random samples of size $n_1 = 10$ and $n_2 = 12$ from normal populations have the standard deviations $s_1 = 9.2$ and $s_2 = 15.8$. Use the 0.02 level of significance to test whether or not the two populations have equal standard deviations.

11.23 Test runs with six models of an experimental engine showed that they operated for 25, 31, 19, 32, 27, and 24 minutes with a gallon of a certain kind of fuel. Estimate the standard deviation of the population sampled using
   (a) the sample standard deviation;
   (b) the sample range and the method of Exercise 11.9.

11.24 In a random sample of 50 tax returns, the amount due after an audit averaged $450.50 with a standard deviation of $38.40. Construct a 99 percent confidence interval for the true standard deviation of such amounts.

11.25 With reference to Exercise 10.62 on page 302, test at the 0.10 level of significance whether $\sigma_A = \sigma_B$ or $\sigma_A \neq \sigma_B$.

11.26 In a random sample, 40 lawyers had an average annual income of $72,460 with a standard deviation of $8,290. Letting the probability of a Type I error be at most 0.05, test the null hypothesis $\sigma \geq \$9,500$ against the alternative hypothesis $\sigma < \$9,500$ for the population sampled.

11.6
REFERENCES

Theoretical discussions of the chi-square and F distributions may be found in most textbooks on mathematical statistics; for instance, in

FREUND, J. E., and WALPOLE, R. E., *Mathematical Statistics, 3rd ed.* Englewood Cliffs, N.J.: Prentice-Hall, Inc., 1980.

In principle, the work of this chapter is very similar to that of Chapters 10 and 11. In problems of estimation we shall again construct confidence intervals, or assess the maximum error. In tests of hypotheses we shall again formulate null hypotheses and alternative hypotheses, choose a level of significance, construct one-tailed tests or two-tailed tests, and so forth. The main difference is that we will be concerned with other parameters; instead of population means or standard deviations, we will deal with population proportions, percentages, or probabilities.

Sections 12.1 and 12.2 deal with the estimation of proportions; Section 12.3 deals with tests concerning proportions; Sections 12.4 and 12.5 deal with tests concerning two or more proportions; in Section 12.6 we shall learn how to analyze data tallied into two-way classifications; and in Section 12.7 we shall learn how to judge whether the differences between an observed frequency distribution and corresponding expectations can be attributed to chance.

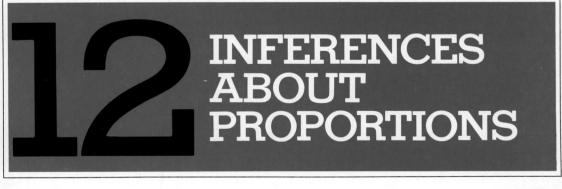

12 INFERENCES ABOUT PROPORTIONS

12.1
THE ESTIMATION OF PROPORTIONS

The information that is usually available for the estimation of a true proportion (percentage, or probability) is a **sample proportion** $\frac{x}{n}$, where x is the number of times that an event has occurred in n trials. For instance, if a study shows that 54 of 120 cheerleaders, presumably a random sample, suffered what auditory experts called "moderate to severe damage" to their voices, then $\frac{x}{n} = \frac{54}{120} = 0.45$, and we can use this figure as a point estimate of the true proportion of cheerleaders who are afflicted in this way. Similarly, a supermarket chain might estimate the proportion of its shoppers who regularly use cents-off coupons as 0.68, if a random sample of 300 shoppers included 204 who regularly use cents-off coupons.

Throughout this section it will be assumed that the situations satisfy (at least approximately) the conditions underlying the binomial distribution; that is, our information will consist of the number of successes observed in a given number of independent trials, and it will be assumed that for each trial the probability of a success—the parameter we want to estimate—has the constant value p. Thus, the sampling distribution of the counts on which our methods will be based is the binomial distribution with the mean $\mu = np$ and the standard deviation $\sigma = \sqrt{np(1-p)}$, and we know that this distribution can be approximated with a normal distribution when n is large.[†] It follows that, for large values of n, the statistic

$$z = \frac{x - np}{\sqrt{np(1-p)}}$$

has approximately the standard normal distribution. If we substitute this expression for z into the inequality $-z_{\alpha/2} < z < z_{\alpha/2}$ (as on page 269) and use some simple algebra, we arrive at

$$\frac{x}{n} - z_{\alpha/2}\sqrt{\frac{p(1-p)}{n}} < p < \frac{x}{n} + z_{\alpha/2}\sqrt{\frac{p(1-p)}{n}}$$

[†] On page 223 we said that the normal approximation to the binomial distribution may be used when np and $n(1-p)$ are both greater than 5. So, when $n = 50$, for example, the normal-curve methods discussed here and later in this chapter may be used if p lies between 0.10 and 0.90; when $n = 100$ they may be used if p lies between 0.05 and 0.95; and when $n = 200$ they may be used if p lies between 0.025 and 0.975. This illustrates what we mean here by "n being large."

This may look like a confidence interval formula for p [indeed, if it is repeatedly used the inequalities should be satisfied $(1 - \alpha)100$ percent of the time] but observe that the unknown parameter p appears not only in the middle, but also in $\sqrt{\dfrac{p(1 - p)}{n}}$ to the left of the first inequality sign and to the right of the other. The quantity $\sqrt{\dfrac{p(1 - p)}{n}}$ is called the **standard error of a proportion,** as it is, in fact, the standard deviation of the sampling distribution of a sample proportion (see Exercise 12.23 on page 332).

To get around this difficulty, we substitute $\dfrac{x}{n}$ for p in $\sqrt{\dfrac{p(1 - p)}{n}}$, and we thus arrive at the following $(1 - \alpha)100$ **percent large-sample confidence interval for** p:

*Large-sample
confidence
interval for p*

$$\frac{x}{n} - z_{\alpha/2}\sqrt{\frac{\dfrac{x}{n}\left(1 - \dfrac{x}{n}\right)}{n}} < p < \frac{x}{n} + z_{\alpha/2}\sqrt{\frac{\dfrac{x}{n}\left(1 - \dfrac{x}{n}\right)}{n}}$$

EXAMPLE   If 400 persons, constituting a random sample, were given a flu vaccine and 136 of them experienced some discomfort, construct a 95 percent confidence interval for the true proportion of persons who experience discomfort from the vaccine.

Solution   Substituting $n = 400$, $\dfrac{x}{n} = \dfrac{136}{400} = 0.34$, and $z_{0.025} = 1.96$ into the formula, we get

$$0.34 - 1.96\sqrt{\frac{(0.34)(0.66)}{400}} < p < 0.34 + 1.96\sqrt{\frac{(0.34)(0.66)}{400}}$$

$$0.29 < p < 0.39$$

Clearly, this interval either contains the true proportion p or it does not, and we really don't know which. However, the 95 percent confidence implies that the interval was obtained by a method which leads to correct results 95 percent of the time.

The large-sample theory presented here can also be used to judge the maximum size of the error when we use a sample proportion as a point estimate of a population proportion p. Looking at the problem in this way, we can assert with $(1 - \alpha)100$ percent confidence that the size of our error is at most

$$E = z_{\alpha/2} \sqrt{\frac{p(1 - p)}{n}}$$

or, approximately, that it is at most

Maximum error of estimate

$$E = z_{\alpha/2} \sqrt{\frac{\dfrac{x}{n}\left(1 - \dfrac{x}{n}\right)}{n}}$$

where we again substituted $\dfrac{x}{n}$ for p in $\sqrt{\dfrac{p(1 - p)}{n}}$.

EXAMPLE In a random sample of 150 vacationers interviewed at a resort, 108 said that they chose the resort mainly because of its climate. With 99 percent confidence, what can we say about the maximum error, if we use $\dfrac{x}{n} = \dfrac{108}{150} = 0.72$ as an estimate of the true proportion of vacationers who choose the resort mainly because of its climate.

Solution Substituting $n = 150$, $\dfrac{x}{n} = 0.72$, and $z_{0.005} = 2.575$ into the formula for E, we get

$$E = 2.575 \sqrt{\frac{(0.72)(0.28)}{150}} = 0.094$$

As in the estimation of means, we can use the expression for the maximum error to determine how large a sample is needed to attain a desired degree of precision. Suppose that we want to use a sample proportion to estimate a population proportion p, and that we want to be able to assert with $(1 - \alpha)100$ percent confidence that the error of this estimate is not greater than some prescribed quantity E. As before, we write

$$E = z_{\alpha/2} \sqrt{\frac{p(1 - p)}{n}}$$

and upon solving this equation for n, we get

Sample size

$$n = p(1 - p) \left[\frac{z_{\alpha/2}}{E}\right]^2$$

Since this formula involves p, it cannot be used unless we have some information about the possible values p might assume. Without such information, we make use of the fact that $p(1 - p)$ equals $\frac{1}{4}$ when $p = \frac{1}{2}$ and it is smaller than $\frac{1}{4}$ for all other values of p. Hence, if we use the formula

Sample size

$$n = \frac{1}{4}\left[\frac{z_{\alpha/2}}{E}\right]^2$$

our sample may be larger than necessary, and we will be able to say that our confidence is at least $(1 - \alpha)100$ percent that the error does not exceed E.

When we do have some information about the values that p might assume, we can take this into account by substituting into the first of the two formulas for n whichever of these values of p is closest to $\frac{1}{2}$. For instance, if it is reasonable to suppose that the true proportion we are trying to estimate lies on the interval from 0.20 to 0.40, we substitute $p = 0.40$.

EXAMPLE Suppose that we want to estimate what proportion of the adult population of the United States has high blood pressure, and that we want to be at least 99 percent confident that the error of our estimate is at most 0.05. How large a sample will we need if

(a) we have no idea what the true proportion might be;

(b) we know that the true proportion may be anywhere from 0.05 to 0.20?

Solution (a) Substituting $E = 0.05$ and $z_{0.005} = 2.575$ into the second of the two formulas for n, we get

$$n = \tfrac{1}{4}\left[\frac{2.575}{0.05}\right]^2 = 664$$

rounded up to the nearest integer; (b) substituting these same values together with $p = 0.20$ into the first formula, we get

$$n = (0.20)(0.80)\left[\frac{2.575}{0.05}\right]^2 = 425$$

rounded up to the nearest integer. This shows how some knowledge about the values p might assume can substantially reduce the sample size needed to attain a desired degree of precision.

The methods which we have discussed in this section are all large-sample techniques. For small samples, confidence intervals for proportions may be found from special tables, such as the ones given in earlier editions of this book; nowadays, they are rarely used.

12.2
THE ESTIMATION OF PROPORTIONS
(A Bayesian Method) ⋆

In the preceding section we looked upon the true proportion p (which we tried to estimate) as an unknown constant; in Bayesian estimation this parameter is looked upon as a random variable having a prior distribution which reflects one's belief about the values it can assume. As in the Bayesian estimation of means, we are thus faced with the problem of combining prior information with direct sample evidence.

To give an example, consider a large company which routinely pays thousands of invoices submitted by its suppliers. Of course, it is of interest to know what proportion of these invoices might contain errors, and interviews of three of the company's executives reveal the following information: Mr. Martin feels that only 0.005 (or a half of 1 percent) of the invoices contain errors; Mr. Green, who is generally regarded to be as reliable in his estimates as Mr. Martin, feels that 0.01 (or 1 percent) of the invoices contain errors; and Mr Jones, who is generally regarded to be twice as reliable in his estimates as either Mr. Martin or Mr. Green, feels that 0.02 (or 2 percent) of the invoices contain errors. Assuming that $p = 0.005$, $p = 0.01$, and $p = 0.02$ are the only possibilities, we thus have the following prior probabilities for the proportion of invoices containing errors:

p	Prior probability
0.005	0.25
0.010	0.25
0.020	0.50

(The prior probabilities assigned to Mr. Martin's and Mr. Green's estimates are the same, and that assigned to Mr. Jones's estimate is twice as large.)

Now suppose that a random sample of 200 invoices is carefully checked, and that only one of them contains an error. The probabilities of this happening when $p = 0.005$, $p = 0.010$, or $p = 0.020$ are

$$\binom{200}{1}(0.005)^1(0.995)^{199} = 0.37$$

$$\binom{200}{1}(0.010)^1(0.990)^{199} = 0.27$$

and

$$\binom{200}{1}(0.020)^1(0.980)^{199} = 0.07$$

where we used the formula for the binomial distribution, and logarithms to perform the calculations. Combining these probabilities with the prior probabilities by using the formula for Bayes' theorem (see Section 5.7), we find that the **posterior probability** of $p = 0.005$ is

$$\frac{(0.25)(0.37)}{(0.25)(0.37) + (0.25)(0.27) + (0.50)(0.07)} = 0.47$$

while the corresponding posterior probabilities of $p = 0.010$ and $p = 0.020$ are

$$\frac{(0.25)(0.27)}{(0.25)(0.37) + (0.25)(0.27) + (0.50)(0.07)} = 0.35$$

and

$$\frac{(0.50)(0.07)}{(0.25)(0.37) + (0.25)(0.27) + (0.50)(0.07)} = 0.18$$

We have thus arrived at the following posterior distribution for the proportion of invoices containing errors:

p	Posterior probability
0.005	0.47
0.010	0.35
0.020	0.18

This distribution reflects the prior judgments as well as the direct sample evidence, and it should not come as a surprise that the highest posterior probability goes to $p = 0.005$—after all, the sample proportion actually equaled $\frac{x}{n} = \frac{1}{200} = 0.005$.

To continue, we could use the mean of the posterior distribution, namely, $(0.005)(0.47) + (0.010)(0.35) + (0.020)(0.18) = 0.009$ as a point estimate of the true proportion of invoices containing errors. In contrast to the mean of the prior distribution, which was $(0.005)(0.25) + (0.010)(0.25) + (0.020)(0.50) = 0.014$, the mean of the posterior distribution reflects the prior judgments as well as the direct sample evidence.

In the preceding example it was assumed that p had to be 0.005, 0.010, or 0.020 to simplify the calculations. The method of analysis would have been the same, however, if we had considered ten different values of p or even a hundred. In fact, there exist Bayesian techniques, similar to that of

Section 10.4, in which p can take on any value on the continuous interval from 0 to 1, and its prior distribution is a continuous curve.

EXERCISES

12.1 In a random sample of 200 persons interviewed in a large city, 108 said that they oppose the construction of any more freeways. Construct a 95 percent confidence interval for the corresponding proportion for the population sampled.

12.2 With reference to the preceding exercise, what can we say with 99 percent confidence about the maximum error, if we use the sample proportion as an estimate of the proportion of persons in the given city who oppose the construction of any more freeways?

12.3 Among 100 fish caught in a certain lake, 18 were inedible as a result of the chemical pollution of the environment. Construct a 99 percent confidence interval for the corresponding true proportion.

12.4 With reference to the preceding exercise, what can we say with 95 percent confidence about the maximum error, if we use the sample proportion as an estimate of the true proportion of the fish in the lake that is inedible as a result of chemical pollution?

12.5 A random sample of 300 shoppers at a supermarket includes 204 who regularly use cents-off coupons. Construct a 98 percent confidence interval for the true proportion of shoppers at the supermarket who use cents-off coupons.

12.6 In a random sample of 120 cheerleaders, 54 had suffered moderate to severe damage to their voices. With 90 percent confidence, what can we say about the maximum error, if we use the sample proportion, $\frac{54}{120} = 0.45$, as an estimate of the true proportion of cheerleaders who are afflicted in this way?

12.7 In a random sample of 1,200 voters interviewed nationwide, only 324 felt that the salaries of certain government officials should be raised. Construct a 95 percent confidence interval of the corresponding true proportion.

12.8 In a random sample of 400 television viewers interviewed in a certain area, 152 had seen a certain controversial program. With 99 percent confidence, what can we say about the maximum error, if we use the sample proportion $\frac{152}{400} = 0.38$, as an estimate of the corresponding true proportion?

12.9 In a random sample of 240 high school seniors in a Western state, 168 said that they expect to continue their education at an in-state college or university. Construct a 95 percent confidence interval for the corresponding true percentage.

12.10 With reference to the preceding exercise, what can we say with 98 percent confidence about the maximum error, if we use $\frac{168}{240} \cdot 100 = 70$ as an estimate of the actual percentage of high school seniors in the state who expect to continue their education at an in-state college or university?

12.11 In a random sample of 140 supposed UFO sightings, 119 could easily be explained in terms of natural phenomena. Construct a 99 percent confidence interval for the probability that a supposed UFO sighting will easily be explained in terms of natural phenomena.

12.12 In a random sample of 80 persons convicted in U.S. District Courts on narcotics charges, 36 received probation. With 95 percent confidence, what can we say about the maximum error, if we use $\frac{36}{80} = 0.45$ as an estimate of the probability that a person convicted in a U.S. District Court on narcotics charges will receive probation?

12.13 If a sample constitutes at least 5 percent of a finite population, and the sample itself is large, we can use the finite population correction factor to reduce the width of confidence intervals for p. If we make this correction, the large-sample confidence limits for p become

Large-sample confidence limits for p (finite population)

$$\frac{x}{n} \pm z_{\alpha/2} \sqrt{\frac{\frac{x}{n}\left(1 - \frac{x}{n}\right)}{n}} \cdot \sqrt{\frac{N - n}{N - 1}}$$

where N is, as before, the size of the population sampled.

(a) Among the 360 families in an apartment complex a random sample of 100 is interviewed, and it is found that 34 of the families have children of college age. Construct a 95 percent confidence interval for the actual proportion of all the families living in the complex who have children of college age.

(b) A shoe manufacturer feels that unless its employees agree to a 5 percent wage reduction it cannot stay in business. If in a random sample of 80 of the 400 employees only 24 felt "kindly disposed" to the reduction, find a 99 percent confidence interval for the corresponding proportion for all the employees.

12.14 A private opinion poll is engaged by a politician to estimate what proportion of her constituents favor the decriminalization of certain narcotics violations. How large a sample will the poll have to take to be at least 95 percent confident that the sample proportion is off by at most 0.02?

12.15 Rework the preceding exercise, given that the poll has reason to believe that the true proportion does not exceed 0.30.

12.16 Suppose that we want to estimate what proportion of all drivers exceed the 55-mph speed limit on a stretch of road between Los Angeles and Bakersfield. How large a sample will we need to be at least 99 percent confident that the error of our estimate, the sample proportion, is at most 0.04?

12.17 Rework the preceding exercise, given that we have good reason to believe that the proportion we are trying to estimate is at least 0.65.

12.18 A national manufacturer wants to determine what percentage of purchases of razor blades for use by men is actually made by women. How large a sample will the manufacturer need to be at least 98 percent confident that the sample percentage is not off by more than 2.5 percent?

12.19 A large finance company wants to estimate from a sample the probability that any one of its many customers will make a major purchase on credit during the coming year. How large a sample will it need to be at least 99 percent confident that the difference between the sample proportion and the actual probability does not exceed 0.03?

★12.20 In planning the operation of a new school, one school board member claims that 4 out of 5 newly hired teachers will stay with the school for more than a year, while another school board member claims that it would be correct to say 7 out of 10. In the past, the two board members have been about equally reliable in their predictions, so that in the absence of direct information we would assign their judgments equal weight. What posterior probabilities would we assign to their claims if it were found that 11 of 12 newly hired teachers stayed with the school for more than a year?

★12.21 The landscaping plans for a new hotel call for a row of palm trees along the driveway. The landscape designer tells the owner that if he plants *Washingtonia filifera*, 20 percent of the trees will fail to survive the first heavy frost, the manager of the nursery which supplies the trees tells the owner that 10 percent of the trees will fail to survive the first heavy frost, and the owner's wife tells him that 30 percent of the trees will fail to survive the first heavy frost.

    (a) If the owner feels that in this matter the landscape designer is 10 times as reliable as his wife and the manager of the nursery is 9 times as reliable as his wife, what prior probabilities should he assign to these percentages?

    (b) If 13 of these palm trees are planted and 2 fail to survive the first heavy frost, what posterior probabilities should the manager assign to the three percentages?

★12.22 The method of Section 12.2 can also be used to find the posterior distribution of the number of "successes" in a finite population. Suppose, for instance, that a coin dealer receives a shipment of five ancient coins from abroad, and that, on the basis of past experience, he feels that the probabilities that 0, 1, 2, 3, 4, or all 5 of them are counterfeits are 0.74, 0.11, 0.02, 0.01, 0.02, and 0.10. Since modern methods of counterfeiting have become greatly refined, the cost of authentication has risen sharply and the dealer decides to select one of the coins at random and send it away for authentication. If it turns out that the coin is a forgery, find

    (a) the posterior probabilities that 0, 1, 2, 3, or all 4 of the remaining coins are counterfeits;

    (b) the mean of the posterior distribution obtained in part (a) and use it to estimate what proportion of the remaining coins are counterfeits.

12.23 Since the proportion of successes is simply the number of successes divided by n, the mean and the standard deviation of the sampling distribution of the proportion of successes may be obtained by dividing by n the mean and the standard deviation of the sampling distribution of the number of successes. Use this argument to verify the standard error formula given on page 325.

12.3
TESTS CONCERNING PROPORTIONS

In this section we shall be concerned with tests of hypotheses which enable us to decide, on the basis of sample data, whether the true value of a proportion (percentage, or probability) equals, is greater than, or is less than a given constant. These tests will make it possible, for example, to determine whether the true proportion of fifth graders who can name the governor of their state is 0.35, whether it is true that 10 percent of the answers which the IRS gives to taxpayers' telephone inquiries are in error, or whether the true probability is 0.25 that the downtime of a new computer will exceed two hours in any given week.

Questions of this kind are usually decided on the basis of the observed number of successes in n trials, or the observed proportion of successes, and it will be assumed throughout this section that these trials are independent and that the probability of a success is the same for each trial. In other words, we shall assume that we can use the binomial distribution and that we are, in fact, testing hypotheses about the parameter p of binomial populations.

When n is small, tests concerning true proportions can be based directly on tables of binomial probabilities such as Table V at the end of the book. Suppose, for instance, that we want to investigate the claim that at least 60 percent of the students attending a large university are opposed to a plan to increase student fees in order to build new parking facilities. Specifically, we want to test the null hypothesis $p = 0.60$ against the alternative hypothesis $p < 0.60$ on the basis of a random sample of size $n = 14$ and at the 0.05 level of significance.

In the test criteria of the two preceding chapters, where we always dealt with continuous sampling distributions, we were able to draw the dividing lines so that the probabilities associated with the tails, or regions of rejection, were exactly α or $\alpha/2$. Since this cannot always be done when we deal with binomial or other discrete random variables, we modify the criteria as follows:

> We draw the dividing lines so that the probability of getting a value in the "tail" is as close as possible to the level of significance α (or to $\alpha/2$ in a two-tailed test) without exceeding it.

Thus, for our example we observe from Table V that for $p = 0.60$ and $n = 14$ the probability of 4 or fewer successes is

$$0.001 + 0.003 + 0.014 = 0.018$$

and that the probability of 5 or fewer successes is

$$0.001 + 0.003 + 0.014 + 0.041 = 0.059$$

Since the first of these probabilities is less than $\alpha = 0.05$ and the second probability exceeds it, we reject the null hypothesis $p = 0.60$ (and accept the alternative hypothesis $p < 0.60$) when in a random sample of $n = 14$ students interviewed there are at most 4 who are opposed to the plan to increase student fees in order to build new parking facilities (see also Figure 12.1).

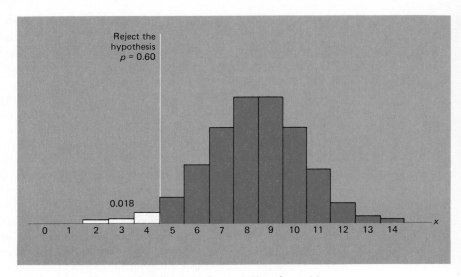

FIGURE 12.1 *Binomial distribution with p = 0.60 and n = 14.*

In applications where we are given x, the observed number of successes, we do not actually have to find the dividing lines of the test criteria. Instead, we proceed as in the following example, which concerns a two-tailed test about a proportion:

EXAMPLE It has been claimed that 40 percent of all shoppers can identify a highly advertised trade mark. If $x = 4$ shoppers can identify the trade mark in a random sample of size $n = 20$, test at the 0.05 level of significance whether to accept the null hypothesis $p = 0.40$ or the alternative hypothesis $p \neq 0.40$.

Solution 1. *Null hypothesis*: $p = 0.40$
   Alternative hypothesis: $p \neq 0.40$
 2. *Level of significance*: $\alpha = 0.05$
 3. *Criterion*: Reject the null hypothesis if the probability of $x \leq 4$, or that of $x \geq 4$, is less than or equal to $\alpha/2 = 0.025$; otherwise, accept it.
 4. *Calculations*: Table V shows that the probability of $x \leq 4$ is

$$0.003 + 0.012 + 0.035 = 0.050$$

and that the probability of $x \geq 4$ is

$$0.035 + 0.075 + \cdots + 0.005 + 0.001 = 0.984$$

 5. *Decision*: Since both of these probabilities exceed 0.025, the null hypothesis cannot be rejected; we conclude that 40 percent of all shoppers can identify the trade mark.

When n is large, tests concerning population proportions (percentages, or probabilities) are usually based on the normal-curve approximation to the binomial distribution. Using again the z-statistic which led to the large-sample confidence interval for p, we base the test of the null hypothesis $p = p_0$ on the statistic

Statistic for large-sample test concerning proportion

$$z = \frac{x - np_0}{\sqrt{np_0(1 - p_0)}}$$

which has approximately the standard normal distribution.[†] The test criteria are again those of Figure 10.9 on page 293 with p and p_0 substituted for μ and μ_0. For the one-sided alternative $p < p_0$ we reject the null hypothesis when $z < -z_\alpha$; for the one-sided alternative $p > p_0$ we reject the null hypothesis when $z > z_\alpha$; and for the two-sided alternative $p \neq p_0$ we reject the null hypotheses when $z < -z_{\alpha/2}$ or $z > z_{\alpha/2}$.

EXAMPLE A nutritionist claims that at least 75 percent of the preschool children in a certain country have protein-deficient diets. If a sample survey reveals that 206 of 300 preschool children in that country have protein-deficient diets and the probability of a Type I error is to be at most 0.01, should we accept the null hypothesis $p \geq 0.75$ or the alternative hypothesis $p < 0.75$?

Solution
1. *Null hypothesis*: $p \geq 0.75$
   Alternative hypothesis: $p < 0.75$
2. *Level of significance*: $\alpha \leq 0.01$
3. *Criterion*: Since the probability of a Type I error is greatest when $p = 0.75$, we proceed as if we were testing the null hypothesis $p = 0.75$ against the alternative hypothesis $p < 0.75$. Thus, the null hypothesis must be rejected if $z < -2.33$, where

$$z = \frac{x - np_0}{\sqrt{np_0(1 - p_0)}}$$

and otherwise it must be accepted.
4. *Calculations*:

$$z = \frac{206 - 300(0.75)}{\sqrt{300(0.75)(0.25)}} = -2.53$$

[†] Many statisticians make a continuity correction here by substituting $x - \frac{1}{2}$ or $x + \frac{1}{2}$ for x in the formula for z, whichever makes z smaller. However, when n is large the effect of this correction is usually negligible.

5. *Decision*: Since $z = -2.53$ is less than -2.33, the null hypothesis must be rejected; in other words, we conclude that fewer than 75 percent of the preschool children in the given country have protein-deficient diets.

EXAMPLE A television critic claims that 80 percent of all viewers find the noise level of a certain commercial objectionable. If a random sample of 320 television viewers includes 245 who find the noise level of the commercial objectionable, use the 0.05 level of significance to test whether the difference between $\dfrac{x}{n} = \dfrac{245}{320} = 0.766$ and $p = 0.80$ is significant.

Solution
1. *Null hypothesis*: $p = 0.80$
   Alternative hypothesis: $p \neq 0.80$
2. *Level of significance*: $\alpha = 0.05$
3. *Criterion*: Reject the null hypothesis if $z < -1.96$ or $z > 1.96$, where

$$z = \frac{x - np_0}{\sqrt{np_0(1 - p_0)}}$$

Otherwise, state that the difference is not significant.
4. *Calculations*:

$$z = \frac{245 - 320(0.80)}{\sqrt{320(0.80)(0.20)}} = -1.54$$

5. *Decision*: Since $z = -1.54$ falls on the interval from -1.96 to 1.96, the null hypothesis cannot be rejected; in other words, we find that the difference between $\frac{245}{320} = 0.766$ and $p = 0.80$ is not significant.

EXERCISES

12.24 A doctor claims that only 10 percent of all persons exposed to a certain amount of radiation will feel any ill effects. If 18 persons are checked, at least how many of them will have to feel some ill effects so that the null hypothesis $p = 0.10$ can be rejected against the alternative hypothesis $p > 0.10$ at the 0.05 level of significance?

12.25 Suppose that we want to test the null hypothesis $p = 0.30$ against the alternative hypothesis $p \neq 0.30$ at the 0.05 level of significance. How many successes must be observed in $n = 20$ trials for the null hypothesis to be rejected?

12.26 A coin is flipped 15 times to check whether it is balanced, namely, whether the probability for heads is 0.50. How many heads do we have to get so that we can reject the null hypothesis $p = 0.50$ against the alternative hypothesis $p \neq 0.50$
   (a) at the 0.05 level of significance;
   (b) at the 0.01 level of significance?

12.27 The manufacturer of a spot remover claims that his product removes 90 percent of all spots. If, in a random sample, the spot remover removes 11 of 16 spots, test the null hypothesis $p = 0.90$ against the alternative hypothesis $p < 0.90$ at the 0.05 level of significance.

12.28 In a random sample of 12 undergraduate business students, 6 say that they will take advanced work in accounting. Test the claim that 20 percent of all undergraduate business students will take advanced work in accounting, using the alternative hypothesis $p > 0.20$ and the 0.01 level of significance.

12.29 A food processor wants to know whether the probability is really 0.60 that a customer will prefer a new kind of packaging to the old kind. If, in a random sample, 7 of 18 customers prefer the new kind of packaging to the old kind, test the null hypothesis $p = 0.60$ against the alternative hypothesis $p \neq 0.60$ at the 0.05 level of significance.

12.30 In a random sample of 14 industrial accidents, 12 were due to unsafe working conditions. Test the claim that 50 percent of all industrial accidents are due to unsafe working conditions, using the 0.05 level of significance and the alternative hypothesis $p \neq 0.50$.

12.31 In a random sample of 200 automobile accidents, it was found that 64 were due at least in part to driver fatigue. Use the 0.05 level of significance to test whether or not this supports the claim that 35 percent of all automobile accidents are due at least in part to driver fatigue.

12.32 In a study of aviophobia, a psychologist claims that 30 percent of all women are afraid of flying. If, in a random sample, 41 of 150 women are afraid of flying, test the null hypothesis $p = 0.30$ against the alternative hypothesis $p \neq 0.30$ at the 0.05 level of significance.

12.33 An airline claims that at most 6 percent of all lost luggage is never found. If, in a random sample, 17 of 200 pieces of lost luggage are not found and the probability of a Type I error is not to exceed 0.01, test the null hypothesis $p \leq 0.06$ against the alternative hypothesis $p > 0.06$.

12.34 In a random sample of 600 cars making a right turn at a certain intersection, 157 pulled into the wrong lane. Test the null hypothesis that actually 30 percent of all drivers make this mistake at the given intersection, using the alternative hypothesis $p \neq 0.30$ and the level of significance
    (a) $\alpha = 0.05$;
    (b) $\alpha = 0.01$.

12.35 To check an ambulance service's claim that 40 percent of its calls are life-threatening emergencies, a random sample was taken from its files, and it was found that 49 of 150 calls were life-threatening emergencies. Test the null hypothesis $p = 0.40$ against the alternative hypothesis $p < 0.40$ at the 0.05 level of significance.

12.36 In the construction of tables of random numbers there are various ways of testing for possible departures from randomness. For instance, there should be about as many even digits (0, 2, 4, 6, or 8) as there are odd digits (1, 3, 5, 7, or 9). Count the number of even digits among the 350 digits constituting the first ten rows of the table on page 518, and test at the 0.05 level of significance whether, on the basis of this criterion, there is any reason to be concerned about the possibility that the random numbers are, in fact, not random.

12.4
DIFFERENCES BETWEEN PROPORTIONS

There are many problems in which we must decide whether an observed difference between two sample proportions can be attributed to chance, or whether it is indicative of the fact that the corresponding population proportions are not equal. For instance, we may want to decide on the basis of sample data whether there is a difference between the actual proportions of persons with and without flu shots who catch the disease, or we may want to test on the basis of samples whether two manufacturers of electronic equipment ship equal proportions of defectives.

The method we shall use to test whether an observed difference between two sample proportions is significant, or whether it can be attributed to chance, is based on the following theory: If x_1 and x_2 are the numbers of successes obtained in n_1 trials of one kind and n_2 of another, the trials are all independent, and the corresponding probabilities of a success are, respectively, p_1 and p_2, then the sampling distribution of $\dfrac{x_1}{n_1} - \dfrac{x_2}{n_2}$ has the mean $p_1 - p_2$ and the standard deviation

$$\sqrt{\frac{p_1(1 - p_1)}{n_1} + \frac{p_2(1 - p_2)}{n_2}}$$

called the **standard error of the difference between two proportions**.

When we test the null hypothesis $p_1 = p_2 (= p)$ against an appropriate alternative hypothesis, the mean of the sampling distribution of the difference between two sample proportions is $p_1 - p_2 = 0$, and the standard error formula can be written

$$\sqrt{p(1 - p)\left(\frac{1}{n_1} + \frac{1}{n_2}\right)}$$

where p is usually estimated by **pooling** the data and substituting for p the combined sample proportion $\hat{p} = \dfrac{x_1 + x_2}{n_1 + n_2}$, which reads "$p$-hat." Then, since for large samples the sampling distribution of the difference between two proportions can be approximated closely with a normal distribution, we base the test on the statistic

Statistic for test concerning difference between two proportions

$$z = \frac{\dfrac{x_1}{n_1} - \dfrac{x_2}{n_2}}{\sqrt{\hat{p}(1 - \hat{p})\left(\dfrac{1}{n_1} + \dfrac{1}{n_2}\right)}} \quad with \quad \hat{p} = \frac{x_1 + x_2}{n_1 + n_2}$$

which has approximately the standard normal distribution. The test criteria are again those of Figure 10.9 with $p_1 - p_2$ substituted for μ and 0 substituted for μ_0. For the one-sided alternative hypothesis $p_1 < p_2$ we reject the null hypothesis if $z < -z_\alpha$, for the one-sided alternative hypothesis $p_1 > p_2$ we reject the null hypothesis if $z > z_\alpha$, and for the two-sided alternative hypothesis $p_1 \neq p_2$ we reject the null hypothesis if $z < -z_{\alpha/2}$ or $z > z_{\alpha/2}$.

EXAMPLE To test the effectiveness of a new pain-relieving drug, 80 patients at a clinic were given a pill containing the drug and 80 others were given a placebo containing only powdered sugar. If 56 of the patients in the first group and 38 of those in the second group felt a beneficial effect, what can we conclude about the effectiveness of the new drug at the 0.01 level of significance?

Solution
1. *Null hypothesis*: $p_1 = p_2$
   Alternative hypothesis: $p_1 > p_2$
2. *Level of significance*: $\alpha = 0.01$
3. *Criterion*: Reject the null hypothesis if $z > 2.33$, where z is given by the formula on page 338; otherwise, state that the difference between the two sample proportions is not significant.
4. *Calculations*: Substituting $x_1 = 56, x_2 = 38, n_1 = 80, n_2 = 80$, and $\hat{p} = \dfrac{56 + 38}{80 + 80} = 0.5875$ into the formula for z, we get

$$z = \frac{\dfrac{56}{80} - \dfrac{38}{80}}{\sqrt{(0.5875)(0.4125)\left(\dfrac{1}{80} + \dfrac{1}{80}\right)}} = 2.89$$

5. *Decision*: Since $z = 2.89$ exceeds 2.33, the null hypothesis must be rejected; in other words, we conclude that the new pain-relieving drug is effective.

12.5
DIFFERENCES AMONG PROPORTIONS

There are also many problems in which we must decide whether differences among more than two sample proportions are significant, or whether they can be attributed to chance. For instance, if 24 of 200 brand A tires, 21 of 200 brand B tires, 18 of 200 brand C tires, and 33 of 200 brand D tires failed to last 20,000 miles, we may want to decide whether the differences among $\frac{24}{200} = 0.120$, $\frac{21}{200} = 0.105$, $\frac{18}{200} = 0.090$, and $\frac{33}{200} = 0.165$ are significant, or whether they may be due to chance.

To illustrate the method we use to analyze this kind of data, suppose that a survey in which independent random samples of 80 single persons, 120 married persons, and 100 widowed or divorced persons were asked whether "friends and social life" or "job or primary activity" contributes more to their general happiness, yielded the results shown in the following table:

	Single	Married	Widowed or divorced
Friends and social life	47	59	56
Job or primary activity	33	61	44
Total	80	120	100

The proportions of persons choosing "friends and social life" are $\frac{47}{80} = 0.59$, $\frac{59}{120} = 0.49$, and $\frac{56}{100} = 0.56$ (rounded to two decimals) for the three groups, and we want to decide at the 0.05 level of significance whether the differences among them can be attributed to chance.

If we let p_1, p_2, and p_3 denote the true proportions of persons who would choose "friends and social life" in the three populations sampled, the null hypothesis we want to test is $p_1 = p_2 = p_3$ and the alternative hypothesis is that p_1, p_2, and p_3 are not all equal. If the null hypothesis is true, the three samples come from populations having a common proportion p (of persons who would choose "friends and social life"), and we can combine the three samples and look on them as one sample from one population. Thus, we can estimate the common proportion p as

$$\hat{p} = \frac{47 + 59 + 56}{80 + 120 + 100} = \frac{162}{300} = 0.54$$

With this estimate we would expect $80(0.54) = 43.2$ of the single persons, $120(0.54) = 64.8$ of the married persons, and $100(0.54) = 54.0$ of the widowed or divorced persons to choose "friends and social life." Subtracting these figures from the respective sizes of the three samples, we find that $80 - 43.2 = 36.8$ of the single persons, $120 - 64.8 = 55.2$ of the married persons, and $100 - 54.0 = 46.0$ of the widowed or divorced persons would be expected to choose "job or primary activity." These results are summarized in the following table, where the **expected frequencies** are shown in parentheses below the **observed frequencies**:

	Single	Married	Widowed or divorced
Friends and social life	47 (43.2)	59 (64.8)	56 (54.0)
Job or primary activity	33 (36.8)	61 (55.2)	44 (46.0)

To test the null hypothesis that the p's are all equal in a problem like this, we compare the frequencies which were actually observed with the frequencies we would expect if the null hypothesis is true. It stands to reason that the null hypothesis should be accepted if the discrepancies between the observed and expected frequencies are small. On the other hand, if the discrepancies between the two sets of frequencies are large, this suggests that the null hypothesis must be false.

Using the letter o for the observed frequencies and the letter e for the expected frequencies, we base their comparison on the following χ^2 (chi-square) statistic:

Statistic for test concerning differences among proportions

$$\chi^2 = \sum \frac{(o - e)^2}{e}$$

In words, χ^2 is the sum of the quantities obtained by dividing $(o - e)^2$ by e separately for each **cell** of the table, and for our example we get

$$\chi^2 = \frac{(47 - 43.2)^2}{43.2} + \frac{(59 - 64.8)^2}{64.8} + \frac{(56 - 54.0)^2}{54.0}$$

$$+ \frac{(33 - 36.8)^2}{36.8} + \frac{(61 - 55.2)^2}{55.2} + \frac{(44 - 46.0)^2}{46.0}$$

$$= 2.016$$

It remains to be seen whether this value is large enough to reject the null hypothesis $p_1 = p_2 = p_3$.

In a problem like this, if the null hypothesis that the population proportions are all equal is true, the sampling distribution of the χ^2 statistic

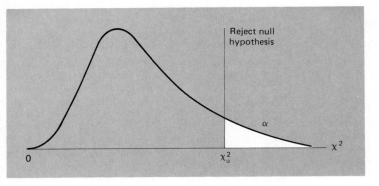

FIGURE 12.2 *Test criterion.*

is approximately the chi-square distribution, which we introduced in Section 11.1. Since the null hypothesis will be rejected only when the value obtained for χ^2 is too large to be accounted for by chance, we base our decision on the criterion of Figure 12.2, where χ_α^2 is such that the area under the chi-square distribution to its right equals α. The parameter of the chi-square distribution, the number of degrees of freedom (or simply the degrees of freedom), equals $k - 1$ when we compare k sample proportions. Intuitively, we can justify this formula with the argument that once we have calculated $k - 1$ of the expected frequencies in either row of the table, all of the other expected frequencies can be obtained by subtraction from the totals of the rows and columns.

Returning to our example and supposing that the level of significance is to be 0.05, we find that $\chi_{0.05}^2 = 5.991$ for $3 - 1 = 2$ degrees of freedom, and that the null hypothesis cannot be rejected since $\chi^2 = 2.016$ does not exceed 5.991. Consequently, we either reserve judgment or we accept the null hypothesis that equal proportions of single, married, and widowed or divorced persons would choose "friends and social life" (as contributing more than "job or primary activity" to their general happiness).

In general, if we want to test the null hypothesis $p_1 = p_2 = \cdots = p_k$ on the basis of random samples from k populations, we combine the data, as in our example, and thus get the following pooled estimate of the common population proportion p:

Estimate of common population proportion

$$\hat{p} = \frac{x_1 + x_2 + \cdots + x_k}{n_1 + n_2 + \cdots + n_k}$$

where the n's are the sample sizes, and the x's the numbers of successes, in the k samples. We then multiply the n's by $\hat{p}$ to get the expected frequencies for the first row of the table; after that we subtract these values from the respective sample sizes to get the expected frequencies for the second row of the table. An alternative way of getting the expected frequency for any one cell is to multiply the total of the row to which it belongs by the total of the column to which it belongs, and then divide by the **grand total**, $n_1 + n_2 + \cdots + n_k$, for the entire table (see also page 350). Next, we calculate χ^2 as just defined, with $\dfrac{(o - e)^2}{e}$ determined separately for each of the $2k$ cells of the table, and reject the null hypothesis $p_1 = p_2 = \cdots = p_k$ if this value of χ^2 exceeds χ_α^2 for $k - 1$ degrees of freedom.

When we calculate the expected frequencies, we usually round them to the nearest integer or to one decimal. The entries in Table III are given to three decimals, in some instances more, but there is seldom any need to carry more than two decimals in calculating the value of the χ^2 statistic itself. Also, the test we have been discussing is only an approximate test since the sampling distribution of the χ^2 statistic is only approximately the chi-square distribution, and it should not be used when one (or more) of the expected frequencies is less than 5. If this is the case, we can sometimes combine two or more of the samples in such a way that none of the e's is less than 5.

EXAMPLE   Given that 24 of 200 brand A tires, 21 of 200 brand B tires, 18 of 200 brand C tires, and 33 of 200 brand D tires failed to last 20,000 miles, use the 0.05 level of significance to test the null hypothesis that there is no difference in the durability of the four kinds of tires.

Solution   1. *Null hypothesis*: $p_1 = p_2 = p_3 = p_4$
   Alternative hypothesis: p_1, p_2, p_3, and p_4 are not all equal
   2. *Level of significance*: $\alpha = 0.05$
   3. *Criterion*: Reject the null hypothesis if $\chi^2 > 7.815$, the value of $\chi_{0.05}^2$ for $4 - 1 = 3$ degrees of freedom, where

$$\chi^2 = \sum \frac{(o - e)^2}{e}$$

and otherwise accept it.
   4. *Calculations*

$$\hat{p} = \frac{24 + 21 + 18 + 33}{200 + 200 + 200 + 200} = 0.12$$

and we get $200(0.12) = 24$ for each of the expected frequencies for the first row and $200 - 24 = 176$ for each of the expected frequencies for the second row; then, substitution into the formula for χ^2 yields

$$\chi^2 = \frac{(24-24)^2}{24} + \frac{(21-24)^2}{24} + \frac{(18-24)^2}{24} + \frac{(33-24)^2}{24}$$

$$+ \frac{(176-176)^2}{176} + \frac{(179-176)^2}{176} + \frac{(182-176)^2}{176}$$

$$+ \frac{(167-176)^2}{176}$$

$$= 5.97$$

5. *Decision*: Since $\chi^2 = 5.97$ does not exceed 7.815, the null hypothesis cannot be rejected; we conclude that there is no difference in the durability of the four kinds of tires.

The method we have discussed here serves only to test the null hypothesis $p_1 = p_2 = \cdots = p_k$ against the alternative hypothesis that the p's are not all equal. However, in the special case where $k = 2$ we can use instead the method of Section 12.4 and test also against either of the alternative hypotheses $p_1 < p_2$ or $p_1 > p_2$. Indeed, for $k = 2$ the two methods are equivalent, as it can be shown that the χ^2 statistic equals the square of the z statistic given on page 338 (see Exercises 12.48 and 12.49 on page 347).

EXERCISES

12.37 One method of seeding clouds was successful in 57 of 150 attempts, while another method was successful in 33 of 100 attempts. At the 0.05 level of significance, can we conclude that the first method is better than the second?

12.38 One mail solicitation for a charity brought 412 responses to 5,000 letters and another, more expensive, mail solicitation brought 311 responses to 3,000 letters. Use the 0.01 level of significance to test the null hypothesis that the two solicitations are equally effective against the alternative that the more expensive one is more effective.

12.39 In a random sample of 250 persons who skipped breakfast, 102 reported that they experienced midmorning fatigue, and in a random sample of 250 persons who ate breakfast, 73 reported that they experienced midmorning fatigue. Use the 0.01 level of significance to test the null hypothesis that there is no difference between the corresponding population proportions against the alternative hypothesis that midmorning fatigue is more prevalent among persons who skip breakfast.

12.40 A study showed that 74 of 200 persons who saw a deodorant advertised during the telecast of a baseball game and 86 of 200 other persons who saw it advertised on a variety show remembered two hours later the name of the deodorant. Use the 0.05 level of significance to test whether or not the difference between the two sample proportions, $\frac{74}{200} = 0.37$ and $\frac{86}{200} = 0.43$, is significant.

12.41 A random sample of 150 high school students was asked whether they would turn to their father or their mother for help with a homework assignment in mathematics, and another random sample of 150 high school students was asked the same question with regard to a homework assignment in English. Use the results shown in the following table and the 0.01 level of significance to test whether or not there is a difference between the true proportions of high school students who turn to their fathers rather than their mothers for help in these two subjects:

	Mathematics	English
Mother	59	85
Father	91	65

12.42 Among 500 marriage license applications, chosen at random in 1971, 48 of the women were at least one year older than the men, and among 500 marriage license applications, chosen at random in 1977, 85 of the women were at least one year older than the men. Use the 0.05 level of significance to test whether or not there was an increase from 1971 to 1977 in the proportion of women on marriage license applications who are at least one year older than the men.

★12.43 If we want to test the null hypothesis that the difference between two population proportions equals some constant δ (*delta*), we can base our decision on the statistic

Statistic for test concerning difference between two proportions

$$z = \frac{\dfrac{x_1}{n_1} - \dfrac{x_2}{n_2} - \delta}{\sqrt{\dfrac{\dfrac{x_1}{n_1}\left(1 - \dfrac{x_1}{n_1}\right)}{n_1} + \dfrac{\dfrac{x_2}{n_2}\left(1 - \dfrac{x_2}{n_2}\right)}{n_2}}}$$

whose sampling distribution is approximately the standard normal distribution when n_1 and n_2 are both large. The criteria for tests based on this statistic are again those of Figure 10.9 on page 293.

(a) In a true–false test, a test item is considered to be good if it discriminates between well-prepared and poorly prepared students. If 205 of 250 well-prepared students and 137 of 250 poorly prepared students answered a certain test item correctly, test at the 0.05 level of significance whether to accept the null hypothesis $p_1 - p_2 = 0.20$ or the alternative hypothesis $p_1 - p_2 > 0.20$.

(b) Two groups of 80 patients each took part in an experiment in which one group received pills containing an anti-allergy drug, while the other group received a placebo (a pill containing no drug). If in the group given the drug 23 exhibited allergic symptoms while in the group given the placebo 51 exhibited such symptoms, test at the 0.01 level of significance whether the true percentage of patients exhibiting allergic symptoms is at least 15 percent less for those who received the drug.

12.44 A market research study shows that among 100 men, 100 women, and 200 children interviewed, there were 44, 53, and 133 who did not like the flavor of a new toothpaste. Use the 0.05 level of significance to decide whether or not the differences among the three sample proportions, $\frac{44}{100} = 0.44$, $\frac{53}{100} = 0.53$, and $\frac{133}{200} = 0.665$, are significant.

12.45 The following table shows the results of a survey in which random samples of the parents of 12th graders in three school districts were asked whether they are for or against the removal of certain controversial books from a required reading list:

	District 1	District 2	District 3
For the removal of the books	8	13	12
Against the removal of the books	52	67	48

Use the 0.05 level of significance to test whether the differences among the sample proportions for the three school districts are significant.

12.46 The data shown in the following table were obtained in a study of methods of suicide attempt:

	White		Black	
	Female	*Male*	*Female*	*Male*
Ingestion or inhilation of drug	13	10	14	4
Violent self-injury	9	9	5	9

Use the 0.05 level of significance to test whether the differences among the four sample proportions (of suicide by violent self-injury) are significant.

12.47 The following table shows the results of a study in which random samples of 100 members of each of five large unions were asked whether they are for or against a certain piece of legislation:

	Union 1	Union 2	Union 3	Union 4	Union 5
For it	74	81	69	75	91
Against it	26	19	31	25	9

Use the 0.01 level of significance to test whether or not the corresponding true proportions for the legislation are all equal.

12.48 The manager of a motel, in trying to decide which of two supposedly equally good cigarette-vending machines to install, tests each machine 500 times, and he finds that machine I fails to work (neither delivers the cigarettes nor returns the money) 26 times and machine II fails to work 12 times. Based on the χ^2 statistic of Section 12.5, can he conclude at the 0.05 level of significance that the two machines are not equally good?

12.49 Rework the preceding exercise using the z statistic of Section 12.4, and verify that the value of the χ^2 statistic equals the square of the z statistic.

12.50 Verify that the sum of the expected numbers of successes, $n_1\hat{p}, n_2\hat{p}, \ldots$, and $n_k\hat{p}$, equals the sum of the observed numbers of successes.

12.6
CONTINGENCY TABLES

The χ^2 statistic plays an important role in many other problems where information is obtained by counting or enumerating, rather than measuring. The method we shall describe here for analyzing such **count data** is an extension of the method of the preceding section, and it applies to two kinds of problems, which differ conceptually but are analyzed the same way.

In the first kind of problem we deal with trials permitting more than two possible outcomes. For instance, the weather can get better, remain the same, or get worse, an undergraduate can be a freshman, a sophomore, a junior, or a senior, and a movie may be rated G, PG, R, or X. Also, in the example on page 340 each person might have been asked whether "friends and social life," "job or primary activity," or "health and physical condition" contributes most to his or her general happiness, and this might have resulted in the following table:

	Single	Married	Widowed or divorced
Friends and social life	41	49	42
Job or primary activity	27	50	33
Health and physical condition	12	21	25
Total	80	120	100

We refer to this kind of table as a 3×3 table (where 3×3 reads "3 by 3"), because it contains 3 rows and 3 columns; more generally, when there are c samples and each trial permits r alternatives, we refer to the resulting table as an $r \times c$ **table.** Here, as in the examples of the preceding section, the column totals, which are the sample sizes, are fixed. On the other hand, the row totals $(41 + 49 + 42 = 132, 27 + 50 + 33 = 110,$ and $12 + 21 + 25 = 58)$ depend on the responses of the persons interviewed, and hence on chance.

In the second kind of problem where the method of this section applies, the column totals as well as the row totals are left to chance. Suppose, for instance, that we want to investigate whether there is a relationship between the intelligence of persons who have gone through a certain job-training program and their subsequent performance on the job. Suppose, furthermore, that a random sample of 400 cases taken from very extensive files yielded the following results:

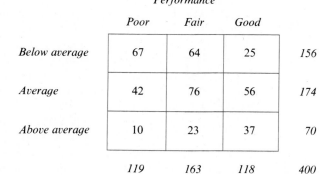

		Performance			
		Poor	Fair	Good	
	Below average	67	64	25	156
IQ	Average	42	76	56	174
	Above average	10	23	37	70
		119	163	118	400

This is also a 3 × 3 table, and it is mainly in connection with problems like this that $r \times c$ tables are referred to as **contingency tables**.

Before we demonstrate how $r \times c$ tables are analyzed, let us examine what null hypotheses we want to test. In the problem dealing with the different factors which contribute to one's happiness, we want to test the null hypothesis that the probabilities of the three alternatives are the same for each group. In other words, we want to test the null hypothesis that a person's choice of one of the three alternatives is independent of his or her being single, married, or widowed or divorced. In the problem directly above we are also concerned with a null hypothesis of independence; namely, the null hypothesis that the on the job performance of persons who have gone through the training program is independent of their IQ.

To show how an $r \times c$ table is analyzed, let us refer to the second example and begin by calculating the expected cell frequencies. If the null hypothesis of independence is true, the probability of randomly choosing a person whose IQ is below average and whose on the job performance is poor is given by the product of the probability of choosing a person whose IQ is below average and the probability of choosing a person whose on the job performance is poor. Using the total of the first row, the total of the first column, and the grand total for the entire table to estimate these probabilities, we get

$$\frac{67 + 64 + 25}{400} = \frac{156}{400}$$

for the probability of choosing a person whose IQ is below average, and

$$\frac{67 + 42 + 10}{400} = \frac{119}{400}$$

for the probability of choosing a person whose on the job performance is poor. Hence, we estimate the probability of choosing a person whose IQ is below average and whose on the job performance is poor as $\frac{156}{400} \cdot \frac{119}{400}$, and in a sample of size 400 we would expect to find

$$400 \cdot \frac{156}{400} \cdot \frac{119}{400} = \frac{156 \cdot 119}{400} = 46.4$$

persons who fit this description.

In this last result, $\dfrac{156 \cdot 119}{400}$ is just the product of the total of the first row and the total of the first column divided by the grand total for the entire table. Indeed, the argument which led to this result can be used to show that in general

> **The expected frequency for any cell of a contingency table may be obtained by multiplying the total of the row to which it belongs by the total of the column to which it belongs and then dividing by the grand total for the entire table.**

With this rule we get an expected frequency of $\dfrac{156 \cdot 163}{400} = 63.6$ for the second cell of the first row, and $\dfrac{174 \cdot 119}{400} = 51.8$ and $\dfrac{174 \cdot 163}{400} = 70.9$ for the first two cells of the second row.

It is not necessary to calculate all the expected cell frequencies in this way, as it can be shown that the sum of the expected frequencies for any row or column must equal the sum of the corresponding observed frequencies. Therefore, we can get some of the expected cell frequencies by subtraction from row or column totals. For instance, for our example we get

$$156 - 46.4 - 63.6 = 46.0$$

for the third cell of the first row,

$$174 - 51.8 - 70.9 = 51.3$$

for the third cell of the second row, and

$$119 - 46.4 - 51.8 = 20.8$$

$$163 - 63.6 - 70.9 = 28.5$$

and

$$118 - 46.0 - 51.3 = 20.7$$

for the three cells of the third row. These results are summarized in the following table, where the expected frequencies are shown in parentheses below the corresponding observed frequencies:

		Performance		
		Poor	Fair	Good
	Below average	67 (46.4)	64 (63.6)	25 (46.0)
IQ	Average	42 (51.8)	76 (70.9)	56 (51.3)
	Above average	10 (20.8)	23 (28.5)	37 (20.7)

From here on the work is like that of the preceding section; we calculate the χ^2 statistic according to the formula

$$\chi^2 = \sum \frac{(o - e)^2}{e}$$

with $\dfrac{(o - e)^2}{e}$ calculated separately for each cell of the table. Then we reject the null hypothesis at the level of significance α if the value obtained for χ^2 exceeds χ^2_α for $(r - 1)(c - 1)$ degrees of freedom. In connection with this formula for the number of degrees of freedom observe that after $c - 1$ of the expected frequencies have been calculated for each of $r - 1$ of the rows by means of the rule on page 350, all the other expected frequencies may be obtained by subtraction from row or column totals. For our example the number of degrees of freedom is $(3 - 1)(3 - 1) = 4$, and it should be observed that after we had calculated four of the expected frequencies, two for each of the first two rows, by means of the rule on page 350, all the others were obtained by subtraction from row or column totals. Thus, if the level of significance is to be 0.01 in our example, we reject the null hypothesis if the value we obtain for χ^2 exceeds $\chi^2_{0.01} = 13.277$ for 4 degrees of freedom. Returning to the table on page 350, we find that

$$\chi^2 = \frac{(67 - 46.4)^2}{46.4} + \frac{(64 - 63.6)^2}{63.6} + \frac{(25 - 46.0)^2}{46.0}$$

$$+ \frac{(42 - 51.8)^2}{51.8} + \frac{(76 - 70.9)^2}{70.9} + \frac{(56 - 51.3)^2}{51.3}$$

$$+ \frac{(10 - 20.8)^2}{20.8} + \frac{(23 - 28.5)^2}{28.5} + \frac{(37 - 20.7)^2}{20.7}$$

$$= 40.89$$

and since $\chi^2 = 40.89$ exceeds 13.277, the null hypothesis must be rejected; in other words, we conclude that there is a relationship between IQ and on the job performance.

The method which we have used here to analyze the contingency table applies also when the column totals are fixed sample sizes, as in the "happiness" example, and do not depend on chance. In that case, the rule by which we multiply the row total by the column total and then divide by the grand total must be justified in a different way, but this is of no consequence—the expected cell frequencies are determined in exactly the same way (see Exercise 12.57 on page 357).

EXAMPLE   With reference to the 3 × 3 table on page 348, use the 0.05 level of significance to test whether the probabilities of the three alternatives are the same for persons who are single, married, or widowed or divorced.

Solution   1. *Null hypothesis*: The probabilities of the three alternatives are the same for persons who are single, married, or widowed or divorced.
Alternative hypothesis: The probabilities of the three alternatives are not the same for the three kinds of persons.
2. *Level of significance*: $\alpha = 0.05$
3. *Criterion*: Reject the null hypothesis if $\chi^2 > 9.488$, the value of $\chi^2_{0.05}$ for $(3 - 1)(3 - 1) = 4$ degrees of freedom, where

$$\chi^2 = \sum \frac{(o - e)^2}{e}$$

and otherwise accept it.
4. *Calculations*: The expected frequencies for the first two cells of the first row are $\frac{132 \cdot 80}{300} = 35.2$ and $\frac{132 \cdot 120}{300} = 52.8$; those for the first two cells of the second row are $\frac{110 \cdot 80}{300} = 29.3$ and $\frac{110 \cdot 120}{300} = 44$; and by subtraction, those for the third cells of the first two rows are 44.0 and 36.7, and those for the third row are 15.5, 23.2, and 19.3. Thus, substitution into the formula for χ^2 yields

$$\chi^2 = \frac{(41 - 35.2)^2}{35.2} + \frac{(49 - 52.8)^2}{52.8} + \frac{(42 - 44.0)^2}{44.0}$$

$$+ \frac{(27 - 29.3)^2}{29.3} + \frac{(50 - 44.0)^2}{44.0} + \frac{(33 - 36.7)^2}{36.7}$$

$$+ \frac{(12 - 15.5)^2}{15.5} + \frac{(21 - 23.2)^2}{23.2} + \frac{(25 - 19.3)^2}{19.3}$$

$$= 5.37$$

5. *Decision*: Since $\chi^2 = 5.37$ does not exceed 9.488, the null hypothesis cannot be rejected; we conclude that the probabilities of the three alternatives are the same for persons who are single, married, or widowed or divorced.

Since the sampling distribution of the χ^2 statistic we are using here is only approximately a chi-square distribution, it should not be used when any of the expected cell frequencies are less than 5. When there are expected cell frequencies less than 5, it may be possible to combine some of the cells, subtract 1 degree of freedom for each cell eliminated, and then perform the test just as it has been described.

12.7
GOODNESS OF FIT

In this section we shall treat a further application of the χ^2 criterion, in which we compare an observed frequency distribution with a distribution we might expect according to theory or assumptions. We refer to such a comparison as a test of **goodness of fit.**

To illustrate, let us consider Table XI, the table of random numbers, which is supposed to have been constructed in such a way that each digit is a value of a random variable which takes on the values 0, 1, 2, 3, 4, 5, 6, 7, 8, and 9 with equal probabilities of 0.10. To see whether it is reasonable to maintain that this is, indeed, the case, we might count how many times each digit appears in the table or part of the table; specifically, let us take the 250 digits in the first five columns of the table on page 517. This yields the values shown in the "observed frequency" column of the following table:

Digit	Probability	Observed frequency o	Expected frequency e
0	0.10	22	25
1	0.10	17	25
2	0.10	23	25
3	0.10	26	25
4	0.10	27	25
5	0.10	31	25
6	0.10	26	25
7	0.10	23	25
8	0.10	29	25
9	0.10	26	25
		250	250

The expected frequencies in the right-hand column were obtained by multiplying each of the probabilities of 0.10 by 250, the total number of digits counted.

To test whether the discrepancies between the observed and expected frequencies can be attributed to chance, we use the same chi-square statistic as in Sections 12.5 and 12.6:

Statistic for test of goodness of fit

$$\chi^2 = \sum \frac{(o - e)^2}{e}$$

calculating $\frac{(o - e)^2}{e}$ separately for each class of the distribution. Then, if the value we get for χ^2 exceeds χ^2_α, we reject the null hypothesis on which the expected frequencies are based at the level of significance α. The number of degrees of freedom is $k - m$, where k is the number of terms $\frac{(o - e)^2}{e}$ added in the formula for χ^2, and m is the number of quantities we must determine from the observed data to calculate the expected frequencies.

EXAMPLE   Based on the observed frequencies in the table on page 353, test at the 0.05 level of significance whether there is any indication that the digits in Table XI are not truly random.

Solution   1. *Null hypothesis*: The probability of each digit is 0.10;
Alternative hypothesis: The probabilities are not all 0.10.
2. *Level of significance*: $\alpha = 0.05$
3. *Criterion*: Reject the null hypothesis if $\chi^2 > 16.919$, the value of $\chi^2_{0.05}$ for $k - m = 10 - 1 = 9$ degrees of freedom, where

$$\chi^2 = \sum \frac{(o - e)^2}{e}$$

and otherwise accept it. (The number of degrees of freedom is $10 - 1 = 9$, since we compare ten observed frequencies with the corresponding expected frequencies, and only one quantity, the total frequency of 250, is needed from the observed data to calculate the expected frequencies.)
4. *Calculations*: Since each of the expected frequencies is $250(0.10) = 25$, substitution into the formula for χ^2 yields

$$\chi^2 = \frac{(22 - 25)^2}{25} + \frac{(17 - 25)^2}{25} + \frac{(23 - 25)^2}{25} + \frac{(26 - 25)^2}{25}$$
$$+ \frac{(27 - 25)^2}{25} + \frac{(31 - 25)^2}{25} + \frac{(26 - 25)^2}{25} + \frac{(23 - 25)^2}{25}$$
$$+ \frac{(29 - 25)^2}{25} + \frac{(26 - 25)^2}{25}$$
$$= 5.60$$

5. *Decision*: Since $\chi^2 = 5.60$ does not exceed 16.919, the null hypothesis cannot be rejected; in other words, there is no indication of a lack of randomness.

The method we have illustrated here is used quite generally to test how well distributions, expected on the basis of theory or assumptions, fit, or describe, observed data. Thus, in the exercises which follow we shall test whether observed distributions have (at least approximately) the shape of binomial, Poisson, and normal distributions. As in the tests of the two preceding sections, the sampling distribution of the χ^2 statistic is only approximately a chi-square distribution when it is used for tests of goodness of fit. So, if any one of the expected frequencies is less than 5, we must again combine some of the data; that is, we combine adjacent classes of the distribution.

EXERCISES

12.51 Suppose that for 120 mental patients who did not receive psychotherapy and 120 mental patients who received psychotherapy a panel of psychiatrists determined after six months whether their condition had deteriorated, remained unchanged, or improved. Based on the results shown in the following table, test at the 0.05 level of significance whether the therapy is effective:

	No therapy	Therapy
Deteriorated	6	11
Unchanged	65	31
Improved	49	78

12.52 The following sample data pertain to shipments received by a large firm from three different vendors:

	Vendor A	Vendor B	Vendor C
Rejected	12	8	20
Not perfect but acceptable	23	12	30
Perfect	85	60	110

Use the 0.01 level of significance to test whether the three vendors ship products of equal quality.

12.53 A sample survey, designed to show how students attending a large university get to their classes, yielded the following results:

	Freshman	Sophomore	Junior	Senior
Walk	104	87	89	72
Automobile	22	29	35	43
Bicycle	46	34	37	32
Other	28	50	39	53
Totals	200	200	200	200

Use the 0.05 level of significance to test the null hypothesis that the same proportions of freshmen, sophomores, juniors, and seniors use these means of transportation to get to class.

12.54 In a study of students' parents' feelings about a required course in sex education, a random sample of 360 parents are classified according to whether they have one, two, or three or more children in the school system, and also whether they feel that the course is poor, adequate, or good. Supposing that the results are as shown in the following table, test at the 0.05 level of significance whether there is a relationship between parents' reaction to the course and the number of children they have in school:

	Number of children		
	1	2	3 or more
Poor	48	40	12
Adequate	55	53	29
Good	57	46	20

12.55 A market research organization wants to determine, on the basis of the following information, whether there is a relationship between the size of a tube of toothpaste which a shopper buys and the number of persons in the shopper's household:

		Number of persons in household			
		1–2	*3–4*	*5–6*	*7 or more*
Size of tube bought	*Giant*	23	116	78	43
	Large	54	25	16	11
	Small	31	68	39	8

At the 0.01 level of significance, is there a relationship?

12.56 In a study to determine whether there is a relationship between bank employees' standard of dress and their professional advancement, a random sample of size $n = 500$ yielded the results shown in the following table:

	Speed of advancement		
	Slow	*Average*	*Fast*
Very well dressed	38	135	129
Well dressed	32	68	43
Poorly dressed	13	25	17

Test at the 0.05 level of significance whether there is a real relationship between standard of dress and speed of professional advancement.

12.57 Use an argument similar to that on page 351 to show that the rule for calculating the expected cell frequencies (multiplying the row total by the column total and then dividing by the grand total) applies also when the column totals are fixed sample sizes and do not depend on chance.

★12.58 If the analysis of a contingency table shows that there is a relationship between the two variables under consideration, the strength of the relationship can be measured by the **contingency coefficient**

Contingency coefficient

$$C = \sqrt{\frac{\chi^2}{\chi^2 + n}}$$

where χ^2 is the value of the chi-square statistic obtained for the table and n is the grand total of all the frequencies. This coefficient takes on values between 0 (corresponding to independence) and a maximum value less than 1 depending on the size of the table; for instance, it can be shown that for a 3×3 table the maximum value of C is $\sqrt{2/3} = 0.82$.

   (a) Calculate C for the example on page 348, which concerns the relationship between IQ and on the job performance.
   (b) Calculate C for the contingency table of Exercise 12.55.

12.59 To see whether a die is balanced, it is rolled 720 times and the following results are obtained: 1 showed 129 times, 2 showed 107 times, 3 showed 98 times, 4 showed 132 times, 5 showed 136 times, and 6 showed 118 times. At the 0.05 level of significance, do these results support the hypothesis that the die is balanced?

12.60 Ten years' data show that in a given city there were no bank robberies in 57 months, one bank robbery in 36 months, two bank robberies in 15 months, and three or more bank robberies in 12 months. At the 0.05 level of significance, does this substantiate the claim that the probabilities of 0, 1, 2, or 3 or more bank robberies in any one month are 0.40, 0.30, 0.20, and 0.10?

12.61 A quality control engineer takes daily samples of four tractors coming off an assembly line and on 200 consecutive working days he obtains the data summarized in the following table:

Number of tractors requiring adjustments	Number of days
0	102
1	78
2	19
3	1
4	0

Use the 0.01 level of significance to test the null hypothesis that the data may be looked upon as random samples from a binomial population with $n = 4$ and $p = 0.10$.

12.62 The following is the distribution of the number of calls received at the switchboard of a government building during 400 five-minute intervals:

Number of calls	Frequency
0	95
1	116
2	112
3	47
4 or more	30

Use the 0.05 level of significance to test whether the number of calls received by the switchboard in a five-minute interval is a random variable having a Poisson distribution with $\lambda = 1.5$, namely, that the probabilities for 0, 1, 2, 3, or 4 or more calls are 0.22, 0.33, 0.25, 0.13, and 0.07.

12.63 The following is the distribution of the number of minutes it took 80 persons to complete a certain tax form:

Time required to complete form (minutes)	Frequency
10–14	8
15–19	28
20–24	27
25–29	12
30–34	4
35–39	1
	80

As can easily be verified, the mean of this distribution is $\bar{x} = 20.7$ and its standard deviation is $s = 5.4$.

(a) Find the probabilities that a random variable having a normal distribution with $\mu = 20.7$ and $\sigma = 5.4$ will take on a value less than 14.5, between 14.5 and 19.5, between 19.5 and 24.5, between 24.5 and 29.5, between 29.5 and 34.5, and greater than 34.5.

(b) Changing the first and last classes of the distribution to "14 or less" and "35 or more," find the expected normal curve frequencies corresponding to the six classes of the distribution by multiplying the probabilities obtained in part (a) by the total frequency.

(c) Use the 0.05 level of significance to test the null hypothesis that the data constitute a random sample from a normal population. (In making the χ^2 test, the number of degrees of freedom is $k - 3$ since the total frequency, the mean, and the standard deviation are determined from the observed data.)

12.8
CHECKLIST OF KEY TERMS
(with page references to their definitions)

12.9
REVIEW EXERCISES

12.64 The following table shows how many times, Monday through Friday, a bus was late arriving at a given stop in 40 weeks:

Number of times bus was late	Number of weeks
0	4
1	11
2	15
3 or more	10

Use the 0.05 level of significance to test the null hypothesis that the bus is late 30 percent of the time, namely, that the data constitute a random sample from a binomial population with $n = 5$ and $p = 0.30$.

12.65 In a random sample of 90 persons eating dinner by themselves in a French restaurant, 63 had wine with their dinner.

(a) Find a 95 percent confidence interval for the true proportions of persons eating by themselves in that restaurant who have wine with their dinner.

(b) If we use the sample proportion, $\frac{63}{90} = 0.70$, to estimate this true proportion, what can we say with 99 percent confidence about the maximum error?

12.66 The following table shows how samples of the residents of three federally financed housing projects replied to the question whether they would continue to live there if they had the choice:

	Project 1	Project 2	Project 3
Yes	63	84	69
No	37	16	31

Use the 0.05 level of significance to test whether the differences among the three sample proportions (of "yes" answers) may be attributed to chance.

12.67 Based on the result of $n = 12$ trials, we want to test the null hypothesis $p = 0.20$ against the alternative hypothesis $p > 0.20$. If we reject the null hypothesis when the number of successes is 5 or more and otherwise we accept it, find
  (a) the probability of a Type I error;
  (b) the probability of a Type II error when $p = 0.30$;
  (c) the probability of a Type II error when $p = 0.50$.

12.68 In a study conducted at a large airport, 81 of 300 persons who had just gotten off a plane and 32 of 200 persons who were about to board a plane admitted that they were afraid of flying. Use the z statistic to test at the 0.01 level of significance whether the difference between the corresponding sample proportions is significant.

12.69 Use the χ^2 statistic to rework the preceding exercise, and verify that the value obtained for χ^2 equals the square of the value obtained for z.

12.70 In a random sample of 200 retired persons, 137 stated that they prefer living in an apartment to living in a one-family home. At the 0.05 level of significance, does this refute the claim that 60 percent of all retired persons prefer living in an apartment to living in a one-family home?

12.71 A sample check reveals that 186 of 200 of a professional football team's season-ticket holders intend to renew their tickets for the next season. Construct a 95 percent confidence interval for the true proportion of the team's season-ticket holders who intend to renew their tickets for the next season.

12.72 A researcher wants to determine what percentage of the farm workers in a certain area are illegal aliens. How large a sample will she need if she wants to be at least 99 percent confident that her estimate is not off by more than 2.5 percent, and she feels that the actual percentage is at most 10 percent?

12.73 A mental health study yielded the results shown in the following table:

	Men	Women
Never been in therapy	125	144
Therapy for six months or less	54	42
Therapy for more than six months	21	14

Analyze this 3×2 table by performing an appropriate chi-square test at the 0.05 level of significance.

12.74 In $n = 11$ trials, a certain weed killer was effective eight times. At the 0.05 level of significance, test the null hypothesis $p = 0.50$ against the alternative hypothesis $p > 0.50$.

12.75 In an experiment, an interviewer of job applicants is asked to write down his first impression (favorable or unfavorable) after two minutes and his final impression at the end of the interview. Use the following data and the 0.01 level of significance to test the interviewer's claim that his first and final impressions are the same 90 percent of the time:

| | | *First impression* | |
		Favorable	*Unfavorable*
Final impression	*Favorable*	186	33
	Unfavorable	54	127

12.76 To see whether a newly discovered manuscript can be attributed to a famous nineteenth-century historian, his literary style was analyzed statistically, and it was found, among other things, that in many samples of 1,000 words of text material he used the word "from" 0 to 4 times 43 percent of the time, 5 to 7 times 32 percent of the time, and 8 or more times 25 percent of the time. Now, if in 20 samples of 1,000 words of text material from the newly discovered manuscript, the word "from" is used 0 to 4 times, 5 to 7 times, and 8 or more times in 15, 3, and 2 of the samples, is this evidence at the 0.05 level of significance that the manuscript is not the work of this historian?

12.77 Among 120 baseball players, 64 improved their batting average after eating a certain breakfast food regularly for several weeks, the other 56 did not. Is this evidence at the 0.05 level of significance that the probability is 0.70 that the breakfast food will improve the performance of a baseball player?

12.78 The following is the distribution of the sulfur oxides emission data obtained on page 16:

Tons of sulfur oxides	Frequency
5.0–8.9	3
9.0–12.9	10
13.0–16.9	14
17.0–20.9	25
21.0–24.9	17
25.0–28.9	9
29.0–32.9	2
	80

As we showed on page 66, the mean and the standard deviation of this distribution are $\bar{x} = 18.85$ and $s = 5.55$.

(a) Find the probabilities that a random variable having a normal distribution with $\mu = 18.85$ and $\sigma = 5.55$ takes on a value less than 8.95, between 8.95 and 12.95, between 12.95 and 16.95, between 16.95 and 20.95, between 20.95 and 24.95, between 24.95 and 28.95, and greater than 28.95.

(b) Change the first and last classes of the distribution to "8.9 or less" and "29.0 or more," and find the expected normal curve frequencies corresponding to the seven classes of the distribution.

(c) Use the 0.05 level of significance to test the null hypothesis that the data constitute a random sample from a normal population.

12.79 Tests of the fidelity and the selectivity of 190 radios produced the results shown in the following table:

		Fidelity		
		Low	Average	High
Selectivity	Low	7	12	31
	Average	35	59	18
	High	15	13	0

Use the 0.01 level of significance to test the null hypothesis that fidelity is independent of selectivity.

12.80 Tests are made on the proportion of defective castings produced by two molds. If in a random sample of 100 castings from mold A there were 14 defectives and in a random sample of 200 castings from mold B there were 37 defectives, can the null hypothesis $p_1 = p_2$ be rejected against the alternative hypothesis $p_1 < p_2$ at the 0.05 level of significance?

12.81 In a random sample of 500 voters in a large city, 285 favor a new issue of municipal bonds. If we use $\frac{285}{500} = 0.57$ as an estimate of the true proportion of voters in this city who favor the bonds, what can we say with 99 percent confidence about the maximum error of this estimate?

12.82 A political leader hires a private poll to estimate what percentage of residents of his precinct believe a rumor about him spread by his opponent. How large a sample will the poll have to take to be able to assert with a probability of at least 0.90 that the sample percentage will be within 5 percent of the true value?

12.83 Suppose that we want to test the null hypothesis $p = 0.30$ on the basis of the number of successes, x, in 14 trials. For what values of x would we reject the null hypothesis if the level of significance and the alternative hypothesis are, respectively,

(a) $\alpha = 0.05$ and $p \neq 0.30$;
(b) $\alpha = 0.05$ and $p > 0.30$;
(c) $\alpha = 0.05$ and $p < 0.30$;
(d) $\alpha = 0.10$ and $p \neq 0.30$?

In each case state the *actual* level of significance.

12.84 Among 210 persons with alcohol problems admitted to the psychiatric emergency room of a hospital, 26 were admitted on a Monday, 23 on a Tuesday, 19 on a Wednesday, 25 on a Thursday, 33 on a Friday, 44 on a Saturday, and 40 on a Sunday. Use the 0.05 level of significance to test the null hypothesis that the psychiatric emergency room can expect equally many persons with alcohol problems on each day of the week.

12.85 When asked to identify Jean Jacques Rousseau, 28 of 120 persons said that he is a deep-sea diver. If the probability of a Type I error is to be at most 0.05, does this substantiate the claim that at least 30 percent of all persons make this mistake?

★12.86 The method of Section 12.2 can also be used to find the posterior distribution of the parameter λ of the Poisson distribution. Suppose, for instance, that several brokers are discussing the sale of a large estate, and that one of them feels that a newspaper ad should produce three serious inquiries, a second feels that it should produce five serious inquiries, and a third feels that it should produce six.

(a) If in the past the second broker has been twice as reliable as the first and the first has been three times as reliable as the third, what prior probabilities should we assign to their claims?

(b) How would these probabilities be affected if the ad actually produced only one serious inquiry and it can be assumed that the number of serious inquiries is a random variable having the Poisson distribution with either $\lambda = 3$, $\lambda = 5$, or $\lambda = 6$ corresponding to the three claims?

12.87 Three coins are tossed 240 times and 0, 1, 2, or 3 heads showed 51, 104, 69, and 16 times. Use the 0.01 level of significance to test whether it is reasonable to suppose that the coins are balanced and randomly tossed.

12.10
REFERENCES

The theory which underlies the various tests of this chapter is discussed in most textbooks on mathematical statistics; for instance, in Chapters 8 and 13 of the book by Freund and Walpole listed on page 263. Details about contingency tables may be found in

EVERITT, B. S., *The Analysis of Contingency Tables*. New York: John Wiley & Sons, Inc., 1977.

In this chapter we shall consider the problem of deciding whether observed differences among more than two sample means can be attributed to chance, or whether there are real differences among the means of the populations sampled. For instance, we may want to decide on the basis of sample data whether there really is a difference in the effectiveness of three methods of teaching a foreign language, we may want to compare the average yields per acre of several varieties of wheat, we may want to see whether there really is a difference in the average mileage obtained with four kinds of gasoline, we may want to judge whether there really is a difference in the durability of five kinds of carpets, and so on. The method we shall introduce for this purpose is a powerful statistical tool called analysis of variance, ANOVA for short.

Following an example in Section 13.1, we shall present the one-way analysis of variance in Section 13.3, the two-way analysis of variance in Section 13.5, and a generalization in Section 13.7. Some related aspects of experimental design are discussed in Sections 13.2, 13.4, 13.6, and 13.8.

13 ANALYSIS OF VARIANCE

13.1
DIFFERENCES AMONG k MEANS: AN EXAMPLE

Suppose that we want to compare the cleansing action of three detergents on the basis of the following whiteness readings made on fifteen swatches of white cloth, which were first soiled with India ink and then washed in an agitator-type machine with the respective detergents:

$$Detergent\ A:\quad 77, 81, 71, 76, 80$$

$$Detergent\ B:\quad 72, 58, 74, 66, 70$$

$$Detergent\ C:\quad 76, 85, 82, 80, 77$$

The means of these three samples are 77, 68, and 80, and we want to know whether the differences among them are significant or whether they can be attributed to chance.

In general, in a problem like this, if μ_1, $\mu_2, \ldots$, and μ_k are the means of the k populations from which the samples are drawn, we want to test the null hypothesis $\mu_1 = \mu_2 = \cdots = \mu_k$ against the alternative hypothesis that these means are not all equal.[†] Evidently, this null hypothesis would be supported if the differences among the sample means are small, and the alternative hypothesis would be supported if at least some of the differences among the sample means are large. Thus, we need a precise measure of the discrepancies among the $\bar{x}$'s, and with it a rule which tells us when the discrepancies are so large that the null hypothesis should be rejected. Possible choices for such a measure are the standard deviation of the $\bar{x}$'s or their variance and it is the latter which we shall use here. For the three detergents, the mean of the $\bar{x}$'s is

$$\frac{77 + 68 + 80}{3} = 75$$

[†] In connection with work later in this chapter, it is desirable to write the means as $\mu_1 = \mu + \alpha_1, \mu_2 = \mu + \alpha_2, \ldots,$ and $\mu_k = \mu + \alpha_k$. Here

$$\mu = \frac{\mu_1 + \mu_2 + \cdots + \mu_k}{k}$$

is called the **grand mean** and the α's, whose sum is zero (see Exercise 13.11 on page 379) are called the **treatment effects**. In this notation, the null hypothesis $\mu_1 = \mu_2 = \cdots = \mu_k$ becomes $\alpha_1 = \alpha_2 = \cdots = \alpha_k = 0$, and the alternative hypothesis is that the α's are not all equal to zero.

and their variance is

$$s_{\bar{x}}^2 = \frac{(77 - 75)^2 + (68 - 75)^2 + (80 - 75)^2}{3 - 1}$$

$$= 39$$

where the subscript $\bar{x}$ is used to show that this is the variance of the sample means.

Let us now make two assumptions which are critical to the method of analysis we shall use. First, it will be assumed that the populations we are sampling can be approximated closely with normal distributions; second, it will be assumed that these populations all have the same variance σ^2. With these assumptions, we note that, if the null hypothesis $\mu_1 = \mu_2 = \cdots = \mu_k$ is true, we can look upon the k samples as if they came from one and the same (normal) population and, hence, upon the variance of their means, $s_{\bar{x}}^2$, as an estimate of $\sigma_{\bar{x}}^2$, the square of the standard error of the mean. Now, since $\sigma_{\bar{x}} = \dfrac{\sigma}{\sqrt{n}}$ for samples from infinite populations, we can look upon $s_{\bar{x}}^2$ as an estimate of $\sigma_{\bar{x}}^2 = \left(\dfrac{\sigma}{\sqrt{n}}\right)^2 = \dfrac{\sigma^2}{n}$ and, therefore, upon $n \cdot s_{\bar{x}}^2$ as an estimate of σ^2. For instance, for our example we have $n \cdot s_{\bar{x}}^2 = 5 \cdot 39 = 195$ as an estimate of σ^2, the common variance of the three populations sampled.

If σ^2 were known, we could compare $n \cdot s_{\bar{x}}^2$ with σ^2 and reject the null hypothesis that the population means are all equal if this value is much larger than σ^2. However, in most practical problems σ^2 is not known and we have no choice but to estimate it on the basis of the sample data. Having assumed under the null hypothesis that the k samples do, in fact, come from identical populations, we could use any one of their variances, $s_1^2, s_2^2, \ldots,$ or s_k^2, as a second estimate of σ^2, and we can also use their mean. Averaging, or **pooling**, the three sample variances in our example, we get

$$\frac{s_1^2 + s_2^2 + s_3^2}{3} = \frac{1}{3}\left[\frac{(77-77)^2 + (81-77)^2 + (71-77)^2 + (76-77)^2 + (80-77)^2}{5-1} \right.$$

$$+ \frac{(72-68)^2 + (58-68)^2 + (74-68)^2 + (66-68)^2 + (70-68)^2}{5-1}$$

$$\left. + \frac{(76-80)^2 + (85-80)^2 + (82-80)^2 + (80-80)^2 + (77-80)^2}{5-1} \right]$$

$$= 23$$

and we now have two estimates of σ^2,

$$n \cdot s_{\bar{x}}^2 = 195 \quad \text{and} \quad \frac{s_1^2 + s_2^2 + s_3^2}{3} = 23$$

If the first of two such estimates of σ^2 (which is based on the variation among the sample means) is much larger than the second estimate (which is based on the variation within the samples and, hence, measures variation due to chance), it stands to reason that the null hypothesis should be rejected. After all, in that case the variation among the sample means would be greater than we would expect it to be if it were due only to chance. To put the comparison of the two estimates of σ^2 on a rigorous basis, we use the statistic

Statistic for test concerning differences among means

$$F = \frac{\textit{estimate of } \sigma^2 \textit{ based on the variation among the } \bar{x}\textit{'s}}{\textit{estimate of } \sigma^2 \textit{ based on the variation within the samples}}$$

which is appropriately called a **variance ratio**.

If the null hypothesis is true and if the assumptions made are valid, the sampling distribution of this statistic is the F distribution, which we introduced in Chapter 11. Since the null hypothesis will be rejected only when F is large (that is, when the variation among the $\bar{x}$'s is too great to be attributed to chance), we base our decision on the criterion of Figure 13.1. For $\alpha = 0.05$ or 0.01, the values of F_α may be looked up in Table IV at the end of the book, and if we compare the means of k random samples of size n, we have $k - 1$ degrees of freedom for the numerator and $k(n - 1)$ degrees of freedom for the denominator.[†]

Returning to our example, we find that for $k = 3$ and $n = 5$ the numerator and denominator degrees of freedom are $k - 1 = 3 - 1 = 2$ and $k(n - 1) = 3(5 - 1) = 12$, and that for these degrees of freedom $F_{0.01} = 6.93$. Since

$$F = \tfrac{195}{23} = 8.48$$

exceeds this critical value, the null hypothesis $\mu_1 = \mu_2 = \mu_3$ must be rejected; in other words, we conclude that the differences among the $\bar{x}$'s are too large to be attributed entirely to chance.

The technique we have just described is the simplest form of an **analysis of variance**. Although we could go ahead and perform F tests for differences among k means without further discussion, it will be instructive to look at the problem from an analysis-of-variance point of view, and we shall do so in Section 13.3.

[†] In connection with the numerator degrees of freedom, the numerator of F is $n \cdot s_{\bar{x}}^2$ and $s_{\bar{x}}^2$ is the variance of k means and, hence, has $k - 1$ degrees of freedom in accordance with the terminology introduced in the footnote to page 271. As for the denominator degrees of freedom, $k(n - 1)$, the denominator of F is the mean of k sample variances each of which has $n - 1$ degrees of freedom.

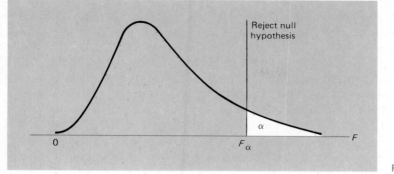

FIGURE 13.1 *F distribution*.

13.2
THE DESIGN OF EXPERIMENTS:
RANDOMIZATION

In the example of the preceding section, it may seem perfectly natural to conclude that the three detergents are not equally effective; and yet, a moment's reflection will show that this conclusion is not so "natural" at all. For all we know, the swatches cleaned with detergent *B* may have been more soiled than the others, the washing times may have been longer for detergent *C*, there may have been differences in water hardness or water temperature, and even the instruments used to make the whiteness readings may have gone out of adjustment after the readings for detergents *A* and *C* were made.

It is entirely possible, of course, that the differences among the three sample means, 77, 68, and 80, are due largely to differences in the quality of the three detergents, but we have just listed several other factors which could be responsible. It is important to remember that a significance test may show that differences among sample means are too large to be attributed to chance, but it cannot say why the differences occurred.

In general, if we want to show that one factor (among various others) can be considered the cause of an observed phenomenon, we must somehow make sure that none of the other factors can reasonably be held responsible. There are various ways in which this can be done; for instance, we can conduct a **rigorously controlled experiment** in which all variables except the one of concern are held fixed. To do this in the example dealing with the three detergents, we might soil the swatches with exactly equal amounts of India ink, always use exactly the same washing time, water of exactly the same temperature and hardness, and inspect the measuring instruments after each use. Under these rigid conditions, significant differences among the sample means cannot be due to differently soiled swatches, or differences in washing time, water temperature, water hardness, or measuring instruments. On the positive side, the differences show that the detergents are not all equally effective if they are used in this narrowly restricted way. Of course, we cannot say whether the same differences would exist if the washing time is longer or shorter, if the water has a different temperature or hardness, and so on.

In most cases, "overcontrolled" experiments like the one just described do not really provide us with the kind of information we want. Also, such experiments are rarely possible in actual practice; for example, it would have been difficult in our illustration to be sure that the instruments really were measuring identically on repeated washings or that some other factor, not thought of or properly controlled, was not responsible for the observed differences in whiteness. So, we look for alternatives. At the other extreme we can conduct an experiment in which none of the extraneous factors is controlled, but in which we protect ourselves against their effects by **randomization.** That is, we design, or plan, the experiment in such a way that the variations caused by these extraneous factors can all be combined under the general heading of "chance."

In our example dealing with the three detergents, we could accomplish this by numbering the swatches (which need not be equally soiled) from 1 to 15, specifying the random order in which they are to be washed and measured, and randomly selecting the five swatches which are to be washed with each of the three detergents. When all the variations due to uncontrolled extraneous factors can thus be included under the heading of chance variation, we refer to the design of the experiment as a **completely randomized design.**

As should be apparent, randomization protects against the effects of the extraneous factors only in a probabilistic sort of way. For instance, in our example it is possible, though very unlikely, that detergent A will be randomly assigned to the five swatches which happen to be the least soiled, or that the water happens to be coldest when we wash the five swatches with detergent B. It is partly for this reason that we often try to control some of the factors and randomize the others, and thus use designs that are somewhere between the two extremes which we have described.

13.3
ONE-WAY ANALYSIS OF VARIANCE

In analysis of variance, the basic idea is to express a measure of the total variation of a set of data as a sum of terms, which can be attributed to specific sources, or causes, of variation; in its simplest form, it applies to experiments which are planned as completely randomized designs. In the example of Section 13.1, two such sources of variation are (1) actual differences in the cleansing action of the three detergents, and (2) chance, which in problems of this kind is usually called the **experimental error.**

The measure of the total variation which we shall use here is the **total sum of squares**[†]

$$SST = \sum_{i=1}^{k} \sum_{j=1}^{n} (x_{ij} - \bar{x}..)^2$$

where x_{ij} is the jth observation of the ith sample ($i = 1, 2, \ldots, k$ and $j = 1, 2, \ldots, n$), and $\bar{x}..$ is the **grand mean,** the mean of all the kn measurements or observations. Note that if we divide the total sum of squares SST by $kn - 1$, we get the variance of the combined data.

If we let $\bar{x}_{i}.$ denote the mean of the ith sample (for $i = 1, 2, \ldots, k$), we can now write the following identity, which forms the basis of a **one-way analysis of variance:**[‡]

Identity for one-way analysis of variance

$$SST = n \cdot \sum_{i=1}^{k} (\bar{x}_{i}. - \bar{x}..)^2 + \sum_{i=1}^{k} \sum_{j=1}^{n} (x_{ij} - \bar{x}_{i}.)^2$$

Looking closely at the two terms into which the total sum of squares SST has been partitioned, we find that the first term is a measure of the variation among the sample means; in fact, if we divide it by $k - 1$ we get the quantity which we denoted by $n \cdot s_x^2$ on page 367. Similarly, the second term is a measure of the variation within the individual samples, and if we divide this term by $k(n - 1)$, we get the quantity which we put into the denominator of F in Section 13.1.

It is customary to refer to the first term, the quantity which measures the variation among the sample means, as the **treatment sum of squares** $SS(Tr)$, and to the second term, which measures the variation within the samples, as the **error sum of squares** SSE. This terminology is explained by the fact that most analysis-of-variance techniques were originally developed in connection with agricultural experiments where different fertilizers, for

---

[†] The use of double subscripts and double summations is explained briefly in Section 3.11.

[‡] This identity may be derived by writing the total sum of squares as

$$SST = \sum_{i=1}^{k} \sum_{j=1}^{n} (x_{ij} - \bar{x}..)^2$$

$$= \sum_{i=1}^{k} \sum_{j=1}^{n} [(\bar{x}_{i}. - \bar{x}..) + (x_{ij} - \bar{x}_{i}.)]^2$$

and then expanding the squares $[(\bar{x}_{i}. - \bar{x}..) + (x_{ij} - \bar{x}_{i}.)]^2$ by means of the binomial theorem and simplifying algebraically.

example, were regarded as different **treatments** applied to the soil. So, we shall refer to the three detergents in our example as three different treatments, and in other problems we may refer to four nationalities as four different treatments, five kinds of advertising campaigns as five different treatments, and so on. The word "error" in "error sum of squares" pertains to the experimental error, or chance.

Before we go any further, let us verify the identity $SST = SS(Tr) + SSE$ with reference to the example of Section 13.1, which deals with the three detergents. Substituting into the formulas for the different sums of squares, we get

$$SST = (77 - 75)^2 + (81 - 75)^2 + (71 - 75)^2$$
$$+ (76 - 75)^2 + (80 - 75)^2 + (72 - 75)^2$$
$$+ (58 - 75)^2 + (74 - 75)^2 + (66 - 75)^2$$
$$+ (70 - 75)^2 + (76 - 75)^2 + (85 - 75)^2$$
$$+ (82 - 75)^2 + (80 - 75)^2 + (77 - 75)^2$$
$$= 666$$

$$SS(Tr) = 5[(77 - 75)^2 + (68 - 75)^2 + (80 - 75)^2]$$
$$= 390$$

$$SSE = (77 - 77)^2 + (81 - 77)^2 + (71 - 77)^2 + (76 - 77)^2$$
$$+ (80 - 77)^2 + (72 - 68)^2 + (58 - 68)^2 + (74 - 68)^2$$
$$+ (66 - 68)^2 + (70 - 68)^2 + (76 - 80)^2 + (85 - 80)^2$$
$$+ (82 - 80)^2 + (80 - 80)^2 + (77 - 80)^2$$
$$= 276$$

and it can be seen that

$$SS(Tr) + SSE = 390 + 276 = 666 = SST$$

To test the null hypothesis $\mu_1 = \mu_2 = \cdots = \mu_k$ (or $\alpha_1 = \alpha_2 = \cdots = \alpha_k = 0$ in the notation of the footnote to page 366) against the alternative hypothesis that the treatment means are not all equal (or equivalently that the treatment effects are not all zero), we now proceed as in Section 13.1 and compare $SS(Tr)$ with SSE by means of an F statistic. In practice, we usually exhibit the necessary work as follows in an **analysis-of-variance table:**

Source of variation	Degrees of freedom	Sum of squares	Mean square	F
Treatments	$k - 1$	$SS(Tr)$	$MS(Tr) = \dfrac{SS(Tr)}{k - 1}$	$\dfrac{MS(Tr)}{MSE}$
Error	$k(n - 1)$	SSE	$MSE = \dfrac{SSE}{k(n - 1)}$	
Total	$kn - 1$	SST		

Here the second column lists the degrees of freedom (the number of independent deviations from the mean on which the sums of squares are based), the fourth column lists the **mean squares** $MS(Tr)$ and MSE, which are obtained by dividing the corresponding sums of squares by their degrees of freedom, and the right-hand column gives the value of the F-statistic as the ratio of the two mean squares. These two mean squares are, in fact, the two estimates of σ^2 referred to on page 367; also, the numerator and denominator degrees of freedom for the F test, $k - 1$ and $k(n - 1)$, are shown opposite "Treatments" and "Error" in the "Degrees of freedom" column. The significance test is the same as before; we compare F with F_α for $k - 1$ and $k(n - 1)$ degrees of freedom.

EXAMPLE   Construct an analysis-of-variance table for the example dealing with the three detergents.

Solution   The degrees of freedom for the treatment, error, and total sums of squares are $k - 1 = 3 - 1 = 2$, $k(n - 1) = 3(5 - 1) = 12$, and $kn - 1 = 3 \cdot 5 - 1 = 14$ (which provides a check as it equals the sum of the other two). The sums of squares are $SS(Tr) = 390$, $SSE = 276$, and $SST = 666$, so that $MS(Tr) = \frac{390}{2} = 195$, $MSE = \frac{276}{12} = 23$, and $F = \frac{195}{23} = 8.48$. All these results are summarized in the following table:

Source of variation	Degrees of freedom	Sum of squares	Mean square	F
Treatments	2	390	195	8.48
Error	12	276	23	
Total	14	666		

From here on, the significance test is as on page 368. Since $F = 8.48$ exceeds 6.93, the value of $F_{0.01}$ for 2 and 12 degrees of freedom, the null hypothesis that the three detergents are equally effective must be rejected.

The numbers which we used in our illustration were intentionally chosen so that the calculations would be relatively easy. In actual practice, the calculation of the sums of squares can be quite tedious unless we use the following computing formulas, in which $T_i.$ denotes the total of the observations for the ith treatment (that is, the sum of the values in the ith sample), and $T..$ denotes the grand total of all the data:

Computing formulas for sums of squares

$$SST = \sum_{i=1}^{k} \sum_{j=1}^{n} x_{ij}^2 - \frac{1}{kn} \cdot T_{..}^2$$

$$SS(Tr) = \frac{1}{n} \cdot \sum_{i=1}^{k} T_{i.}^2 - \frac{1}{kn} \cdot T_{..}^2$$

and by subtraction

$$SSE = SST - SS(Tr)$$

EXAMPLE Use these computing formulas to verify the sums of squares obtained on page 372 for the example dealing with the three detergents.

Solution Substituting $k = 3$, $n = 5$, $T_1. = 385$, $T_2. = 340$, $T_3. = 400$, $T.. = 1,125$, and $\sum \sum x^2 = 85,041$ into the formulas, we get

$$SST = 85,041 - \tfrac{1}{15}(1,125)^2$$

$$= 85,041 - 84,375$$

$$= 666$$

$$SS(Tr) = \tfrac{1}{5}[385^2 + 340^2 + 400^2] - 84,375$$

$$= 390$$

and

$$SSE = 666 - 390 = 276$$

As can be seen, these results are identical with the ones obtained before.

The method we have discussed here applies only when each sample has the same number of observations, but minor modifications make it applicable also to situations where the sample sizes are not all equal. If there are n_i observations for the ith treatment (in the ith sample), the computing formulas for the sums of squares become

Computing formulas for sums of squares (unequal sample sizes)

$$SST = \sum_{i=1}^{k} \sum_{j=1}^{n_i} x_{ij}^2 - \frac{1}{N} \cdot T_{..}^2$$

$$SS(Tr) = \sum_{i=1}^{k} \frac{T_{i\cdot}^2}{n_i} - \frac{1}{N} \cdot T_{..}^2$$

$$SSE = SST - SS(Tr)$$

where $N = n_1 + n_2 + \cdots + n_k$. The only other change is that the total number of degrees of freedom is $N - 1$, and the degrees of freedom for treatments and error are $k - 1$ and $N - k$.

EXAMPLE A laboratory technician wants to compare the breaking strength of three kinds of thread and originally he had planned to repeat each determination six times. Not having enough time, however, he has to base his analysis on the following results (in ounces):

> *Thread 1*: 18.0, 16.4, 15.7, 19.6, 16.5, 18.2
>
> *Thread 2*: 21.1, 17.8, 18.6, 20.8, 17.9, 19.0
>
> *Thread 3*: 16.5, 17.8, 16.1

Perform an analysis of variance to test at the 0.05 level of significance whether the differences among the sample means are significant.

Solution Using the same steps as in previous chapters for tests of hypotheses, we get

1. *Null hypothesis*: $\mu_1 = \mu_2 = \mu_3$
   Alternative hypothesis: The μ's are not all equal
2. *Level of significance*: $\alpha = 0.05$
3. *Criterion*: Reject the null hypothesis if $F > 3.89$, the value of $F_{0.05}$ for $k - 1 = 3 - 1 = 2$ and $N - k = 15 - 3 = 12$ degrees of freedom, where F is to be determined by an analysis of variance; otherwise, accept it.

4. *Calculations*: Substituting $n_1 = 6$, $n_2 = 6$, $n_3 = 3$, $N = 15$, $T_1. = 104.4$, $T_2. = 115.2$, $T_3. = 50.4$, $T.. = 270.0$, and $\sum\sum x^2 = 4{,}897.46$ into the computing formulas for the sums of squares, we get

$$SST = 4{,}897.46 - \frac{1}{15}(270.0)^2 = 37.46$$

$$SS(Tr) = \frac{104.4^2}{6} + \frac{115.2^2}{6} + \frac{50.4^2}{3} - \frac{1}{15}(270.0)^2$$

$$= 15.12$$

and

$$SSE = 37.46 - 15.12 = 22.34$$

The remainder of the work is shown in the following analysis-of-variance table:

Source of variation	Degrees of freedom	Sum of squares	Mean square	F
Treatments	2	15.12	7.56	4.06
Error	12	22.34	1.86	
Total	14	37.46		

5. *Decision*: Since $F = 4.06$ exceeds 3.89, the value of $F_{0.05}$ for 2 and 12 degrees of freedom, the null hypothesis must be rejected; in other words, we conclude that there is a difference in the mean strength of the three kinds of thread.

EXERCISES　　**13.1** The following are the mileages which a test driver got with four gallons each of five brands of gasoline:

> *Brand A*:　30, 25, 27, 26
>
> *Brand B*:　29, 26, 29, 28
>
> *Brand C*:　32, 32, 35, 37
>
> *Brand D*:　29, 34, 32, 33
>
> *Brand E*:　32, 26, 31, 27

(a) Use the method of Section 13.1 to test at the 0.01 level of significance whether the differences among the five sample means are significant.

(b) Perform an analysis of variance, using the computing formulas for the required sums of squares, and compare the resulting value of F with that obtained in part (a).

13.2 An agronomist planted three test plots each with four varieties of wheat and obtained the following yields (in pounds per plot):

$$Variety\ A:\quad 57, 62, 61$$

$$Variety\ B:\quad 52, 53, 60$$

$$Variety\ C:\quad 53, 56, 56$$

$$Variety\ D:\quad 56, 59, 59$$

(a) Use the method of Section 13.1 to test at the 0.05 level of significance whether the differences among the four sample means can be attributed to chance.
(b) Perform an analysis of variance, using the computing formulas for the required sums of squares, and compare the resulting value of F with that obtained in part (a).

13.3 An experiment is performed to determine which of three different golf-ball designs, A, B, and C, will give the greatest distance when driven. Criticize the experiment if
(a) one golf pro hits all the design A balls, another all the design B balls, and a third all the design C balls;
(b) all the balls are hit with drivers of the same manufacture;
(c) all the design A balls are hit first, the design B balls next, and the design C balls last;
(d) all the design A balls are hit from a tee, while the design B and design C balls are hit without tees;
(e) all the design A balls are driven from the No. 1 tee, all the design B balls are driven from the No. 2 tee, and all the design C balls are driven from the No. 3 tee.

13.4 The following are eight consecutive weeks' earnings (in dollars) of three door-to-door cosmetics salespersons employed by a firm:

$$Salesperson\ 1:\quad 176, 212, 188, 206, 200, 184, 193, 209$$

$$Salesperson\ 2:\quad 187, 193, 184, 198, 210, 199, 180, 195$$

$$Salesperson\ 3:\quad 164, 203, 180, 187, 223, 196, 189, 211$$

Use the 0.05 level of significance to test the null hypothesis that on the average the three salespersons' weekly earnings are the same.

13.5 Samples of peanut butter produced by three different manufacturers are tested for aflatoxin content (ppb), with the following results:

$$Brand\ 1:\quad 0.5, 6.3, 1.1, 2.7, 5.5, 4.3$$

$$Brand\ 2:\quad 2.5, 1.8, 3.6, 5.2, 1.2, 0.7$$

$$Brand\ 3:\quad 3.3, 1.5, 0.4, 4.8, 2.2, 1.0$$

Use the 0.05 level of significance to test whether the differences among the three sample means are significant.

13.6 The following are the numbers of mistakes made on five occasions by four compositors setting the type for a technical report:

Compositor 1: 10, 13, 9, 11, 12

Compositor 2: 11, 13, 8, 16, 12

Compositor 3: 10, 15, 13, 11, 15

Compositor 4: 15, 7, 11, 12, 9

At the 0.05 level of significance, can we conclude that there is a real difference in the numbers of mistakes made in general by the four compositors?

13.7 The following data show the yields of soybeans (in bushels per acre) planted two inches apart on essentially similar plots with the rows 20, 24, 28, and 32 inches apart:

20 in	24 in	28 in	32 in
23.1	21.7	21.9	19.8
22.8	23.0	21.3	20.4
23.2	22.4	21.6	19.3
23.4	21.1	20.2	18.5
23.6	21.9	21.6	19.1
21.7	23.4	23.8	21.9

Test at the 0.05 level of significance whether the differences among the four sample means can be attributed to chance.

13.8 The following are the numbers of words per minute which a secretary typed on several occasions on four different typewriters:

Typewriter C: 71, 75, 69, 77, 61, 72, 71, 78

Typewriter D: 68, 71, 74, 66, 69, 67, 70, 62

Typewriter E: 75, 70, 81, 73, 78, 72

Typewriter F: 62, 59, 71, 68, 63, 65, 72, 60, 64

Use the 0.01 level of significance to test whether the differences among the four sample means can be attributed to chance.

13.9 The following are the weight losses of certain machine parts due to friction (in milligrams) when used with three different lubricants:

Lubricant X: 10, 13, 12, 10, 14, 8, 12, 13

Lubricant Y: 9, 8, 12, 9, 8, 11, 7, 6, 8, 11, 9

Lubricant Z: 6, 7, 7, 5, 9, 8, 4, 10

Test at the 0.01 level of significance whether the differences among the three sample means are significant.

13.10 To study its performance, a newly designed motorboat was timed over a marked course under various wind and water conditions. Use the following data (in minutes) to test, at the 0.05 level of significance, whether the differences among the three sample means are significant:

$$Calm\ conditions:\quad 26, 19, 16, 22$$

$$Moderate\ conditions:\quad 25, 27, 25, 20, 18, 23$$

$$Choppy\ conditions:\quad 23, 25, 28, 31, 26$$

13.11 With reference to the notation introduced in the footnote to page 366, show that the sum of the α's, the treatment effects, is equal to zero.

13.4
THE DESIGN OF EXPERIMENTS: BLOCKING

To introduce another important concept in the design of experiments in addition to controlling and randomizing, suppose that a reading comprehension test is given to random samples of three eighth graders from each of four schools, and that the results are

$$School\ A:\quad 87, 70, 92$$

$$School\ B:\quad 43, 75, 56$$

$$School\ C:\quad 70, 66, 50$$

$$School\ D:\quad 67, 85, 70$$

The means of these four samples are 83, 58, 62, and 77, and since the differences among them are very large, it would seem reasonable to conclude that there are some real differences in the average reading comprehension of eighth graders in the four schools. This does not follow, however, from a one-way analysis of variance. We get

Source of variation	Degrees of freedom	Sum of squares	Mean square	F
Treatments	3	1,278	426	2.90
Error	8	1,176	147	
Total	11	2,454		

and since $F = 2.90$ does not exceed 4.07, the value of $F_{0.05}$ for 3 and 8 degrees of freedom, the null hypothesis (that the population means are all equal) cannot be rejected at the 0.05 level of significance.

The reason for this is that there are not only considerable differences among the four means, but also very large differences among the values within the samples. In the first sample they range from 70 to 92, in the second sample from 43 to 75, in the third sample from 50 to 70, and in the fourth sample from 67 to 85. Giving this some thought, it would seem reasonable to conclude that these differences within the samples may well be due to differences in intelligence, an extraneous factor (we might call it a "nuisance" factor) which was randomized by taking a random sample of eighth graders from each school. Thus, variations due to differences in intelligence were included in the experimental error; this "inflated" the error sum of squares which went into the denominator of the F statistic, and the results were not significant.

To avoid this kind of situation, we could hold the extraneous factor fixed, but this will seldom give us the information we want. In our example, we could limit the study to eighth graders with an IQ of, say, 105, but then the results would apply only to eighth graders with IQ's of 105. Another possibility is to vary the known source of variability (the extraneous factor) deliberately over as wide a range as necessary, and to do it in such a way that the variability it causes can be measured and, hence, eliminated from the experimental error. This means that we should plan the experiment in such a way that we can perform a **two-way analysis of variance,** in which the total variability of the data is partitioned into three components attributed, respectively, to treatments (in our example, the four schools), the extraneous factor, and experimental error.

As we shall see later, this can be accomplished in our example by randomly selecting from each school one eighth grader with a low IQ, one eighth grader with an average IQ, and one eighth grader with a high IQ, where "low," "average," and "high" are presumably defined in a rigorous way. Suppose, then, that we proceed in this way and get the results shown in the following table:

	Low IQ	Average IQ	High IQ
School A	71	92	89
School B	44	51	85
School C	50	64	72
School D	67	81	86

What we have done here is called **blocking,** and the three levels of intelligence are called **blocks.** In general, blocks are the levels at which we hold an extraneous factor fixed, so that we can measure its contribution to the total variability of the data by means of a two-way analysis of variance. The design which we used for our example is called a **complete block design**; it is "complete" in the sense that each treatment appears the same number of times in each block. In our example, there is one eighth grader from each school in each block.

13.5
TWO-WAY ANALYSIS OF VARIANCE

In this section we shall present the theory of a **two-way analysis of variance** in connection with a complete block design, and refer to the two variables (factors) under consideration as "treatments" and "blocks." As we shall indicate later, the same kind of analysis applies also to **two-factor experiments**, where both variables are of material concern.

Before we go into any details, let us point out that there are essentially two ways of analyzing such two-variable experiments, and they depend on whether the two variables are independent, or whether they **interact**. Suppose, for instance, that a tire manufacturer is experimenting with different kinds of treads, and that he finds that one kind is especially good for use on dirt roads while another kind is especially good for use on hard pavement. If this is the case, we say that there is an interaction between road conditions and tread design. On the other hand, if each of the treads is affected equally by the different road conditions, we would say that there is no interaction and that the two variables (road conditions and tread design) are independent. In this book we shall study only the case where there is no interaction.

To formulate the hypotheses to be tested in the two-variable case, let us write μ_{ij} for the population mean which corresponds to the ith treatment and the jth block (in our numerical example, the average reading comprehension score of eighth graders with the intelligence level j in the ith school) and express it as

$$\mu_{ij} = \mu + \alpha_i + \beta_j$$

As in the footnote to page 366, μ is the grand mean (the average of all the population means μ_{ij}) and the α_i are the treatment effects (whose sum is zero). Correspondingly, we refer to the β_j as the **block effects** (whose sum is also zero), and write the two null hypotheses we want to test as

$$\alpha_1 = \alpha_2 = \cdots = \alpha_k = 0$$

and

$$\beta_1 = \beta_2 = \cdots = \beta_n = 0$$

The alternative to the first null hypothesis (which in our illustration amounts to the hypothesis that the average reading comprehension of eighth graders is the same in all four schools) is that the treatment effects α_i are not all zero; the alternative to the second null hypothesis (which in our illustration amounts to the hypothesis that the average reading comprehension of eighth graders is the same for all three levels of intelligence) is that the block effects β_j are not all zero.

To test the second of the null hypotheses, we need a quantity, similar to the treatment sum of squares, which measures the variation among the block means (58, 72, and 83 for the data on page 380). So, if we let $T_{\cdot j}$ denote the total of all the values in the jth block, substitute it for $T_{i\cdot}$ in the computing formula for $SS(Tr)$ on page 374, sum on j instead of i, and interchange n and k, we obtain, analogous to $SS(Tr)$ the **block sum of squares**

Computing formula for block sum of squares

$$SSB = \frac{1}{k} \cdot \sum_{j=1}^{n} T_{\cdot j}^2 - \frac{1}{kn} \cdot T_{\cdot\cdot}^2$$

In a two-way analysis of variance (with no interaction) we compute SST and $SS(Tr)$ according to the formulas on page 374, SSB according to the formula immediately above, and then we get SSE by subtraction. Since

$$SST = SS(Tr) + SSB + SSE$$

we have

Error sum of squares (Two-way analysis of variance)

$$SSE = SST - [SS(Tr) + SSB]$$

Observe that the error sum of squares for a two-way analysis of variance does not equal the error sum of squares for a one-way analysis of variance performed on the same data, even though we denote both with the symbol SSE. In fact, we are now partitioning the error sum of squares for the one-way analysis of variance into two terms: the block sum of squares, SSB, and the remainder which is the new error sum of squares, SSE.

We can now construct the following analysis-of-variance table for a two-way analysis of variance (with no interaction):

Source of variation	Degrees of freedom	Sum of squares	Mean square	F
Treatments	$k - 1$	$SS(Tr)$	$MS(Tr) = \dfrac{SS(Tr)}{k - 1}$	$\dfrac{MS(Tr)}{MSE}$
Blocks	$n - 1$	SSB	$MSB = \dfrac{SSB}{n - 1}$	$\dfrac{MSB}{MSE}$
Error	$(k - 1)(n - 1)$	SSE	$MSE = \dfrac{SSE}{(k - 1)(n - 1)}$	
Total	$kn - 1$	SST		

The mean squares are again the sums of squares divided by their respective degrees of freedom, and the two F values are the mean squares for treatments and blocks divided by the mean square for error. Also, the degrees of freedom for blocks are $n - 1$ (like those for treatments with n substituted for k), and the degrees of freedom for error are found by subtracting the degrees of freedom for treatments and blocks from $kn - 1$, the total number of degrees of freedom:

$$(kn - 1) - (k - 1) - (n - 1) = kn - k - n + 1$$
$$= (k - 1)(n - 1)$$

Thus, in the significance test for treatments the numerator and denominator degrees of freedom for F are $k - 1$ and $(k - 1)(n - 1)$, and in the significance test for blocks the numerator and denominator degrees of freedom for F are $n - 1$ and $(k - 1)(n - 1)$.

EXAMPLE   Based on the data on page 380, namely,

	Low IQ	Average IQ	High IQ
School A	71	92	89
School B	44	51	85
School C	50	64	72
School D	67	81	86

test at the 0.05 level of significance whether the differences among the means obtained for the four schools (treatments) are significant, and also whether the differences among the means obtained for the three levels of IQ (blocks) are significant.

Solution

1. *Null hypotheses:* $\alpha_1 = \alpha_2 = \alpha_3 = \alpha_4 = 0$ and

$$\beta_1 = \beta_2 = \beta_3 = 0$$

Alternative hypotheses: The α's are not all equal to zero; the β's are not all equal to zero.

2. *Levels of significance:* $\alpha = 0.05$ for both tests

3. *Criteria:* For treatments, reject the null hypothesis if $F > 4.76$, the value of $F_{0.05}$ for $k - 1 = 4 - 1 = 3$ and $(k - 1)(n - 1) = (4 - 1)(3 - 1) = 6$ degrees of freedom; for blocks, reject the null hypothesis if $F > 5.14$, the value of $F_{0.05}$ for $n - 1 = 3 - 1 = 2$ and $(k - 1)(n - 1) = (4 - 1)(3 - 1) = 6$ degrees of freedom. Otherwise, reserve judgments.

4. *Calculations:* Substituting $k = 4$, $n = 3$, $T_1. = 252$, $T_2. = 180$, $T_3. = 186$, $T_4. = 234$, $T._1 = 232$, $T._2 = 288$, $T._3 = 332$, $T.. = 852$, and $\sum\sum x^2 = 63{,}414$ into the computing formulas for the sums of squares, we get

$$
\begin{aligned}
SST &= 63{,}414 - \tfrac{1}{12}(852)^2 \\
&= 63{,}414 - 60{,}492 \\
&= 2{,}922
\end{aligned}
$$

$$
\begin{aligned}
SS(Tr) &= \tfrac{1}{3}[252^2 + 180^2 + 186^2 + 234^2] - 60{,}492 \\
&= 1{,}260
\end{aligned}
$$

$$
\begin{aligned}
SSB &= \tfrac{1}{4}[232^2 + 288^2 + 332^2] - 60{,}492 \\
&= 1{,}256
\end{aligned}
$$

and

$$
\begin{aligned}
SSE &= 2{,}922 - [1{,}260 + 1{,}256] \\
&= 406
\end{aligned}
$$

The remainder of the work is shown in the following analysis-of-variance table:

Source of variation	Degrees of freedom	Sum of squares	Mean square	F
Treatments	3	1,260	420	6.21
Blocks	2	1,256	628	9.28
Error	6	406	67.67	
Total	11	2,922		

5. *Decisions*: For treatments, since $F = 6.21$ exceeds 4.76, the value of $F_{0.05}$ for 3 and 6 degrees of freedom, the null hypothesis must be rejected; for blocks, since $F = 9.28$ exceeds 5.14, the value of $F_{0.05}$ for 2 and 6 degrees of freedom, the null hypothesis must be rejected. In other words, we conclude that the average reading comprehension of eighth graders is not the same in the four schools, and that it is not the same for the three levels of intelligence. Observe that by blocking we were able to show that the differences among the means obtained for the four schools are significant, whereas without blocking, in the experiment described on page 379, the differences among the means obtained for the four schools were not significant.

As we already pointed out on page 381, a two-way analysis of variance can also be used in connection with **two-factor experiments,** where both variables (factors) are of material concern. It could be used, for example, in the analysis of the following data collected in an experiment designed to test whether the range of a missile flight (in miles) is affected by differences among launchers and also by differences among fuels (see Exercise 13.16 on page 387):

	Fuel 1	Fuel 2	Fuel 3	Fuel 4
Launcher X	45.9	57.6	52.2	41.7
Launcher Y	46.0	51.0	50.1	38.8
Launcher Z	45.7	56.9	55.3	48.1

Note that we used a different format for the table to distinguish between two-factor experiments and complete block designs.

Also, when a two-way analysis of variance is used in this way, we usually call the two variables **factor *A*** and **factor *B*** (instead of treatments and blocks) and write *SSA* instead of *SS(Tr)*; we still write *SSB*, but now *B* stands for the factor *B* instead of blocks.

EXERCISES

13.12 The following are the cholesterol contents (in milligrams per package) which four laboratories obtained for 6-ounce packages of three very similar diet foods:

	Laboratory 1	Laboratory 2	Laboratory 3	Laboratory 4
Diet food A	3.7	2.8	3.1	3.4
Diet food B	3.1	2.6	2.7	3.0
Diet food C	3.5	3.4	3.0	3.3

perform a two-way analysis of variance, using the 0.05 level of significance for both tests.

13.13 Four different, although supposedly equivalent, forms of a standardized achievement test in science were given to each of five students, and the following are the scores which they obtained:

	Student C	Student D	Student E	Student F	Student G
Form 1	77	62	52	66	68
Form 2	85	63	49	65	76
Form 3	81	65	46	64	79
Form 4	88	72	55	60	66

Perform a two-way analysis of variance, using the 0.01 level of significance for both tests. If each student separately randomized the order in which he or she took the four tests (by using random numbers or some kind of gambling device), we refer to the design of this experiment as a **randomized block design.** The purpose of this randomization is to take care of such possible extraneous factors as fatigue or, perhaps, the experience gained from repeatedly taking the test.

13.14 A laboratory technician measures the breaking strength of each of five kinds of linen threads by using four different measuring instruments, I_1, I_2, I_3, and I_4, and obtains the following results (in ounces):

	I_1	I_2	I_3	I_4
Thread 1	20.9	20.4	19.9	21.9
Thread 2	25.0	26.2	27.0	24.8
Thread 3	25.5	23.1	21.5	24.4
Thread 4	24.8	21.2	23.5	25.7
Thread 5	19.6	21.2	22.1	21.1

Perform a two-way analysis of variance, using the 0.05 level of significance for both tests.

13.15 Regarding the weeks as blocks, reanalyze the data of Exercise 13.4 by means of a two-way analysis of variance. Use the 0.05 level of significance for both tests.

13.16 With reference to the missile-range data on page 385, perform a two-way analysis of variance, using the 0.05 level of significance for both tests.

13.17 The following are the numbers of defectives produced by four workmen operating, in turn, three different machines:

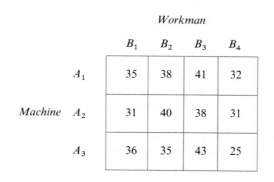

Workman

	B_1	B_2	B_3	B_4
A_1	35	38	41	32
A_2	31	40	38	31
A_3	36	35	43	25

Machine

Perform a two-way analysis of variance, using the 0.05 level of significance for both tests.

13.6
THE DESIGN OF EXPERIMENTS: REPLICATION

In Section 13.4 we showed how we can increase the amount of information to be gained from an experiment by blocking, that is, by eliminating the effect of an extraneous factor. Another way to increase the amount of information to be gained from an experiment is to increase the volume of the data. For instance, in the example on page 379 we might increase the size of the samples and give the reading comprehension test to twenty eighth graders from each school instead of three. For more complicated designs, the same thing can be accomplished by executing the entire experiment more than once, and this is called **replication**. With reference to the example on page 380, we might conduct the experiment (select and test twelve eighth graders) in one week, and then replicate (repeat) the entire experiment in the next week.

Conceptually, replication does not present any difficulties, but computationally it does, and that is why we shall not go into this any further. Indeed, if an experiment requiring a two-way analysis of variance is replicated, it will then require a three-way analysis of variance, since replication, itself, may be a source of variation in the data. For instance, this might be the case in our example if it got very hot and humid during the second week, making it difficult for the students to concentrate.

13.7
LATIN SQUARES ★

We have seen how blocking can be used to eliminate the variability due to one extraneous factor from the experimental error, and, in principle, two or more extraneous sources of variation can be handled in the same way. The only real problem is that this may inflate the size of an experiment beyond practical bounds. Suppose, for instance, that in the example dealing with the reading comprehension of eighth graders we would also like to eliminate whatever variability there may be due to differences in age (12, 13, or 14) and in sex. Allowing for all possible combinations of intelligence, age, and sex, we will have to use $3 \cdot 3 \cdot 2 = 18$ different blocks, and if there is to be one eighth grader from each school in each block, we will have to select and test $18 \cdot 4 = 72$ eighth graders in all. If we also wanted to eliminate whatever variability there may be due to ethnic background, for which we might consider five categories, this would raise the required number of eighth graders to $72 \cdot 5 = 360$.

In this section we will show how problems like this can sometimes be resolved, at least in part, by using a **Latin square design**; at the same time, we hope to impress upon the reader that it is through proper design that experiments can be made to yield a wealth of information. To give an example, suppose that a market research organization wants to compare four ways of packaging a breakfast food, but it is concerned about possible regional differences in the popularity of the breakfast food, and also about the effects of promoting the breakfast food in different ways. So, it decides to test market the different kinds of packaging in the northeastern, southeastern, northwestern, and southwestern parts of the United States and to promote them with discounts, lotteries, coupons, and two-for-one sales. Thus, there are $4 \cdot 4 = 16$ blocks (combinations of regions and methods of promotion) and it would take $16 \cdot 4 = 64$ market areas (cities) to promote each kind of packaging once within each block. It is of interest to note, however, that with proper planning 16 market areas (cities) will suffice. To illustrate, let us consider the following arrangement, called a **Latin square**, in which the letters A, B, C, and D represent the four kinds of packaging:

	Discounts	Lotteries	Coupons	Two-for-one sales
Northeast	A	B	C	D
Southeast	B	C	D	A
Northwest	C	D	A	B
Southwest	D	A	B	C

In general, a Latin square is a square array of the letters A, B, C, D, ..., of the English (Latin) alphabet, which is such that each letter occurs once and only once in each row and in each column.

The above Latin square, looked upon as an experimental design, requires that discounts be used with packaging A in a city in the Northeast, with packaging B in a city in the Southeast, with packaging C in a city in the Northwest, and with packaging D in a city in the Southwest; that lotteries be used with packaging B in a city in the Northeast, with packaging C in a city in the Southeast, with packaging D in a city in the Northwest, and with packaging A in a city in the Southwest; and so on. Note that each kind of promotion is used once in each region and once with each kind of packaging; each kind of packaging is used once in each region and once with each kind of promotion; and each region is used once with each kind of packaging and once with each kind of promotion. As we shall see, this will enable us to perform an analysis of variance leading to significance tests for all three variables.

The analysis of an $r \times r$ Latin square is very similar to a two-way analysis of variance. The total sum of squares and the sums of squares for rows and columns are calculated in the same way in which we previously calculated SST, $SS(Tr)$, and SSB, but we must find an extra sum of squares which measures the variability due to the variable represented by the letters $A, B, C, D, \ldots$, namely, a new treatment sum of squares. The formula for this sum of squares is

Treatment sum of squares for Latin square

$$SS(Tr) = \frac{1}{r} \cdot (T_A^2 + T_B^2 + T_C^2 + \cdots) - \frac{1}{r^2} \cdot T_{..}^2$$

where T_A is the total of the observations corresponding to treatment A, T_B is the total of the observations corresponding to treatment B, and so forth. Finally, the error sum of squares is again obtained by subtraction:

Error sum of squares for Latin square

$$SSE = SST - [SSR + SSC + SS(Tr)]$$

where SSR and SSC are the sums of squares for rows and columns.

We can now construct the following analysis-of-variance table for the analysis of an $r \times r$ Latin square:

Source of variation	Degrees of freedom	Sum of squares	Mean square	F
Rows	$r - 1$	SSR	$MSR = \dfrac{SSR}{r-1}$	$\dfrac{MSR}{MSE}$
Columns	$r - 1$	SSC	$MSC = \dfrac{SSC}{r-1}$	$\dfrac{MSC}{MSE}$
Treatments	$r - 1$	$SS(Tr)$	$MS(Tr) = \dfrac{SS(Tr)}{r-1}$	$\dfrac{MS(Tr)}{MSE}$
Error	$(r-1)(r-2)$	SSE	$MSE = \dfrac{SSE}{(r-1)(r-2)}$	
Total	$r^2 - 1$	SST		

The mean squares are again the sums of squares divided by their respective degrees of freedom, and the three F-values are the mean squares for rows, columns, and treatments divided by the mean square for error. The degrees of freedom for rows, columns, and treatments are all $r - 1$, and, by subtraction, the degree of freedom for error is

$$(r^2 - 1) - (r - 1) - (r - 1) - (r - 1) = r^2 - 3r + 2$$
$$= (r - 1)(r - 2)$$

Thus, for each of the three significance tests the numerator and denominator degrees of freedom for F are $r - 1$ and $(r - 1)(r - 2)$.

EXAMPLE   Suppose that in the breakfast-food example the market research organization actually gets the data shown in the following table, where the figures are one week's sales in $10 thousand:

	Discounts	Lotteries	Coupons	Two-for-one sales
Northeast	A 48	B 38	C 42	D 53
Southeast	B 39	C 43	D 50	A 54
Northwest	C 42	D 50	A 47	B 44
Southwest	D 46	A 48	B 46	C 52

Analyze this Latin square, using the 0.05 level of significance for each test.

Solution

1. *Null hypotheses*: The row, column, and treatment effects (defined in the same way as the others) are all equal to zero.
   Alternative hypotheses: The row effects are not all equal to zero, the column effects are not all equal to zero, and the treatment effects are not all equal to zero.
2. *Levels of significance*: $\alpha = 0.05$ for each test.
3. *Criteria*: For rows, columns, or treatments, reject the null hypothesis if $F > 4.76$, the value of $F_{0.05}$ for $r - 1 = 4 - 1 = 3$ and $(r - 1)(r - 2) = (4 - 1)(4 - 2) = 6$ degrees of freedom. Otherwise, accept the alternative hypotheses or reserve judgment.
4. *Calculations*: Substituting $r = 4$, $T_{1.} = 181$, $T_{2.} = 186$, $T_{3.} = 183$, $T_{4.} = 192$, $T_{.1} = 175$, $T_{.2} = 179$, $T_{.3} = 185$, $T_{.4} = 203$, $T_A = 197$, $T_B = 167$, $T_C = 179$, $T_D = 199$, $T_{..} = 742$, and $\sum\sum x^2 = 34{,}756$ into the computing formulas for the sums of squares, we get

$$SST = 34{,}756 - \tfrac{1}{16}(742)^2 = 34{,}756 - 34{,}410.25$$
$$= 345.75$$

$$SSR = \tfrac{1}{4}[181^2 + 186^2 + 183^2 + 192^2] - 34{,}410.25$$
$$= 17.25$$

$$SSC = \tfrac{1}{4}[175^2 + 179^2 + 185^2 + 203^2] - 34{,}410.25$$
$$= 114.75$$

$$SS(Tr) = \tfrac{1}{4}[197^2 + 167^2 + 179^2 + 199^2] - 34{,}410.25$$
$$= 174.75$$

and

$$SSE = 345.75 - [17.25 + 114.75 + 174.75]$$
$$= 39.00$$

The remainder of the work is shown in the following analysis-of-variance table:

Source of variation	Degrees of freedom	Sum of squares	Mean square	F
Rows	3	17.25	$\frac{17.25}{3} = 5.75$	$\frac{5.75}{6.5} = 0.9$
Columns	3	114.75	$\frac{114.75}{3} = 38.25$	$\frac{38.25}{6.5} = 5.9$
Treatments	3	174.75	$\frac{174.75}{3} = 58.25$	$\frac{58.25}{6.5} = 9.0$
Error	6	39.00	$\frac{39.00}{6} = 6.5$	
Total	15	345.75		

5. *Decisions*: For rows, since $F = 0.9$ does not exceed 4.76, the null hypothesis cannot be rejected; for columns, since $F = 5.9$ exceeds 4.76, the null hypothesis must be rejected; for treatments, since $F = 9.0$ exceeds 4.76, the null hypothesis must be rejected. In other words, we conclude that differences in promotion and packaging, but not the different regions, affect the breakfast food's sales.

The construction of Latin squares (that is, square arrays in which each of the letters appears once and only once in each row and in each column) is a problem of pure mathematics; so far as applied work in statistics is concerned, Latin square patterns may be looked up in tables (for instance, the one listed among the references on page 400).

13.8
THE DESIGN OF EXPERIMENTS: SOME FURTHER CONSIDERATIONS

There are many other experimental designs besides the ones we have discussed in this chapter, and they serve a great variety of special purposes. Widely used, for example, are the **incomplete block designs,** which apply when it is impossible to have each treatment in each block.

The need for such a design arises, for instance, when we want to compare 13 kinds of tires but cannot put them all on a test car at the same time. Numbering the tires from 1 to 13, we might use the following experimental design

Test run	Kinds of tires			
1	1	2	4	10
2	2	3	5	11
3	3	4	6	12
4	4	5	7	13
5	5	6	8	1
6	6	7	9	2
7	7	8	10	3
8	8	9	11	4
9	9	10	12	5
10	10	11	13	6
11	11	12	1	7
12	12	13	2	8
13	13	1	3	9

where there are 13 test runs, or blocks, and since each kind of tire appears together with each other kind of tire once within the same block, the design is referred to as a **balanced incomplete block design**. The fact that each kind of tire appears together with each other kind of tire once within the same block is important; it facilitates the statistical analysis because it assures that we have the same amount of information for comparing each pair of tires. In general, the analysis of incomplete block designs is fairly complicated, and we shall not go into it here, as it has been our purpose only to demonstrate what can be accomplished by the careful design of an experiment.

EXERCISES

★ 13.18 Suppose that we want to compare the number of defective pieces produced by four machine operators working on five different machine parts (1, 2, 3, 4, and 5) in three different shifts (I, II, and III).
   (a) If the machine operators are to be regarded as the "treatments," list the blocks (combinations of machine parts and shifts) that would be required if each part is to be produced in each shift.
   (b) How many observations would be required if each machine operator is to work once on each machine part during each shift?

★ 13.19 An agronomist wants to compare the yield of 16 varieties of corn, and at the same time study the effect of five different fertilizers and two methods of irrigation. How many test plots must he plant if each variety of corn is to be grown once with each possible combination of fertilizers and methods of irrigation?

★ 13.20 A manufacturer of pharmaceuticals wants to market a new cold remedy which is actually a combination of four medications, and he wants to experiment first with two dosages for each medication. If A_L and A_H denote the low and high dosage of medication A, B_L and B_H the low and high dosage of medication B, C_L and C_H the low and high dosage of medication C, and D_L and D_H the low and high dosage of medication D, list the 16 preparations he has to test if each dosage of each medication is to be used once in combination with each dosage of each of the other medications.

★ 13.21 Making use of the fact that each of the letters must occur once and only once in each row and each column, complete the following Latin squares:

(a)
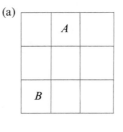

(b)
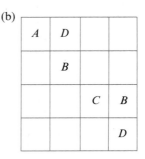

★13.22 The sample data in the following 3 × 3 Latin square are the grades in an American history test obtained by nine college students of various ethnic backgrounds and of various professional interests, who were taught by instructors A, B, and C:

| | Ethnic background | | |
	Mexican	German	Polish
Law	A 75	B 86	C 69
Medicine	B 95	C 79	A 86
Engineering	C 70	A 83	B 93

Analyze this Latin square, using the 0.05 level of significance for each test.

★13.23 To compare four different golf-ball designs, A, B, C, and D, each kind was driven by each of four golf pros, P_1, P_2, P_3, and P_4, using once each the four different drivers, D_1, D_2, D_3, and D_4. The distances from the tee to the points where the balls came to rest (in yards) were as shown in the following table:

	D_1	D_2	D_3	D_4
P_1	D 231	B 215	A 261	C 199
P_2	C 234	A 300	B 280	D 266
P_3	A 301	C 208	D 247	B 255
P_4	B 253	D 258	C 210	A 290

Analyze this Latin square, using the 0.05 level of significance for each test.

13.24 To test their ability to make decisions under pressure, the nine senior executives of a company are to be interviewed by each of four psychologists. As it takes a psychologist a full day to interview three of the executives, the schedule for the interviews is arranged as follows, where the nine executives are denoted A, B, C, D, E, F, G, H, and I:

Day	Psychologist	Executives		
March 2	I	B	C	?
March 3	I	E	F	G
March 4	I	H	I	A
March 5	II	C	?	H
March 6	II	B	F	A
March 9	II	D	E	?
March 10	III	D	G	A
March 11	III	C	F	?
March 12	III	B	E	H
March 13	IV	B	?	I
March 16	IV	C	?	A
March 17	IV	D	F	H

Replace the six question marks with the appropriate letters, given that each of the nine executives is to be interviewed together with each of the other executives once and only once on the same day. Note that this will make the arrangement a balanced incomplete block design, which may be important because each executive is tested together with each other executive once under identical conditions.

13.25 A newspaper regularly prints the columns of seven writers but has room for only three in each edition. Complete the following schedule, in which the writers are numbered 1–7, so that each writer's column appears three times per week, and a column of each writer appears together with a column of each other writer once per week:

	Writers		
Monday	1	2	3
Tuesday	4		
Wednesday	1	4	5
Thursday	2		
Friday	1	6	7
Saturday	5		
Sunday	2	4	6

13.9
CHECKLIST OF KEY TERMS
(with page references to their definitions)

13.10
REVIEW EXERCISES

13.26 To find the best arrangement of instruments on a control panel of an airplane, three different arrangements were tested by simulating emergency conditions and observing the reaction time required to correct the condition. The reaction times (in tenths of a second) of twelve pilots (randomly assigned to the different arrangements) were as follows:

> *Arrangement 1*: 8, 15, 10, 11
>
> *Arrangement 2*: 16, 11, 14, 19
>
> *Arrangement 3*: 12, 7, 13, 8

At the 0.01 level of significance, can the differences among the means of the three samples be attributed to chance?

13.27 A school has seven department heads who are assigned to seven different committees as shown in the following table:

	Department heads
Textbooks	Dodge, Fleming, Griffith, Anderson
Athletics	Bowman, Evans, Griffith, Anderson
Band	Bowman, Carlson, Fleming, Anderson
Dramatics	Bowman, Carlson, Dodge, Griffith
Tenure	Carlson, Evans, Fleming, Griffith
Salaries	Bowman, Dodge, Evans, Fleming
Discipline	Carlson, Dodge, Evans, Anderson

    (a) Verify that this arrangement is a balanced incomplete block design.

    (b) If Dodge, Bowman, and Carlson are (in that order) appointed chairperson of the first three committees, how will the chairpersons of the other four committees have to be chosen so that each of the department heads is chairperson of one of the committees?

★**13.28** The figures in the following 5×5 Latin square are the numbers of minutes engines $E_1, E_2, E_3, E_4,$ and E_5, tuned up by mechanics $M_1, M_2, M_3, M_4,$ and M_5, ran with a gallon of fuel $A, B, C, D,$ or E:

	E_1	E_2	E_3	E_4	E_5
M_1	A 31	B 24	C 20	D 20	E 18
M_2	B 21	C 27	D 23	E 25	A 31
M_3	C 21	D 27	E 25	A 29	B 21
M_4	D 21	E 25	A 33	B 25	C 22
M_5	E 21	A 37	B 24	C 24	D 20

Analyze this Latin square, using the 0.01 level of significance for each of the tests of significance.

13.29 The sample data in the following table are the grades in a statistics test obtained by nine college students from three majors who were taught by three different instructors:

	Instructor A	Instructor B	Instructor C
Marketing	77	88	71
Finance	88	97	81
Insurance	85	95	72

Analyze this two-factor experiment, using the 0.05 level of significance.

13.30 A food processor considers three spices for use individually or together in a new cake mix. If A, B, and C denote that the three spices are used, and a, b, and c denote that they are not used, then AbC, for example, denotes that the first and third spices are used and the second spice is not used. List the eight possible ways in which the food processor may or may not use the three spices.

13.31 The manager of a restaurant wants to determine whether the sales of chicken dinners depend on how this entree is described on the menu. He has three kinds of menus printed, listing chicken dinners among the other entrees, featuring them as "Chef's Special," or as "Gourmet's Delight," and he intends to use each kind of menu on six different Sundays. Actually, the manager collects only the following data, showing the number of chicken dinners sold on twelve Sundays:

Listed among other entrees	76, 94, 85, 77
Featured as Chef's Special	109, 117, 102, 92, 115
Featured as Gourmet's Delight	100, 83, 102

Perform a one-way analysis of variance at the 0.05 level of significance.

★**13.32** Among the nine persons interviewed in a poll, three are Easterners, three are Southerners, and three are Westerners. By profession, three of them are teachers, three are lawyers, and three are doctors, and no two of the same profession come from the same part of the United States. Also, three are Democrats, three are Republicans, and three are Independents, and no two of the same political affiliation are of the same profession or come from the same part of the United States. If one of the teachers is an Easterner and an Independent, another teacher is a Southerner and a Republican, and one of the lawyers is a Southerner and a Democrat, what is the political affiliation of the doctor who is a Westerner?

(*Hint*: Construct a 3 × 3 Latin square; this exercise is a simplified version of a famous problem posed by R. A. Fisher in his classical work *The Design of Experiments.*)

13.33 The following data are the amounts of time (in minutes) that it took a person to drive to work, Monday through Friday, along four different routes:

	Monday	*Tuesday*	*Wednesday*	*Thursday*	*Friday*
Route 1	27	26	26	32	23
Route 2	28	29	27	30	26
Route 3	30	34	31	34	27
Route 4	29	28	31	31	27

Perform a two-way analysis of variance, using the 0.05 level of significance for both tests.

13.11
REFERENCES

The following are some of the many books that have been written on the subject of analysis of variance:

GUENTHER, W. C., *Analysis of Variance*. Englewood Cliffs, N.J.: Prentice-Hall, Inc., 1964.

SNEDECOR, G. W., and COCHRAN, W. G., *Statistical Methods, 6th ed.* Ames, Iowa: Iowa State University Press, 1973.

Problems relating to the design of experiments are also treated in the above books and in

ANDERSON, V. L., and McLEAN, R. A., *Design of Experiments: A Realistic Approach*. New York: Marcel Dekker, Inc., 1974.

BOX, G. E. P., HUNTER, W. G., and HUNTER, J. S., *Statistics for Experimenters*. New York: John Wiley & Sons, Inc., 1978.

COCHRAN, W. G., and COX, G. M., *Experimental Design, 2nd ed.* New York: John Wiley & Sons, Inc., 1957.

FINNEY, D. J., *An Introduction to the Theory of Experimental Design*. Chicago: The University of Chicago Press, 1960.

HICKS, C. R., *Fundamental Concepts in the Design of Experiments*. New York: Holt, Rinehart and Winston, 1964.

ROMANO, A., *Applied Statistics for Science and Industry*. Boston: Allyn and Bacon, Inc., 1977.

A table of Latin squares of size $r = 3, 4, 5, \ldots$, and 12 may be found in the above-mentioned book by W. G. Cochran and G. M. Cox.

The main goal of many statistical investigations is to establish relationships which make it possible to predict one or more variables in terms of others. For instance, studies are made to predict the future sales of a product in terms of its price, a person's weight loss in terms of the number of weeks he or she has been on an 800-calories per-day diet, family expenditures on medical care in terms of family income, the per capita consumption of certain food items in terms of their nutritional value and the amount of money spent advertising them on television, and so forth.

Of course, it would be ideal if we could predict one quantity exactly in terms of another, but this is seldom possible. In most instances we must be satisfied with predicting averages or expected values. For instance, we cannot predict exactly how much money a specific college graduate will earn ten years after graduation, but given suitable data we can predict the average earnings of all college graduates ten years after graduation. Similarly, we can predict the average yield of a variety of wheat in terms of the total rainfall in July, and we can predict the expected grade-point average of a student starting law school in terms of his or her IQ. This problem of predicting the average value of one variable in terms of the known value of another variable (or the known values of other variables) is called the problem of regression. This term dates back to Francis Galton (1822–1911), who used it first in connection with a study of the relationship between the heights of fathers and sons.

The case where predictions are based on the known value of one variable is treated in Sections 14.1 through 14.4, and the case where predictions are based on the known values of several variables is treated in Section 14.5.

14 REGRESSION

14.1
CURVE FITTING

Whenever possible, we try to express, or approximate, relationships between known quantities and quantities that are to be predicted in terms of mathematical equations. This has been very successful in the natural sciences, where it is known, for instance, that at a constant temperature the relationship between the volume, y, and the pressure, x, of a gas is given by the formula

$$y = \frac{k}{x}$$

where k is a numerical constant. Also, it has been shown that the relationship between the size of a culture of bacteria, y, and the length of time, x, it has been exposed to certain environmental conditions is given by the formula

$$y = a \cdot b^x$$

where a and b are numerical constants. More recently, equations like these have also been used to describe relationships in the behavioral sciences, the social sciences, and other fields. For instance, the first of the equations above is often used in economics to describe the relationship between price and demand, and the second has been used to describe the growth of one's vocabulary or the accumulation of wealth.

Whenever we use observed data to arrive at a mathematical equation which describes the relationship between two variables—a procedure known as **curve fitting**—we must face three kinds of problems: (1) We must decide what kind of "predicting" equation we want to use; (2) we must find the particular equation which is "best" in some sense; and (3) we must investigate certain questions regarding the merits of the particular equation, and of predictions made from it.

With respect to the first of these problems of curve fitting, there are many different kinds of curves (and their equations) that can be used for predictive purposes. The choice of one of them is sometimes decided for us by theoretical considerations, but usually it is decided by direct inspection of the data. We plot the data on ordinary (arithmetic) graph paper, sometimes on special graph paper with special scales (see Section 14.4), and we decide by visual inspection upon the kind of curve (a straight line, a parabola, . . .)

which best describes the overall pattern of the data. There exist various methods for putting this decision on a more objective basis, but they will not be discussed in this book. So far as the second and third problems of curve fitting are concerned, we shall study the second in some detail in Section 14.2, and the third in Section 14.3.

14.2
LINEAR REGRESSION

Of the many equations that can be used to predict values of one variable, y, from given values of another variable, x, simplest and most widely used is the **linear equation in two unknowns,** which is of the form

$$y = a + bx$$

where a is the y-intercept (the value of y for $x = 0$) and b is the slope of the line (namely, the change in y which accompanies a change of one unit in x). Ordinarily, the values of a and b are estimated from given data, and once they have been determined, we can substitute a value of x into the equation and calculate the corresponding predicted value of y. Linear equations are useful and important not only because many relationships are actually of this form, but also because they often provide close approximations to relationships which would otherwise be difficult to describe in mathematical terms.

The term "linear equation" arises from the fact that, when plotted on ordinary (arithmetic) graph paper, all pairs of values of x and y which satisfy an equation of the form $y = a + bx$ yield points which fall on a straight line. Suppose, for instance, that we want to predict the bushels-per-acre yield of wheat in a certain Midwestern county, y, in terms of the county's annual rainfall (in inches, measured from September through August), x. Based on past experience, we have the predicting equation

$$y = 0.23 + 4.42x$$

whose graph is shown in Figure 14.1, and any pair of values of x and y which are such that $y = 0.23 + 4.42x$ forms a point (x, y) that falls on the line. Substituting $x = 6$, for instance, we find that when there is an annual rainfall of 6 inches we can expect a yield of $0.23 + 4.42 \cdot 6 = 26.75$ bushels per acre; and when there is an annual rainfall of 12 inches we can expect a yield of $0.23 + 4.42 \cdot 12 = 53.27$ bushels per acre.

Once we have decided to fit a straight line, we are faced with the problem of finding the equation of the particular line which in some sense provides the best possible fit. To show how this is done, let us consider the following

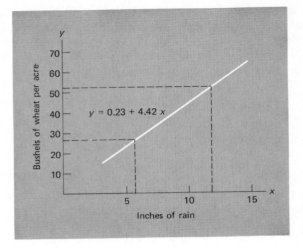

FIGURE 14.1 *Graph of linear equation.*

sample data obtained in a study of the relationship between the number of years that applicants for certain foreign service jobs studied German in high school or college and the scores which they received in a proficiency test in that language:

Number of years x	Score in test y
3	57
4	78
4	72
2	58
5	89
3	63
4	73
5	84
3	75
2	48

If we plot the points which correspond to these ten pairs of values as in Figure 14.2, we observe that, even though the points do not all fall on a straight line, the overall pattern of the relationship is reasonably well described by the white line. There is no noticeable departure from linearity in the scatter of the points, so we feel justified in deciding that a straight line is a suitable description of the underlying relationship.

We now face the problem of finding the equation of the line which in some sense provides the best fit to the data and which, it is hoped, will later yield the best possible predictions of y from x. Logically speaking, there is

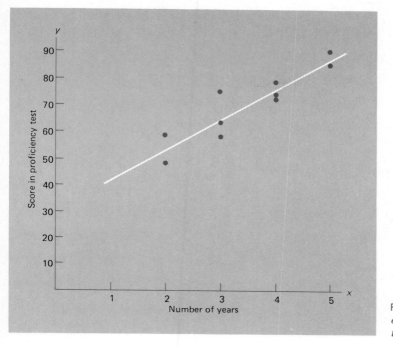

FIGURE 14.2 *Data on number of years studied and scores in test.*

no limit to the number of straight lines which can be drawn on a piece of graph paper. Some of these lines would fit the data so poorly that we could not consider them seriously, but many others would seem to provide more or less "good" fits, and the problem is to find the one line which fits the data "best" in some well-defined sense. If all the points actually fall on a straight line there is no problem, but this is an extreme case which we rarely encounter in practice. In general, we have to be satisfied with a line having certain desirable properties, short of perfection.

The criterion which, today, is used almost exclusively for defining a "best" fit dates back to the early part of the nineteenth century and the work of the French mathematician Adrien Legendre; it is known as the **method of least squares.** As it will be used here, this method requires that the line which we fit to our data be such that the sum of the squares of the vertical deviations of the points from the line is a minimum.

For the problem concerning the proficiency scores in German, the method of least squares requires that the sum of the squares of the distances represented by the solid-line segments of Figure 14.3 be as small as possible. To explain why this is done, let us consider one of the applicants for the foreign service jobs, say, the one who studied German for two years and received a 48 in the test. If we mark $x = 2$ on the horizontal scale and read the corresponding value of y off the line of Figure 14.3, we find that the corresponding score is about 53; therefore, the error in the prediction based on the

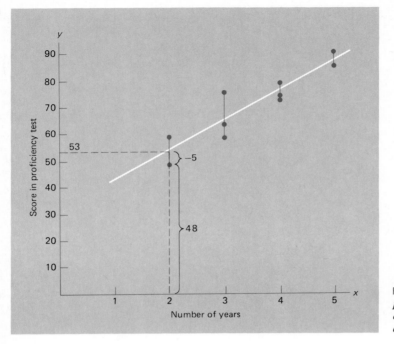

FIGURE 14.3 *Error of prediction based on line fit to data on number of years studied and score in proficiency test.*

line, represented by the vertical distance from the point to the line, is $48 - 53 = -5$. Altogether, there are ten such errors in our example, and the least-squares criterion requires that we minimize the sum of their squares. We minimize the sum of the squares of the deviations and not the sum of the deviations, themselves, for the same reason why we squared the deviations from the mean in the definition of the standard deviation. We are interested in the magnitude of the vertical deviations from the line, not in their signs.

To show how a **least-squares line** is actually fitted to a set of data, let us consider n pairs of numbers (x_1, y_1), $(x_2, y_2), \ldots$, and (x_n, y_n), which might represent such things as the thrust and the speed of n rockets, the height and weight of n persons, the reading rate and reading comprehension of n students, or the number of persons unemployed in two cities in n years. If we write the equation of the line as $\hat{y} = a + bx$, where the symbol $\hat{y}$ (y-hat) is used to distinguish between an observed value of y and the corresponding value $\hat{y}$ on the line, the least-squares criterion requires that we minimize the sum of the squares of the differences between the y's and the $\hat{y}$'s (see Figure 14.4). This means that we must find the numerical values of the constants a and b appearing in the equation $\hat{y} = a + bx$ for which

$$\sum (y - \hat{y})^2 = \sum [y - (a + bx)]^2$$

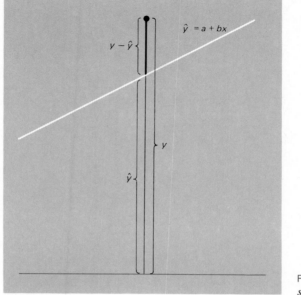

FIGURE 14.4 *Least-squares line.*

is as small as possible. As it takes calculus or fairly tedious algebra to find the expressions for a and b which minimize $\sum (y - \hat{y})^2$, let us merely state the result that they are given by the solutions for a and b of the following system of two linear equations:

Normal equations

$$\sum y = na + b(\sum x)$$
$$\sum xy = a(\sum x) + b(\sum x^2)$$

In these equations, called the **normal equations,** n is the number of pairs of observations, $\sum x$ and $\sum y$ are the sums of the observed x's and y's, $\sum x^2$ is the sum of the squares of the x's, and $\sum xy$ is the sum of the products obtained by multiplying each x by the corresponding y.

EXAMPLE Fit a least-squares line to the data on page 404, which pertain to the numbers of years that certain applicants for foreign service jobs have studied German in high school or college and the scores which they received in a proficiency test in that language.

Solution We get the sums needed for substitution into the normal equations by performing the calculations shown in the following table:

Number of years x	Test score y	x^2	xy
3	57	9	171
4	78	16	312
4	72	16	288
2	58	4	116
5	89	25	445
3	63	9	189
4	73	16	292
5	84	25	420
3	75	9	225
2	48	4	96
35	697	133	2,554

(There are many kinds of desk calculators, or hand-held calculators, on which the various sums can be accumulated directly, so that there is no need to fill in all the details.) Substituting $n = 10$ and the four column totals into the two normal equations, we get

$$697 = 10a + 35b$$

$$2{,}554 = 35a + 133b$$

and we must now solve these two simultaneous linear equations for a and b. There are several ways in which this can be done; from elementary algebra we can use either the method of elimination or the method of determinants. Using the first, we get $a = 31.55$ and $b = 10.90$.

As an alternative to these procedures we can use the following formulas, which result when we symbolically solve the two normal equations for a and b:

Solutions of normal equations

$$a = \frac{(\sum y)(\sum x^2) - (\sum x)(\sum xy)}{n(\sum x^2) - (\sum x)^2}$$

$$b = \frac{n(\sum xy) - (\sum x)(\sum y)}{n(\sum x^2) - (\sum x)^2}$$

EXAMPLE   Rework the preceding example, using the above formulas for a and b.

Solution   Substituting $n = 10$ and the various sums shown in the above table, we get

$$a = \frac{(697)(133) - (35)(2,554)}{10(133) - (35)^2} = \frac{3,311}{105} = 31.53$$

and

$$b = \frac{10(2,554) - (35)(697)}{105} = \frac{1,145}{105} = 10.90$$

The difference between the values obtained for a by the two methods, 31.55 and 31.53, is due to rounding.

There is still another way of finding a and b which is often used because of its convenience. It consists of using the second of the above formulas to calculate b, and then substituting the result into the first of the two normal equations to get the value of a. Symbolically solving the first of the two normal equations for a, we get

Alternative formula for a

$$a = \frac{\sum y - b(\sum x)}{n}$$

and we use this formula in conjunction with the original formula for b. In our example we would thus have obtained

$$a = \frac{697 - 10.90(35)}{10} = 31.55$$

Once we have determined the equation of a least-squares line, we can use it to make predictions.

EXAMPLE   Use the least-squares line $\hat{y} = 31.53 + 10.90x$ to predict the proficiency score of an applicant who has studied German in high school or college for two years.

Solution   Substituting $x = 2$ into the equation, we get

$$\hat{y} = 31.53 + 10.90(2) = 53.33$$

and this is the best prediction we can make in the least-squares sense.

When we make a prediction like this, we cannot really expect that we will always hit the answer right on the nose; in fact, we cannot possibly be right when the answer has to be a whole number, as in our illustration, and our prediction is 53.33. With reference to this example, it would be very unreasonable to expect that every applicant who has studied German for a given number of years will get the same score in the proficiency test; indeed, the data on page 404 show that this is not the case. Thus, to make meaningful predictions based on least-squares lines, we must look upon the values of $\hat{y}$ obtained by substituting given values of x as averages, or expected values. Interpreted in this way, we refer to least-squares lines as **regression lines,** or better as **estimated regression lines,** since the values of a and b are estimated on the basis of sample data. Questions relating to the goodness of these estimates will be discussed in Section 14.3.

In the discussion of this section we have considered only the problem of fitting a straight line to paired data. More generally, the method of least squares can also be used to fit other kinds of curves and to derive predicting equations in more than two unknowns. The problem of fitting some curves other than straight lines by the method of least squares will be discussed briefly in Section 14.4, and a simple example of a predicting equation in more than two unknowns will be given in Section 14.5.

EXERCISES

14.1 The following table shows how many weeks six persons have worked at an automobile inspection station and the number of cars each one inspected between noon and 2 P.M. on a given day:

Number of weeks employed x	Number of cars inspected y
2	13
7	21
9	23
1	14
5	15
12	21

(a) Solve the normal equations to find the equation of the least-squares line which will enable us to predict y in terms of x.
(b) Use the formulas on page 408 to check the values of a and b obtained in part (a).

14.2 Use the result of part (a) of the preceding exercise to estimate how many cars someone who has been working at the inspection station for eight weeks can be expected to inspect during the given two-hour period.

14.3 The following data pertain to the chlorine residual in a swimming pool at various times after it has been treated with chemicals:

Number of hours	Chlorine residual (parts per million)
2	1.8
4	1.5
6	1.4
8	1.1
10	1.1
12	0.9

(a) Fit a least-squares line from which we can predict the chlorine residual in terms of the number of hours since the pool has been treated with chemicals.

(b) Use the equation of the least-squares line to estimate the chlorine residual in the pool five hours after it has been treated with chemicals.

14.4 The following sample data show the demand for a product (in thousands of units) and its price (in cents) charged in six different market areas:

Price	18	10	14	11	16	13
Demand	9	125	57	90	22	79

(a) Fit a least-squares line from which we can predict the demand for the product in terms of its price.

(b) Estimate the demand for the product in a market area where it is priced at 15 cents.

14.5 Verify that the equation of the example on page 403 can be obtained by fitting a least-squares line to the following data:

Rainfall (inches)	Yield of wheat (bushels per acre)
12.9	62.5
7.2	28.7
11.3	52.2
18.6	80.6
8.8	41.6
10.3	44.5
15.9	71.3
13.1	54.4

14.6 The following are the midterm and final examination grades of eight students in a course in European history:

Midterm	Final examination
75	81
66	57
92	95
86	77
65	71
44	62
60	63
79	84

(a) Find the equation of the least-squares line which will enable us to predict final examination grades in this course from midterm grades.

(b) Use the least-squares equation obtained in part (a) to predict the final examination grade of a student who got a 68 on the midterm test.

14.7 Raw material used in the production of a synthetic fiber is stored in a place which has no humidity control. Measurements of the relative humidity in the storage place and the moisture content of a sample of the raw material (both in percentages) on 12 days yielded the following results:

Humidity	Moisture content
46	12
53	14
37	11
42	13
34	10
29	8
60	17
44	12
41	10
48	15
33	9
40	13

(a) Fit a least-squares line from which we can predict the moisture content in terms of the humidity.

(b) Use the result of part (a) to estimate the moisture content when the relative humidity is 38 percent.

★14.8 Suppose that in the preceding exercise we had wanted to estimate what humidity will yield a moisture content of 10 percent. We could substitute $\hat{y} = 10$ into the equation obtained in part (a) of the preceding exercise and solve for x, but this would not provide an estimate in the least-squares sense. To make the best possible least-squares predictions, or estimates, of humidity in terms of moisture content, we denote the moisture contents by x, the humidity readings by y, and then fit a least-squares line to these data. Find the equation of this line and use it to estimate the humidity which will yield a moisture content of 10 percent.

14.9 When the x's are equally spaced (that is, when the differences between successive values of x are all equal), the calculation of a and b can be simplified by coding the x's by assigning them the values $\ldots, -3, -2, -1, 0, 1, 2, 3, \ldots$, when n is odd, or the values $\ldots, -5, -3, -1, 1, 3, 5, \ldots$, when n is even. With this coding, the sum of the x's is zero, and the formulas for a and b on page 408 become

$$a = \frac{\sum y}{n} \quad \text{and} \quad b = \frac{\sum xy}{\sum x^2}$$

Of course, the equation of the resulting least-squares line expresses y in terms of the coded x's, and we have to account for this when we use the equation to make predictions.

(a) During its first five years of operation, a company's gross income from sales was 1.4, 2.1, 2.6, 3.5, and 3.7 million dollars. Fit a least-squares line and, assuming that the trend continues, predict the company's gross income from sales during its sixth year of operation.

(b) For the years 1971–1976, the average weekly earnings of production workers in manufacturing in the United States were $142.04, $154.69, $165.65, $176.40, $189.51, and $207.60. Fit a least-squares line and, assuming that the trend continues, predict the average weekly earnings of production workers in manufacturing in the United States for 1984.

14.10 Use the method of the preceding exercise to rework Exercise 14.3.

★14.11 Sometimes it is known that the line we fit to a set of data must pass through the origin (namely, that $a = 0$), and in that case the method of least squares yields

$$b = \frac{\sum xy}{\sum x^2}$$

Fit this kind of line to the data of Exercise 14.5 and plot it together with the line obtained in that exercise.

14.3
REGRESSION ANALYSIS

In the preceding section we used a least-squares line to predict that someone who has studied German in high school or college for two years will score 53.33 in the proficiency test, but even if we interpret the line correctly as a regression line (that is, treat the predictions made from it as averages or expected values), several questions remain to be answered.

How good are the values we found for the constants a and b in the equation $\hat{y} = a + bx$? After all, $a = 31.53$ and $b = 10.90$ are only estimates based on sample data, and if we base our work on a sample of ten other applicants for the foreign service jobs, the method of least squares would probably yield different values of a and b.

How good an estimate is 53.33 of the average score of persons who have had two years of German in high school or college?

Also, we might ask

How can we obtain limits (two numbers) for which we can assert with a given degree of confidence that they will contain the score of a person who has studied German in high school or college for two years?

When in the first question above we said that $a = 31.53$ and $b = 10.90$ are only estimates based on sample data, we implied the existence of corresponding true values, usually denoted by α and β, and therefore of a true regression line $\mu_{y|x} = \alpha + \beta x$, where $\mu_{y|x}$ is the true mean of y for a given value of x. It is customary to refer to α and β as **regression coefficients,** and to a and b as the corresponding **estimated regression coefficients.†**

To clarify the idea of a true regression line, let us consider Figure 14.5, in which we have drawn the distributions of y for several values of x. With reference to our example, these curves should be looked upon as the distributions of the proficiency scores of persons who have had one, two, or three years of German in high school or college; to complete the picture, we can visualize similar distributions for all other values of x within the range of values under consideration. Note that the means of all the distributions of Figure 14.5 lie on the true regression line $\mu_{y|x} = \alpha + \beta x$.

In **linear regression analysis** we assume that the x's are constants, not values of random variables, and that for each value of x the variable to be predicted, y, has a certain distribution (as shown in Figure 14.5) whose mean is $\alpha + \beta x$. In **normal regression analysis** we assume, furthermore, that these distributions are all normal distributions with the same standard deviation σ. In other words, the distributions pictured in Figure 14.5, as well as those we imagine, are normal distributions with the means $\alpha + \beta x$ and the standard deviation σ.

---

† The coefficients α and β should not be confused with the probabilities α and β of committing Type I and Type II errors.

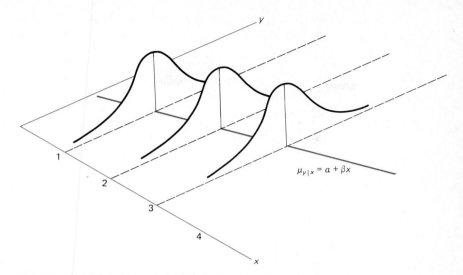

$$\mu_{y|x} = a + \beta x$$

FIGURE 14.5 *Distributions of y for given values of x.*

If we make all these assumptions, inferences about the regression co-efficients α and β can be based on the statistics

Statistics for inferences about regression coefficients

$$t = \frac{a - \alpha}{s_e\sqrt{\dfrac{1}{n} + \dfrac{n \cdot \bar{x}^2}{n(\sum x^2) - (\sum x)^2}}}$$

$$t = \frac{b - \beta}{s_e}\sqrt{\frac{n(\sum x^2) - (\sum x)^2}{n}}$$

whose sampling distributions are t distributions with $n - 2$ degrees of freedom. Here α and β are the regression coefficients we want to estimate or test, and a and b are their estimates calculated from a given set of data by the method of least squares. Also, n, $\bar{x}$, $\sum x$, and $\sum x^2$ come from the original data, and s_e is an estimate of σ, the common standard deviation of the normal distributions pictured in Figure 14.5, given by

$$s_e = \sqrt{\frac{\sum (y - \hat{y})^2}{n - 2}}$$

Here again, the y's are the observed values of y and the $\hat{y}$'s are the corresponding values on the least-squares line. We call s_e the **standard error of estimate,** and it should be observed that its square, s_e^2, is the sum of the squares of the vertical deviations of the points from the line in Figure 14.3 divided by $n - 2$. (We lose two degrees of freedom, so to speak, because

α and β are replaced by estimates, a and b, in the formula for s_e.) In practice, we determine the value of s_e by means of the computing formula

Standard error of estimate

$$s_e = \sqrt{\frac{\sum y^2 - a(\sum y) - b(\sum xy)}{n - 2}}$$

The following example illustrates how the two t statistics are used to make inferences about the regression coefficients.

EXAMPLE   Suppose that someone claims that $\beta = 12.5$ in the example dealing with the applicants for the foreign service jobs and their proficiency in German, and that the purpose of the study is to test this claim. Since β, the slope of the regression line, is the average change in y associated with a change of one unit in x, the hypothesis asserts that each additional year of German studied in high school or college adds 12.5 on the average to a person's proficiency score. Test this hypothesis against the alternative hypothesis $\beta \neq 12.5$ at the 0.05 level of significance.

Solution   1. *Null hypothesis*: $\beta = 12.5$
 Alternative hypothesis: $\beta \neq 12.5$
 2. *Level of significance*: $\alpha = 0.05$
 3. *Criterion*: Reject the null hypothesis if $t < -2.306$ or $t > 2.306$, where 2.306 is the value of $t_{0.025}$ for $10 - 2 = 8$ degrees of freedom, and t is given by

$$t = \frac{b - \beta}{s_e} \sqrt{\frac{n(\sum x^2) - (\sum x)^2}{n}}$$

Otherwise, reserve judgment or accept it.
 4. *Calculations*: Substituting $n = 10$, $\sum y = 697$, $\sum xy = 2{,}554$, $\sum y^2 = 50{,}085$ (from the data on page 404), $a = 31.53$, and $b = 10.90$ into the formula for s_e, we get

$$s_e = \sqrt{\frac{50{,}085 - (31.53)(697) - (10.90)(2{,}554)}{10 - 2}}$$

$$= 5.81$$

Then, substituting this value together with $b = 10.90$, $\beta = 12.5$, $n = 10$, $\sum x = 35$, and $\sum x^2 = 133$ into the formula for t, we get

$$t = \frac{10.90 - 12.5}{5.81} \sqrt{\frac{10(133) - (35)^2}{10}}$$

$$= -0.89$$

5. *Decision*: Since $t = -0.89$ falls on the interval from -2.306 to 2.306, the null hypothesis cannot be rejected; in other words, the difference between $\beta = 12.5$ and $b = 10.90$ may well be due to chance.

Tests concerning the regression coefficient α are performed in the same way, except that we use the first, instead of the second, of the two t statistics. In most practical applications, however, the regression coefficient α is not of much interest—it is just the y-intercept, namely, the value of y which corresponds to $x = 0$. In many cases it has no real meaning.

To construct confidence intervals for the regression coefficients α and β, we substitute into the middle term of $-t_{\alpha/2} < t < t_{\alpha/2}$ the appropriate t statistic from page 415. Then, simple algebra leads to the formulas

Confidence limits for regression coefficients

$$a \pm t_{\alpha/2} \cdot s_e \sqrt{\frac{1}{n} + \frac{n \cdot \bar{x}^2}{n(\sum x^2) - (\sum x)^2}}$$

and

$$b \pm \frac{t_{\alpha/2} \cdot s_e}{\sqrt{\dfrac{n(\sum x^2) - (\sum x)^2}{n}}}$$

where the degree of confidence is $(1 - \alpha)100$ percent and $t_{\alpha/2}$ is the entry in Table II for $n - 2$ degrees of freedom.

EXAMPLE The following data show the average numbers of hours which six students spent on homework per week and their grade-point indexes for the courses they took:

Hours spent on homework x	Grade-point index y
15	2.0
28	2.7
13	1.3
20	1.9
4	0.9
10	1.7

Construct a 95 percent confidence interval for β, the amount by which a student in the population sampled can expect to raise his grade-point index if he studies an extra hour per week.

Solution   Substituting $n = 6$, $\sum x = 90$, $\sum x^2 = 1{,}694$, $\sum y = 10.5$, and $\sum xy = 181.1$ into the formulas for a and b on page 408, we get

$$a = \frac{(10.5)(1{,}694) - (90)(181.1)}{6(1{,}694) - (90)^2} = 0.721$$

and

$$b = \frac{6(181.1) - (90)(10.5)}{6(1{,}694) - (90)^2} = 0.0686$$

Then, since $\sum y^2 = 20.29$, we get

$$s_e = \sqrt{\frac{20.29 - (0.721)(10.5) - (0.0686)(181.1)}{6 - 2}}$$

$$= 0.272$$

and it follows that the 95 percent confidence limits for β are

$$0.0686 \pm \frac{(2.776)(0.272)}{\sqrt{\dfrac{6(1{,}694) - (90)^2}{6}}}$$

or 0.0686 ± 0.0407. Rounding to three decimals, we can write the 95 percent confidence interval for β as

$$0.028 < \beta < 0.109$$

This confidence interval is rather wide, and this is due to two things—the very small size of the sample and the variation measured by s_e, namely, the variation among the grade-point indexes of students doing the same amount of home work.

To answer the second question asked on page 414, the one concerning the estimation, or prediction, of the average value of y for a given value of x, we use a method that is very similar to the one just discussed. Basing our argument on another t statistic, we arrive at the following $(1 - \alpha)100$ percent confidence limits for $\mu_{y|x_0}$, the mean of y when $x = x_0$:

Confidence limits
for mean of y
when $x = x_0$

$$(a + bx_0) \pm t_{\alpha/2} \cdot s_e \sqrt{\frac{1}{n} + \frac{n(x_0 - \bar{x})^2}{n(\sum x^2) - (\sum x)^2}}$$

As before, the number of degrees of freedom is $n - 2$ and $t_{\alpha/2}$ may be read from Table II.

EXAMPLE Referring again to the data on page 404, suppose that we want to estimate the mean proficiency score of applicants who have had two years of German in high school or college. Construct a 99 percent confidence interval for this mean.

Solution Copying $\sum x = 35$ and $\sum x^2 = 133$ from page 408, $a + bx_0 = 31.53 + 10.90(2) = 53.33$ from page 409, and $s_e = 5.81$ from page 416, and substituting these values together with $n = 10$, $\bar{x} = \frac{35}{10} = 3.5$, and $t_{0.005} = 3.355$ (for $10 - 2 = 8$ degrees of freedom) into the confidence interval formula, we get

$$53.33 \pm (3.355)(5.81)\sqrt{\frac{1}{10} + \frac{10(2 - 3.5)^2}{10(133) - (35)^2}}$$

or

$$53.33 \pm 10.93$$

Hence, we can write the 99 percent confidence interval for the mean proficiency score of applicants who have had two years of German in high school or college as

$$42.20 < \mu_{y|2} < 64.26$$

EXERCISES 14.12 With reference to Exercise 14.1 on page 410, test the null hypothesis $\beta = 1.2$ (namely, the hypothesis that each additional week on the job adds 1.2 to the number of cars a person can be expected to inspect in the given period of time) against the alternative hypothesis $\beta < 1.2$. Use the 0.05 level of significance.

14.13 With reference to Exercise 14.3 on page 411, test the null hypothesis $\beta = -0.12$ against the alternative hypothesis $\beta > -0.12$ at the 0.05 level of significance. Also, state in words the hypothesis being tested.

14.14 The following table shows the assessed values and the selling prices of eight houses, constituting a random sample of all the houses sold recently in a metropolitan area:

Assessed value (thousands of dollars)	Selling price (thousands of dollars)
40.3	63.4
72.0	118.3
32.5	55.2
44.8	74.0
27.9	48.8
51.6	81.1
80.4	123.2
58.0	92.5

Fit a least-squares line which will enable us to predict the selling price of a house in terms of its assessed value and test the null hypothesis $\beta = 1.30$ against the alternative hypothesis $\beta > 1.30$ at the 0.05 level of significance.

14.15 With reference to Exercise 14.3 on page 411, test the null hypothesis $\alpha = 2.1$ against the alternative hypothesis $\alpha < 2.1$ at the 0.05 level of significance. Also explain in words what hypothesis is being tested.

14.16 The following data show the advertising expenses (expressed as a percentage of total expenses) and the net operating profits (expressed as a percentage of total sales) in a random sample of six drugstores:

Advertising expenses	Net operating profits
1.5	3.6
1.0	2.8
2.8	5.4
0.4	1.9
1.3	2.9
2.0	4.3

Test the null hypothesis $\alpha = 0.8$ against the alternative hypothesis $\alpha > 0.8$ at the 0.01 level of significance.

14.17 With reference to Exercise 14.4 on page 411, construct a 95 percent confidence interval for β.

14.18 With reference to Exercise 14.7 on page 412, construct a 99 percent confidence interval for β.

14.19 With reference to Exercise 14.16, construct a 98 percent confidence interval for β. Also, explain in words what quantity is being estimated.

14.20 With reference to Exercise 14.3 on page 411, construct a 95 percent confidence interval for α.

14.21 With reference to Exercise 14.16, construct a 99 percent confidence interval for α. Also, explain in words what quantity is being estimated.

14.22 With reference to Exercise 14.1 on page 410, construct a 95 percent confidence interval for the average number of cars inspected in the given period of time by a person who has worked at the inspection station for eight weeks.

14.23 With reference to Exercise 14.14, construct a 99 percent confidence interval for the average selling price of a house in the given metropolitan area which has an assessed value of $45,000.

14.24 With reference to Exercise 14.16, construct a 95 percent confidence interval for the mean net operating profits (expressed as a percentage of total sales) when the advertising expenses are 2.0 percent of total expenses.

★14.25 The third of the questions asked in the beginning of Section 14.3 differs from the other two in that it does not concern the estimation of a population parameter, but the prediction of a single future observation. Limits (two values) for which we can assert with a given degree of confidence that they will contain such an observation are called **limits of prediction.** Appropriate limits for predicting with $(1 - \alpha)100$ percent confidence a value of y when $x = x_0$ are given by

Limits of
prediction

$$(a + bx_0) \pm t_{\alpha/2} \cdot s_e \sqrt{1 + \frac{1}{n} + \frac{n(x_0 - \bar{x})^2}{n(\sum x^2) - (\sum x)^2}}$$

Again, $t_{\alpha/2}$ may be read from Table II, and the number of degrees of freedom is $n - 2$. To illustrate, let us refer again to the language proficiency example, and let us find 99 percent limits of prediction for the proficiency score of an applicant who has studied German in high school or college for two years. Noting that the only difference between the above limits and the corresponding confidence limits for $\mu_{y|x_0}$ is that we add 1 to the quantity under the radical, we can immediately write the limits of prediction as

$$53.33 \pm (3.355)(5.81)\sqrt{1 + \frac{1}{10} + \frac{10(2 - 3.5)^2}{10(133) - (35)^2}}$$

or

$$53.33 \pm 22.35$$

Thus, the 99 percent limits of prediction are 30.98 and 75.68.
  (a) With reference to Exercise 14.1 on page 410, find 95 percent limits of prediction for the number of cars that will be inspected in the given period of time by a person who has worked at the inspection station for eight weeks.
  (b) With reference to Exercise 14.14, construct 99 percent limits of prediction for the selling price of a house in the given metropolitan area which has an assessed value of $45,000.

14.4
NONLINEAR REGRESSION ★

When data depart more or less widely from linearity, we must consider fitting some curve other than a straight line. In this section we shall first give two cases where the relationship is not linear but the method of Section 14.2 can nevertheless be employed; then we shall give an example of **polynomial curve fitting** by fitting a **parabola** whose equation is

$$y = a + bx + cx^2$$

It is common practice to plot paired data on various kinds of graph paper, to see whether there are scales for which the points fall close to a straight line. Of course, when this is the case for ordinary (arithmetic) graph paper, we proceed as in Section 14.2. If it is the case when we use **semilog paper** (with equal subdivisions for x and a logarithmic scale for y, as shown in Figure 14.7), this indicates that an **exponential curve** will provide a good fit. The equation of such a curve is

$$y = a \cdot b^x$$

or in logarithmic form

$$\log y = \log a + x(\log b)$$

which is a linear equation in x and $\log y$, where "log" stands for logarithm to the base 10. Observe that if we write A, B, and Y for $\log a$, $\log b$, and $\log y$, the equation becomes $Y = A + Bx$, which is the usual equation of a straight line.

For fitting an exponential curve, the two normal equations on page 407 become

Normal equations for fitting exponential curve

$$\sum \log y = n(\log a) + (\log b)(\sum x)$$
$$\sum x(\log y) = (\log a)(\sum x) + (\log b)(\sum x^2)$$

and for any set of paired data they can be solved for $\log a$ and $\log b$, and hence for a and b.

EXAMPLE   The following are data on a company's net profits during the first six years that it has been in business:

Year	Net profits (thousands of dollars)
1	112
2	149
3	238
4	354
5	580
6	867

Figure 14.6, where these data are plotted on ordinary graph paper, shows that the relationship is not linear; in Figure 14.7, where a logarithmic scale is used for the y's, the overall pattern is remarkably well "straightened out." Thus, fit an exponential curve to the given data.

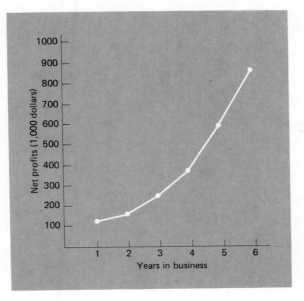

FIGURE 14.6 *Net profits plotted on ordinary graph paper.*

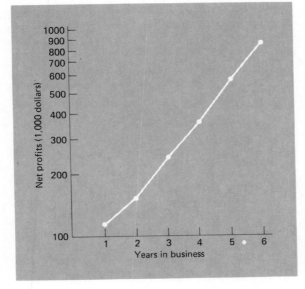

FIGURE 14.7 *Net profits plotted on semi-log paper.*

Solution Looking up the required logarithms in Table XIV, we get the sums needed for substitution into the two normal equations by performing the calculations shown in the following table:

x	y	$\log y$	$x \cdot \log y$	x^2
1	112	2.0492	2.0492	1
2	149	2.1732	4.3464	4
3	238	2.3766	7.1298	9
4	354	2.5490	10.1960	16
5	580	2.7634	13.8170	25
6	867	2.9380	17.6280	36
21		14.8494	55.1664	91

Substituting the column totals and $n = 6$ into the two normal equations, we get

$$14.8494 = 6(\log a) + 21(\log b)$$

$$55.1664 = 21(\log a) + 91(\log b)$$

To solve these equations, we substitute into formulas analogous to those for a and b on page 408, and we find that

$$\log a = \frac{14.8494(91) - 21(55.1664)}{6(91) - (21)^2} = 1.8362$$

$$\log b = \frac{6(55.1664) - 21(14.8494)}{6(91) - (21)^2} = 0.1825$$

and hence that $a = 68.6$ and $b = 1.52$. Thus, the equation of the exponential curve which best describes the relationship between the company's net profits and the number of years it has been in business is given by

$$\hat{y} = 68.6(1.52)^x$$

where $\hat{y}$ is in thousands of dollars.

To estimate (predict) the company's net profit, say, for the eighth year it will have been in business, it is more convenient to use the logarithmic form of the equation; namely,

$$\log \hat{y} = 1.8362 + x(0.1825)$$

Substituting $x = 8$, we get

$$\log \hat{y} = 1.8362 + 8(0.1825)$$

$$= 3.2962$$

and hence $\hat{y} = 1,980$ (or \$1,980,000).

If points representing paired data fall close to a straight line when plotted on **log-log paper** (with logarithmic scales for both x and y), this indicates that an equation of the form

$$y = a \cdot x^b$$

will provide a good fit. In the logarithmic form, the equation of such a **power function** is

$$\log y = \log a + b(\log x)$$

which is a linear equation in $\log x$ and $\log y$. (Writing A, X, and Y for $\log a$, $\log x$, and $\log y$, the equation becomes $Y = A + bX$, which is the usual equation of a straight line.) For fitting a power function, the two normal equations on page 407 become

Normal equations for fitting power function

$$\sum \log y = n(\log a) + b(\sum \log x)$$
$$\sum (\log x)(\log y) = (\log a)(\sum \log x) + b(\sum \log^2 x)$$

and for any set of paired data they can be solved for $\log a$ and b, and hence for a and b. In these equations, $\sum (\log x)(\log y)$ is the sum of the products obtained by multiplying the logarithm of each observed value of x by the logarithm of the corresponding value of y, and $\sum \log^2 x$ is the sum of the squares of the logarithms of the x's. Since the work required to fit a power function is very similar to that required to fit an exponential curve, we shall not illustrate it, but in Exercises 14.29 and 14.30 on page 429 the reader will find data to which the method can be applied.

When the values of y first increase and then decrease, or first decrease and then increase, a **parabola** having the equation

$$y = a + bx + cx^2$$

will often provide a good fit. In fitting a parabola by the method of least squares, we must determine a, b, and c so that $\sum [y - (a + bx + cx^2)]^2$ is a minimum. Again stating merely the results, we must solve the following set of normal equations for a, b, and c:

Normal equations for fitting parabola

$$\sum y = na + b(\sum x) + c(\sum x^2)$$
$$\sum xy = a(\sum x) + b(\sum x^2) + c(\sum x^3)$$
$$\sum x^2 y = a(\sum x^2) + b(\sum x^3) + c(\sum x^4)$$

In these equations, $\sum x^2y$ is the sum of the products obtained by multiplying the square of each value of x by the corresponding observed value of y, and $\sum x^3$ and $\sum x^4$ are the sums of the third and fourth powers of the x's.

EXAMPLE   The following are data on the drying time of a varnish and the amount of a certain chemical additive:

Amount of additive (grams) x	Drying time (hours) y
1	7.2
2	6.7
3	4.7
4	3.7
5	4.7
6	4.2
7	5.2
8	5.7

Since the values of y first decrease and then increase, as can be seen from the data and also from Figure 14.8, fit a parabola.

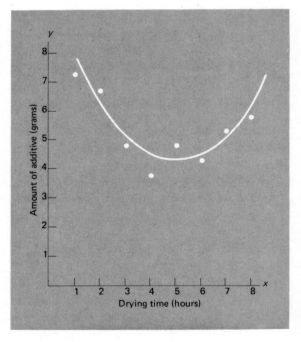

FIGURE 14.8   *Parabola fitted to varnish-drying-time data.*

Solution To calculate a, b, and c from the normal equations requires that we find n, $\sum x$, $\sum x^2$, $\sum x^3$, $\sum x^4$, $\sum y$, $\sum xy$, and $\sum x^2 y$. The work needed to find these quantities is shown in the following table:

x	y	x^2	x^3	x^4	xy	x^2y
1	7.2	1	1	1	7.2	7.2
2	6.7	4	8	16	13.4	26.8
3	4.7	9	27	81	14.1	42.3
4	3.7	16	64	256	14.8	59.2
5	4.7	25	125	625	23.5	117.5
6	4.2	36	216	1,296	25.2	151.2
7	5.2	49	343	2,401	36.4	254.8
8	5.7	64	512	4,096	45.6	364.8
36	42.1	204	1,296	8,772	180.2	1,023.8

Substituting the appropriate column totals and $n = 8$ into the three normal equations, we get

$$42.1 = 8a + 36b + 204c$$

$$180.2 = 36a + 204b + 1{,}296c$$

$$1{,}023.8 = 204a + 1{,}296b + 8{,}772c$$

and the solution of this system of equations (rounded to one decimal) is $a = 9.2$, $b = -2.0$, and $c = 0.2$, as can be verified by the method of elimination or by using determinants. Thus, the equation of the parabola is

$$\hat{y} = 9.2 - 2.0x + 0.2x^2$$

Having obtained this equation, let us use it to determine the expected drying time of the varnish when 6.5 grams of the chemical are added. Substituting $x = 6.5$ into the equation of the parabola, we get

$$\hat{y} = 9.2 - 2.0(6.5) + 0.2(6.5)^2$$

$$= 4.65 \text{ hours}$$

We have introduced the parabola, also called a **second-degree polynomial equation,** as a curve which bends once; that is, its values first increase and then decrease, or first decrease and then increase. As curves which bend more than once, polynomial equations of higher degree in x than two, such as $y = a + bx + cx^2 + dx^3$ and $y = a + bx + cx^2 + dx^3 + ex^4$, can also be fitted by the method of least squares. In practice, we also use parts of such curves, especially parts of parabolas, when there is only a slight curvature in the pattern we want to describe (see Exercise 14.31 on page 429).

EXERCISES

★ 14.26 Fit an exponential curve to the following data on the percentage of the radial tires made by a certain manufacturer that are still usable after having been driven the given numbers of miles:

Miles driven (thousands) x	Percentage usable y
1	97.2
2	91.8
5	82.5
10	64.4
20	41.0
30	29.9
40	17.6
50	11.3

Also estimate the percentage of the tires we can expect to be usable after they have been driven for 25,000 miles.

★ 14.27 The following data pertain to the growth of a colony of bacteria in a culture medium:

Days since inoculation x	Bacteria count (thousands) y
2	112
4	148
6	241
8	363
10	585

Fit an exponential curve and use it to estimate the bacteria count at the end of the fifth day.

★ 14.28 When the x's are equally spaced, the calculations that are needed to fit an exponential curve can be simplified by using the coding of Exercise 14.9 on page 413;

that is, by assigning the x's the values $\ldots, -3, -2, -1, 0, 1, 2, 3, \ldots$ when n is odd, or the values $\ldots, -5, -3, -1, 1, 3, 5, \ldots$ when n is even.

(a) Use this kind of coding to rework the example on page 422. What coded value of x do we have to substitute into the least-squares equation to predict the company's profit for the eighth year that it will have been in business? Make this substitution and compare the result with that obtained on page 424.

(b) In the years 1970–1974, Argentina's Index of Consumer Prices (with 1967 = 100) was 142, 191, 303, 486, and 604. Suitably coding the years, fit an exponential curve to these data.

★14.29 The following data pertain to the demand for a product (in thousands of units) and its price (in cents) charged in five different market areas:

Price	20	16	10	11	14
Demand	22	41	120	89	56

Fit a power function and use it to estimate the demand when the price of the product is 12 cents.

★14.30 The following data pertain to the volume of a gas (in cubic inches) and its pressure (in pounds per square inch), when the gas is compressed at a constant temperature:

Volume x	Pressure y
50	16.0
30	40.1
20	78.0
10	190.5
5	532.2

Fit a power function and use it to estimate the pressure of this gas when it is compressed to a volume of 15 cubic inches.

★14.31 Fit a parabola to the data of Exercise 14.29 and use it to estimate the demand when the price of the product is 12 cents. Explain why the parabola cannot be used, for example, to estimate the demand when the price of the product is 40 cents.

★14.32 When the x's are equally spaced, the calculations that are needed to fit a parabola can be simplified by using the coding of Exercise 14.9 on page 413; that is, by assigning the x's the values $\ldots, -3, -2, -1, 0, 1, 2, 3, \ldots$ when n is odd, or $\ldots, -5, -3, -1, 1, 3, 5, \ldots$ when n is even. Since this kind of coding makes $\sum x$ and $\sum x^3$ equal to 0, b can be found directly with the formula

$$b = \frac{\sum xy}{\sum x^2}$$

and a and c can be found by solving the system of equations

$$\sum y = na + c\left(\sum x^2\right)$$
$$\sum x^2 y = a\left(\sum x^2\right) + c\left(\sum x^4\right)$$

(a) Use this kind of coding to rework the example on page 426. What coded value of x do we have to substitute into the least-squares equation to estimate the average drying time of the varnish when 6.5 grams of the chemical are added. Make this substitution and compare the result with that obtained on page 427.

(b) In the years 1968–1972 the installed capacity of nuclear electric energy in the United States was 2,817, 3,980, 6,493, 8,687, and 15,301 thousands of kilowatts. Fit a parabola to these data and plot it together with the original data on ordinary (arithmetic) graph paper.

14.5
MULTIPLE REGRESSION ⋆

Although there are many problems in which one variable can be predicted quite accurately in terms of another, it stands to reason that predictions should improve if one considers additional relevant information. For instance, we should be able to make better predictions of the performance of newly hired teachers if we consider not only their education, but also their years of experience and their personality. Also, we should be able to make better predictions of a new textbook's success if we consider not only the quality of the work, but also the potential demand and the competition.

Many mathematical formulas can serve to express relationships among more than two variables, but most commonly used in statistics (partly for reasons of convenience) are linear equations of the form

$$y = b_0 + b_1 x_1 + b_2 x_2 + \cdots + b_k x_k$$

Here y is the variable which is to be predicted, $x_1, x_2, \ldots,$ and x_k are the k known variables on which predictions are to be based, and $b_0, b_1, b_2, \ldots,$ and b_k are numerical constants which must be determined from observed data.

To illustrate, consider the following equation which was obtained in a study of the demand for different meats:

$$\hat{y} = 3.489 - 0.090 x_1 + 0.064 x_2 + 0.019 x_3$$

Here y denotes the total consumption of federally inspected beef and veal in millions of pounds, x_1 denotes a composite retail price of beef in cents per pound, x_2 denotes a composite retail price of pork in cents per pound, and x_3 denotes income as measured by a certain payroll index. With this equation, we can predict the total consumption of federally inspected beef and veal on the basis of specified values of x_1, x_2, and x_3.

The main problem in deriving a linear equation in more than two variables which best describes a given set of data is that of finding numerical values for b_0, b_1, b_2,..., and b_k. This is usually done by the method of least squares; that is, we minimize the sum of squares $\sum (y - \hat{y})^2$, where as before the y's are the observed values and the $\hat{y}$'s are the values calculated by means of the linear equation. In principle, the problem of determining the values of b_0, b_1, b_2,..., and b_k is the same as it is in the two-variable case, but manual solutions may be very tedious because the method of least squares requires that we solve as many normal equations as there are unknown constants b_0, b_1, b_2,..., and b_k. For instance, when there are two independent variables x_1 and x_2, and we want to fit the equation $y = b_0 + b_1 x_1 + b_2 x_2$, we must solve the three normal equations

Normal equations (two independent variables)

$$\sum y = n \cdot b_0 + b_1(\sum x_1) + b_2(\sum x_2)$$
$$\sum x_1 y = b_0(\sum x_1) + b_1(\sum x_1^2) + b_2(\sum x_1 x_2)$$
$$\sum x_2 y = b_0(\sum x_2) + b_1(\sum x_1 x_2) + b_2(\sum x_2^2)$$

Here $\sum x_1 y$ is the sum of the products obtained by multiplying each given value of x_1 by the corresponding value of y, $\sum x_1 x_2$ is the sum of the products obtained by multiplying each given value of x_1 by the corresponding value of x_2, and so on.

EXAMPLE   The following data show the number of bedrooms, the number of baths, and the prices at which eight one-family houses sold recently in a certain community:

Number of bedrooms x_1	Number of baths x_2	Price (dollars) y
3	2	48,800
2	1	44,300
4	3	53,800
2	1	44,200
3	2	49,700
2	2	44,900
5	3	58,400
4	2	52,900

Find a linear equation which will enable us to predict the average sales price of a one-family house in the given community in terms of the number of bedrooms and the number of baths.

Solution   To get the sums needed for substitution into the three normal equations, we perform the calculations shown in the following table:

x_1	x_2	y	x_1y	x_2y	x_1^2	x_1x_2	x_2^2
3	2	48,800	146,400	97,600	9	6	4
2	1	44,300	88,600	44,300	4	2	1
4	3	53,800	215,200	161,400	16	12	9
2	1	44,200	88,400	44,200	4	2	1
3	2	49,700	149,100	99,400	9	6	4
2	2	44,900	89,800	89,800	4	4	4
5	3	58,400	292,000	175,200	25	15	9
4	2	52,900	211,600	105,800	16	8	4
25	16	397,000	1,281,100	817,700	87	55	36

Then, substituting the column totals and $n = 8$ into the normal equations, we get

$$397,000 = 8b_0 + 25b_1 + 16b_2$$

$$1,281,100 = 25b_0 + 87b_1 + 55b_2$$

$$817,700 = 16b_0 + 55b_1 + 36b_2$$

and the solution of this system of linear equations is $b_0 = 35,197$, $b_1 = 4,149$, and $b_2 = 731$ (see Exercise 14.37 on page 433). Thus, the least-squares equation is

$$\hat{y} = 35,197 + 4,149x_1 + 731x_2$$

and this tells us that, in this study, each extra bedroom adds on the average $4,149, and each bath $731, to the sales price of a house. To estimate (predict) the average sales price of three-bedroom houses with two baths, for instance, we substitute $x_1 = 3$ and $x_2 = 2$, and get

$$\hat{y} = 35,197 + (4,149)(3) + (731)(2) = \$49,106$$

or approximately $49,100.

EXERCISES   ★14.33 The following are data on the ages and incomes of a random sample of five executives working for a large multinational company, and the number of years each went to college:

Age x_1	Years college x_2	Income (dollars) y
38	4	31,700
46	0	27,300
39	5	35,500
43	2	30,800
32	4	25,900

Fit an equation of the form $y = b_0 + b_1 x_1 + b_2 x_2$ to these data, and use it to estimate the average income of 39-year-old executives with four years of college working for this company.

★14.34 The following data were collected to determine the relationship between two processing variables and the hardness of a certain kind of steel:

Hardness (Rockwell 30-T) y	Copper content (percent) x_1	Annealing temperature (degrees F) x_2
78.9	0.02	1,000
55.2	0.02	1,200
80.9	0.10	1,000
57.4	0.10	1,200
85.3	0.18	1,000
60.7	0.18	1,200

Fit an equation of the form $y = b_0 + b_1 x_1 + b_2 x_2$ to these data.

★14.35 Rework the preceding exercise after coding the three x_1 values -1, 0, and 1, and the two x_2 values -1 and 1.

★14.36 The following are sample data provided by a moving company on the weights of six shipments, the distances they were moved, and the damage that was incurred:

Weight (1,000 lb) x_1	Distance (1,000 miles) x_2	Damage (dollars) y
4.0	1.5	160
3.0	2.2	112
1.6	1.0	69
1.2	2.0	90
3.4	0.8	123
4.8	1.6	186

(a) Fit an equation of the form $y = b_0 + b_1 x_1 + b_2 x_2$ to the given data.

(b) Use the equation obtained in part (a) to predict the damage when a shipment weighing 2,400 pounds is moved 1,200 miles.

★14.37 Use the method of elimination or determinants to verify that the solution of the three normal equations on page 432 is $b_0 = 35{,}197$, $b_1 = 4{,}149$, and $b_2 = 731$. (It is likely that there will be a difference in the results due to rounding.)

14.6
CHECKLIST OF KEY TERMS
(with page references to their definitions)

14.7
REVIEW EXERCISES

14.38 The following are the numbers of hours which ten persons (interviewed as part of a sample survey) spent watching television, x, and reading books or magazines, y, per week:

x	y
18	7
25	5
19	1
12	5
12	10
27	2
15	3
9	9
12	8
18	4

Fit a least-squares line which will enable us to predict y in terms of x, and construct a 95 percent confidence interval for the regression coefficient β.

14.39 With reference to the preceding exercise, construct a 99 percent confidence interval for the number of hours a person who watches 20 hours of television can be expected to read books or magazines per week.

★14.40 In the years 1969–1973 the total value of New Zealand's exports were 1,194, 1,250, 1,380, 1,636, and 2,094 million SDR's (Special Drawing Rights, or "paper gold" issued by the International Monetary Fund for the use by its members). Use the coding of Exercise 14.9 to fit a parabola to these data.

★14.41 The following are data on the average weekly profits (in $1,000) of five restaurants, their seating capacities, and the average daily traffic (in thousands of cars) which passes their locations:

Seating capacity x_1	Traffic count x_2	Weekly net profit y
120	19	23.8
200	8	24.2
150	12	22.0
180	15	26.2
240	16	33.5

(a) Fit an equation of the form $y = b_0 + b_1 x_1 + b_2 x_2$ to these data.
(b) Use the equation obtained in part (a) to predict the average weekly net profit of a restaurant with a seating capacity of 210 at a location where the daily traffic count averages 14,000 cars.

★14.42 The following data pertain to the cosmic-ray doses measured at various altitudes:

Altitude (100 feet) x	Dose rate (mrem/year) y
0.5	28
4.5	30
7.8	32
12.0	36
48.0	58
53.0	69

Fit an exponential curve and use it to estimate the cosmic radiation at 6,000 feet.

14.43 The following are the high school averages, x, and first-year college grade-point indexes, y, of ten students:

x	y
3.0	2.6
2.7	2.4
3.8	3.9
2.6	2.1
3.2	2.6
3.4	3.3
2.8	2.2
3.1	3.2
3.5	2.8
3.3	2.5

Fit a least-squares line which will enable us to predict first-year college grade-point indexes in terms of high school averages.

14.44 With reference to the preceding exercise, test the null hypothesis $\beta = 1.0$ against the alternative hypothesis $\beta > 1.0$ at the 0.05 level of significance.

14.45 The following are the processing times (minutes), x, and hardness readings, y, of certain machine parts:

x	y
20	282
34	275
19	171
10	142
24	145
31	340
25	282
13	105
29	233

Fit a least-squares line and use it to predict the hardness reading of a part which has been processed for 25 minutes.

14.46 With reference to the preceding exercise, test the null hypothesis $\alpha = 55.0$ against the alternative hypothesis $\alpha \neq 55.0$ at the 0.01 level of significance.

14.47 With reference to Exercise 14.45, find 95 percent limits of prediction for the hardness reading of a machine part which has been processed for 25 minutes.

★**14.48** The following data show the grades which ten students got in an examination, their IQ's, and the numbers of hours they studied for the examination:

Number of hours studied x_1	IQ x_2	Grade in examination y
9	116	70
3	101	38
16	102	90
19	99	99
6	118	59
11	111	75
14	96	77
12	120	84
6	101	49
9	100	61

Fit an equation of the form $y = b_0 + b_1x_1 + b_2x_2$ to the given data, and use it to predict the grade of a student who has an IQ of 107 and studies 13 hours for the examination.

14.8
REFERENCES

Methods of deciding which kind of curve to fit to a given set of paired data may be found in books on numerical analysis and in more advanced texts in statistics. Further information about the material of this chapter may be found in

CHATTERJEE, S., and PRICE, B., *Regression Analysis by Example.* New York: John Wiley & Sons, Inc., 1977.

DRAPER, N. R., and SMITH, H., *Applied Regression Analysis,* 2nd ed. New York: John Wiley & Sons, Inc., 1981.

EZEKIEL, M., and FOX, K. A., *Methods of Correlation and Regression Analysis,* 3rd ed. New York: John Wiley & Sons, Inc., 1959.

HARRIS, R. J., *A Primer of Multivariate Statistics.* New York: Academic Press, Inc., 1971.

MOSTELLER, F., and TUKEY, J. W., *Data Analysis and Regression.* Reading, Mass.: Addison-Wesley Publishing Company, Inc., 1977.

aving learned how to fit a least-squares line to paired data, we turn now to the problem of determining how well such a line actually fits the data. Of course, we can get some idea of this by inspecting a diagram which shows the line together with the data, but to describe how we can be more objective, let us refer back to the original data of the example dealing with the foreign service job applicants' proficiency in German, namely,

Years studied German x	Score in test y
3	57
4	78
4	72
2	58
5	89
3	63
4	73
5	84
3	75
2	48

As can be seen from this table, there are substantial differences among the scores, the smallest being 48 and the largest being 89. However, we also see that the 48 was obtained by a person who had studied German for two years, while the 89 was obtained by a person who had studied German for five years, and this suggests that the differences among the scores may well be due, at least in part, to differences in the length of time that the applicants had studied German. This observation raises the following question, which we shall answer in this chapter: Of the total variation among the y's, how much can be attributed to the relationship between the two variables x and y (that is, to the fact that the observed y's correspond to different values of x), and how much can be attributed to chance.

In Section 15.1 we shall introduce the coefficient of correlation as a measure of the strength of the linear relationship between two variables, in Section 15.2 we shall learn how to interpret it, and in Section 15.3 we shall study related problems of inference; the problems of multiple and partial correlation will be touched upon lightly in Section 15.4.

15 CORRELATION

15.1
THE COEFFICIENT OF CORRELATION

With regard to the question raised in the chapter opening, we are faced here with an analysis-of-variance problem—Figure 15.1 shows what we mean. Referring to this figure, we see that $y - \bar{y}$, the deviation of any observed value of y from the mean of all the y's, can be written as the sum of two parts: $\hat{y} - \bar{y}$, the deviation of the value on the line (corresponding to an observed value of x) from the mean of the y's, and $y - \hat{y}$, the deviation of the observed value of y from the corresponding value on the line. Symbolically, we write

$$y - \bar{y} = (\hat{y} - \bar{y}) + (y - \hat{y})$$

for any observed value y, and if we square the expressions on both sides of this identity and sum over all n values of y, we find that algebraic simplifications lead to

$$\sum (y - \bar{y})^2 = \sum (\hat{y} - \bar{y})^2 + \sum (y - \hat{y})^2$$

As our measure of the total variation of the y's we use the quantity on the left above, $\sum (y - \bar{y})^2$, called the **total sum of squares**; it is just $n - 1$ times

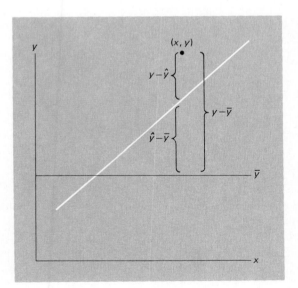

FIGURE 15.1 *Illustration that*
$$y - \bar{y} = (\hat{y} - \bar{y}) + (y - \hat{y}).$$

the variance of the y's, and, as the equation shows, it has been partitioned into two additive components. The first of these, $\sum (\hat{y} - \bar{y})^2$, called the **regression sum of squares,** consists of the sum of the squares of the deviations of the values on the line from the mean $\bar{y}$; it measures that portion of the total variation of the y's which would exist if differences in x were the only cause of differences among the y's (that is, if all the y's lay directly on the line).

This is hardly ever the case, though, in practice, and the fact that all the points do not lie on a least-squares line is an indication that there are other factors than differences among the x's which affect the values of y. It is customary to combine all these other factors under the general heading of "chance." Chance variation is thus measured by the amounts by which the points deviate from the line; specifically, it is measured by $\sum (y - \hat{y})^2$, called the **residual sum of squares,** which is the second of the two components into which we partitioned the total sum of squares.

To calculate the various sums of squares for our numerical example, where $\bar{y} = \frac{697}{10} = 69.7$, we find by substitution into the formula (but see Exercise 15.22 on page 452) that for the proficiency scores in German the total sum of squares is

$$\sum (y - \bar{y})^2 = (57 - 69.7)^2 + (78 - 69.7)^2 + \cdots + (48 - 69.7)^2$$
$$= 1,504.10$$

To find the residual sum of squares by substitution into the formula (but again see Exercise 15.22 on page 452), we must first determine the value of $\hat{y} = 31.53 + 10.90x$ for each of the given values of x, and we get $31.53 + 10.90(3) = 64.23$, $31.53 + 10.90(4) = 75.13, \ldots,$ and $31.53 + 10.90(2) = 53.33$. Then,

$$\sum (y - \hat{y})^2 = (57 - 64.23)^2 + (78 - 75.13)^2 + \cdots + (48 - 53.33)^2$$
$$= 255.51$$

and we can say that

$$\frac{\sum (y - \hat{y})^2}{\sum (y - \bar{y})^2} = \frac{255.51}{1,504.10} = 0.17$$

is the proportion of the total variation of the scores which can be attributed to chance, while

$$1 - \frac{\sum (y - \hat{y})^2}{\sum (y - \bar{y})^2} = 1 - 0.17 = 0.83$$

is the proportion of the total variation of the scores which can be attributed to the relationship with x, namely, to the differences in the number of years which the applicants have studied German.

If we take the square root of the last proportion (namely, the proportion of the total variation of the y's which can be attributed to the relationship with x), we obtain the statistical measure which is called the **coefficient of correlation**. It is denoted by the letter r, and its sign is chosen so that it is the same as that of the estimated regression coefficient b. Thus, for our example, where $b = 10.90$, we get

$$r = \sqrt{0.83} = 0.91$$

rounded to two decimals.

It follows from the rule for the sign of r that the correlation coefficient is positive when the least-squares line has an upward slope, namely, when the relationship between x and y is such that small values of y tend to go with small values of x and large values of y tend to go with large values of x. Also, the correlation coefficient is negative when the least-squares line has a downward slope, namely, when large values of y tend to go with small values of x and small values of y tend to go with large values of x. Geometrically, examples of **positive** and **negative correlations** are shown in the first two diagrams of Figure 15.2.

Since part of the variation of the y's cannot exceed their total variation, $\sum (y - \hat{y})^2$ cannot exceed $\sum (y - \bar{y})^2$, and it follows from the formula defining r that correlation coefficients must lie on the interval from -1 to $+1$. If all the points actually fall on a straight line, the residual sum of squares, $\sum (y - \hat{y})^2$, is zero, and the resulting value of r, -1 or $+1$, is indicative of the perfect fit. If, however, the scatter of the points is such that the least-squares line is a horizontal line coincident with $\bar{y}$ (that is, a line with slope 0 which intersects the y-axis at $a = \bar{y}$), then $\sum (y - \hat{y})^2$ equals $\sum (y - \bar{y})^2$ and $r = 0$. In that case none of the variation of the y's can be

FIGURE 15.2 *Types of correlation.*

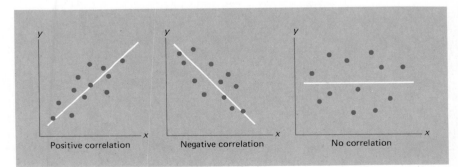

Positive correlation   Negative correlation   No correlation

attributed to their relationship with x, and the fit is so poor that knowledge of x is of no help in predicting y. The predicted value of y is $\bar{y}$ regardless of x. An example of this is shown in the third diagram of Figure 15.2.

The formula which defines r shows clearly the nature, or essence, of the coefficient of correlation, but in actual practice it is seldom used to determine its value. Instead, we use the computing formula[†]

Computing formula for coefficient of correlation

$$r = \frac{n(\sum xy) - (\sum x)(\sum y)}{\sqrt{n(\sum x^2) - (\sum x)^2}\sqrt{n(\sum y^2) - (\sum y)^2}}$$

which automatically gives r the correct sign. This formula may look imposing but, with the exception of $\sum y^2$, the quantities needed for substitution are the same ones which were required to calculate the regression coefficients a and b.

EXAMPLE   Use the computing formula for r to verify the value $r = 0.91$, which we calculated on page 441.

Solution   Copying $n = 10$, $\sum x = 35$, $\sum x^2 = 133$, $\sum y = 697$, and $\sum xy = 2,554$ from page 408, and $\sum y^2 = 50,085$ from page 416, we find that substitution into the computing formula for r yields

$$r = \frac{10(2,554) - (35)(697)}{\sqrt{10(133) - (35)^2}\sqrt{10(50,085) - (697)^2}}$$

$$= 0.91$$

This agrees, as it should, with the result obtained on page 441.

[†] This computing formula follows directly from an alternative, but equivalent, definition of the coefficient of correlation based on the **sample covariance**

$$s_{xy} = \frac{\sum (x - \bar{x})(y - \bar{y})}{n - 1}$$

In this formula, we add the products obtained by multiplying the deviation of each x from $\bar{x}$ by the deviation of the corresponding y from $\bar{y}$, and divide by $n - 1$. In this way we literally measure the way in which the values of x and y vary together. If the relationship between the x's and the y's is such that large values of x tend to go with large values of y, and small values of x with small values of y, the deviations $x - \bar{x}$ and $y - \bar{y}$ tend to be both positive or both negative, so that most of the products $(x - \bar{x})(y - \bar{y})$ and, hence, the covariance, are positive. On the other hand, if the relationship between the x's and the y's is such that large values of x tend to go with small values of y and vice versa, the deviations $x - \bar{x}$ and $y - \bar{y}$ tend to be of opposite sign, so most of the products and, hence, the covariance, are negative. Using the covariance and the standard deviations, s_x and s_y, of the x's and the y's, we can define the coefficient of correlation as

$$r = \frac{s_{xy}}{s_x \cdot s_y}$$

The sum of products $\sum (x - \bar{x})(y - \bar{y})$ divided by n is called a **product moment**, and this explains the term "product-moment coefficient of correlation."

15.2
THE INTERPRETATION OF r

When r equals $+1$, -1, or 0, there is no problem about the interpretation of the coefficient of correlation. As we have already indicated, it is $+1$ or -1 when all the points actually fall on a straight line, and it is zero when the fit of the least-squares line is so poor that knowledge of x does not help in the prediction of y. In general, $100r^2$ gives the percentage of the total variation of the y's which is explained by, or is due to, their relationship with x. This itself is an important measure of the relationship between two variables; beyond this, it permits valid comparisons of several coefficients of correlation. For instance, if $r = 0.80$ in one study and $r = 0.40$ in another study, it would be misleading to say that the correlation of 0.80 is "twice as good" or "twice as strong" as the correlation of 0.40. When $r = 0.80$, then 64 percent of the variation of the y's is accounted for by the relationship with x, and when $r = 0.40$, only 16 percent of the variation of the y's is accounted for by the relationship with x. Thus, in the sense of "percentage of variation accounted for" we can say that a correlation of 0.80 is four times as strong as a correlation of 0.40. In the same way, we say that a correlation of 0.60 is nine times as strong as a correlation of 0.20.

There are several pitfalls in the interpretation of the coefficient of correlation. First, it is often overlooked that r measures only the strength of linear relationships; second, a strong correlation does not necessarily imply a cause–effect relationship.

If r is calculated indiscriminately, for instance, for the three sets of data of Figure 15.3, we get $r = 0.75$ in each case, but it is a meaningful

FIGURE 15.3 *Three sets of paired data for which $r = 0.75$.*

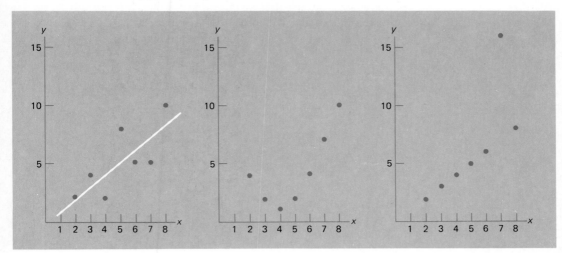

measure of the strength of the relationship only in the first case. In the second case there is a very strong curvilinear relationship between the two variables, and in the third case six of the seven points actually fall on a straight line, but the seventh point is so far off that it suggests the possibility of a gross error of measurement or an error in recording the data. Thus, before we calculate r, we should always plot the data to see whether there is reason to believe that the relationship is, in fact, linear.

The fallacy of interpreting a high value of r (that is, a value close to $+1$ or -1) as an indication of a cause–effect relationship, is best explained with a few examples. Frequently used as an illustration, is the high positive correlation between the annual sales of chewing gum and the incidence of crime in the United States. Obviously, one cannot conclude that crime might be reduced by prohibiting the sale of chewing gum; both variables depend upon the size of the population, and it is this mutual relationship with a third variable (population size) which produces the positive correlation. Another example is the strong positive correlation which was observed between the number of storks seen nesting in English villages and the number of children born in the same villages. We leave it to the reader's ingenuity to explain why there might be a strong correlation in this case in the absence of any cause–effect relationship.

15.3
CORRELATION ANALYSIS

When r is calculated on the basis of sample data, we may get a strong positive or negative correlation purely by chance, even though there is actually no relationship whatever between the two variables under consideration.

Suppose, for instance, that we take a pair of dice, one red and one green, roll them five times, and get the following results:

Red die x	Green die y
4	5
2	2
4	6
2	1
6	4

Presumably, there is no relationship between x and y, the numbers of points showing on the two dice. It is hard to see why large values of x should go with large values of y and small values of x with small values of y, but calculating r, we get the surprisingly high value $r = 0.66$. This raises the question of whether something is wrong with the assumption that there is no relationship between x and y, and to answer it we shall have to see whether the high value of r may be attributed to chance.

When a correlation coefficient is calculated from sample data, as in the above example, the value we get for r is only an estimate of a corresponding parameter, the **population correlation coefficient,** which we denote by ρ (the Greek letter *rho*). What r measures for a sample, ρ measures for a population.

To make inferences about ρ on the basis of r, we shall have to make several assumptions about the distributions of the random variables whose values we observe. In **normal correlation analysis** we make the same assumptions as in normal regression analysis (see page 414), except that the x's are not constants, but values of a random variable having a normal distribution.

Since the sampling distribution of r is rather complicated under these assumptions, it is common practice to base inferences about ρ on the **Fisher Z transformation,** a change of scale from r to Z, which is given by

$$Z = \frac{1}{2} \cdot \ln \frac{1 + r}{1 + r}$$

Here ln denotes "natural logarithm," that is, logarithm to the base e, where $e = 2.71828\ldots$. This transformation is named after R. A. Fisher, a prominent statistician, who showed that under the assumptions of normal correlation analysis and for any value of ρ, the distribution of Z is approximately normal with

$$\mu_Z = \frac{1}{2} \cdot \ln \frac{1 + \rho}{1 - \rho} \qquad \text{and} \qquad \sigma_Z = \frac{1}{\sqrt{n - 3}}$$

Hence,

Statistic for inferences about ρ

$$z = \frac{Z - \mu_Z}{\sigma_Z} = (Z - \mu_Z)\sqrt{n - 3}$$

has approximately the standard normal distribution. The application of this theory is greatly facilitated by the use of Table IX at the end of the book, which gives the values of Z corresponding to $r = 0.00, 0.01, 0.02, 0.03, \ldots$, and 0.99. Observe that only positive values are given in this table; if r is negative, we simply look up $-r$ and take the negative of the corresponding Z. Note also that the formula for μ_Z is like that for Z with r replaced by ρ; therefore, Table IX can be used to look up values of μ_Z corresponding to given values of ρ.

EXAMPLE Use the 0.05 level of significance to test the null hypothesis of no correlation for the example in which we rolled a pair of dice five times and obtained $r = 0.66$.

Solution
1. *Null hypothesis*: $\rho = 0$
   Alternative hypothesis: $\rho \neq 0$
2. *Level of significance*: $\alpha = 0.05$
3. *Criterion*: Since $\mu_z = 0$ for $\rho = 0$, reject the null hypothesis if $z < -1.96$ or $z > 1.96$, where

$$z = Z \cdot \sqrt{n-3}$$

and otherwise state that the value of r is not significant.

4. *Calculations*: The value of Z corresponding to $r = 0.66$ is 0.793 according to Table IX, so that

$$z = 0.793\sqrt{5 - 3}$$
$$= 1.12$$

5. *Decision*: Since $z = 1.12$ falls between -1.96 and 1.96, the null hypothesis of no correlation cannot be rejected; in other words, $r = 0.66$ is not significant at $\alpha = 0.05$ for five rolls of a pair of dice. (An alternative way of handling this kind of problem is given in Exercise 15.13 on page 450.)

EXAMPLE With reference to the proficiency scores of the applicants for foreign service jobs, where $n = 10$ and $r = 0.91$, test the null hypothesis $\rho = 0.70$ against the alternative hypothesis $\rho > 0.70$ at the 0.05 level of significance.

Solution
1. *Null hypothesis*: $\rho = 0.70$
   Alternative hypothesis: $\rho > 0.70$
2. *Level of significance*: $\alpha = 0.05$
3. *Criterion*: Reject the null hypothesis if $z > 1.645$, where

$$z = (Z - \mu_z)\sqrt{n - 3}$$

and otherwise accept the null hypothesis or reserve judgment.

4. *Calculations*: The values of Z corresponding to $r = 0.91$ and $\rho = 0.70$ are 1.528 and 0.867 according to Table IX, so that

$$z = (1.528 - 0.867)\sqrt{10 - 3}$$
$$= 1.75$$

5. *Decision*: Since $z = 1.75$ exceeds 1.645, the null hypothesis must be rejected; we conclude that $\rho > 0.70$ for the proficiency scores of applicants to the foreign service jobs and the number of years that they have studied German in high school or college.

To construct confidence intervals for ρ, we first construct confidence intervals for μ_Z, and then convert to r and ρ by means of Table IX. A confidence interval formula for μ_Z may be obtained by substituting

$$z = (Z - \mu_Z)\sqrt{n - 3}$$

for the middle term of the double inequality $-z_{\alpha/2} < z < z_{\alpha/2}$, and then manipulating the terms algebraically so that the middle term is μ_Z. This leads to the following $(1 - \alpha)100$ percent confidence interval for μ_Z:

Confidence interval for μ_Z

$$Z - \frac{z_{\alpha/2}}{\sqrt{n - 3}} < \mu_Z < Z + \frac{z_{\alpha/2}}{\sqrt{n - 3}}$$

EXAMPLE   If $r = 0.62$ for the cost estimates of two mechanics for a random sample of 30 repair jobs, construct a 95 percent confidence interval for the population correlation coefficient.

Solution   Reading the value of Z which corresponds to $r = 0.62$ from Table IX, and substituting it together with $n = 30$ and $z_{0.025} = 1.96$ into the above confidence interval formula for μ_Z, we get

$$0.725 - \frac{1.96}{\sqrt{27}} < \mu_Z < 0.725 + \frac{1.96}{\sqrt{27}}$$

or

$$0.348 < \mu_Z < 1.102$$

Finally, looking up the values of r which come closest to $Z = 0.348$ and $Z = 1.102$ in Table IX, we get the 95 percent confidence interval

$$0.33 < \rho < 0.80$$

for the true strength of the linear relationship between cost estimates made by the two mechanics.

The following is an example which involves negative values of r and Z:

EXAMPLE If $r = 0.20$ for a random sample of $n = 40$ paired data, construct a 95 percent confidence interval for ρ.

Solution Since $Z = 0.203$ corresponds to $r = 0.20$ in Table IX, we first get the 95 percent confidence interval

$$0.203 - \frac{1.96}{\sqrt{37}} < \mu_Z < 0.203 + \frac{1.96}{\sqrt{37}}$$

or

$$-0.119 < \mu_Z < 0.525$$

Then, looking up in Table IX the values of r which come closest to $Z = 0.119$ and $Z = 0.525$, we get the 95 percent confidence interval

$$-0.12 < \rho < 0.48$$

for the population correlation coefficient.

EXERCISES

15.1 The following are the numbers of minutes it took 12 mechanics to assemble a piece of machinery in the morning, x, and in the late afternoon, y:

x	y
12	14
11	11
9	14
13	11
10	12
11	15
12	12
14	13
10	16
9	10
11	10
12	14

Calculate r.

15.2 With reference to the preceding exercise, test the null hypothesis of no correlation at the 0.05 level of significance.

15.3 The following table shows the percentages of the vote predicted by a poll for eight candidates for the U.S. Senate in different states, x, and the percentages of the vote which they actually received, y:

x	y
43	50
46	42
51	57
59	55
41	46
53	48
52	53
62	56

Calculate r.

15.4 Since r does not depend on the scales of x and y, its calculation can often be simplified by adding a suitable positive or negative number to each x, each y, or both. Rework the preceding exercise after subtracting 41 from each x and 42 from each y.

15.5 The following data were obtained in a study of the relationship between the resistance (ohms) and the failure time (minutes) of certain overloaded resistors:

Resistance	Failure time
48	45
28	25
33	39
40	45
36	36
39	35
46	36
40	45
30	34
42	39
44	51
48	41
39	38
34	32
47	45

Calculate r and determine what percentage of the variation in failure time is due to differences in resistance.

15.6 With reference to the preceding exercise, test the null hypothesis $\rho = 0$ against the alternative hypothesis $\rho \neq 0$ at the 0.01 level of significance.

15.7 With reference to Exercise 14.1 on page 410,
   (a) calculate r for the given data;
   (b) determine what percentage of the variation in y is accounted for by differences in x;
   (c) construct a 95 percent confidence interval for ρ.

15.8 With reference to Exercise 14.7 on page 412,
   (a) calculate r for the given data;
   (b) test whether r is significant at $\alpha = 0.05$.

15.9 If we calculate r for each of the following sets of data, should we be surprised if we get $r = 1$ and $r = -1$? Explain your answers.

(a)	x	y	(b)	x	y
	6	9		12	5
	14	11		8	15

15.10 With reference to Exercise 14.33, calculate r for each of the following pairs of variables:
   (a) age and income;
   (b) age and years college;
   (c) years college and income.

15.11 If $r = 0.41$ for one set of paired data and $r = 0.29$ for another set of paired data, compare the strengths of the two relationships.

15.12 State in each case whether you would expect a positive correlation, a negative correlation, or no correlation:
   (a) the ages of husbands and wives;
   (b) the amount of rubber on tires and the number of miles they have been driven;
   (c) the number of hours that golfers practice and their scores;
   (d) shoe size and IQ;
   (e) pollen count and the sale of anti-allergy drugs;
   (f) income and education;
   (g) the number of sunny days in August in Detroit and the attendance at the Detroit Zoo;
   (h) shirt size and sense of humor;
   (i) number of persons getting flu shots and number of persons catching the flu.

15.13 Under the assumptions of normal correlation analysis, the test of the null hypothesis $\rho = 0$ may also be based on the statistic

$$t = \frac{r\sqrt{n-2}}{\sqrt{1 - r^2}}$$

which has the t distribution with $n - 2$ degrees of freedom. Use this statistic to test in each case whether the value of r is significant at the 0.05 level of significance:
   (a) $n = 12$ and $r = 0.50$;
   (b) $n = 20$ and $r = 0.62$.

15.14 Use the t statistic of the preceding exercise to test in each case whether the value of r is significant at the 0.01 level of significance:

(a) $n = 14$ and $r = 0.78$;

(b) $n = 16$ and $r = 0.51$.

15.15 In a study of the relationship between the annual production of cotton and citrus fruits in Arizona, data for $n = 16$ years yielded $r = -0.56$. At the 0.05 level of significance, test the null hypothesis $\rho = -0.40$ against the alternative hypothesis $\rho \neq -0.40$.

15.16 In a study of the relationship between the death rate from lung cancer and the per capita consumption of cigarettes twenty years earlier, data for $n = 9$ countries yielded $r = 0.73$. At the 0.05 level of significance, test the null hypothesis $\rho = 0.50$ against the alternative hypothesis $\rho > 0.50$.

15.17 In a study of the relationship between the available heat (per cord) of green wood and air-dried wood, data for $n = 13$ kinds of wood yielded $r = 0.94$. Use the 0.01 level of significance to test the null hypothesis $\rho = 0.75$ against the alternative hypothesis $\rho \neq 0.75$.

15.18 If $n = 18$ and $r = -0.64$ for certain paired data, test the null hypothesis $\rho = -0.30$ against the alternative hypothesis $\rho < -0.30$ at the 0.05 level of significance.

15.19 Assuming that the conditions underlying normal correlation analysis are met, use the Fisher Z transformation to construct approximate 95 percent confidence intervals for ρ when

(a) $r = 0.80$ and $n = 15$;

(b) $r = -0.22$ and $n = 30$;

(c) $r = 0.64$ and $n = 100$.

15.20 Assuming that the conditions underlying normal correlation analysis are met, use the Fisher Z transformation to construct approximate 99 percent confidence intervals for ρ when

(a) $r = -0.87$ and $n = 19$;

(b) $r = 0.39$ and $n = 24$;

(c) $r = 0.16$ and $n = 40$.

★**15.21** Correlation methods are sometimes used to study the relationship between two (time) series of data which are recorded annually, monthly, weekly, daily, and so on. Suppose, for instance, that in the years 1963–1976 a large textile manufacturer spent 0.8, 0.5, 0.8, 1.0, 1.0, 0.9, 0.8, 1.2, 1.0, 0.9, 0.8, 1.0, 1.0, and 0.8 million dollars on research and development, and that in these years its share of the market was 20.4, 18.6, 19.1, 18.0, 18.2, 19.6, 20.0, 20.4, 19.2, 20.5, 20.8, 18.9, 19.0, and 19.8 percent. To see whether and how the company's share of the market in a given year may be related to its expenditure on research and development in prior years, let x_t denote the company's research and development expenditures and y_t its market share in the year t, and calculate

(a) the correlation coefficient for y_t and x_{t-1};

(b) the correlation coefficient for y_t and x_{t-2};

(c) the correlation coefficient for y_t and x_{t-3};

(d) the correlation coefficient for y_t and x_{t-4}.

For instance, in part (a) calculate r after pairing the 1964 percentage share of the market with the 1963 expenditures on research and development, the 1965 market share with the 1964 expenditures, and so on, . . . , and in part (d) calculate

r after pairing the 1967 percentage share of the market with the 1963 expenditures on research and development, the 1968 market share with the 1964 expenditures, and so on. These time-lag correlations are called **cross correlations**. To continue,

   (e) test, at the 0.05 level of significance, whether the correlation coefficients obtained in parts (a) through (d) are significant:

   (f) discuss the apparent duration of the effect of expenditures on research and development on the company's share of the market.

15.22 On page 440 we calculated for the proficiency scores the total sum of squares and the residual sum of squares by direct substitution into $\sum (y - \bar{y})^2$ and $\sum (y - \hat{y})^2$. However, it can be shown that

$$\sum (y - \bar{y})^2 = \sum y^2 - n \cdot \bar{y}^2$$

and

$$\overset{4}{\sum} (y - \hat{y})^2 = \sum y^2 - a(\sum y) - b(\sum xy)$$

where a and b are the y-intercept and the slope of the least-squares line, and in most problems these two formulas will greatly simplify the calculation of these sums of squares. Use these computing formulas to recalculate the two sums of squares for the proficiency scores. The answer for $\sum (y - \bar{y})^2$ should be the same as on page 440, but the answer for $\sum (y - \hat{y})^2$ will differ somewhat from that obtained before due to the rounding of a and b to two decimals.

15.4
MULTIPLE AND PARTIAL CORRELATION ⋆

In the beginning of this chapter we defined the correlation coefficient as a measure of the goodness of the fit of a least-squares line to a set of paired data. If predictions are to be made with equations of the form

$$\hat{y} = b_0 + b_1 x_1 + b_2 x_2 + \cdots + b_k x_k$$

as in Section 14.5, we define the **multiple correlation coefficient** in the same way in which we originally defined r. We take the square root of the quantity

$$1 - \frac{\sum (y - \hat{y})^2}{\sum (y - \bar{y})^2}$$

which is the proportion of the total variation of the y's that can be attributed to the relationship with the x's. The only difference is that we now calculate $\hat{y}$ by means of the multiple regression equation instead of the equation $\hat{y} = a + bx$.

EXAMPLE In the example on page 432, where we derived the equation $\hat{y} = 35,197 + 4,149x_1 + 731x_2$, it can be shown that $\sum(y - \hat{y})^2 = 686,719$ and $\sum(y - \bar{y})^2 = 185,955,000$. What is the value of the multiple correlation coefficient?

Solution Since

$$1 - \frac{\sum(y - \hat{y})^2}{\sum(y - \bar{y})^2} = 1 - \frac{686,719}{185,955,000} = 0.9963$$

it follows that the multiple correlation coefficient is $\sqrt{0.9963} = 0.998$.

This example also serves to illustrate that adding more independent variables in a correlation study is not always sufficiently productive to justify the extra work. The value of r for y and x_1 alone is 0.996, so that very little is gained by considering x_2. In other words, predictions based on the number of bedrooms alone are virtually as good as predictions which account also for the number of baths. The situation is quite different, though, in Exercise 15.24, where two independent variables together account for a much higher proportion of the total variation in y than does either x_1 or x_2 alone.

When we discussed the problem of correlation and causation, we showed that a strong correlation between two variables may be due entirely to their dependence on a third variable. We illustrated this with the examples of chewing gum sales and the crime rate, and child births and the number of storks. To give another example, let us consider the two variables x_1, the weekly amount of hot chocolate sold by a refreshment stand at a summer resort, and x_2, the weekly number of visitors to the resort. If, on the basis of suitable data, we get $r = -0.30$ for these variables, this should come as a surprise—after all, we would expect more sales of hot chocolate when there are more visitors, and vice versa, and hence a positive correlation.

However, if we think for a moment, we may surmise that the negative correlation of -0.30 may well be due to the fact that the variables x_1 and x_2 are both related to a third variable x_3, the average weekly temperature at the resort. If the temperature is high, there will be more visitors, but they will prefer cold drinks to hot chocolate; if the temperature is low, there will be fewer visitors, but they will prefer hot chocolate to cold drinks. So, let us suppose that further data yield $r = -0.70$ for x_1 and x_3, and $r = 0.80$ for x_2 and x_3. These values seem reasonable since low sales of hot chocolate should go with high temperatures and vice versa, while the number of visitors should be high when the temperature is high, and low when the temperature is low.

In the preceding example, we should really have investigated the relationship between x_1 and x_2 (hot chocolate sales and the number of visitors to the resort) when all other factors, primarily temperature, are held fixed. As it is seldom possible to control matters to such an extent, it has been found that a statistic called the **partial correlation coefficient** does a fair job of eliminating the effects of other variables. If we write the ordinary correlation coefficients for x_1 and x_2, x_1 and x_3, and x_2 and x_3, as r_{12}, r_{13}, and r_{23}, the partial correlation coefficient for x_1 and x_2 with x_3 fixed is given by

Partial correlation coefficient

$$r_{12.3} = \frac{r_{12} - r_{13} \cdot r_{23}}{\sqrt{1 - r_{13}^2}\sqrt{1 - r_{23}^2}}$$

EXAMPLE   Calculate $r_{12.3}$ for the above example.

Solution   Substituting $r_{12} = -0.30$, $r_{13} = -0.70$, and $r_{23} = 0.80$ into the formula, we get

$$r_{12.3} = \frac{(-0.30) - (-0.70)(0.80)}{\sqrt{1 - (-0.70)^2}\sqrt{1 - (0.80)^2}} = 0.61$$

This shows that, as we expected, there is a positive relationship between the sales of hot chocolate and the number of visitors to the resort when the effect of differences in temperature is eliminated.

We have given this example mainly to illustrate what we mean by partial correlation, but it also serves as a reminder that correlation coefficients can be very misleading unless they are interpreted with care.

EXERCISES   ★15.23 In a multiple regression problem, the residual sum of squares is $\sum (y - \hat{y})^2 = 75{,}240$ and the total sum of squares is $\sum (y - \bar{y})^2 = 112{,}550$. Find the value of the multiple correlation coefficient.

★15.24 Use the least-squares equation obtained in Exercise 14.33 on page 432 to calculate $\hat{y}$ for each of the five executives, determine the two sums of squares $\sum (y - \hat{y})^2$ and $\sum (y - \bar{y})^2$, and calculate the multiple correlation coefficient. Also compare the result with the values obtained for r in parts (a) and (c) of Exercise 15.10 on page 450.

★15.25 Use the least-squares equation obtained in Exercise 14.36 on page 433 to calculate $\hat{y}$ for each of the six shipments, determine the two sums of squares $\sum (y - \hat{y})^2$ and $\sum (y - \bar{y})^2$, and calculate the multiple correlation coefficient.

★15.26 Use the results of Exercise 15.10 on page 450 to find
  (a) the partial correlation coefficient for age and income when years college is held fixed;
  (b) the partial correlation coefficient for years college and income when age is held fixed.

★15.27 With reference to the example on page 453, find the partial correlation coefficient for x_1 and x_3 (sales of hot chocolate and temperature) when x_2 (number of visitors) is held fixed.

15.5
CHECKLIST OF KEY TERMS
(with page references to their definitions)

Coefficient of correlation, 441
Fisher Z transformation, 445
★Multiple correlation coefficient, 452
Negative correlation, 441
Normal correlation analysis, 445
★Partial correlation coefficient, 454
Population correlation coefficient, 445
Positive correlation, 441
Product moment, 442
Product-moment coefficient of correlation, 442
Regression sum of squares, 440
Residual sum of squares, 440
Total sum of squares, 439

15.6
REVIEW EXERCISES

15.28 If $r = 0.56$ for one set of paired data and $r = -0.97$ for another, compare the strengths of the two relationships.

★15.29 Use the least-squares equation obtained in Exercise 14.41 on page 435 to calculate $\hat{y}$ for each of the five restaurants, determine the two sums of squares $\sum (y - \hat{y})^2$ and $\sum (y - \bar{y})^2$, and calculate the multiple correlation coefficient.

15.30 If a set of $n = 20$ paired observations yields $r = 0.38$, test the null hypothesis $\rho = 0$ at the 0.05 level of significance using
  (a) the method described in the text;
  (b) the method of Exercise 15.13.

15.31 The following data pertain to a study of the effects of environmental pollution on wildlife; in particular, the effect of DDT on the thickness of the eggshells of certain birds:

DDT residue in yolk lipids (parts per million)	Thickness of eggshells (millimeters)
117	0.49
65	0.52
303	0.37
98	0.53
122	0.49
150	0.42

Calculate r for these data.

15.32 With reference to the preceding exercise, test whether the value obtained for r is significant at $\alpha = 0.05$.

15.33 If $r = 0.25$ for the ages of a group of college students and their knowledge of foreign affairs, what percentage of the variation of their knowledge of foreign affairs can be attributed to differences in age?

15.34 Assuming that the conditions underlying normal correlation analysis are met, use the Fisher Z transformation to construct approximate 95 percent confidence intervals for ρ when
  (a) $r = 0.44$ and $n = 20$;
  (b) $r = -0.32$ and $n = 38$.

★**15.35** In a multiple regression problem, the residual sum of squares is $\sum (y - \hat{y})^2 = 926$ and the total sum of squares is $\sum (y - \bar{y})^2 = 1,702$. Find the value of the multiple correlation coefficient.

15.36 Calculate r for the data of Exercise 14.38 on page 434, and test the null hypothesis of no correlation at the 0.01 level of significance.

15.37 The following are the numbers of inquiries which a real estate agency received in eight weeks about houses for rent, x, and houses for sale, y:

x	y
60	82
72	85
47	62
38	53
17	29
45	50
33	69
57	88

Calculate r.

★**15.38** With reference to Exercise 14.48 on page 436, find the partial correlation coefficient for number of hours studied and grade in the test when IQ is held fixed.

★15.39 Correlation methods are sometimes used to study the internal structure (or systematic patterns) in series of data (time series) which are recorded on an annual, monthly, weekly, ..., basis. Consider, for example, the diagram of Figure 15.4, which shows a line chart of a company's annual sales for the years 1958–1977. There is an obvious linear trend in the series as shown by the least-squares line, and to look for further patterns, we can study the deviations from the line, $y - \hat{y}$, which are $-2, 6, 0, 3, -2, -13, -5, -10, 1, 18, 6, 10, 1, -8, -15, 0, 3, -7, 15,$ and 0 million dollars for the years 1958–1977. Letting y_t denote the deviation from the line in the year t, calculate

   (a) the correlation coefficient for y_t and y_{t-1};
   (b) the correlation coefficient for y_t and y_{t-2};
   (c) the correlation coefficient for y_t and y_{t-3};
   (d) the correlation coefficient for y_t and y_{t-4};
   (e) the correlation coefficient for y_t and y_{t-5}.

For instance, in part (a) calculate r after pairing the 1959 deviation from the line with the 1958 deviation from the line, the 1960 deviation from the line with the 1959 deviation from the line, and so on, ..., and in part (e) calculate r after pairing the 1963 deviation from the line with the 1958 deviation from the line, the 1964 deviation from the line with the 1959 deviation from the line, and, finally, the 1977 deviation from the line with the 1972 deviation from the line. These lag-time correlations are also called **autocorrelations**.

   (f) Test at the 0.05 level of significance whether the correlation coefficients obtained in parts (a) through (e) are significant.
   (g) Discuss the possible existence of a cyclical (or repeating) pattern in the series of data.

FIGURE 15.4 *Diagram for Exercise* 15.39.

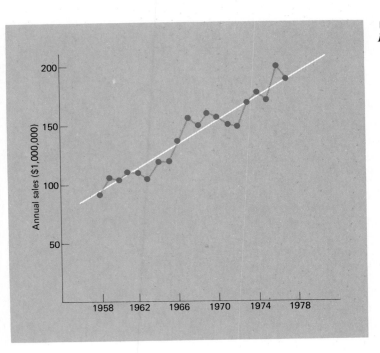

15.40 Use the method of Exercise 15.13 to rework the example on page 444 which pertained to five rolls of a pair of dice.

15.7
REFERENCES

More detailed information about multiple and partial correlation may be found in

EZEKIEL, M., and FOX, K. A., *Methods of Correlation and Regression Analysis, 3rd ed.* New York: John Wiley & Sons, Inc., 1959.

HARRIS, R. J., *A Primer of Multivariate Statistics.* New York: Academic Press, Inc., 1975.

and an advanced theoretical treatment is given in Volume 2 of

KENDALL, M. G., and STUART, A., *The Advanced Theory of Statistics, 3rd ed.* New York: Hafner Press, 1973.

Volume 1 of this book provides the theoretical foundation of significance tests for r.

Most of the tests discussed in Chapters 10 through 13 require specific assumptions about the population, or populations, sampled. In many cases we must assume that the populations have roughly the shape of normal distributions, that their variances are known or are known to be equal, or that the samples are independent. Since there are many situations where it is doubtful whether these assumptions can be met, statisticians have developed alternative techniques based on less stringent assumptions, which have become known as nonparametric tests.

Aside from the fact that nonparametric tests can be used under more general conditions than the standard tests which they replace, they are often easier to explain and understand; moreover, in many nonparametric tests the computational burden is so light that they come under the heading of "quick and easy" or "shortcut" techniques. For all these reasons, nonparametric tests have become quite popular, and extensive literature is devoted to their theory and application.

In Sections 16.1 through 16.3 we shall present the sign test as a nonparametric alternative to tests concerning means and tests concerning differences between means based on paired data; in Sections 16.4 through 16.8 we shall study methods based on rank sums, which serve as a nonparametric alternative to tests concerning differences between means based on independent samples, as an alternative to the sign test which is less wasteful of information, and as a nonparametric alternative to the one-way analysis of variance; in Sections 16.9 through 16.11 we shall learn how to test the randomness of a sample after the data have been obtained; and in Section 16.12 we shall present a nonparametric test of the significance of a relationship between paired data.

16 NON-PARAMETRIC TESTS

16.1
THE ONE-SAMPLE SIGN TEST

Except for the large-sample tests, all the standard tests concerning means of Chapter 10 are based on the assumption that the populations sampled have roughly the shape of normal distributions. When this assumption is untenable in practice, these standard tests can be replaced by any one of several nonparametric alternatives, among them the **sign test,** which we shall study in this section, and in Sections 16.2 and 16.3.

The **one-sample sign test** applies when we sample a continuous symmetrical population, so that the probability of getting a sample value less than the mean and the probability of getting a sample value greater than the mean are both $\frac{1}{2}$. To test the null hypothesis $\mu = \mu_0$ against an appropriate alternative on the basis of a random sample of size n, we replace each sample value greater than μ_0 with a plus sign and each sample value less than μ_0 with a minus sign, and then we test the null hypothesis that these plus and minus signs are values of a random variable having the binomial distribution with $p = \frac{1}{2}$.[†] (If a sample value equals μ_0, which may well happen when we deal with rounded data, we simply discard it.)

To perform a one-sample sign test when the sample is small, we refer directly to a table of binomial probabilities such as Table V at the end of the book; when the sample is large, we use the normal approximation to the binomial distribution, as will be illustrated in Section 16.3.

EXAMPLE   The following data constitute a random sample of 15 measurements of the octane rating of a certain kind of gasoline:

97.5	95.2	97.3	96.0	96.8	100.3	97.4	95.3
93.2	99.1	96.1	97.6	98.2	98.5	94.9	

Use the one-sample sign test to test the null hypothesis $\mu = 98.5$ against the alternative hypothesis $\mu < 98.5$ at the 0.01 level of significance.

[†] If we cannot assume that the population is symmetrical, we can still use the one-sample sign test, but we have to test the null hypothesis $\tilde{\mu} = \tilde{\mu}_0$, where $\tilde{\mu}$ is the population median.

Solution Since one of the sample values equals 98.5, it must be discarded and the sample size for the one-sample sign test is only $n = 14$.

1. *Null hypothesis*: $\mu = 98.5$
   Alternative hypothesis: $\mu < 98.5$
2. *Level of significance*: $\alpha = 0.01$
3. *Criterion*: The criterion may be based on the number of plus signs or the number of minus signs. Using the number of plus signs, denoted by x, reject the null hypothesis if the probability of getting x or fewer plus signs is less than or equal to 0.01; otherwise, accept the null hypothesis or reserve judgment.
4. *Calculations*: Replacing each value greater than 98.5 with a plus sign and each value less than 98.5 with a minus sign, the 14 sample values yield

$$- \; - \; - \; - \; - \; + \; - \; - \; - \; + \; - \; - \; - \; -$$

Thus, $x = 2$ and Table V shows that for $n = 14$ and $p = 0.50$ the probability of $x \leq 2$ is $0.001 + 0.006 = 0.007$.
5. *Decision*: Since 0.007 is less than 0.01, the null hypothesis must be rejected; we conclude that the mean octane rating of the given kind of gasoline is less than 98.5.

16.2
THE PAIRED-SAMPLE SIGN TEST

The sign test can also be used when we deal with paired data as in Exercises 10.64 and 10.65 on pages 303 and 304. In such problems, each pair of sample values is replaced with a plus sign if the first value is greater than the second, with a minus sign if the first value is smaller than the second, or it is discarded if the two values are equal. Then, we proceed as in Section 16.1.

EXAMPLE To determine the effectiveness of a new traffic control system, the numbers of accidents that occurred at a random sample of ten dangerous intersections during the four weeks before and the four weeks following the installation of the new system were observed and the following data were obtained:

3 and 1,	4 and 2,	2 and 3,	5 and 2,	3 and 3,
2 and 0,	3 and 2,	6 and 3,	1 and 2,	1 and 0

Use the sign test at the 0.05 level of significance to test whether the new traffic control system is only as good as the old system or whether the new system is more effective.

Solution   Since one of the pairs, 3 and 3, has to be discarded, the sample size for the sign test is only $n = 9$.

1. *Null hypothesis*: $\mu_1 = \mu_2$, where μ_1 and μ_2 are the mean numbers of accidents in four weeks at a dangerous intersection with the old and the new control systems.
   Alternative hypothesis: $\mu_1 > \mu_2$
2. *Level of significance*: $\alpha = 0.05$
3. *Criterion*: If x is the number of plus signs, reject the null hypothesis if the probability of getting x or more plus signs is less than or equal to 0.05; otherwise, accept it.
4. *Calculations*: Replacing each pair of values with a plus sign if the first value is greater than the second or with a minus sign if the first value is smaller than the second, the nine unequal sample pairs yield

$$+ \ + \ - \ + \ + \ + \ + \ - \ +$$

   Thus, $x = 7$ and Table V shows that for $n = 9$ and $p = 0.50$ the probability of $x \geq 7$ is $0.070 + 0.018 + 0.002 = 0.090$.
5. *Decision*: Since 0.090 exceeds 0.05, the null hypothesis cannot be rejected; we conclude that the new control system is only as good as the old system.

16.3
THE SIGN TEST (Large Samples)

When np and $n(1 - p)$ are both greater than 5, so that we can use the normal approximation to the binomial distribution, the sign test may be based on the large-sample test of Section 12.3; namely, on the statistic

$$z = \frac{x - np_0}{\sqrt{np_0(1 - p_0)}}$$

with $p_0 = 0.50$, which has approximately the standard normal distribution.

EXAMPLE   The following are the ratings of two supervisors of the performance of 20 employees, a random sample, on the scale from 1 to 30:

Supervisor A	Supervisor B
28	23
25	25
19	12
17	15
20	23
19	19
22	18
30	27
25	22
17	20
23	21
26	26
15	17
20	15
28	24
21	23
28	26
27	22
24	24
25	23

At the 0.05 level of significance, does this support the claim that on the average supervisor A gives higher ratings than supervisor B?

Solution Since four pairs of equal ratings have to be discarded, the sample size for the sign test is only $n = 16$.

1. *Null hypothesis*: $\mu_A = \mu_B$
   Alternative hypothesis: $\mu_A > \mu_B$
2. *Level of significance*: $\alpha = 0.05$
3. *Criterion*: If x is the number of plus signs (positive differences between the ratings of supervisors A and B), reject the null hypothesis if $z > 1.645$, where

$$z = \frac{x - np_0}{\sqrt{np_0(1 - p_0)}}$$

with $p_0 = 0.50$; otherwise, accept it.
4. *Calculations*: Replacing each positive difference with a plus sign and each negative difference with a minus sign, we get

$$+ \; + \; + \; - \; + \; + \; - \; + \; - \; + \; + \; + \; - \; + \; + \; +$$

Thus, $x = 12$ and

$$z = \frac{12 - 16(0.50)}{\sqrt{16(0.50)(0.50)}} = 2.00$$

5. *Decision*: Since $z = 2.00$ exceeds 1.645, the null hypothesis must be rejected; in other words, the data support the claim that on the average supervisor A gives higher performance ratings than supervisor B.

EXERCISES

16.1 On 15 occasions, a random sample, a city employee had to wait 4, 8, 7, 7, 2, 6, 8, 5, 9, 6, 1, 5, 6, 5, and 9 minutes for the bus he takes to work. Use the sign test based on Table V and the 0.05 level of significance to test the null hypothesis $\mu = 5$ (that on the average he has to wait five minutes) against the alternative hypothesis $\mu \neq 5$.

16.2 A random sample of nine women buying new eyeglasses tried on 12, 11, 14, 15, 10, 14, 11, 8, and 12 different frames. Use the sign test at the 0.05 level of significance to test whether $\mu = 10$ (namely, that on the average a woman buying new eyeglasses tries on 10 different frames) or whether $\mu > 10$.

16.3 The following random-sample data are the weights (in grams) of 14 packages of a certain kind of candy: 100.8, 100.0, 102.6, 100.3, 98.2, 101.0, 100.5, 102.5, 100.0, 97.1, 103.6, 100.9, 99.8, and 101.0. Use the sign test based on Table V and the 0.01 level of significance to test the null hypothesis $\mu = 100.0$ (that on the average a package of that kind of candy weighs 100.0 grams) against the alternative hypothesis $\mu \neq 100.0$.

16.4 Playing four rounds of golf at a country club, a random sample of ten professionals totaled 280, 282, 279, 278, 283, 280, 279, 280, 287, and 283. Use the sign test at the 0.05 level of significance to test the null hypothesis $\mu = 284$ (that on the average golf professionals total 284 for four rounds at the golf course) against the alternative hypothesis $\mu < 284$.

16.5 The following are the grades which 15 students received on the midterm and final examination in a course in European history: 66 and 73, 88 and 91, 75 and 78, 90 and 86, 63 and 69, 58 and 67, 75 and 75, 82 and 80, 73 and 76, 84 and 89, 85 and 81, 93 and 96, 70 and 76, 85 and 82, and 90 and 97. Use the sign test based on Table V and the 0.05 level of significance to test the null hypothesis $\mu_1 = \mu_2$ (that on the average students score equally well on the midterm and final examinations) against the alternative hypothesis $\mu_1 < \mu_2$.

16.6 The following are the numbers of passengers carried on flights No. 136 and No. 137 between Chicago and Phoenix on 12 days:

232 and 189, 265 and 230, 249 and 236, 250 and 261,

255 and 249, 236 and 218, 270 and 258, 266 and 253,

249 and 251, 240 and 233, 257 and 254, 239 and 249

Use the sign test based on Table V and the 0.01 level of significance to test the null hypothesis $\mu_1 = \mu_2$ (that on the average the two flights carry equally many passengers) against the alternative hypothesis $\mu_1 > \mu_2$.

16.7 Use the sign test based on Table V to rework Exercise 10.65 on page 304.

16.8 The following are the miles per gallon obtained with 40 tankfuls of a certain kind of gas

24.1	25.0	24.8	24.3	24.2	25.3	24.2	23.6	24.5	24.4
24.5	23.2	24.0	23.8	23.8	25.3	24.5	24.6	24.0	25.2
25.2	24.4	24.7	24.1	24.6	24.9	24.1	25.8	24.2	24.2
24.8	24.1	25.6	24.5	25.1	24.6	24.3	25.2	24.7	23.3

Use the sign test at the 0.01 level of significance to test the null hypothesis $\mu = 24.2$ (that on the average the gasoline yields 24.2 miles per gallon) against the alternative hypothesis $\mu > 24.2$.

16.9 Use the normal approximation to the binomial distribution to rework Exercise 16.3.

16.10 Use the normal approximation to the binomial distribution to rework Exercise 16.6.

16.11 The following are the numbers of employees absent from two departments of a large firm on 25 days: 4 and 3, 2 and 5, 6 and 6, 3 and 6, 1 and 4, 2 and 4, 5 and 2, 1 and 4, 3 and 4, 6 and 5, 2 and 5, 7 and 1, 4 and 6, 1 and 3, 2 and 5, 0 and 3, 6 and 5, 4 and 6, 1 and 2, 4 and 1, 2 and 4, 0 and 1, 5 and 3, 2 and 3, and 2 and 4. Use the sign test at the 0.05 level of significance to test the null hypothesis $\mu_1 = \mu_2$ (that on the average equally many employees are absent from the two departments) against the alternative hypothesis $\mu_1 < \mu_2$.

16.4
RANK SUMS: THE U TEST

In this section we shall present a nonparametric alternative to the two-sample t test for the difference between two means. It is called the U test, the **Wilcoxon test**, or the **Mann–Whitney test**, named after the statisticians who contributed to its development. With this test we will be able to test the null hypothesis that the two samples come from identical populations without having to assume that the populations sampled have roughly the shape of normal distributions; in fact, the test requires only that the populations be continuous (to avoid ties), and in practice it does not matter whether this assumption is satisfied.

To illustrate how the U test is performed, suppose that we want to compare the grain size of sand obtained from two different locations on the moon on the basis of the following diameters (in millimeters):

Location 1:   0.37, 0.70, 0.75, 0.30, 0.45, 0.16, 0.62, 0.73, 0.33

Location 2:   0.86, 0.55, 0.80, 0.42, 0.97, 0.84, 0.24, 0.51, 0.92, 0.69

The means of these two random samples are 0.49 and 0.68, and their difference seems large, but we shall have to see whether it is significant.

We begin the U test by arranging the data jointly, as if they comprise one sample, in an increasing order of magnitude. For our data we get

0.16	0.24	0.30	0.33	0.37	0.42	0.45	0.51	0.55	0.62
1	2	1	1	1	2	1	2	2	1

0.69	0.70	0.73	0.75	0.80	0.84	0.86	0.92	0.97
2	1	1	1	2	2	2	2	2

where we indicated for each value whether it came from location 1 or location 2. Assigning the data, in this order, the ranks $1, 2, 3, \ldots$, and 19, we find that the values of the first sample (location 1) occupy ranks 1, 3, 4, 5, 7, 10, 12, 13, and 14, while those of the second sample (location 2) occupy ranks, 2, 6, 8, 9, 11, 15, 16, 17, 18, and 19. There are no ties here, but if there were, we would assign each of the tied observations the mean of the ranks which they jointly occupy. For instance, if the third and fourth values were the same, we would assign each the rank $\frac{3+4}{2} = 3.5$, and if the ninth, tenth, and eleventh values were the same, we would assign each the rank $\frac{9+10+11}{3} = 10$.

Now, if there is an appreciable difference between the means of the two populations, most of the lower ranks are likely to go to the values of one sample while most of the higher ranks are likely to go to the values of the other sample. The test of the null hypothesis that the two samples come from identical populations may thus be based on W_1, the sum of the ranks of the values of the first sample, or on W_2, the sum of the ranks of the values of the second sample. In practice, it does not matter which sample we refer to as sample 1 and which sample we refer to as sample 2, and whether we base the test on W_1 or W_2. (When the sample sizes are unequal, we usually let the smaller of the two samples be sample 1; however, this is not required for the work in this book.) If the sample sizes are n_1 and n_2, the sum of W_1 and W_2 is simply the sum of the first $n_1 + n_2$ positive integers, which is known to be

$$\frac{(n_1 + n_2)(n_1 + n_2 + 1)}{2}$$

This formula enables us to find W_2 if we know W_1, and vice versa. For our illustration we get

$$W_1 = 1 + 3 + 4 + 5 + 7 + 10 + 12 + 13 + 14$$

$$= 69$$

and since the sum of the first 19 positive integers is $\dfrac{19 \cdot 20}{2} = 190$, it follows that $W_2 = 190 - 69 = 121$. (This value is the sum of the ranks 2, 6, 8, 9, 11, 15, 16, 17, 18, and 19.)

When the use of **rank sums** was first proposed as a nonparametric alternative to the two-sample t test, the decision was based on W_1 or W_2, but now the decision is based on either of the related statistics

<table>
<tr><td rowspan="3" style="vertical-align: middle;">U₁ and U₂
statistics</td></tr>
</table>

U₁ and U₂ statistics

$$U_1 = n_1 n_2 + \frac{n_1(n_1 + 1)}{2} - W_1$$

or

$$U_2 = n_1 n_2 + \frac{n_2(n_2 + 1)}{2} - W_2$$

or on the statistic U, which always equals the smaller of the two. The resulting tests are equivalent to those based on W_1 or W_2, but they have the advantage that they lend themselves more readily to the construction of tables of critical values. Not only do U_1 and U_2 take on values on the interval from 0 to $n_1 n_2$—indeed, their sum is always equal to $n_1 n_2$—but their sampling distributions are symmetrical about $\dfrac{n_1 n_2}{2}$.

When the alternative hypothesis is $\mu_1 \neq \mu_2$, we base the test on the statistic U, which has the advantage that the resulting test is one-tailed, and hence easier to tabulate. Accordingly,

> We reject the null hypothesis that the two samples come from identical populations and accept the alternative hypothesis $\mu_1 \neq \mu_2$, if the value we obtain for U is less than or equal to the critical value in Table VI for $\alpha = 0.05$ or $\alpha = 0.01$.

In the construction of Table VI, the critical values are the largest values of U for which the probability of getting a value less than or equal to them is less than or equal to α; the blank spaces in the table indicate that the null hypothesis cannot be rejected regardless of the value we obtain for U. More extensive tables may be found in handbooks of statistical tables, but when n_1 and n_2 are both greater than 8, we can use instead the large-sample test described in Section 16.5.

EXAMPLE   With reference to the grain-size data on page 465, use the U test at the 0.05 level of significance to test the null hypothesis that the two samples come from identical populations against the alternative hypothesis that the two populations have unequal means.

Solution

1. *Null hypothesis*: Populations are identical
   Alternative hypothesis: $\mu_1 \neq \mu_2$
2. *Level of significance*: $\alpha = 0.05$
3. *Criterion*: Reject the null hypothesis if $U \leq 20$, which is the critical value in Table VI for $n_1 = 9$, $n_2 = 10$, and $\alpha = 0.05$; otherwise, reserve judgment.
4. *Calculations*: Having already shown that $W_1 = 69$ and $W_2 = 121$, we get

$$U_1 = 9 \cdot 10 + \frac{9 \cdot 10}{2} - 69 = 66$$

$$U_2 = 9 \cdot 10 + \frac{10 \cdot 11}{2} - 121 = 24$$

and, hence, $U = 24$. Note that $U_1 + U_2 = 66 + 24 = 90$, which equals $n_1 n_2 = 9 \cdot 10$.

5. *Decision*: Since $U = 24$ is greater than 20, the null hypothesis cannot be rejected; in other words, we cannot conclude that there is a difference in the mean grain size of sand from the two locations on the moon.

When the alternative hypothesis is $\mu_1 > \mu_2$ or $\mu_1 < \mu_2$, we base the test on U_1 or U_2 instead of U, so that in either case the null hypothesis is rejected for values in the left-hand tail of the sampling distribution of the test statistic. Accordingly,

> **We reject the null hypothesis that the two samples come from identical populations and accept the alternative hypothesis $\mu_1 > \mu_2$, if the value we obtain for U_1 is less than or equal to the critical value in Table VI for $\alpha = 0.05$ or $\alpha = 0.01$; when the alternative hypothesis is $\mu_1 < \mu_2$, we reject the null hypothesis if the value we obtain for U_2 is less than or equal to the critical value in Table VI.**

EXAMPLE

The following are the burning times (rounded to the nearest tenth of a minute) of random samples of two kinds of emergency flares:

Brand 1: 17.2, 18.1, 19.3, 21.1, 14.4, 13.7, 18.8, 15.2, 20.3, 17.5

Brand 2: 13.6, 19.1, 11.8, 14.6, 14.3, 22.5, 12.3, 13.5, 10.9, 14.8

Use the U test at the 0.05 level of significance to test whether it is reasonable to say that on the average brand 1 flares are better (last longer) than brand 2 flares.

Solution

1. *Null hypothesis*: Populations are identical
   Alternative hypothesis: $\mu_1 > \mu_2$
2. *Level of significance*: $\alpha = 0.05$
3. *Criterion*: Reject the null hypothesis if $U_1 \leq 27$, which is the critical value in Table VI for $n_1 = 10$, $n_2 = 30$, and $\alpha = 0.05$; otherwise, reserve judgment.
4. *Calculations*: Ranking the data jointly according to size, we find that the values of the first sample (brand 1) occupy ranks 12, 14, 17, 19, 8, 6, 15, 11, 18, and 13, so that

$$W_1 = 12 + 14 + 17 + 19 + 8 + 6 + 15 + 11 + 18 + 13$$
$$= 133$$

and $U_1 = 10 \cdot 10 + \dfrac{10 \cdot 11}{2} - 133 = 22.$

5. *Decision*: Since $U_1 = 22$ is less than 27, the null hypothesis must be rejected; we conclude that brand 1 flares are, indeed, better than brand 2 flares.

When there is no difference between the means of two populations, the U test can also be used to test the null hypothesis that two samples come from identical populations against the alternative that the two populations have unequal dispersions. The test is performed in the same way as before, except for a modification in the way in which the data are ranked. Again, the data are arranged in an increasing order of magnitude, but now they are ranked from both ends toward the middle. We assign rank 1 to the smallest value, ranks 2 and 3 to the largest and second largest, ranks 4 and 5 to the second and third smallest, ranks 6 and 7 to the third and fourth largest, and so on. Subsequently, the test is performed exactly as before. With the new kind of ranking, however, a small rank sum for one sample suggests that the population from which it came has a greater variation than the other population since the sample values occupy the more extreme positions. We shall not illustrate this technique with reference to the above example, because this procedure loses its sensitivity for detecting differences in variability when the population means are not equal.

16.5
RANK SUMS: THE U TEST (Large Samples)

The large-sample U test may be based on either U_1 or U_2 as defined in the preceding section, but since the resulting tests are equivalent and it does not matter how we number the samples, we shall use here the statistic U_1.

Under the null hypothesis that the two samples come from identical populations, it can be shown that the mean and the standard deviation of the sampling distribution of U_1 are[†]

<table>
<tr><td rowspan="2">Mean and standard deviation of U_1 statistic</td><td>

$$\mu_{U_1} = \frac{n_1 n_2}{2}$$

and

$$\sigma_{U_1} = \sqrt{\frac{n_1 n_2 (n_1 + n_2 + 1)}{12}}$$

</td></tr>
</table>

Furthermore, if n_1 and n_2 are both greater than 8, the sampling distribution of U_1 can be approximated closely with a normal curve. Thus, we base the test of the null hypothesis that the two samples come from identical populations on the statistic

<table>
<tr><td>Statistic for large-sample U test</td><td>

$$z = \frac{U_1 - \mu_{U_1}}{\sigma_{U_1}}$$

</td></tr>
</table>

which has approximately the standard normal distribution. When the alternative hypothesis is $\mu_1 \neq \mu_2$, we reject the null hypothesis if $z < -z_{\alpha/2}$ or $z > z_{\alpha/2}$; when the alternative hypothesis is $\mu_1 > \mu_2$, we reject the null hypothesis if $z < -z_\alpha$ since small values of U_1 correspond to large values of W_1; and when the alternative hypothesis is $\mu_1 < \mu_2$, we reject the null hypothesis if $z > z_\alpha$ since large values of U_1 correspond to small values of W_1.

EXAMPLE The following are the weight gains (in pounds) of two random samples of young turkeys fed two different diets but otherwise kept under identical conditions:

Diet 1: 16.3, 10.1, 10.7, 13.5, 14.9, 11.8, 14.3, 10.2, 12.0, 14.7, 23.6, 15.1, 14.5, 18.4, 13.2, 14.0

Diet 2: 21.3, 23.8, 15.4, 19.6, 12.0, 13.9, 18.8, 19.2, 15.3, 20.1, 14.8, 18.9, 20.7, 21.1, 15.8, 16.2

Use the large-sample U test at the 0.01 level of significance to test the null hypothesis that the two populations sampled are identical

---

[†] If there are ties in rank, these formulas provide only approximations, but if the number of ties is small, there is usually no need to make any corrections.

against the alternative hypothesis that on the average the second diet produces a greater gain in weight.

Solution 1. *Null hypothesis*: Populations are identical
   Alternative hypothesis: $\mu_1 < \mu_2$
2. *Level of significance*: $\alpha = 0.01$
3. *Criterion*: Reject the null hypothesis if $z > 2.33$, where

$$z = \frac{U_1 - \mu_{U_1}}{\sigma_{U_1}}$$

and otherwise accept the null hypothesis or reserve judgment.
4. *Calculations*: Ranking the data jointly according to size, we find that the values of the first sample occupy ranks 21, 1, 3, 8, 15, 4, 11, 2, 5.5, 13, 31, 16, 12, 22, 7, and 10. (The 5th and 6th values are both equal to 12.0, so we assign each the rank 5.5.) Thus,

$$W_1 = 1 + 2 + 3 + 4 + 5.5 + 7 + 8 + 10 + 11 + 12 + 13$$
$$+ 15 + 16 + 21 + 22 + 31$$
$$= 181.5$$

and

$$U_1 = 16 \cdot 16 + \frac{16 \cdot 17}{2} - 181.5$$
$$= 210.5$$

Since $\mu_{U_1} = \dfrac{16 \cdot 16}{2} = 128$ and $\sigma_{U_1} = \sqrt{\dfrac{16 \cdot 16 \cdot 33}{12}} = 26.53$, it follows that

$$z = \frac{210.5 - 128}{26.53} = 3.11$$

5. *Decision*: Since $z = 3.11$ exceeds 2.33, the null hypothesis must be rejected; we conclude that on the average the second diet produces a greater gain in weight.

EXERCISES   16.12 The following are the scores which random samples of students from two minority groups obtained on a current events test:

Minority group 1: 73, 82, 39, 68, 91, 75, 89, 67, 50, 86, 57, 65

Minority group 2: 51, 42, 36, 53, 88, 59, 49, 66, 25, 64, 18, 76

Use the U test based on Table VI and the 0.05 level of significance to test whether or not students from the two minority groups can be expected to score equally well on this test.

16.13 The following are the numbers of mistakes counted on pages randomly selected from reports typed by a firm's two secretaries:

$$Secretary\ X:\quad 5, 13, 9, 9, 4, 10$$

$$Secretary\ Y:\quad 3, 5, 6, 4, 1, 8$$

Use the U test at the 0.01 level of significance to test whether or not the two secretaries average equally many mistakes per page.

16.14 The following are the numbers of minutes it took random samples of 15 men and 12 women to complete a written test given for the renewal of their driver's licenses:

$$Men:\quad 9.9, 7.4, 8.9, 9.1, 7.7, 9.7, 11.8, 9.2, 10.0, 10.2, 9.5, 10.8, 8.0, 11.0, 7.5$$

$$Women:\quad 8.6, 10.9, 9.8, 10.7, 9.4, 10.3, 7.3, 11.5, 7.6, 9.3, 8.8, 9.6$$

Use the U test based on Table VI at the 0.05 level of significance to decide whether to accept the null hypothesis $\mu_1 = \mu_2$ or the alternative hypothesis $\mu_1 \neq \mu_2$, where μ_1 and μ_2 are the average amounts of time it takes men and women to complete the test.

16.15 The following are figures on the numbers of burglaries committed in a city in random samples of six days in the spring and six days in the fall:

$$Spring:\quad 36, 25, 32, 38, 28, 35$$

$$Fall:\quad 27, 20, 15, 29, 18, 22$$

Use the U test at the 0.01 level of significance to test the claim that on the average there are equally many burglaries per day in the spring as in the fall against the alternative that there are fewer in the fall.

16.16 The following are the Rockwell hardness numbers obtained for six aluminum die castings randomly selected from production lot A and eight from production lot B:

$$Production\ lot\ A:\quad 75, 56, 63, 70, 58, 74$$

$$Production\ lot\ B:\quad 63, 85, 77, 80, 86, 76, 72, 82$$

Use the U test at the 0.05 level of significance to test whether the castings of production lot B are on the average harder than those of production lot A.

16.17 The following are the weekly food expenditures of families with two children chosen at random from two suburbs of a large city:

$$Suburb\ F:\quad 76.19, 84.00, 79.89, 82.78, 98.45, 87.38, 82.50, 90.60$$

$$Suburb\ G;\quad 75.76, 78.51, 87.91, 75.12, 90.19, 67.36, 87.16$$

Use the U test at the 0.01 level of significance to test the claim that on the average such weekly food expenditures are higher in suburb F than in suburb G.

16.18 Use the large-sample U test to rework the example on page 465, which deals with the grain size of sand from two locations on the moon.

16.19 Use the large-sample U test to rework the example on page 468, which deals with the burning times of two kinds of emergency flares.

16.20 An examination designed to measure basic knowledge of American history was given to random samples of freshmen at two major universities, and their grades were

University A: 77, 72, 58, 92, 87, 93, 97, 91, 70, 98,
76, 90, 62, 69, 90, 78, 96, 84, 73, 80

University B: 89, 74, 45, 56, 71, 74, 94, 88, 66, 62,
88, 63, 88, 37, 63, 75, 78, 34, 75, 68

Apply the U test at the 0.05 level of significance to test the null hypothesis that there is no difference in the average knowledge of American history between freshmen entering the two universities.

16.21 The following are data on the breaking strength (in pounds) of random samples of two kinds of 2-inch cotton ribbons:

Type I ribbon: 144, 181, 200, 187, 169, 171, 186, 194,
176, 182, 133, 183, 197, 165, 180, 198

Type II ribbon: 175, 164, 172, 194, 176, 198, 154, 134,
169, 164, 185, 159, 161, 189, 170, 164

Use the U test at the 0.05 level of significance to test the claim that Type I ribbon is on the average stronger than Type II ribbon.

16.6
RANK SUMS: THE SIGNED-RANK TEST ★

The sign test is very easy to perform and has intuitive appeal, but it is wasteful of information because it utilizes only the signs of the differences between the observations and μ_0 in the one-sample case, or the signs of the differences between the pairs of observations in the paired-sample case. It is for this reason that an alternative nonparametric test, the **signed-rank test** also called the **Wilcoxon signed-rank test,** is often preferred.

In this test, we rank the differences without regard to their signs, assigning rank 1 to the smallest numerical difference (that is, to the smallest difference in absolute value), rank 2 to the second smallest numerical difference,..., and rank n to the largest numerical difference. Zero differences are again discarded, and if two or more differences are numerically equal, we assign each one the mean of the ranks which they jointly occupy.

To illustrate this procedure, let us refer to the example of Section 16.1, where we were given 15 measurements of the octane rating of a certain kind of gasoline and used the one-sample sign test to test the null hypothesis $\mu = 98.5$ against the alternative hypothesis $\mu < 98.5$ at the 0.01 level of significance. These measurements are shown in the first column of the following table:

Measurements	Differences	Ranks
97.5	-1.0	4
95.2	-3.3	12
97.3	-1.2	6
96.0	-2.5	10
96.8	-1.7	7
100.3	1.8	8
97.4	-1.1	5
95.3	-3.2	11
93.2	-5.3	14
99.1	0.6	2
96.1	-2.4	9
97.6	-0.9	3
98.2	-0.3	1
98.5	0.0	
94.9	-3.6	13

The differences between the measurements and $\mu = 98.5$ are given in the second column, and since the smallest numerical difference is 0.3 (after we discarded to zero difference), the next smallest is 0.6, ..., and the largest is 5.3, the ranks are as shown in the third column.

If the null hypothesis $\mu = \mu_0$ (or the null hypothesis $\mu_1 = \mu_2$ in the paired-sample case) is true, the sum of the ranks corresponding to the positive differences should be about equal to the sum of the ranks corresponding to the negative differences. Indeed, we can base the signed-rank test on T^+, the sum of the ranks corresponding to the positive differences, T^-, the sum of the ranks corresponding to the negative differences, or T, which is always the smaller of the two.

When the alternative hypothesis is $\mu \neq \mu_0$ (or $\mu_1 \neq \mu_2$ in the paired-sample case), we base the test on the statistic T, which has the advantage that the resulting test is one-tailed, and hence easier to tabulate. Accordingly,

> We reject the null hypothesis $\mu = \mu_0$ (or $\mu_1 = \mu_2$) and accept the alternative hypothesis $\mu \neq \mu_0$ (or $\mu_1 \neq \mu_2$), if the value we obtain for T is less than or equal to the critical value in Table VII for $\alpha = 0.05$ or $\alpha = 0.01$.

In the construction of Table VII, the critical values are the largest values of T for which the probability of getting a value less than or equal to them is less than or equal to α; the blank spaces in the table indicate that the null hypothesis cannot be rejected regardless of the value we obtain for T.

More extensive tables may be found in handbooks of statistical tables, but when n is 15 or more we can use the large-sample test described in Section 16.7.

EXAMPLE In a random sample taken at a public playground, it took 36, 29, 44, 28, 40, 50, 39, 47, and 33 minutes to play a set of tennis. Use the signed-rank test at the 0.05 level of significance to test whether or not it takes on the average 35 minutes to play a set of tennis at the public playground.

Solution
1. *Null hypothesis*: $\mu = 35$
   Alternative hypothesis: $\mu \neq 35$
2. *Level of significance*: $\alpha = 0.05$
3. *Criterion*: Reject the null hypothesis if $T \leq 6$, which is the critical value in Table VII for $n = 9$ and $\alpha = 0.05$; otherwise, accept it.
4. *Calculations*: Subtracting 35 from each value we get 1, -6, 9, -7, 5, 15, 4, 12, and -2, and the ranks of the magnitudes of these differences are 1, 5, 7, 6, 4, 9, 3, 8, and 2; thus, $T^- = 5 + 6 + 2 = 13$, $T^+ = 1 + 7 + 4 + 9 + 3 + 8 = 32$, and $T = 13$.
5. *Decision*: Since $T = 13$ is greater than 6, the null hypothesis cannot be rejected; the data tend to support the 35 minutes average.

When the alternative hypothesis is one-sided, we base the signed-rank test on T^+ or T^-, and modify the procedure as we did in connection with the U test in Section 16.4. Specifically, when the alternative hypothesis is $\mu > \mu_0$ (or $\mu_1 > \mu_2$), we base the test on the statistic T^- and reject the null hypothesis if the value we obtain for T^- is less than or equal to the critical value in Table VII; when the alternative hypothesis is $\mu < \mu_0$ (or $\mu_1 < \mu_2$), we base the test on the statistic T^+ and reject the null hypothesis if the value we obtain for T^+ is less than or equal to the critical value in Table VII. In Exercise 16.25 on page 479 the reader will be asked to apply this technique to the octane ratings on page 460.

16.7
RANK SUMS: THE SIGNED-RANK TEST
(Large Samples) ★

The large-sample signed-rank test may be based on either T^+ or T^- as defined in the preceding section, but since $T^+ + T^-$ always equals $1 + 2 + \cdots + n = \dfrac{n(n + 1)}{2}$, the resulting tests are equivalent and it does not matter which one we choose. In what follows, we shall base the tests on the statistic T^+.

Under the assumption that each difference is as likely to be positive as negative, it can be shown that the mean and the standard deviation of the sampling distribution of T^+ are

Mean and standard deviation of T^+ statistic

$$\mu_{T^+} = \frac{n(n+1)}{4}$$

and

$$\sigma_{T^+} = \sqrt{\frac{n(n+1)(2n+1)}{24}}$$

Furthermore, if $n \geq 15$, the sampling distribution of T^+ can be approximated closely with a normal curve. Thus, we base the large-sample signed-rank test on the statistic

Statistic for large-sample signed-rank test

$$z = \frac{T^+ - \mu_{T^+}}{\sigma_{T^+}}$$

which has approximately the standard normal distribution. When the alternative hypothesis is $\mu \neq \mu_0$ (or $\mu_1 \neq \mu_2$), we reject the null hypothesis if $z < -z_{\alpha/2}$ or $z > z_{\alpha/2}$; when the alternative hypothesis is $\mu > \mu_0$ (or $\mu_1 > \mu_2$), we reject the null hypothesis if $z > z_\alpha$; and when the alternative hypothesis is $\mu < \mu_0$ (or $\mu_1 < \mu_2$), we reject the null hypothesis if $z < -z_\alpha$.

EXAMPLE The following are the weights in pounds, before and after, of 16 persons who stayed on a certain weight-reducing diet for two weeks: 169.0 and 159.9, 188.6 and 181.3, 222.1 and 209.0, 160.1 and 162.3, 187.5 and 183.5, 202.5 and 197.6, 167.8 and 171.4, 214.3 and 202.1, 143.8 and 145.1, 198.2 and 185.5, 166.9 and 158.6, 142.9 and 145.4, 160.5 and 159.5, 198.7 and 190.6, 149.7 and 149.0, and 181.6 and 183.1. Use the large-sample signed-rank test at the 0.05 level of significance to test whether the weight-reducing diet is effective.

Solution
1. *Null hypothesis*: $\mu_1 = \mu_2$
   Alternative hypothesis: $\mu_1 > \mu_2$
2. *Level of significance*: $\alpha = 0.05$
3. *Criterion*: Reject the null hypothesis if $z > 1.645$, where

$$z = \frac{T^+ - \mu_{T^+}}{\sigma_{T^+}}$$

and otherwise reserve judgment.

4. *Calculations*: The differences and their ranks are shown in the following table

Before	After	Differences	Ranks
169.0	159.9	9.1	13
188.6	181.3	7.3	10
222.1	209.0	13.1	16
160.1	162.3	−2.2	5
187.5	183.5	4.0	8
202.5	197.6	4.9	9
167.8	171.4	−3.6	7
214.3	202.1	12.2	14
143.8	145.1	−1.3	3
198.2	185.5	12.7	15
166.9	158.6	8.3	12
142.9	145.4	−2.5	6
160.5	159.5	1.0	2
198.7	190.6	8.1	11
149.7	149.0	0.7	1
181.6	183.1	−1.5	4

and it can be seen that

$$T^+ = 13 + 10 + 16 + 8 + 9 + 14 + 15 + 12 + 2 + 11 + 1$$
$$= 111$$

Since $\mu_{T^+} = \dfrac{16 \cdot 17}{4} = 68$ and $\sigma_{T^+} = \sqrt{\dfrac{16 \cdot 17 \cdot 33}{24}} = 19.34$, we get

$$z = \frac{111 - 68}{19.34} = 2.22$$

5. *Decision*: Since $z = 2.22$ is greater than 1.645, the null hypothesis must be rejected; we conclude that the weight-reducing diet is effective.

16.8
RANK SUMS: THE *H* TEST ★

The **H test,** or **Kruskal–Wallis test,** is a rank-sum test which serves to test the null hypothesis that k independent random samples come from identical populations against the alternative hypothesis that the means of these populations are not all equal. Unlike the standard test which it replaces, the one-way analysis of variance of Section 13.3, it does not require the assumption that the populations sampled have, at least approximately, normal distributions.

As in the U test, the data are ranked jointly from low to high as though they constitute a single sample. Then, if R_i is the sum of the ranks assigned to the n_i values of the ith sample and $n = n_1 + n_2 + \cdots + n_k$, the H test is based on the statistic

Statistic for H test

$$H = \frac{12}{n(n + 1)} \sum_{i=1}^{k} \frac{R_i^2}{n_i} - 3(n + 1)$$

If the null hypothesis is true and each sample has at least five observations, the sampling distribution of H can be approximated closely with a chi-square distribution with $k - 1$ degrees of freedom. Consequently, we reject the null hypothesis that the populations sampled are identical and accept the alternative hypothesis that the means of these populations are not all equal, if the value we get for H exceeds χ_α^2 for $k - 1$ degrees of freedom.

EXAMPLE Students are randomly assigned to groups which are taught German by three different methods—classroom instruction and language laboratory, only classroom instruction, and only self-study in language laboratory—and the following are the final examination scores of samples of students from the three groups:

Method 1: 94, 88, 91, 74, 86, 97

Method 2: 85, 82, 79, 84, 61, 72, 80

Method 3: 89, 67, 72, 76, 69

Use the H test at the 0.05 level of significance to test the null hypothesis that the populations sampled are identical against the alternative hypothesis that their means are not all equal.

Solution
1. *Null hypothesis*: Populations are identical
   Alternative hypothesis: Populations means are not all equal
2. *Level of significance*: $\alpha = 0.05$
3. *Criterion*: Reject the null hypothesis if $H > 5.991$, which is the value of $\chi_{0.05}^2$ for $3 - 1 = 2$ degrees of freedom; otherwise, reserve judgment.
4. *Calculations*: Arranging the data jointly according to size, we get 61, 67, 69, 72, 72, 74, 76, 79, 80, 82, 84, 85, 86, 88, 89, 91, 94,

and 97. Assigning the data, in this order, the ranks 1, 2, 3, ..., and 18, we find that

$$R_1 = 6 + 13 + 15 + 16 + 17 + 18 = 84$$

$$R_2 = 1 + 4.5 + 8 + 9 + 10 + 11 + 12 = 55.5$$

$$R_3 = 2 + 3 + 4.5 + 7 + 15 = 31.5$$

and it follows that

$$H = \frac{12}{18 \cdot 19} \left(\frac{84^2}{6} + \frac{55.5^2}{7} + \frac{31.5^2}{5} \right) - 3 \cdot 19$$

$$= 6.67$$

5. *Decision*: Since $H = 6.67$ exceeds 5.991, the null hypothesis must be rejected; we conclude that the three methods of instruction are not all equally effective.

EXERCISES

★16.22 In a random sample of ten issues, a newspaper listed 32, 27, 41, 36, 52, 31, 22, 38, 45, and 34 apartments for rent. Use the signed-rank test at the 0.05 level of significance to test the null hypothesis $\mu = 40$ against the alternative hypothesis $\mu \neq 40$.

★16.23 Use the signed-rank test to rework Exercise 16.3 on page 464.

★16.24 The following are the amounts of money (in dollars) which a sample of twelve persons spent at a cafeteria: 3.56, 5.12, 4.85, 4.22, 3.26, 4.73, 5.55, 4.91, 3.67, 3.95, 4.80, and 4.52. Use the signed-rank test at the 0.01 level of significance to test whether the mean of the population sampled is $\mu = 4.75$ or $\mu < 4.75$.

★16.25 With reference to the octane ratings in the table on page 460, use the signed-rank test at the 0.01 level of significance to test the null hypothesis $\mu = 98.5$ against the alternative hypothesis $\mu < 98.5$.

★16.26 In a random sample of ten summer days, two cities in Arizona reported high temperatures of 108 and 105, 104 and 109, 109 and 110, 116 and 109, 111 and 106, 113 and 112, 114 and 108, 112 and 107, 102 and 104, and 112 and 108 degrees Fahrenheit. Use the signed-rank test at the 0.05 level of significance to test whether or not in the summer the mean daily high temperature in the two cities is the same.

★16.27 The following are the numbers of 3-month and 6-month certificates of deposit which a major bank sold on a sample of 16 business days: 35 and 30, 27 and 17, 24 and 26, 44 and 30, 29 and 28, 30 and 23, 28 and 22, 36 and 24, 51 and 31, 33 and 43, 42 and 34, 41 and 25, 18 and 33, 29 and 26, 33 and 22, and 37 and 32. Use the signed-rank test based on Table VII to test at the 0.05 level of significance whether the bank averages equally many 3-month and 6-month certificates per day.

★16.28 Use the signed-rank test based on Table VII to rework Exercise 16.6 on page 464.

★16.29 Use the signed-rank test based on Table VII to rework Exercise 10.64 on page 303.

★16.30 A sample of 24 suitcases carried by an airline on transoceanic flights weighed 32.0, 46.3, 48.1, 27.7, 35.5, 52.6, 66.0, 41.3, 49.9, 36.1, 50.0, 44.7, 48.2, 36.9, 40.8, 35.1, 63.3, 42.5, 52.4, 40.8, 38.6, 43.2, 41.7, and 35.6 pounds. Use the large-sample signed-rank test at the 0.05 level of significance to test whether or not the mean weight of suitcases carried by the airline on such flights is 37.0 pounds.

★16.31 Use the large-sample signed-rank test to rework Exercise 16.25.

★16.32 Use the large-sample signed-rank test to rework Exercise 16.27.

★16.33 The following is a random sample of the IQ's of husbands and wives: 108 and 103, 104 and 116, 103 and 106, 112 and 104, 99 and 99, 105 and 94, 102 and 110, 112 and 128, 119 and 106, 106 and 103, 125 and 120, 96 and 98, 107 and 117, 115 and 130, 101 and 100, 110 and 101, 103 and 96, 105 and 99, 124 and 120, and 113 and 116. Use the large-sample signed-rank test at the 0.05 level of significance to test whether or not husbands and wives are on the average equally intelligent in the population sampled.

★16.34 The following are the miles per gallon which a test driver got in random samples of six tankfuls of each of three kinds of gasoline:

Gasoline 1:   30, 15, 32, 27, 24, 29

Gasoline 2:   17, 28, 20, 33, 32, 22

Gasoline 3:   19, 23, 32, 22, 18, 25

Use the H test at the 0.05 level of significance to test the claim that there is no difference in the true average mileage yield of the three kinds of gasoline.

★16.35 Use the H test to rework Exercise 13.4 on page 377.

★16.36 Use the H test to rework Exercise 13.8 on page 378.

★16.37 To compare four bowling balls, a professional bowler bowls five games with each ball and gets the following scores:

Bowling ball D: 221, 232, 207, 198, 212

Bowling ball E: 202, 225, 252, 218, 226

Bowling ball F: 210, 205, 189, 196, 216

Bowling ball G: 229, 192, 247, 220, 208

Use the H test at the 0.05 level of significance to test the null hypothesis that on the average the bowler performs equally well with the four bowling balls.

★16.38 Three groups of guinea pigs were injected, respectively, with 0.5, 1.0, and 1.5 mg of a tranquilizer, and the following are the numbers of seconds it took them to fall asleep:

0.5-mg dose: 8.2, 10.0, 10.2, 13.7, 14.0, 7.8, 12.7, 10.9

1.0-mg dose: 9.7, 13.1, 11.0, 7.5, 13.3, 12.5, 8.8, 12.9, 7.9, 10.5

1.5-mg dose: 12.0, 7.2, 8.0, 9.4, 11.3, 9.0, 11.5, 8.5

Use the H test at the 0.01 level of significance to test the null hypothesis that the differences in dosage have no effect on the length of time it takes guinea pigs to fall asleep.

16.9
TESTS OF RANDOMNESS: RUNS

All the methods of inference which we have discussed are based on the assumption that the samples are random; yet, there are many applications where it is difficult to decide whether this assumption is justifiable. This is true, particularly, when we have little or no control over the selection of the data, as is the case, for example, when we rely on whatever records are available to make long-range predictions of the weather, when we use whatever data are available to estimate the mortality rate of a disease, or when we use sales records for past months to make predictions of a department store's future sales. None of this information constitutes a random sample in the strict sense.

There are several methods of judging the randomness of a sample on the basis of the order in which the observations are obtained; they enable us to decide, after the data have been collected, whether patterns that look suspiciously nonrandom may be attributed to chance. The technique we shall describe here and in the next two sections is based on the **theory of runs.**

A **run** is a succession of identical letters (or other kinds of symbols) which is followed and preceded by different letters or no letters at all. To illustrate, consider the following arrangement of healthy, H, and diseased, D, elm trees that were planted many years ago along a country road:

$$\underline{H\ H\ H\ H}\ \underline{D\ D\ D}\ \underline{H\ H\ H\ H\ H\ H\ H}\ \underline{D\ D}\ \underline{H\ H}\ \underline{D\ D\ D\ D}$$

Using underlines to combine the letters which constitute the runs, we find that there is first a run of four H's, then a run of three D's, then a run of seven H's, then a run of two D's, then a run of two H's, and finally a run of four D's.

The **total number of runs** appearing in an arrangement of this kind is often a good indication of a possible lack of randomness. If there are too few runs we might suspect a definite grouping or clustering, or perhaps a trend; if there are too many runs, we might suspect some sort of repeated alternating, or cyclical pattern. In the example above there seems to be a definite clustering— the diseased trees seem to come in groups—but it remains to be seen whether this is significant or whether it can be attributed to chance.

If there are n_1 letters of one kind, n_2 letters of another kind, and u runs, we base this kind of decision on the following criterion:

Reject the null hypothesis of randomness if

$$u \leq u'_{\alpha/2} \quad \text{or} \quad u \geq u_{\alpha/2}$$

where $u'_{\alpha/2}$ and $u_{\alpha/2}$ may be read from Table VIII for values of n_1 and n_2 through 15, and $\alpha = 0.05$ and $\alpha = 0.01$.

In the construction of Table VIII, $u'_{\alpha/2}$ is the largest value of u for which the probability of $u \leq u'_{\alpha/2}$ is less than or equal to $\alpha/2$, $u_{\alpha/2}$ is the smallest value of u for which the probability of $u \geq u_{\alpha/2}$ is less than or equal to $\alpha/2$, and the blank spaces indicate that the null hypothesis of randomness cannot be rejected for values in that tail of the sampling distribution of u regardless of the value of u. More extensive tables for the u test may be found in handbooks of statistical tables.

EXAMPLE With reference to the arrangement of healthy and diseased elm trees given on page 481, use the u test at the 0.05 level of significance to test the null hypothesis of randomness against the alternative hypothesis that the arrangement is not random.

Solution
1. *Null hypothesis*: Arrangement is random
   Alternative hypothesis: Arrangement is not random
2. *Level of significance*: $\alpha = 0.05$
3. *Criterion*: Since $n_1 = 13$, $n_2 = 9$, and $\alpha = 0.05$, we get $u'_{0.025} = 6$ and $u_{0.025} = 17$ from Table VIII; thus, the null hypothesis must be rejected if $u \leq 6$ or $u \geq 17$; otherwise, reserve judgment.
4. *Calculations*: $u = 6$ by inspection of the data.
5. *Decision*: Since $u = 6$ is less than or equal to 6, the null hypothesis must be rejected; we conclude that the arrangement of healthy and diseased elm trees is not random. It appears that the diseased trees come in clusters.

16.10
TESTS OF RANDOMNESS: RUNS (Large Samples)

Under the null hypothesis that n_1 letters of one kind and n_2 letters of another kind are arranged at random, it can be shown that the mean and the standard deviation of u, the total number of runs, are

Mean and standard deviation of u

$$\mu_u = \frac{2n_1 n_2}{n_1 + n_2} + 1$$

and

$$\sigma_u = \sqrt{\frac{2n_1 n_2 (2n_1 n_2 - n_1 - n_2)}{(n_1 + n_2)^2 (n_1 + n_2 - 1)}}$$

Furthermore, if neither n_1 nor n_2 is less than 10, the sampling distribution of u can be approximated closely with a normal curve. Thus, we base the test of the null hypothesis of randomness on the statistic

Statistic for large-sample u test

$$z = \frac{u - \mu_u}{\sigma_u}$$

which has approximately the standard normal distribution. If the alternative hypothesis is that the arrangement is not random, we reject the null hypothesis for $z < -z_{\alpha/2}$ or $z > z_{\alpha/2}$; if the alternative hypothesis is that there is a clustering or a trend, we reject the null hypothesis for $z < -z_\alpha$; and if the alternative hypothesis is that there is an alternating, or cyclical, pattern, we reject the null hypothesis for $z > z_\alpha$.

EXAMPLE   The following is an arrangement of men, *M*, and women, *W*, lined up to purchase tickets for a rock concert:

M W M W M M M M W M W M M M M W W M M M M W W M W M

(cont.) *M M W M M M W W W M W M M M W M W M M M M M W W M*

Test for randomness at the 0.05 level of significance.

Solution   1. *Null hypothesis*: Arrangement is random
     Alternative hypothesis: Arrangement is not random
 2. *Level of significance*: $\alpha = 0.05$
 3. *Criterion*: Reject the null hypothesis if $z < -1.96$ or $z > 1.96$, where

$$z = \frac{u - \mu_u}{\sigma_u}$$

Otherwise, accept the null hypothesis or reserve judgment.
 4. *Calculations*: Since $n_1 = 30$, $n_2 = 18$, and $u = 27$, we get

$$\mu_u = \frac{2 \cdot 30 \cdot 18}{30 + 18} + 1 = 23.5$$

$$\sigma_u = \sqrt{\frac{2 \cdot 30 \cdot 18(2 \cdot 30 \cdot 18 - 30 - 18)}{(30 + 18)^2(30 + 18 - 1)}} = 3.21$$

and, hence,

$$z = \frac{27 - 23.5}{3.21} = 1.09$$

5. *Decision*: Since $z = 1.09$ falls between -1.96 and 1.96, the null hypothesis cannot be rejected; in other words, there is no real evidence to suggest that the arrangement is not random.

16.11
TESTS OF RANDOMNESS: RUNS ABOVE AND BELOW THE MEDIAN

The tests of the two preceding sections are not limited to testing the randomness of sequences of attributes, such as the H's and D's, or M's and W's, of our examples. Any sample consisting of numerical measurements or observations can be treated similarly by using the letters a and b to denote values falling above and below the median of the sample. Numbers equal to the median are omitted. The resulting sequence of a's and b's (representing the data in their original order) can then be tested for randomness on the basis of the total number of runs of a's and b's, namely, the total number of **runs above and below the median**. Depending on the size of n_1 and n_2, we use Table VIII or the large-sample test of Section 16.10.

EXAMPLE   On 24 successive runs between two cities, a bus carried 24, 19, 32, 28, 21, 23, 26, 17, 20, 28, 30, 24, 13, 35, 26, 21, 19, 29, 27, 18, 26, 14, 21, and 23 passengers. Use the total number of runs above and below the median to test at the 0.01 level of significance whether it is reasonable to treat these data as if they constitute a random sample.

Solution   The median of the data is 23.5, and we get the following arrangement of values above and below it

$$a \quad b \quad a \quad a \quad b \quad b \quad a \quad b \quad b \quad a \quad a \quad a \quad b \quad a \quad a \quad b \quad b \quad a \quad a \quad b \quad a \quad b \quad b \quad b$$

1. *Null hypothesis*: Arrangement is random
Alternative hypothesis: Arrangement is not random
2. *Level of significance*: $\alpha = 0.01$
3. *Criterion*: Since $n_1 = 12$ and $n_2 = 12$, we get $u'_{0.005} = 6$ and $u_{0.005} = 20$ from Table VIII; thus, the null hypothesis must be rejected if $u \leq 6$ or $u \geq 20$; otherwise, accept it.
4. *Calculations*: $u = 14$ by inspection of the sequence of a's and b's.
5. *Decision*: Since $u = 14$ falls between 6 and 20, the null hypothesis cannot be rejected; in other words, there is no real evidence to suggest that the data may not be treated as if they constitute a random sample.

16.39 The following is the order in which a broker received buy, *B*, and sell, *S*, orders for a certain stock:

B B B B B B B S S B S S S S S B B B B B

Test for randomness at the 0.05 level of significance.

16.40 A driver buys gasoline either at a Texaco station, *T*, or at a Mobile station, *M*, and the following arrangement shows the order of the stations from which she bought gasoline over a certain period of time:

T T T M T M T M M T T M T M T M T M M T M T

Test for randomness at the 0.05 level of significance.

16.41 Test at the 0.01 level of significance whether the following arrangement of defective, *D*, and nondefective, *N*, pieces coming off an assembly line may be regarded as random:

N N N N N N N N D D D D N N N D D N N N

16.42 The following arrangement indicates whether 50 consecutive persons interviewed by a pollster are for, *F*, or against, *A*, an increase in the state gasoline tax to build more roads:

A A A F A F A A A A F F A A F A A A A A F A A F F

(cont.) A A F A A A A F A F F A A A A A F A A F A A A A F

Test for randomness at the 0.05 level of significance.

16.43 Representing each 0, 2, 4, 6, and 8 by the letter *E* (for even) and each 1, 3, 5, 7, and 9 by the letter *O* (for odd), test at the 0.05 level of significance whether the arrangement of the 50 digits in the first column of random numbers on page 517 may be regarded as random.

16.44 To test whether a radio signal contains a message or constitutes random noise, an interval of time is subdivided into a number of very short intervals and for each of these it is determined whether the signal strength exceeds, *E*, or does not exceed, *N*, a certain level of background noise. Test at the 0.05 level of significance whether the following arrangement, thus obtained, may be regarded as random, and hence that the signal contains no message and may be regarded as random noise:

E N N N E N E N N N E E N N N E E N E N N N E E N N

(cont.) N N E E N E N N E N N N E E E N N N E N E N N N N E N

16.45 Flip a coin 60 times and test at the 0.05 level of significance whether the resulting sequence of H's and T's may be regarded as random.

16.46 The following are the numbers of students absent from a large school on 24 consecutive days: 29, 25, 31, 28, 30, 28, 33, 31, 35, 29, 31, 33, 35, 28, 36, 30, 33, 26, 30, 28, 32, 31, 38, and 27. Use the 0.01 level of significance to test whether there is any indication that the data should not be treated as if they were a random sample.

16.47 The following are the numbers of lunches that an insurance agent claimed as business deductions in 30 consecutive months: 6, 7, 5, 6, 8, 6, 8, 6, 6, 4, 3, 2, 4, 4, 3, 4, 7, 5, 6, 8, 6, 6, 3, 4, 2, 5, 4, 4, 3, and 7. Test for randomness at the 0.01 level of significance.

16.48 The total number of retail stores opening for business and also quitting business within the calendar years 1948–1980 in a large city were 108, 103, 109, 107, 125, 142, 147, 122, 116, 153, 144, 162, 143, 126, 145, 129, 134, 137, 143, 150, 148, 152, 125, 106, 112, 139, 132, 122, 138, 148, 155, 146, and 158. Making use of the fact that the median is 138, test at the 0.05 level of significance whether there is a significant trend.

16.12
RANK CORRELATION

Since the significance test for r of Section 15.3 is based on very stringent assumptions, we sometimes use a nonparametric alternative which can be applied under much more general conditions. This test of the null hypothesis of no correlation is based on the **rank-correlation coefficient,** often called **Spearman's rank-correlation coefficient** and denoted by r_S.

To calculate the rank-correlation coefficient for a given set of paired data, we first rank the x's among themselves from low to high or high to low; then we rank the y's in the same way, find the sum of the squares of the differences, d, between the ranks of the x's and the y's, and substitute into the formula

Rank correlation coefficient

$$r_S = 1 - \frac{6(\sum d^2)}{n(n^2 - 1)}$$

where n is the number of pairs of x's and y's. When there are ties in rank, we proceed as before and assign to each of the tied observations the mean of the ranks which they jointly occupy.

EXAMPLE The following are the numbers of hours which ten students studied for an examination and the grades which they received:

Number of hours studied x	Grade in examination y
9	56
5	44
11	79
13	72
10	70
5	54
18	94
15	85
2	33
8	65

Calculate r_S.

Solution   Ranking the x's among themselves from low to high, and also the y's, we get the ranks shown in the first two columns of the following table:

Rank of x	Rank of y	d	d^2
5	4	1.0	1.00
2.5	2	0.5	0.25
7	8	−1.0	1.00
8	7	1.0	1.00
6	6	0.0	0.00
2.5	3	−0.5	0.25
10	10	0.0	0.00
9	9	0.0	0.00
1	1	0.0	0.00
4	5	−1.0	1.00
			4.50

Then, determining the d's and their squares, and substituting $n = 10$ and $\sum d^2 = 4.50$ into the formula for r_S, we get

$$r_S = 1 - \frac{6(4.50)}{10(10^2 - 1)} = 0.97$$

If we calculate r for these data (the original x's and y's), we get 0.96, so that the difference between r and r_S is very small in this case.

As can be seen from this example, r_S is easy to compute manually, and this is why it is sometimes used instead of r when no calculator is available.

When there are no ties, r_S actually equals the correlation coefficient r calculated for the two sets of ranks; when ties exist there may be a small (but usually negligible) difference. Of course, by using ranks instead of the original data we lose some information, but this is usually offset by the rank-correlation coefficient's computational ease.

When we use r_S to test the null hypothesis of no correlation between two variables x and y, we do not have to make any assumptions about the nature of the populations sampled. Under the null hypothesis of no correlation—indeed, the null hypothesis that the x's and y's are randomly matched—the sampling distribution of r_S has the mean 0 and the standard deviation

$$\sigma_{r_S} = \frac{1}{\sqrt{n-1}}$$

Since this sampling distribution can be approximated with a normal distribution even for relatively small values of n, we base the test of the null hypothesis on the statistic

Statistic for testing significance of r_S

$$z = \frac{r_S - 0}{1/\sqrt{n-1}} = r_S\sqrt{n-1}$$

which has approximately the standard normal distribution.

EXAMPLE With reference to the preceding example where we had $n = 10$ and $r_S = 0.97$, test the null hypothesis of no correlation at the 0.01 level of significance.

Solution 1. *Null hypothesis*: $\rho = 0$ (no correlation)
   Alternative hypothesis: $\rho \neq 0$
2. *Level of significance*: $\alpha = 0.01$
3. *Criterion*: Reject the null hypothesis if $z < -2.575$ or $z > 2.575$, where

$$z = r_S\sqrt{n-1}$$

and otherwise reserve judgment.
4. *Calculations*: For $n = 10$ and $r_S = 0.97$, we get

$$z = 0.97\sqrt{10-1} = 2.91$$

5. *Decision*: Since $z = 2.91$ exceeds 2.575, the null hypothesis must be rejected; in other words, there is a relationship between study time and grades in the population sampled.

16.49 Calculate r_S for the following sample data representing the numbers of minutes it took 12 mechanics to assemble a piece of machinery in the morning, x, and in the late afternoon, y:

x	y
10.8	15.1
16.6	16.8
11.1	10.9
10.3	14.2
12.0	13.8
15.1	21.5
13.7	13.2
18.5	21.1
17.3	16.4
14.2	19.3
14.8	17.4
15.3	19.0

16.50 Test whether the value obtained for r_S in the preceding exercise is significant at the 0.05 level of significance.

16.51 Calculate r_S for the data of Exercise 15.3 on page 449.

16.52 If a sample of $n = 18$ pairs of data yielded $r_S = 0.39$, is this rank-correlation coefficient significant at the 0.05 level of significance?

16.53 If a sample of $n = 40$ pairs of data yielded $r_S = 0.48$, is this rank correlation coefficient significant at the 0.01 level of significance?

16.54 The following table shows how a panel of nutrition experts and a panel of heads of household ranked 15 breakfast foods on their palatability:

Breakfast food	Nutrition experts	Heads of household
I	7	5
II	3	4
III	11	8
IV	9	14
V	1	2
VI	4	6
VII	10	12
VIII	8	7
IX	5	1
X	13	9
XI	12	15
XII	2	3
XIII	15	10
XIV	6	11
XV	14	13

Calculate r_S as a measure of the consistency of the two rankings.

16.55 The following are the rankings which three judges gave to the work of ten artists:

Judge A	5	8	4	2	3	1	10	7	9	6
Judge B	3	10	1	4	2	5	6	7	8	9
Judge C	8	5	6	4	10	2	3	1	7	9

Calculate r_S for each pair of rankings and decide
(a) which two judges are most alike in their opinions about these artists;
(b) which two judges differ the most in their opinions about these artists.

16.13
SOME FURTHER CONSIDERATIONS

It is important to remember that nonparametric tests which require no (or virtually no) assumptions about the populations sampled are usually **less efficient** than the standard tests which they replace. To illustrate what we mean here by "less efficient," let us refer to the example on page 255, where we showed that the mean of a random sample of size $n = 128$ is as reliable an estimate of the mean of a symmetrical population as the median of a random sample of size $n = 200$. Thus, the median requires a larger sample, and this is what we mean when we say that it is "less efficient."

To put it another way, nonparametric tests tend to be wasteful of information. It is generally true that the more we are willing to assume, the more we can infer from a sample; but at the same time, the more we assume, the more we limit the applicability of our methods.

16.14
CHECKLIST OF KEY TERMS
(with page references to their definitions)

★H test, 477
★Kruskal–Wallis test, 477
  Mann–Whitney test, 465
  Nonparametric tests, 459
  One-sample sign test, 460
  Paired-sample sign test, 461
  Rank correlation coefficient, 486
  Rank sums, 465
  Run, 481
  Runs above and below the median, 484
  Sign test, 460

16.15
REVIEW EXERCISES

16.56 The following are the prices (in dollars) charged for a certain camera in a random sample of 15 discount stores: 57.25, 58.14, 54.19, 56.17, 57.21, 55.38, 54.75, 57.29, 57.80, 54.50, 55.00, 56.35, 54.26, 60.23, and 53.99. Use the sign test at the 0.05 level of significance to test whether or not the mean price of the camera in the population sampled is $55.00.

★**16.57** Use the signed-rank test to rework the preceding exercise.

16.58 The following sequence shows whether a certain member was present, *P*, or absent, *A*, at twenty consecutive meetings of a fraternal organization:

$$P \quad P \quad P \quad P \quad P \quad P \quad P \quad P \quad A \quad P \quad P \quad P \quad P \quad P \quad P \quad P \quad A \quad A \quad A \quad A$$

At the 0.05 level of significance, is there any real indication of a lack of randomness?

★**16.59** The following are the numbers of minutes that patients had to wait for their appointments with four doctors:

Doctor A:	18, 26, 29, 22, 16
Doctor B:	9, 11, 28, 26, 15
Doctor C:	20, 13, 22, 25, 10
Doctor D:	21, 26, 39, 32, 24

Use the *H* test at the 0.05 level of significance to test the null hypothesis that the four samples come from identical populations against the alternative hypothesis that the means of the populations are not all equal.

16.60 The following are data on the percentage kill of two kinds of insecticides used against mosquitos:

Insecticide X: 41.9, 46.9, 44.6, 43.9, 42.0, 44.0, 41.0, 43.1, 39.0, 45.2, 44.6, 42.0

Insecticide Y: 45.7, 39.8, 42.8, 41.2, 45.0, 40.2, 40.2, 41.7, 37.4, 38.8, 41.7, 38.7

Use the *U* test based on Table VI to test at the 0.05 level of significance whether the two samples come from identical populations or whether the two insecticides are on the average not equally effective.

16.61 Use the large-sample U test to rework the preceding exercise.

16.62 The following are the batting averages, x, and home runs hit, y, by a random sample of fifteen major league baseball players during the first half of the season:

x	y
0.252	12
0.305	6
0.299	4
0.303	15
0.285	2
0.191	2
0.283	16
0.272	6
0.310	8
0.266	10
0.215	0
0.211	3
0.272	14
0.244	6
0.320	7

Calculate the rank-correlation coefficient and test whether it is significant at the 0.01 level of significance.

16.63 The following are the numbers of persons who attended a "singles only" dance on twelve Saturdays: 172, 208, 169, 232, 123, 165, 197, 178, 221, 195, 209, and 182. Use the sign test based on Table V to test at the 0.05 level of significance whether or not the true mean of the population sampled is $\mu = 169$.

★16.64 Use the signed-rank test to rework the preceding exercise.

16.65 The following are the numbers of burglaries committed in two cities on twenty days: 83 and 80, 98 and 87, 115 and 86, 112 and 122, 77 and 102, 103 and 94, 116 and 81, 136 and 96, 156 and 158, 83 and 127, 105 and 104, 117 and 102, 86 and 100, 150 and 108, 119 and 124, 111 and 91, 137 and 103, 160 and 153, 121 and 140, and 143 and 105. Use the large-sample sign test at the 0.05 level of significance to test whether or not on the average there are as many burglaries per day in the two cities.

★16.66 Use the large-sample signed-rank test to rework the preceding exercise.

16.67 The following sequence shows whether a television news program had at least 25 percent of a city's viewing audience, A, or less than 25 percent, L, on 36 consecutive weekday evenings:

$$L \quad L \quad L \quad L \quad A \quad A \quad L \quad L \quad L \quad A \quad L \quad L \quad L \quad A \quad A \quad A \quad A \quad L$$

(cont.) $A \quad L \quad L \quad L \quad A \quad A \quad L \quad L \quad L \quad L \quad L \quad A \quad L \quad L \quad L \quad L \quad L \quad A$

Test for randomness at the 0.05 level of significance.

16.68 If $r_S = 0.27$ for $n = 25$ paired data, is this rank-correlation coefficient significant at the 0.05 level of significance?

16.69 The following are the scores of a random sample of twelve golfers on the first two days of a tournament: 68 and 71, 73 and 76, 70 and 73, 74 and 71, 69 and 72, 72 and 74, 67 and 70, 72 and 68, 71 and 72, 73 and 74, 68 and 69, and 70 and 72. Use the sign test at the 0.05 level of significance to test whether on the average golfers scored equally well on the first two days of the tournament or whether they tended to score lower on the first day.

16.70 Use the sign test to rework Exercise 16.30.

16.71 The following are the closing prices of a commodity (in dollars) on twenty consecutive trading days: 378, 379, 379, 378, 377, 376, 374, 374, 373, 373, 374, 375, 376, 376, 376, 375, 374, 374, 373, and 374. Test for randomness at the 0.01 level of significance.

16.72 The following are the numbers of minutes it took two ambulance services to reach the scenes of accidents:

> *Ambulance service 1*: 9.3, 5.5, 13.1, 10.0, 7.6, 9.2, 11.2, 6.4, 14.0, 10.3
>
> *Ambulance service 2*: 12.7, 6.6, 9.1, 4.5, 7.2, 6.4, 7.5

Assuming that the data constitute random samples, use the U test at the 0.05 level of significance to test the null hypothesis that the two samples come from identical populations against the alternative hypothesis that on the average the first ambulance service is slower.

16.16
REFERENCES

Further information about the nonparametric tests discussed in this chapter and many others may be found in

CONOVER, W. J., *Practical Nonparametric Statistics*. New York: John Wiley & Sons, Inc., 1971.

GIBBONS, J. D., *Nonparametric Statistical Inference*. New York: McGraw-Hill Book Company, 1971.

LEHMANN, E. L., *Nonparametrics: Statistical Methods Based on Ranks*. San Francisco: Holden-Day, Inc., 1975.

MOSTELLER, F., and ROURKE, R. E. K., *Sturdy Statistics, Nonparametrics and Order Statistics*. Reading, Mass.: Addison-Wesley Publishing Company, Inc., 1973.

NOETHER, G. E., *Elements of Nonparametric Statistics, 2nd ed.* New York: John Wiley & Sons, Inc., 1976.

SIEGEL, S., *Nonparametric Statistics for the Behavioral Sciences*. New York: McGraw-Hill Book Company, 1956.

The tables that are needed to perform various nonparametric tests for small samples are given in some of the aforementioned books, including the one by S. Siegel, and, among others, in

OWEN, D. B., *Handbook of Statistical Tables*. Reading, Mass.: Addison-Wesley Publishing Company, Inc., 1962.

STATISTICAL TABLES

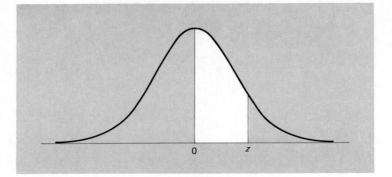

The entries in Table I are the probabilities that a random variable having the standard normal distribution will take on a value between 0 and z; they are given by the area of the white region under the curve in the figure shown above.

TABLE I Normal-Curve Areas

z	.00	.01	.02	.03	.04	.05	.06	.07	.08	.09
0.0	.0000	.0040	.0080	.0120	.0160	.0199	.0239	.0279	.0319	.0359
0.1	.0398	.0438	.0478	.0517	.0557	.0596	.0636	.0675	.0714	.0753
0.2	.0793	.0832	.0871	.0910	.0948	.0987	.1026	.1064	.1103	.1141
0.3	.1179	.1217	.1255	.1293	.1331	.1368	.1406	.1443	.1480	.1517
0.4	.1554	.1591	.1628	.1664	.1700	.1736	.1772	.1808	.1844	.1879
0.5	.1915	.1950	.1985	.2019	.2054	.2088	.2123	.2157	.2190	.2224
0.6	.2257	.2291	.2324	.2357	.2389	.2422	.2454	.2486	.2517	.2549
0.7	.2580	.2611	.2642	.2673	.2704	.2734	.2764	.2794	.2823	.2852
0.8	.2881	.2910	.2939	.2967	.2995	.3023	.3051	.3078	.3106	.3133
0.9	.3159	.3186	.3212	.3238	.3264	.3289	.3315	.3340	.3365	.3389
1.0	.3413	.3438	.3461	.3485	.3508	.3531	.3554	.3577	.3599	.3621
1.1	.3643	.3665	.3686	.3708	.3729	.3749	.3770	.3790	.3810	.3830
1.2	.3849	.3869	.3888	.3907	.3925	.3944	.3962	.3980	.3997	.4015
1.3	.4032	.4049	.4066	.4082	.4099	.4115	.4131	.4147	.4162	.4177
1.4	.4192	.4207	.4222	.4236	.4251	.4265	.4279	.4292	.4306	.4319
1.5	.4332	.4345	.4357	.4370	.4382	.4394	.4406	.4418	.4429	.4441
1.6	.4452	.4463	.4474	.4484	.4495	.4505	.4515	.4525	.4535	.4545
1.7	.4554	.4564	.4573	.4582	.4591	.4599	.4608	.4616	.4625	.4633
1.8	.4641	.4649	.4656	.4664	.4671	.4678	.4686	.4693	.4699	.4706
1.9	.4713	.4719	.4726	.4732	.4738	.4744	.4750	.4756	.4761	.4767
2.0	.4772	.4778	.4783	.4788	.4793	.4798	.4803	.4808	.4812	.4817
2.1	.4821	.4826	.4830	.4834	.4838	.4842	.4846	.4850	.4854	.4857
2.2	.4861	.4864	.4868	.4871	.4875	.4878	.4881	.4884	.4887	.4890
2.3	.4893	.4896	.4898	.4901	.4904	.4906	.4909	.4911	.4913	.4916
2.4	.4918	.4920	.4922	.4925	.4927	.4929	.4931	.4932	.4934	.4936
2.5	.4938	.4940	.4941	.4943	.4945	.4946	.4948	.4949	.4951	.4952
2.6	.4953	.4955	.4956	.4957	.4959	.4960	.4961	.4962	.4963	.4964
2.7	.4965	.4966	.4967	.4968	.4969	.4970	.4971	.4972	.4973	.4974
2.8	.4974	.4975	.4976	.4977	.4977	.4978	.4979	.4979	.4980	.4981
2.9	.4981	.4982	.4982	.4983	.4984	.4984	.4985	.4985	.4986	.4986
3.0	.4987	.4987	.4987	.4988	.4988	.4989	.4989	.4989	.4990	.4990

Also, for $z = 4.0$, 5.0, and 6.0, the areas are 0.49997, 0.4999997, and 0.499999999.

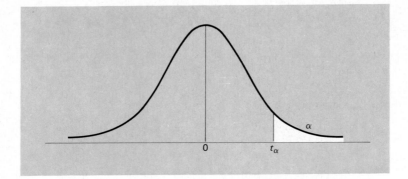

The entries in Table II are values for which the area to their right under the t distribution with given degrees of freedom (the area of the white region under the curve shown above) is equal to α.

TABLE II   Values of t[†]

d.f.	$t_{.100}$	$t_{.050}$	$t_{.025}$	$t_{.010}$	$t_{.005}$	d.f.
1	3.078	6.314	12.706	31.821	63.657	1
2	1.886	2.920	4.303	6.965	9.925	2
3	1.638	2.353	3.182	4.541	5.841	3
4	1.533	2.132	2.776	3.747	4.604	4
5	1.476	2.015	2.571	3.365	4.032	5
6	1.440	1.943	2.447	3.143	3.707	6
7	1.415	1.895	2.365	2.998	3.499	7
8	1.397	1.860	2.306	2.896	3.355	8
9	1.383	1.833	2.262	2.821	3.250	9
10	1.372	1.812	2.228	2.764	3.169	10
11	1.363	1.796	2.201	2.718	3.106	11
12	1.356	1.782	2.179	2.681	3.055	12
13	1.350	1.771	2.160	2.650	3.012	13
14	1.345	1.761	2.145	2.624	2.977	14
15	1.341	1.753	2.131	2.602	2.947	15
16	1.337	1.746	2.120	2.583	2.921	16
17	1.333	1.740	2.110	2.567	2.898	17
18	1.330	1.734	2.101	2.552	2.878	18
19	1.328	1.729	2.093	2.539	2.861	19
20	1.325	1.725	2.086	2.528	2.845	20
21	1.323	1.721	2.080	2.518	2.831	21
22	1.321	1.717	2.074	2.508	2.819	22
23	1.319	1.714	2.069	2.500	2.807	23
24	1.318	1.711	2.064	2.492	2.797	24
25	1.316	1.708	2.060	2.485	2.787	25
26	1.315	1.706	2.056	2.479	2.779	26
27	1.314	1.703	2.052	2.473	2.771	27
28	1.313	1.701	2.048	2.467	2.763	28
29	1.311	1.699	2.045	2.462	2.756	29
inf.	1.282	1.645	1.960	2.326	2.576	inf.

[†] This table is abridged from Table IV of R. A. Fisher, *Statistical Methods for Research Workers, 14th ed.*, by permission of Macmillan Publishing Co., Inc. Copyright © 1970 University of Adelaide.

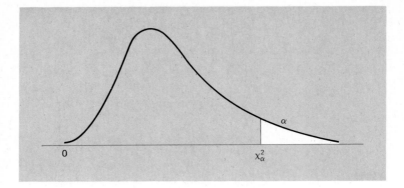

The entries in Table III are values for which the area to their right under the chi-square distribution with given degrees of freedom (the area of the white region under the curve shown above) is equal to α.

TABLE III Values of χ^2†

d.f.	$\chi^2_{.995}$	$\chi^2_{.99}$	$\chi^2_{.975}$	$\chi^2_{.95}$	$\chi^2_{.05}$	$\chi^2_{.025}$	$\chi^2_{.01}$	$\chi^2_{.005}$	d.f.
1	.0000393	.000157	.000982	.00393	3.841	5.024	6.635	7.879	1
2	.0100	.0201	.0506	.103	5.991	7.378	9.210	10.597	2
3	.0717	.115	.216	.352	7.815	9.348	11.345	12.838	3
4	.207	.297	.484	.711	9.488	11.143	13.277	14.860	4
5	.412	.554	.831	1.145	11.070	12.832	15.086	16.750	5
6	.676	.872	1.237	1.635	12.592	14.449	16.812	18.548	6
7	.989	1.239	1.690	2.167	14.067	16.013	18.475	20.278	7
8	1.344	1.646	2.180	2.733	15.507	17.535	20.090	21.955	8
9	1.735	2.088	2.700	3.325	16.919	19.023	21.666	23.589	9
10	2.156	2.558	3.247	3.940	18.307	20.483	23.209	25.188	10
11	2.603	3.053	3.816	4.575	19.675	21.920	24.725	26.757	11
12	3.074	3.571	4.404	5.226	21.026	23.337	26.217	28.300	12
13	3.565	4.107	5.009	5.892	22.362	24.736	27.688	29.819	13
14	4.075	4.660	5.629	6.571	23.685	26.119	29.141	31.319	14
15	4.601	5.229	6.262	7.261	24.996	27.488	30.578	32.801	15
16	5.142	5.812	6.908	7.962	26.296	28.845	32.000	34.267	16
17	5.697	6.408	7.564	8.672	27.587	30.191	33.409	35.718	17
18	6.265	7.015	8.231	9.390	28.869	31.526	34.805	37.156	18
19	6.844	7.633	8.907	10.117	30.144	32.852	36.191	38.582	19
20	7.434	8.260	9.591	10.851	31.410	34.170	37.566	39.997	20
21	8.034	8.897	10.283	11.591	32.671	35.479	38.932	41.401	21
22	8.643	9.542	10.982	12.338	33.924	36.781	40.289	42.796	22
23	9.260	10.196	11.689	13.091	35.172	38.076	41.638	44.181	23
24	9.886	10.856	12.401	13.848	36.415	39.364	42.980	45.558	24
25	10.520	11.524	13.120	14.611	37.652	40.646	44.314	46.928	25
26	11.160	12.198	13.844	15.379	38.885	41.923	45.642	48.290	26
27	11.808	12.879	14.573	16.151	40.113	43.194	46.963	49.645	27
28	12.461	13.565	15.308	16.928	41.337	44.461	48.278	50.993	28
29	13.121	14.256	16.047	17.708	42.557	45.722	49.588	52.336	29
30	13.787	14.953	16.791	18.493	43.773	46.979	50.892	53.672	30

† Based on Table 8 of *Biometrika Tables for Statisticians, Volume I* (Cambridge University Press, 1954), by permission of the *Biometrika* trustees.

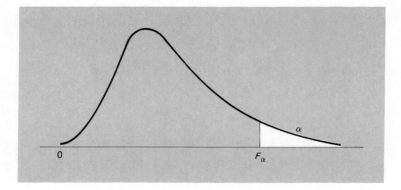

The entries in Table IV are values for which the area to their right under the F distribution with given degrees of freedom (the area of the white region under the curve shown above) is equal to α.

TABLE IV Values of $F_{0.05}$†

Degrees of freedom for numerator

Degrees of freedom for denominator	1	2	3	4	5	6	7	8	9	10	12	15	20	24	30	40	60	120	∞
1	161	200	216	225	230	234	237	239	241	242	244	246	248	249	250	251	252	253	254
2	18.5	19.0	19.2	19.2	19.3	19.3	19.4	19.4	19.4	19.4	19.4	19.4	19.4	19.5	19.5	19.5	19.5	19.5	19.5
3	10.1	9.55	9.28	9.12	9.01	8.94	8.89	8.85	8.81	8.79	8.74	8.70	8.66	8.64	8.62	8.59	8.57	8.55	8.53
4	7.71	6.94	6.59	6.39	6.26	6.16	6.09	6.04	6.00	5.96	5.91	5.86	5.80	5.77	5.75	5.72	5.69	5.66	5.63
5	6.61	5.79	5.41	5.19	5.05	4.95	4.88	4.82	4.77	4.74	4.68	4.62	4.56	4.53	4.50	4.46	4.43	4.40	4.37
6	5.99	5.14	4.76	4.53	4.39	4.28	4.21	4.15	4.10	4.06	4.00	3.94	3.87	3.84	3.81	3.77	3.74	3.70	3.67
7	5.59	4.74	4.35	4.12	3.97	3.87	3.79	3.73	3.68	3.64	3.57	3.51	3.44	3.41	3.38	3.34	3.30	3.27	3.23
8	5.32	4.46	4.07	3.84	3.69	3.58	3.50	3.44	3.39	3.35	3.28	3.22	3.15	3.12	3.08	3.04	3.01	2.97	2.93
9	5.12	4.26	3.86	3.63	3.48	3.37	3.29	3.23	3.18	3.14	3.07	3.01	2.94	2.90	2.86	2.83	2.79	2.75	2.71
10	4.96	4.10	3.71	3.48	3.33	3.22	3.14	3.07	3.02	2.98	2.91	2.85	2.77	2.74	2.70	2.66	2.62	2.58	2.54
11	4.84	3.98	3.59	3.36	3.20	3.09	3.01	2.95	2.90	2.85	2.79	2.72	2.65	2.61	2.57	2.53	2.49	2.45	2.40
12	4.75	3.89	3.49	3.26	3.11	3.00	2.91	2.85	2.80	2.75	2.69	2.62	2.54	2.51	2.47	2.43	2.38	2.34	2.30
13	4.67	3.81	3.41	3.18	3.03	2.92	2.83	2.77	2.71	2.67	2.60	2.53	2.46	2.42	2.38	2.34	2.30	2.25	2.21
14	4.60	3.74	3.34	3.11	2.96	2.85	2.76	2.70	2.65	2.60	2.53	2.46	2.39	2.35	2.31	2.27	2.22	2.18	2.13
15	4.54	3.68	3.29	3.06	2.90	2.79	2.71	2.64	2.59	2.54	2.48	2.40	2.33	2.29	2.25	2.20	2.16	2.11	2.07
16	4.49	3.63	3.24	3.01	2.85	2.74	2.66	2.59	2.54	2.49	2.42	2.35	2.28	2.24	2.19	2.15	2.11	2.06	2.01
17	4.45	3.59	3.20	2.96	2.81	2.70	2.61	2.55	2.49	2.45	2.38	2.31	2.23	2.19	2.15	2.10	2.06	2.01	1.96
18	4.41	3.55	3.16	2.93	2.77	2.66	2.58	2.51	2.46	2.41	2.34	2.27	2.19	2.15	2.11	2.06	2.02	1.97	1.92
19	4.38	3.52	3.13	2.90	2.74	2.63	2.54	2.48	2.42	2.38	2.31	2.23	2.16	2.11	2.07	2.03	1.98	1.93	1.88
20	4.35	3.49	3.10	2.87	2.71	2.60	2.51	2.45	2.39	2.35	2.28	2.20	2.12	2.08	2.04	1.99	1.95	1.90	1.84
21	4.32	3.47	3.07	2.84	2.68	2.57	2.49	2.42	2.37	2.32	2.25	2.18	2.10	2.05	2.01	1.96	1.92	1.87	1.81
22	4.30	3.44	3.05	2.82	2.66	2.55	2.46	2.40	2.34	2.30	2.23	2.15	2.07	2.03	1.98	1.94	1.89	1.84	1.78
23	4.28	3.42	3.03	2.80	2.64	2.53	2.44	2.37	2.32	2.27	2.20	2.13	2.05	2.01	1.96	1.91	1.86	1.81	1.76
24	4.26	3.40	3.01	2.78	2.62	2.51	2.42	2.36	2.30	2.25	2.18	2.11	2.03	1.98	1.94	1.89	1.84	1.79	1.73
25	4.24	3.39	2.99	2.76	2.60	2.49	2.40	2.34	2.28	2.24	2.16	2.09	2.01	1.96	1.92	1.87	1.82	1.77	1.71
30	4.17	3.32	2.92	2.69	2.53	2.42	2.33	2.27	2.21	2.16	2.09	2.01	1.93	1.89	1.84	1.79	1.74	1.68	1.62
40	4.08	3.23	2.84	2.61	2.45	2.34	2.25	2.18	2.12	2.08	2.00	1.92	1.84	1.79	1.74	1.69	1.64	1.58	1.51
60	4.00	3.15	2.76	2.53	2.37	2.25	2.17	2.10	2.04	1.99	1.92	1.84	1.75	1.70	1.65	1.59	1.53	1.47	1.39
120	3.92	3.07	2.68	2.45	2.29	2.18	2.09	2.02	1.96	1.91	1.83	1.75	1.66	1.61	1.55	1.50	1.43	1.35	1.25
∞	3.84	3.00	2.60	2.37	2.21	2.10	2.01	1.94	1.88	1.83	1.75	1.67	1.57	1.52	1.46	1.39	1.32	1.22	1.00

TABLE IV (*Continued*) Values of $F_{0.01}$†

Degrees of freedom for numerator

df (denom)	1	2	3	4	5	6	7	8	9	10	12	15	20	24	30	40	60	120	∞
1	4,052	5,000	5,403	5,625	5,764	5,859	5,928	5,982	6,023	6,056	6,106	6,157	6,209	6,235	6,261	6,287	6,313	6,339	6,366
2	98.5	99.0	99.2	99.2	99.3	99.3	99.4	99.4	99.4	99.4	99.4	99.4	99.4	99.5	99.5	99.5	99.5	99.5	99.5
3	34.1	30.8	29.5	28.7	28.2	27.9	27.7	27.5	27.3	27.2	27.1	26.9	26.7	26.6	26.5	26.4	26.3	26.2	26.1
4	21.2	18.0	16.7	16.0	15.5	15.2	15.0	14.8	14.7	14.5	14.4	14.2	14.0	13.9	13.8	13.7	13.7	13.6	13.5
5	16.3	13.3	12.1	11.4	11.0	10.7	10.5	10.3	10.2	10.1	9.89	9.72	9.55	9.47	9.38	9.29	9.20	9.11	9.02
6	13.7	10.9	9.78	9.15	8.75	8.47	8.26	8.10	7.98	7.87	7.72	7.56	7.40	7.31	7.23	7.14	7.05	6.97	6.88
7	12.2	9.55	8.45	7.85	7.46	7.19	6.99	6.84	6.72	6.62	6.47	6.31	6.16	6.07	5.99	5.91	5.82	5.74	5.65
8	11.3	8.65	7.59	7.01	6.63	6.37	6.18	6.03	5.91	5.81	5.67	5.52	5.36	5.28	5.20	5.12	5.03	4.95	4.86
9	10.6	8.02	6.99	6.42	6.06	5.80	5.61	5.47	5.35	5.26	5.11	4.96	4.81	4.73	4.65	4.57	4.48	4.40	4.31
10	10.0	7.56	6.55	5.99	5.64	5.39	5.20	5.06	4.94	4.85	4.71	4.56	4.41	4.33	4.25	4.17	4.08	4.00	3.91
11	9.65	7.21	6.22	5.67	5.32	5.07	4.89	4.74	4.63	4.54	4.40	4.25	4.10	4.02	3.94	3.86	3.78	3.69	3.60
12	9.33	6.93	5.95	5.41	5.06	4.82	4.64	4.50	4.39	4.30	4.16	4.01	3.86	3.78	3.70	3.62	3.54	3.45	3.36
13	9.07	6.70	5.74	5.21	4.86	4.62	4.44	4.30	4.19	4.10	3.96	3.82	3.66	3.59	3.51	3.43	3.34	3.25	3.17
14	8.86	6.51	5.56	5.04	4.70	4.46	4.28	4.14	4.03	3.94	3.80	3.66	3.51	3.43	3.35	3.27	3.18	3.09	3.00
15	8.68	6.36	5.42	4.89	4.56	4.32	4.14	4.00	3.89	3.80	3.67	3.52	3.37	3.29	3.21	3.13	3.05	2.96	2.87
16	8.53	6.23	5.29	4.77	4.44	4.20	4.03	3.89	3.78	3.69	3.55	3.41	3.26	3.18	3.10	3.02	2.93	2.84	2.75
17	8.40	6.11	5.19	4.67	4.34	4.10	3.93	3.79	3.68	3.59	3.46	3.31	3.16	3.08	3.00	2.92	2.83	2.75	2.65
18	8.29	6.01	5.09	4.58	4.25	4.01	3.84	3.71	3.60	3.51	3.37	3.23	3.08	3.00	2.92	2.84	2.75	2.66	2.57
19	8.19	5.93	5.01	4.50	4.17	3.94	3.77	3.63	3.52	3.43	3.30	3.15	3.00	2.92	2.84	2.76	2.67	2.58	2.49
20	8.10	5.85	4.94	4.43	4.10	3.87	3.70	3.56	3.46	3.37	3.23	3.09	2.94	2.86	2.78	2.69	2.61	2.52	2.42
21	8.02	5.78	4.87	4.37	4.04	3.81	3.64	3.51	3.40	3.31	3.17	3.03	2.88	2.80	2.72	2.64	2.55	2.46	2.36
22	7.95	5.72	4.82	4.31	3.99	3.76	3.59	3.45	3.35	3.26	3.12	2.98	2.83	2.75	2.67	2.58	2.50	2.40	2.31
23	7.88	5.66	4.76	4.26	3.94	3.71	3.54	3.41	3.30	3.21	3.07	2.93	2.78	2.70	2.62	2.54	2.45	2.35	2.26
24	7.82	5.61	4.72	4.22	3.90	3.67	3.50	3.36	3.26	3.17	3.03	2.89	2.74	2.66	2.58	2.49	2.40	2.31	2.21
25	7.77	5.57	4.68	4.18	3.86	3.63	3.46	3.32	3.22	3.13	2.99	2.85	2.70	2.62	2.53	2.45	2.36	2.27	2.17
30	7.56	5.39	4.51	4.02	3.70	3.47	3.30	3.17	3.07	2.98	2.84	2.70	2.55	2.47	2.39	2.30	2.21	2.11	2.01
40	7.31	5.18	4.31	3.83	3.51	3.29	3.12	2.99	2.89	2.80	2.66	2.52	2.37	2.29	2.20	2.11	2.02	1.92	1.80
60	7.08	4.98	4.13	3.65	3.34	3.12	2.95	2.82	2.72	2.63	2.50	2.35	2.20	2.12	2.03	1.94	1.84	1.73	1.60
120	6.85	4.79	3.95	3.48	3.17	2.96	2.79	2.66	2.56	2.47	2.34	2.19	2.03	1.95	1.86	1.76	1.66	1.53	1.38
∞	6.63	4.61	3.78	3.32	3.02	2.80	2.64	2.51	2.41	2.32	2.18	2.04	1.88	1.79	1.70	1.59	1.47	1.32	1.00

Degrees of freedom for denominator

† Reproduced from M. Merrington and C. M. Thompson, "Tables of percentage points of the inverted beta (F) distribution," *Biometrika*, vol. 33 (1943), by permission of the *Biometrika* trustees.

TABLE V Binomial Probabilities

						p						
n	*x*	0.05	0.1	0.2	0.3	0.4	0.5	0.6	0.7	0.8	0.9	0.95
2	0	0.902	0.810	0.640	0.490	0.360	0.250	0.160	0.090	0.040	0.010	0.002
	1	0.095	0.180	0.320	0.420	0.480	0.500	0.480	0.420	0.320	0.180	0.095
	2	0.002	0.010	0.040	0.090	0.160	0.250	0.360	0.490	0.640	0.810	0.902
3	0	0.857	0.729	0.512	0.343	0.216	0.125	0.064	0.027	0.008	0.001	
	1	0.135	0.243	0.384	0.441	0.432	0.375	0.288	0.189	0.096	0.027	0.007
	2	0.007	0.027	0.096	0.189	0.288	0.375	0.432	0.441	0.384	0.243	0.135
	3		0.001	0.008	0.027	0.064	0.125	0.216	0.343	0.512	0.729	0.857
4	0	0.815	0.656	0.410	0.240	0.130	0.062	0.026	0.008	0.002		
	1	0.171	0.292	0.410	0.412	0.346	0.250	0.154	0.076	0.026	0.004	
	2	0.014	0.049	0.154	0.265	0.346	0.375	0.346	0.265	0.154	0.049	0.014
	3		0.004	0.026	0.076	0.154	0.250	0.346	0.412	0.410	0.292	0.171
	4			0.002	0.008	0.026	0.062	0.130	0.240	0.410	0.656	0.815
5	0	0.774	0.590	0.328	0.168	0.078	0.031	0.010	0.002			
	1	0.204	0.328	0.410	0.360	0.259	0.156	0.077	0.028	0.006		
	2	0.021	0.073	0.205	0.309	0.346	0.312	0.230	0.132	0.051	0.008	0.001
	3	0.001	0.008	0.051	0.132	0.230	0.312	0.346	0.309	0.205	0.073	0.021
	4			0.006	0.028	0.077	0.156	0.259	0.360	0.410	0.328	0.204
	5				0.002	0.010	0.031	0.078	0.168	0.328	0.590	0.774
6	0	0.735	0.531	0.262	0.118	0.047	0.016	0.004	0.001			
	1	0.232	0.354	0.393	0.303	0.187	0.094	0.037	0.010	0.002		
	2	0.031	0.098	0.246	0.324	0.311	0.234	0.138	0.060	0.015	0.001	
	3	0.002	0.015	0.082	0.185	0.276	0.312	0.276	0.185	0.082	0.015	0.002
	4		0.001	0.015	0.060	0.138	0.234	0.311	0.324	0.246	0.098	0.031
	5			0.002	0.010	0.037	0.094	0.187	0.303	0.393	0.354	0.232
	6				0.001	0.004	0.016	0.047	0.118	0.262	0.531	0.735
7	0	0.698	0.478	0.210	0.082	0.028	0.008	0.002				
	1	0.257	0.372	0.367	0.247	0.131	0.055	0.017	0.004			
	2	0.041	0.124	0.275	0.318	0.261	0.164	0.077	0.025	0.004		
	3	0.004	0.023	0.115	0.227	0.290	0.273	0.194	0.097	0.029	0.003	
	4		0.003	0.029	0.097	0.194	0.273	0.290	0.227	0.115	0.023	0.004
	5			0.004	0.025	0.077	0.164	0.261	0.318	0.275	0.124	0.041
	6				0.004	0.017	0.055	0.131	0.247	0.367	0.372	0.257
	7					0.002	0.008	0.028	0.082	0.210	0.478	0.698
8	0	0.663	0.430	0.168	0.058	0.017	0.004	0.001				
	1	0.279	0.383	0.336	0.198	0.090	0.031	0.008	0.001			
	2	0.051	0.149	0.294	0.296	0.209	0.109	0.041	0.010	0.001		
	3	0.005	0.033	0.147	0.254	0.279	0.219	0.124	0.047	0.009		
	4		0.005	0.046	0.136	0.232	0.273	0.232	0.136	0.046	0.005	
	5			0.009	0.047	0.124	0.219	0.279	0.254	0.147	0.033	0.005
	6			0.001	0.010	0.041	0.109	0.209	0.296	0.294	0.149	0.051
	7				0.001	0.008	0.031	0.090	0.198	0.336	0.383	0.279
	8					0.001	0.004	0.017	0.058	0.168	0.430	0.663

All values omitted in this table are 0.0005 or less.

TABLE V   Binomial Probabilities (*Continued*)

n	x	0.05	0.1	0.2	0.3	0.4	0.5	0.6	0.7	0.8	0.9	0.95
9	0	0.630	0.387	0.134	0.040	0.010	0.002					
	1	0.299	0.387	0.302	0.156	0.060	0.018	0.004				
	2	0.063	0.172	0.302	0.267	0.161	0.070	0.021	0.004			
	3	0.008	0.045	0.176	0.267	0.251	0.164	0.074	0.021	0.003		
	4	0.001	0.007	0.066	0.172	0.251	0.246	0.167	0.074	0.017	0.001	
	5		0.001	0.017	0.074	0.167	0.246	0.251	0.172	0.066	0.007	0.001
	6			0.003	0.021	0.074	0.164	0.251	0.267	0.176	0.045	0.008
	7				0.004	0.021	0.070	0.161	0.267	0.302	0.172	0.063
	8					0.004	0.018	0.060	0.156	0.302	0.387	0.299
	9						0.002	0.010	0.040	0.134	0.387	0.630
10	0	0.599	0.349	0.107	0.028	0.006	0.001					
	1	0.315	0.387	0.268	0.121	0.040	0.010	0.002				
	2	0.075	0.194	0.302	0.233	0.121	0.044	0.011	0.001			
	3	0.010	0.057	0.201	0.267	0.215	0.117	0.042	0.009	0.001		
	4	0.001	0.011	0.088	0.200	0.251	0.205	0.111	0.037	0.006		
	5		0.001	0.026	0.103	0.201	0.246	0.201	0.103	0.026	0.001	
	6			0.006	0.037	0.111	0.205	0.251	0.200	0.088	0.011	0.001
	7			0.001	0.009	0.042	0.117	0.215	0.267	0.201	0.057	0.010
	8				0.001	0.011	0.044	0.121	0.233	0.302	0.194	0.075
	9					0.002	0.010	0.040	0.121	0.268	0.387	0.315
	10						0.001	0.006	0.028	0.107	0.349	0.599
11	0	0.569	0.314	0.086	0.020	0.004						
	1	0.329	0.384	0.236	0.093	0.027	0.005	0.001				
	2	0.087	0.213	0.295	0.200	0.089	0.027	0.005	0.001			
	3	0.014	0.071	0.221	0.257	0.177	0.081	0.023	0.004			
	4	0.001	0.016	0.111	0.220	0.236	0.161	0.070	0.017	0.002		
	5		0.002	0.039	0.132	0.221	0.226	0.147	0.057	0.010		
	6			0.010	0.057	0.147	0.226	0.221	0.132	0.039	0.002	
	7			0.002	0.017	0.070	0.161	0.236	0.220	0.111	0.016	0.001
	8				0.004	0.023	0.081	0.177	0.257	0.221	0.071	0.014
	9				0.001	0.005	0.027	0.089	0.200	0.295	0.213	0.087
	10					0.001	0.005	0.027	0.093	0.236	0.384	0.329
	11							0.004	0.020	0.086	0.314	0.569
12	0	0.540	0.282	0.069	0.014	0.002						
	1	0.341	0.377	0.206	0.071	0.017	0.003					
	2	0.099	0.230	0.283	0.168	0.064	0.016	0.002				
	3	0.017	0.085	0.236	0.240	0.142	0.054	0.012	0.001			
	4	0.002	0.021	0.133	0.231	0.213	0.121	0.042	0.008	0.001		
	5		0.004	0.053	0.158	0.227	0.193	0.101	0.029	0.003		
	6			0.016	0.079	0.177	0.226	0.177	0.079	0.016		
	7			0.003	0.029	0.101	0.193	0.227	0.158	0.053	0.004	
	8			0.001	0.008	0.042	0.121	0.213	0.231	0.133	0.021	0.002
	9				0.001	0.012	0.054	0.142	0.240	0.236	0.085	0.017
	10					0.002	0.016	0.064	0.168	0.283	0.230	0.099
	11						0.003	0.017	0.071	0.206	0 377	0.341
	12							0.002	0.014	0.069	0.282	0.540

TABLE V   Binomial Probabilities (*Continued*)

n	x	0.05	0.1	0.2	0.3	0.4	0.5	0.6	0.7	0.8	0.9	0.95
13	0	0.513	0.254	0.055	0.010	0.001						
	1	0.351	0.367	0.179	0.054	0.011	0.002					
	2	0.111	0.245	0.268	0.139	0.045	0.010	0.001				
	3	0.021	0.100	0.246	0.218	0.111	0.035	0.006	0.001			
	4	0.003	0.028	0.154	0.234	0.184	0.087	0.024	0.003			
	5		0.006	0.069	0.180	0.221	0.157	0.066	0.014	0.001		
	6		0.001	0.023	0.103	0.197	0.209	0.131	0.044	0.006		
	7			0.006	0.044	0.131	0.209	0.197	0.103	0.023	0.001	
	8			0.001	0.014	0.066	0.157	0.221	0.180	0.069	0.006	
	9				0.003	0.024	0.087	0.184	0.234	0.154	0.028	0.003
	10				0.001	0.006	0.035	0.111	0.218	0.246	0.100	0.021
	11					0.001	0.010	0.045	0.139	0.268	0.245	0.111
	12						0.002	0.011	0.054	0.179	0.367	0.351
	13							0.001	0.010	0.055	0.254	0.513
14	0	0.488	0.229	0.044	0.007	0.001						
	1	0.359	0.356	0.154	0.041	0.007	0.001					
	2	0.123	0.257	0.250	0.113	0.032	0.006	0.001				
	3	0.026	0.114	0.250	0.194	0.085	0.022	0.003				
	4	0.004	0.035	0.172	0.229	0.155	0.061	0.014	0.001			
	5		0.008	0.086	0.196	0.207	0.122	0.041	0.007			
	6		0.001	0.032	0.126	0.207	0.183	0.092	0.023	0.002		
	7			0.009	0.062	0.157	0.209	0.157	0.062	0.009		
	8			0.002	0.023	0.092	0.183	0.207	0.126	0.032	0.001	
	9				0.007	0.041	0.122	0.207	0.196	0.086	0.008	
	10				0.001	0.014	0.061	0.155	0.229	0.172	0.035	0.004
	11					0.003	0.022	0.085	0.194	0.250	0.114	0.026
	12					0.001	0.006	0.032	0.113	0.250	0.257	0.123
	13						0.001	0.007	0.041	0.154	0.356	0.359
	14							0.001	0.007	0.044	0.229	0.488
15	0	0.463	0.206	0.035	0.005							
	1	0.366	0.343	0.132	0.031	0.005						
	2	0.135	0.267	0.231	0.092	0.022	0.003					
	3	0.031	0.129	0.250	0.170	0.063	0.014	0.002				
	4	0.005	0.043	0.188	0.219	0.127	0.042	0.007	0.001			
	5	0.001	0.010	0.103	0.206	0.186	0.092	0.024	0.003			
	6		0.002	0.043	0.147	0.207	0.153	0.061	0.012	0.001		
	7			0.014	0.081	0.177	0.196	0.118	0.035	0.003		
	8			0.003	0.035	0.118	0.196	0.177	0.081	0.014		
	9			0.001	0.012	0.061	0.153	0.207	0.147	0.043	0.002	
	10				0.003	0.024	0.092	0.186	0.206	0.103	0.010	0.001
	11				0.001	0.007	0.042	0.127	0.219	0.188	0.043	0.005
	12					0.002	0.014	0.063	0.170	0.250	0.129	0.031
	13						0.003	0.022	0.092	0.231	0.267	0.135
	14							0.005	0.031	0.132	0.343	0.366
	15								0.005	0.035	0.206	0.463

TABLE V Binomial Probabilities (*Continued*)

n	x	0.05	0.1	0.2	0.3	0.4	0.5	0.6	0.7	0.8	0.9	0.95
16	0	0.440	0.185	0.028	0.003							
	1	0.371	0.329	0.113	0.023	0.003						
	2	0.146	0.275	0.211	0.073	0.015	0.002					
	3	0.036	0.142	0.246	0.146	0.047	0.009	0.001				
	4	0.006	0.051	0.200	0.204	0.101	0.028	0.004				
	5	0.001	0.014	0.120	0.210	0.162	0.067	0.014	0.001			
	6		0.003	0.055	0.165	0.198	0.122	0.039	0.006			
	7			0.020	0.101	0.189	0.175	0.084	0.019	0.001		
	8			0.006	0.049	0.142	0.196	0.142	0.049	0.006		
	9			0.001	0.019	0.084	0.175	0.189	0.101	0.020		
	10				0.006	0.039	0.122	0.198	0.165	0.055	0.003	
	11				0.001	0.014	0.067	0.162	0.210	0.120	0.014	0.001
	12					0.004	0.028	0.101	0.204	0.200	0.051	0.006
	13					0.001	0.009	0.047	0.146	0.246	0.142	0.036
	14						0.002	0.015	0.073	0.211	0.275	0.146
	15							0.003	0.023	0.113	0.329	0.371
	16								0.003	0.028	0.185	0.440
17	0	0.418	0.167	0.023	0.002							
	1	0.374	0.315	0.096	0.017	0.002						
	2	0.158	0.280	0.191	0.058	0.010	0.001					
	3	0.041	0.156	0.239	0.125	0.034	0.005					
	4	0.008	0.060	0.209	0.187	0.080	0.018	0.002				
	5	0.001	0.017	0.136	0.208	0.138	0.047	0.008	0.001			
	6		0.004	0.068	0.178	0.184	0.094	0.024	0.003			
	7		0.001	0.027	0.120	0.193	0.148	0.057	0.009			
	8			0.008	0.064	0.161	0.185	0.107	0.028	0.002		
	9			0.002	0.028	0.107	0.185	0.161	0.064	0.008		
	10				0.009	0.057	0.148	0.193	0.120	0.027	0.001	
	11				0.003	0.024	0.094	0.184	0.178	0.068	0.004	
	12				0.001	0.008	0.047	0.138	0.208	0.136	0.017	0.001
	13					0.002	0.018	0.080	0.187	0.209	0.060	0.008
	14						0.005	0.034	0.125	0.239	0.156	0.041
	15						0.001	0.010	0.058	0.191	0.280	0.158
	16							0.002	0.017	0.096	0.315	0.374
	17								0.002	0.023	0.167	0.418
18	0	0.397	0.150	0.018	0.002							
	1	0.376	0.300	0.081	0.013	0.001						
	2	0.168	0.284	0.172	0.046	0.007	0.001					
	3	0.047	0.168	0.230	0.105	0.025	0.003					
	4	0.009	0.070	0.215	0.168	0.061	0.012	0.001				
	5	0.001	0.022	0.151	0.202	0.115	0.033	0.004				
	6		0.005	0.082	0.187	0.166	·0.071	0.015	0.001			
	7		0.001	0.035	0.138	0.189	0.121	0.037	0.005			
	8			0.012	0.081	0.173	0.167	0.077	0.015	0.001		
	9			0.003	0.039	0.128	0.185	0.128	0.039	0.003		
	10			0.001	0.015	0.077	0.167	0.173	0.081	0.012		
	11				0.005	0.037	0.121	0.189	0.138	0.035	0.001	

TABLE V Binomial Probabilities (*Continued*)

n	x	0.05	0.1	0.2	0.3	0.4	0.5	0.6	0.7	0.8	0.9	0.95
	12			0.001	0.015	0.071	0.166	0.187	0.082	0.005		
	13				0.004	0.033	0.115	0.202	0.151	0.022	0.001	
	14				0.001	0.012	0.061	0.168	0.215	0.070	0.009	
	15					0.003	0.025	0.105	0.230	0.168	0.047	
	16					0.001	0.007	0.046	0.172	0.284	0.168	
	17						0.001	0.013	0.081	0.300	0.376	
	18							0.002	0.018	0.150	0.397	
19	0	0.377	0.135	0.014	0.001							
	1	0.377	0.285	0.068	0.009	0.001						
	2	0.179	0.285	0.154	0.036	0.005						
	3	0.053	0.180	0.218	0.087	0.017	0.002					
	4	0.011	0.080	0.218	0.149	0.047	0.007	0.001				
	5	0.002	0.027	0.164	0.192	0.093	0.022	0.002				
	6		0.007	0.095	0.192	0.145	0.052	0.008	0.001			
	7		0.001	0.044	0.153	0.180	0.096	0.024	0.002			
	8			0.017	0.098	0.180	0.144	0.053	0.008			
	9			0.005	0.051	0.146	0.176	0.098	0.022	0.001		
	10			0.001	0.022	0.098	0.176	0.146	0.051	0.005		
	11				0.008	0.053	0.144	0.180	0.098	0.017		
	12				0.002	0.024	0.096	0.180	0.153	0.044	0.001	
	13				0.001	0.008	0.052	0.145	0.192	0.095	0.007	
	14					0.002	0.022	0.093	0.192	0.164	0.027	0.002
	15					0.001	0.007	0.047	0.149	0.218	0.080	0.011
	16						0.002	0.017	0.087	0.218	0.180	0.053
	17							0.005	0.036	0.154	0.285	0.179
	18							0.001	0.009	0.068	0.285	0.377
	19								0.001	0.014	0.135	0.377
20	0	0.358	0.122	0.012	0.001							
	1	0.377	0.270	0.058	0.007							
	2	0.189	0.285	0.137	0.028	0.003						
	3	0.060	0.190	0.205	0.072	0.012	0.001					
	4	0.013	0.090	0.218	0.130	0.035	0.005					
	5	0.002	0.032	0.175	0.179	0.075	0.015	0.001				
	6		0.009	0.109	0.192	0.124	0.037	0.005				
	7		0.002	0.055	0.164	0.166	0.074	0.015	0.001			
	8			0.022	0.114	0.180	0.120	0.035	0.004			
	9			0.007	0.065	0.160	0.160	0.071	0.012			
	10			0.002	0.031	0.117	0.176	0.117	0.031	0.002		
	11				0.012	0.071	0.160	0.160	0.065	0.007		
	12				0.004	0.035	0.120	0.180	0.114	0.022		
	13				0.001	0.015	0.074	0.166	0.164	0.055	0.002	
	14					0.005	0.037	0.124	0.192	0.109	0.009	
	15					0.001	0.015	0.075	0.179	0.175	0.032	0.002
	16						0.005	0.035	0.130	0.218	0.090	0.013
	17						0.001	0.012	0.072	0.205	0.190	0.060
	18							0.003	0.028	0.137	0.285	0.189
	19								0.007	0.058	0.270	0.377
	20								0.001	0.012	0.122	0.358

TABLE VI Critical Values for the U Test[†]

Two-sided alternative ($\alpha = 0.05$)

n_1 \ n_2	2	3	4	5	6	7	8	9	10	11	12	13	14	15
2							0	0	0	0	1	1	1	1
3			0	1	1	2	2	3	3	4	4	5	5
4			0	1	2	3	4	4	5	6	7	8	9	10
5		0	1	2	3	5	6	7	8	9	11	12	13	14
6		1	2	3	5	6	8	10	11	13	14	16	17	19
7		1	3	5	6	8	10	12	14	16	18	20	22	24
8	0	2	4	6	8	10	13	15	17	19	22	24	26	29
9	0	2	4	7	10	12	15	17	20	23	26	28	31	34
10	0	3	5	8	11	14	17	20	23	26	29	30	36	39
11	0	3	6	9	13	16	19	23	26	30	33	37	40	44
12	1	4	7	11	14	18	22	26	29	33	37	41	45	49
13	1	4	8	12	16	20	24	28	30	37	41	45	50	54
14	1	5	9	13	17	22	26	31	36	40	45	50	55	59
15	1	5	10	14	19	24	29	34	39	44	49	54	59	64

One-sided alternative ($\alpha = 0.05$)

n_1 \ n_2	2	3	4	5	6	7	8	9	10	11	12	13	14	15
2				0	0	0	1	1	1	1	2	2	3	3
3		0	0	1	2	2	3	4	4	5	5	6	7	7
4		0	1	2	3	4	5	6	7	8	9	10	11	12
5	0	1	2	4	5	6	8	9	11	12	13	15	16	18
6	0	2	3	5	7	8	10	12	14	16	17	19	21	23
7	0	2	4	6	8	11	13	15	17	19	21	24	26	28
8	1	3	5	8	10	13	15	18	20	23	26	28	31	33
9	1	4	6	9	12	15	18	21	24	27	30	33	36	39
10	1	4	7	11	14	17	20	24	27	31	34	37	41	44
11	1	5	8	12	16	19	23	27	31	34	38	42	46	50
12	2	5	9	13	17	21	26	30	34	38	42	47	51	55
13	2	6	10	15	19	24	28	33	37	42	47	51	56	61
14	3	7	11	16	21	26	31	36	41	46	51	56	61	66
15	3	7	12	18	23	28	33	39	44	50	55	61	66	72

[†] This table is based on Table 11.4 of D. B. Owen, *Handbook of Statistical Tables,* © 1962, U.S. Department of Energy. Published by Addison-Wesley Publishing Company, Inc., Reading, Mass. Reprinted with permission of the publisher.

TABLE VI Critical Values for the *U* Test (*Continued*)

Two-sided alternative ($\alpha = 0.01$)

n_1 \ n_2	3	4	5	6	7	8	9	10	11	12	13	14	15
3							0	0	0	1	1	1	2
4			0	0	1	1	2	2	3	3	4	5
5		0	1	1	2	3	4	5	6	7	7	8
6	0	1	2	3	4	5	6	7	9	10	11	12
7	0	1	3	4	6	7	9	10	12	13	15	16
8	1	2	4	6	7	9	11	13	15	17	18	20
9	0	1	3	5	7	9	11	13	16	18	20	22	24
10	0	2	4	6	9	11	13	16	18	21	24	26	29
11	0	2	5	7	10	13	16	18	21	24	27	30	33
12	1	3	6	9	12	15	18	21	24	27	31	34	37
13	1	3	7	10	13	17	20	24	27	31	34	38	42
14	1	4	7	11	15	18	22	26	30	34	38	42	46
15	2	5	8	12	16	20	24	29	33	37	42	46	51

One-sided alternative ($\alpha = 0.01$)

n_1 \ n_2	2	3	4	5	6	7	8	9	10	11	12	13	14	15
2												0	0	0
3					0	0	1	1	1	2	2	2	3
4			0	1	1	2	3	3	4	5	5	6	7
5		0	1	2	3	4	5	6	7	8	9	10	11
6		1	2	3	4	6	7	8	9	11	12	13	15
7	0	1	3	4	6	7	9	11	12	14	16	17	19
8	0	2	4	6	7	9	11	13	15	17	20	22	24
9	1	3	5	7	9	11	14	16	18	21	23	26	28
10	1	3	6	8	11	13	16	19	22	24	27	30	33
11	1	4	7	9	12	15	18	22	25	28	31	34	37
12	2	5	8	11	14	17	21	24	28	31	35	38	42
13	0	2	5	9	12	16	20	23	27	31	35	39	43	47
14	0	2	6	10	13	17	22	26	30	34	38	43	47	51
15	0	3	7	11	15	19	24	28	33	37	42	47	51	56

TABLE VII   Critical Values for the Signed-Rank Test[†]

| | Two-sided alternative | | | One-sided alternative | |
| | $\alpha = 0.05$ | $\alpha = 0.01$ | | $\alpha = 0.05$ | $\alpha = 0.01$ |
n			n		
4			4		
5			5	1	
6	1		6	2	
7	2		7	4	0
8	4	0	8	6	2
9	6	2	9	8	3
10	8	3	10	11	5
11	11	5	11	14	7
12	14	7	12	17	10
13	17	10	13	21	13
14	21	13	14	26	16
15	25	16	15	30	20
16	30	19	16	36	24
17	35	23	17	41	28
18	40	28	18	47	33
19	46	32	19	54	38
20	52	37	20	60	43
21	59	43	21	68	49
22	66	49	22	75	56
23	73	55	23	83	62
24	81	61	24	92	69
25	90	68	25	101	77

† From F. Wilcoxon and R. A. Wilcox, *Some Rapid Approximate Statistical Procedures*, American Cyanamid Company, Pearl River, N.Y., 1964. Reproduced with permission of American Cyanamid Company.

TABLE VIII   Critical values of $u^{\dagger}$

n_1 \ n_2	4	5	6	7	8	9	10	11	12	13	14	15
						Values of $u_{0.025}$						
4		9	9									
5	9	10	10	11	11							
6	9	10	11	12	12	13	13	13	13			
7		11	12	13	13	14	14	14	14	15	15	15
8		11	12	13	14	14	15	15	16	16	16	16
9			13	14	14	15	16	16	16	17	17	18
10			13	14	15	16	16	17	17	18	18	18
11			13	14	15	16	17	17	18	19	19	19
12			13	14	16	16	17	18	19	19	20	20
13				15	16	17	18	19	19	20	20	21
14				15	16	17	18	19	20	20	21	22
15				15	16	18	18	19	20	21	22	22

n_1 \ n_2	2	3	4	5	6	7	8	9	10	11	12	13	14	15
							Values of $u'_{0.025}$							
2											2	2	2	2
3					2	2	2	2	2	2	2	2	2	3
4				2	2	2	3	3	3	3	3	3	3	3
5			2	2	3	3	3	3	3	4	4	4	4	4
6		2	2	3	3	3	3	4	4	4	4	5	5	5
7		2	2	3	3	3	4	4	5	5	5	5	5	6
8		2	3	3	3	4	4	5	5	5	6	6	6	6
9		2	3	3	4	4	5	5	5	6	6	6	7	7
10		2	3	3	4	5	5	5	6	6	7	7	7	7
11		2	3	4	4	5	5	6	6	7	7	7	8	8
12	2	2	3	4	4	5	6	6	7	7	7	8	8	8
13	2	2	3	4	5	5	6	6	7	7	8	8	9	9
14	2	2	3	4	5	5	6	7	7	8	8	9	9	9
15	2	3	3	4	5	6	6	7	7	8	8	9	9	10

† This table is adapted, by permission, from F. S. Swed and C. Eisenhart, "Tables for testing randomness of grouping in a sequence of alternatives," *Annals of Mathematical Statistics*, Vol. 14.

TABLE VIII   Critical values of u (*Continued*)

n_1 \ n_2	5	6	7	8	9	10	11	12	13	14	15
					Values of $u_{0.005}$						
5		11									
6	11	12	13	13							
7		13	13	14	15	15	15				
8		13	14	15	15	16	16	17	17	17	
9			15	15	16	17	17	18	18	18	19
10			15	16	17	17	18	19	19	19	20
11			15	16	17	18	19	19	20	20	21
12				17	18	19	19	20	21	21	22
13				17	18	19	20	21	21	22	22
14				17	18	19	20	21	22	23	23
15					19	20	21	22	22	23	24

n_1 \ n_2	3	4	5	6	7	8	9	10	11	12	13	14	15
						Values of $u'_{0.005}$							
3										2	2	2	2
4						2	2	2	2	2	2	2	3
5			2	2	2	2	3	3	3	3	3	3	3
6		2	2	2	3	3	3	3	3	3	3	4	4
7		2	2	3	3	3	3	4	4	4	4	4
8		2	2	3	3	3	3	4	4	4	5	5	5
9		2	2	3	3	3	4	4	5	5	5	5	6
10		2	3	3	3	4	4	5	5	5	5	6	6
11		2	3	3	4	4	5	5	5	6	6	6	7
12	2	2	3	3	4	4	5	5	6	6	6	7	7
13	2	2	3	3	4	5	5	5	6	6	7	7	7
14	2	2	3	4	4	5	5	6	6	7	7	7	8
15	2	3	3	4	4	5	6	6	7	7	7	8	8

TABLE IX   Values of $Z = \frac{1}{2} \cdot \ln \frac{1+r}{1-r}$

r	.00	.01	.02	.03	.04	.05	.06	.07	.08	.09
0.0	0.000	0.010	0.020	0.030	0.040	0.050	0.060	0.070	0.080	0.090
0.1	0.100	0.110	0.121	0.131	0.141	0.151	0.161	0.172	0.182	0.192
0.2	0.203	0.213	0.224	0.234	0.245	0.255	0.266	0.277	0.288	0.299
0.3	0.310	0.321	0.332	0.343	0.354	0.365	0.377	0.388	0.400	0.412
0.4	0.424	0.436	0.448	0.460	0.472	0.485	0.497	0.510	0.523	0.536
0.5	0.549	0.563	0.576	0.590	0.604	0.618	0.633	0.648	0.662	0.678
0.6	0.693	0.709	0.725	0.741	0.758	0.775	0.793	0.811	0.829	0.848
0.7	0.867	0.887	0.908	0.929	0.950	0.973	0.996	1.020	1.045	1.071
0.8	1.099	1.127	1.157	1.188	1.221	1.256	1.293	1.333	1.376	1.422
0.9	1.472	1.528	1.589	1.658	1.738	1.832	1.946	2.092	2.298	2.647

For negative values of r put a minus sign in front of the corresponding Z's, and vice versa.

TABLE X   Binomial Coefficients

n	$\binom{n}{0}$	$\binom{n}{1}$	$\binom{n}{2}$	$\binom{n}{3}$	$\binom{n}{4}$	$\binom{n}{5}$	$\binom{n}{6}$	$\binom{n}{7}$	$\binom{n}{8}$	$\binom{n}{9}$	$\binom{n}{10}$
0	1										
1	1	1									
2	1	2	1								
3	1	3	3	1							
4	1	4	6	4	1						
5	1	5	10	10	5	1					
6	1	6	15	20	15	6	1				
7	1	7	21	35	35	21	7	1			
8	1	8	28	56	70	56	28	8	1		
9	1	9	36	84	126	126	84	36	9	1	
10	1	10	45	120	210	252	210	120	45	10	1
11	1	11	55	165	330	462	462	330	165	55	11
12	1	12	66	220	495	792	924	792	495	220	66
13	1	13	78	286	715	1287	1716	1716	1287	715	286
14	1	14	91	364	1001	2002	3003	3432	3003	2002	1001
15	1	15	105	455	1365	3003	5005	6435	6435	5005	3003
16	1	16	120	560	1820	4368	8008	11440	12870	11440	8008
17	1	17	136	680	2380	6188	12376	19448	24310	24310	19448
18	1	18	153	816	3060	8568	18564	31824	43758	48620	43758
19	1	19	171	969	3876	11628	27132	50388	75582	92378	92378
20	1	20	190	1140	4845	15504	38760	77520	125970	167960	184756

For $r > 10$ it may be necessary to make use of the identity $\binom{n}{r} = \binom{n}{n-r}$.

TABLE XI   Random Numbers[†]

04433	80674	24520	18222	10610	05794	37515
60298	47829	72648	37414	75755	04717	29899
67884	59651	67533	68123	17730	95862	08034
89512	32155	51906	61662	64130	16688	37275
32653	01895	12506	88535	36553	23757	34209
95913	15405	13772	76638	48423	25018	99041
55864	21694	13122	44115	01601	50541	00147
35334	49810	91601	40617	72876	33967	73830
57729	32196	76487	11622	96297	24160	09903
86648	13697	63677	70119	94739	25875	38829
30574	47609	07967	32422	76791	39725	53711
81307	43694	83580	79974	45929	85113	72268
02410	54905	79007	54939	21410	86980	91772
18969	75274	52233	62319	08598	09066	95288
87863	82384	66860	62297	80198	19347	73234
68397	71708	15438	62311	72844	60203	46412
28529	54447	58729	10854	99058	18260	38765
44285	06372	15867	70418	57012	72122	36634
86299	83430	33571	23309	57040	29285	67870
84842	68668	90894	61658	15001	94055	36308
56970	83609	52098	04184	54967	72938	56834
83125	71257	60490	44369	66130	72936	69848
55503	52423	02464	26141	68779	66388	75242
47019	76273	33203	29608	54553	25971	69573
84828	32592	79526	29554	84580	37859	28504
68921	08141	79227	05748	51276	57143	31926
36458	96045	30424	98420	72925	40729	22337
95752	59445	36847	87729	81679	59126	59437
26768	47323	58454	56958	20575	76746	49878
42613	37056	43636	58085	06766	60227	96414
95457	30566	65482	25596	02678	54592	63607
95276	17894	63564	95958	39750	64379	46059
66954	52324	64776	92345	95110	59448	77249
17457	18481	14113	62462	02798	54977	48349
03704	36872	83214	59337	01695	60666	97410
21538	86497	33210	60337	27976	70661	08250
57178	67619	98310	70348	11317	71623	55510
31048	97558	94953	55866	96283	46620	52087
69799	55380	16498	80733	96422	58078	99643
90595	61867	59231	17772	67831	33317	00520
33570	04981	98939	78784	09977	29398	93896
15340	93460	57477	13898	48431	72936	78160
64079	42483	36512	56186	99098	48850	72527
63491	05546	67118	62063	74958	20946	28147
92003	63868	41034	28260	79708	00770	88643
52360	46658	66511	04172	73085	11795	52594
74622	12142	68355	65635	21828	39539	18988
04157	50079	61343	64315	70836	82857	35335
86003	60070	66241	32836	27573	11479	94114
41268	80187	20351	09636	84668	42486	71303

[†] Based on parts of *Tables of 105,000 Random Decimal Digits*. Interstate Commerce Commission, Bureau of Transport Economics and Statistics, Washington D.C.

TABLE XI   Random Numbers (*Continued*)

48611	62866	33963	14045	79451	04934	45576
78812	03509	78673	73181	29973	18664	04555
19472	63971	37271	31445	49019	49405	46925
51266	11569	08697	91120	64156	40365	74297
55806	96275	26130	47949	14877	69594	83041
77527	81360	18180	97421	55541	90275	18213
77680	58788	33016	61173	93049	04694	43534
15404	96554	88265	34537	38526	67924	40474
14045	22917	60718	66487	46346	30949	03173
68376	43918	77653	04127	69930	43283	35766
93385	13421	67957	20384	58731	53396	59723
09858	52104	32014	53115	03727	98624	84616
93307	34116	49516	42148	57740	31198	70336
04794	01534	92058	03157	91758	80611	45357
86265	49096	97021	92582	61422	75890	86442
65943	79232	45702	67055	39024	57383	44424
90038	94209	04055	27393	61517	23002	96560
97283	95943	78363	36498	40662	94188	18202
21913	72958	75637	99936	58715	07943	23748
41161	37341	81838	19389	80336	46346	91895
23777	98392	31417	98547	92058	02277	50315
59973	08144	61070	73094	27059	69181	55623
82690	74099	77885	23813	10054	11900	44653
83854	24715	48866	65745	31131	47636	45137
61980	34997	41825	11623	07320	15003	56774
99915	45821	97702	87125	44488	77613	56823
48293	86847	43186	42951	37804	85129	28993
33225	31280	41232	34750	91097	60752	69783
06846	32828	24425	30249	78801	26977	92074
32671	45587	79620	84831	38156	74211	82752
82096	21913	75544	55228	89796	05694	91552
51666	10433	10945	55306	78562	89630	41230
54044	67942	24145	42294	27427	84875	37022
66738	60184	75679	38120	17640	36242	99357
55064	17427	89180	74018	44865	53197	74810
69599	60264	84549	78007	88450	06488	72274
64756	87759	92354	78694	63638	80939	98644
80817	74533	68407	55862	32476	19326	95558
39847	96884	84657	33697	39578	90197	80532
90401	41700	95510	61166	33757	23279	85523
78227	90110	81378	96659	37008	04050	04228
87240	52716	87697	79433	16336	52862	69149
08486	10951	26832	39763	02485	71688	90936
39338	32169	03713	93510	61244	73774	01245
21188	01850	69689	49426	49128	14660	14143
13287	82531	04388	64693	11934	35051	68576
53609	04001	19648	14053	49623	10840	31915
87900	36194	31567	53506	34304	39910	79630
81641	00496	36058	75899	46620	70024	88753
19512	50277	71508	20116	79520	06269	74173

TABLE XI   Random Numbers (*Continued*)

24418	23508	91507	76455	54941	72711	39406
57404	73678	08272	62941	02349	71389	45605
77644	98489	86268	73652	98210	44546	27174
68366	65614	01443	07607	11826	91326	29664
64472	72294	95432	53555	96810	17100	35066
88205	37913	98633	81009	81060	33449	68055
98455	78685	71250	10329	56135	80647	51404
48977	36794	56054	59243	57361	65304	93258
93077	72941	92779	23581	24548	56415	61927
84533	26564	91583	83411	66504	02036	02922
11338	12903	14514	27585	45068	05520	56321
23853	68500	92274	87026	99717	01542	72990
94096	74920	25822	98026	05394	61840	83089
83160	82362	09350	98536	38155	42661	02363
97425	47335	69709	01386	74319	04318	99387
83951	11954	24317	20345	18134	90062	10761
93085	35203	05740	03206	92012	42710	34650
33762	83193	58045	89880	78101	44392	53767
49665	85397	85137	30496	23469	42846	94810
37541	82627	80051	72521	35342	56119	97190
22145	85304	35348	82854	55846	18076	12415
27153	08662	61078	52433	22184	33998	87436
00301	49425	66682	25442	83668	66236	79655
43815	43272	73778	63469	50083	70696	13558
14689	86482	74157	46012	97765	27552	49617
16680	55936	82453	19532	49988	13176	94219
86938	60429	01137	86168	78257	86249	46134
33944	29219	73161	46061	30946	22210	79302
16045	67736	18608	18198	19468	76358	69203
37044	52523	25627	63107	30806	80857	84383
61471	45322	35340	35132	42163	69332	98851
47422	21296	16785	66393	39249	51463	95963
24133	39719	14484	58613	88717	29289	77360
67253	67064	10748	16006	16767	57345	42285
62382	76941	01635	35829	77516	98468	51686
98011	16503	09201	03523	87192	66483	55649
37366	24386	20654	85117	74078	64120	04643
73587	83993	54176	05221	94119	20108	78101
33583	68291	50547	96085	62180	27453	18567
02878	33223	39199	49536	56199	05993	71201
91498	41673	17195	33175	04994	09879	70337
91127	19815	30219	55591	21725	43827	78862
12997	55013	18662	81724	24305	37661	18956
96098	13651	15393	69995	14762	69734	89150
97627	17837	10472	18983	28387	99781	52977
40064	47981	31484	76603	54088	91095	00010
16239	68743	71374	55863	22672	91609	51514
58354	24913	20435	30965	17453	65623	93058
52567	65085	60220	84641	18273	49604	47418
06236	29052	91392	07551	83532	68130	56970

TABLE XI   Random Numbers (*Continued*)

94620	27963	96478	21559	19246	88097	44926
60947	60775	73181	43264	56895	04232	59604
27499	53523	63110	57106	20865	91683	80688
01603	23156	89223	43429	95353	44662	59433
00815	01552	06392	31437	70385	45863	75971
83844	90942	74857	52419	68723	47830	63010
06626	10042	93629	37609	57215	08409	81906
56760	63348	24949	11859	29793	37457	59377
64416	29934	00755	09418	14230	62887	92683
63569	17906	38076	32135	19096	96970	75917
22693	35089	72994	04252	23791	60249	83010
43413	59744	01275	71326	91382	45114	20245
09224	78530	50566	49965	04851	18280	14039
67625	34683	03142	74733	63558	09665	22610
86874	12549	98699	54952	91579	26023	81076
54548	49505	62515	63903	13193	33905	66936
73236	66167	49728	03581	40699	10396	81827
15220	66319	13543	14071	59148	95154	72852
16151	08029	36954	03891	38313	34016	18671
43635	84249	88984	80993	55431	90793	62603
30193	42776	85611	57635	51362	79907	77364
37430	45246	11400	20986	43996	73122	88474
88312	93047	12088	86937	70794	01041	74867
98995	58159	04700	90443	13168	31553	67891
51734	20849	70198	67906	00880	82899	66065
88698	41755	56216	66852	17748	04963	54859
51865	09836	73966	65711	41699	11732	17173
40300	08852	27528	84648	79589	95295	72895
02760	28625	70476	76410	32988	10194	94917
78450	26245	91763	73117	33047	03577	62599
50252	56911	62693	73817	98693	18728	94741
07929	66728	47761	81472	44806	15592	71357
09030	39605	87507	85446	51257	89555	75520
56670	88445	85799	76200	21795	38894	58070
48140	13583	94911	13318	64741	64336	95103
36764	86132	12463	28385	94242	32063	45233
14351	71381	28133	68269	65145	28152	39087
81276	00835	63835	87174	42446	08882	27067
55524	86088	00069	59254	24654	77371	26409
78852	65889	32719	13758	23937	90740	16866
11861	69032	51915	23510	32050	52052	24004
67699	01009	07050	73324	06732	27510	33761
50064	39500	17450	18030	63124	48061	59412
93126	17700	94400	76075	08317	27324	72723
01657	92602	41043	05686	15650	29970	95877
13800	76690	75133	60456	28491	03845	11507
98135	42870	48578	29036	69876	86563	61729
08313	99293	00990	13595	77457	79969	11339
90974	83965	62732	85161	54330	22406	86253
33273	61993	88407	69399	17301	70975	99129

TABLE XII Random Normal Numbers†

1.801	0.459	1.102	−1.072	−0.336	0.942	−0.290	−0.716	1.396	−0.466
−0.175	−0.754	−0.134	1.231	1.483	−0.149	0.555	1.401	−1.142	0.205
−0.861	−1.460	0.526	0.239	−0.206	2.021	0.313	−0.253	−0.891	1.135
−0.577	0.335	−0.820	0.140	−0.333	0.426	0.209	−0.024	0.323	1.223
0.827	0.802	−0.457	0.560	0.643	−0.729	−0.249	0.338	−0.281	−1.804
−1.344	0.949	−1.459	−1.210	1.016	−0.148	−1.737	0.069	−1.185	0.040
1.476	1.262	−1.428	0.489	−0.523	−0.646	1.721	0.749	0.179	−0.922
0.527	−1.045	0.877	0.646	2.957	−0.972	−1.796	0.309	2.224	−0.070
−0.645	0.117	0.059	−0.080	−1.637	−0.746	1.256	2.520	−0.673	0.994
−0.514	−1.510	−0.714	−1.581	0.905	1.745	1.767	0.682	−0.648	−1.742
−0.656	−0.217	0.287	0.114	1.175	0.791	−0.263	−0.695	−1.348	1.239
−0.778	1.177	0.180	1.156	0.458	1.089	0.339	1.304	0.402	−0.831
0.352	−1.829	−0.645	0.236	0.641	0.920	−1.287	−0.187	−2.339	−0.237
1.352	−0.076	−1.962	0.827	0.252	1.621	0.770	1.324	0.488	−0.037
0.017	0.030	0.211	2.276	0.693	−1.733	0.773	0.652	−0.947	0.148
−0.218	−1.060	−0.553	1.043	2.305	0.380	−0.794	−1.498	1.088	−0.689
1.118	0.816	0.713	0.485	0.185	0.318	−1.050	0.110	0.563	1.177
−1.622	0.436	0.481	0.021	2.070	−0.845	−0.257	−0.680	−0.565	0.024
−1.103	−0.210	−1.088	−0.033	−1.022	0.366	−0.531	2.022	0.210	1.037
−0.677	−0.737	−0.950	−1.517	1.148	0.377	−0.397	−1.902	−0.748	−1.753
1.110	1.120	1.163	1.577	−1.172	−0.133	−0.213	0.154	−0.435	0.218
−0.278	0.569	0.586	1.523	−0.244	−0.170	−1.274	0.874	−1.020	−0.809
0.178	1.314	0.462	−0.253	−0.122	0.108	−1.256	−0.137	1.043	−0.135
0.312	−2.287	−0.655	−1.459	0.075	−0.457	−0.206	−0.326	0.489	−0.149
0.469	−2.066	−0.973	−1.009	−1.410	0.505	0.459	−0.572	−1.186	0.978
−0.730	1.650	0.760	−0.520	−0.671	−0.122	−0.324	−0.202	0.411	−2.103
0.834	0.280	0.744	0.598	0.122	−0.460	−1.310	−1.271	−0.917	0.650
−1.397	−1.053	0.412	1.286	−0.820	−0.371	0.826	−0.666	0.505	0.733
0.238	−0.668	1.861	0.051	0.460	0.079	1.008	−0.487	0.306	−0.061
0.102	−0.907	−0.833	1.103	−0.921	0.145	−0.904	−0.401	0.553	−1.422
−0.160	0.567	−0.638	0.355	0.427	−0.695	−0.846	0.359	1.500	−0.926
0.496	1.179	−0.776	0.511	−1.325	0.275	−0.130	−0.123	1.175	−0.102
0.307	−0.328	−2.474	−0.121	1.371	0.266	1.235	1.827	−0.296	−2.715
−0.559	0.523	1.264	−0.018	−2.791	0.139	1.515	1.976	0.173	−1.728
0.658	−0.261	0.004	−1.296	0.568	−1.215	0.104	0.178	1.126	1.134
−0.856	−2.278	−0.140	−0.164	1.416	−0.043	0.243	−1.399	−0.448	0.120
2.778	0.245	0.282	0.301	−1.506	1.805	1.798	1.078	1.629	−0.648
0.543	0.761	−2.038	−0.533	−0.594	1.742	0.487	1.432	−0.210	−0.358
−0.008	−0.445	−2.551	0.935	1.961	−0.270	−1.557	−1.318	−0.744	−0.860
−1.147	−1.151	−0.522	−2.118	−0.667	0.906	0.639	1.005	−0.480	−1.354
−0.851	0.585	0.672	0.481	−0.888	−0.480	0.041	0.345	−0.537	−0.589
0.023	0.609	0.623	0.356	0.279	−0.051	0.158	−0.353	0.776	0.102
−0.257	0.152	−1.413	0.175	0.149	−1.354	0.286	1.794	−0.571	−0.202
−0.421	−0.344	−0.803	0.832	0.256	−1.296	−1.390	0.379	0.955	0.366
−1.681	2.444	−1.025	1.178	−0.827	−0.200	0.727	0.778	0.169	−1.363
0.717	−1.666	1.071	−2.061	−1.367	−0.450	−0.038	−1.004	−1.240	0.901
−1.266	0.256	−1.312	−0.582	−0.351	−1.002	0.648	0.873	0.015	0.641
0.350	0.552	−1.549	−1.680	1.417	−0.769	−0.514	−1.900	1.017	−1.222
−0.186	0.006	0.148	0.560	−1.081	−0.637	−1.968	−0.623	0.009	−0.369
1.359	1.027	0.740	−2.067	0.543	1.099	0.543	0.064	0.589	−0.016

† Reproduced by permission from RAND Corporation, *A Million Random Digits with 100,000 Normal Deviates* (New York: Macmillan Publishing Co., Inc., third printing, 1966).

TABLE XII   Random Normal Numbers (*Continued*)

0.048	1.040	−0.111	−0.120	1.396	−0.393	−0.220	0.422	0.233	0.197
−0.521	−0.563	−0.116	−0.512	−0.518	−2.194	2.261	0.461	−1.533	−1.836
−1.407	−0.213	0.948	−0.073	−1.474	−0.236	−0.649	1.555	1.285	−0.747
1.822	0.898	−0.691	0.972	−0.011	0.517	0.808	2.651	−0.650	0.592
1.346	−0.137	0.952	1.467	−0.352	0.309	0.578	−1.881	−0.488	−0.329
0.420	−1.085	−1.578	−0.125	1.337	0.169	0.551	−0.745	−0.588	1.810
−1.760	−1.868	0.677	0.545	1.465	0.572	−0.770	0.655	−0.574	1.262
−0.959	0.061	−1.260	−0.573	−0.646	−0.697	−0.026	−1.115	3.591	−0.519
0.561	−0.534	−1.730	−1.172	−0.261	−0.049	0.173	0.027	1.138	0.524
−0.717	0.254	0.421	−1.891	2.592	−1.443	−0.061	−2.520	−0.497	0.909
−2.097	−0.180	−1.298	−0.647	0.159	0.769	−0.735	−0.343	0.966	0.595
0.443	−0.191	0.705	0.420	−0.486	−1.038	−0.396	1.405	0.327	1.198
0.481	0.161	−0.044	−0.864	−0.587	−0.037	−1.304	−1.514	0.946	−0.344
−2.219	−0.123	−0.260	0.680	0.224	−1.217	0.052	0.174	0.692	−1.068
1.723	−0.215	−0.158	0.369	1.073	−2.442	−0.472	2.060	−3.246	−1.020
−0.937	1.253	0.321	−0.541	−0.648	0.265	1.487	−0.554	1.890	0.499
−0.568	−0.146	0.285	1.337	−0.840	0.361	−0.468	0.746	0.470	0.171
−1.717	−1.293	−0.556	−0.545	1.344	0.320	−0.087	0.418	1.076	1.669
−0.151	−0.266	0.920	−2.370	0.484	−1.915	−0.268	0.718	2.075	−0.975
2.278	−1.819	0.245	−0.163	0.980	−1.629	−0.094	−0.573	1.548	−0.896
−0.650	0.669	−0.761	0.154	0.872	0.914	−0.563	−1.434	−0.006	−0.975
−1.086	0.810	0.461	−0.528	2.130	−0.218	0.111	−0.412	−0.580	−1.487
−0.143	−1.196	−1.254	−0.133	0.937	−0.475	−2.348	0.618	−0.057	−0.710
−2.072	0.711	1.241	0.066	−0.341	0.356	1.220	0.431	0.263	−1.623
−0.394	−0.368	−2.108	0.605	0.485	2.068	0.687	−1.474	0.071	−1.196
0.174	−1.131	0.870	2.114	0.201	−0.373	−0.284	−0.234	−2.087	−1.304
0.020	0.102	−1.911	−1.132	1.267	0.420	0.791	1.548	−0.147	−0.453
0.297	0.449	−0.604	−0.858	−1.739	1.143	0.131	0.740	−1.596	0.165
1.160	0.253	0.716	−1.032	−0.595	−1.662	0.632	−0.315	−0.374	0.700
−0.351	−0.490	−0.632	−0.409	−0.116	−1.153	−0.266	−0.125	0.489	−0.366
−0.594	−0.214	−0.461	0.030	−0.595	−0.889	0.638	−0.488	0.418	−0.693
−1.882	1.890	−0.236	0.006	0.966	−0.723	0.229	−2.136	−1.017	−0.008
0.041	2.955	−1.526	2.114	−0.540	1.040	0.753	0.025	0.462	1.221
−0.403	1.237	−1.938	−1.704	−0.103	−0.346	1.214	0.826	0.336	−1.140
−0.068	0.599	0.192	1.503	−0.579	−1.485	−1.645	0.302	−1.348	0.553
−0.361	0.958	0.807	0.787	−0.547	−0.074	−1.378	−0.010	−1.096	0.789
−0.251	0.629	0.459	−0.165	0.016	0.489	−1.205	−0.260	−0.256	−0.399
−1.011	0.893	−0.741	−0.514	−0.576	−0.929	0.478	−0.374	1.950	−0.695
0.780	−2.464	−0.522	0.767	−1.657	−0.983	0.217	−0.529	−0.648	1.454
−0.712	−0.355	−0.564	1.052	−0.169	−0.410	1.543	−2.330	−0.008	−0.955
−0.612	−1.068	−0.644	−0.007	−0.835	0.623	0.093	0.105	−0.318	−0.228
−0.064	0.012	−0.676	0.349	0.303	1.539	0.792	−0.101	−0.344	−0.096
−0.379	1.504	2.375	0.498	−0.996	0.174	−1.268	−1.137	−0.618	0.173
1.145	−1.403	0.770	0.799	0.844	−1.361	−1.059	0.128	1.398	0.277
−0.117	0.585	−1.763	−0.632	0.239	−0.854	1.684	1.024	−0.067	−0.045
1.333	1.374	−0.515	−1.655	0.607	−0.885	−0.902	−1.010	−1.297	−0.139
−0.249	−0.747	1.044	−0.930	0.346	0.575	0.335	−1.159	−1.651	−1.642
−1.022	0.085	−1.441	−0.198	0.844	0.697	0.548	−0.080	0.656	0.443
−0.780	−0.534	−0.339	−0.642	−0.902	−0.827	0.071	−0.678	−0.359	−0.479
−0.687	−0.418	0.991	0.331	−1.003	0.061	−1.416	0.876	0.125	−2.246

TABLE XIII Values of e^{-x}

x	e^{-x}	x	e^{-x}	x	e^{-x}	x	e^{-x}
0.0	1.000	2.5	0.082	5.0	0.0067	7.5	0.00055
0.1	0.905	2.6	0.074	5.1	0.0061	7.6	0.00050
0.2	0.819	2.7	0.067	5.2	0.0055	7.7	0.00045
0.3	0.741	2.8	0.061	5.3	0.0050	7.8	0.00041
0.4	0.670	2.9	0.055	5.4	0.0045	7.9	0.00037
0.5	0.607	3.0	0.050	5.5	0.0041	8.0	0.00034
0.6	0.549	3.1	0.045	5.6	0.0037	8.1	0.00030
0.7	0.497	3.2	0.041	5.7	0.0033	8.2	0.00028
0.8	0.449	3.3	0.037	5.8	0.0030	8.3	0.00025
0.9	0.407	3.4	0.033	5.9	0.0027	8.4	0.00023
1.0	0.368	3.5	0.030	6.0	0.0025	8.5	0.00020
1.1	0.333	3.6	0.027	6.1	0.0022	8.6	0.00018
1.2	0.301	3.7	0.025	6.2	0.0020	8.7	0.00017
1.3	0.273	3.8	0.022	6.3	0.0018	8.8	0.00015
1.4	0.247	3.9	0.020	6.4	0.0017	8.9	0.00014
1.5	0.223	4.0	0.018	6.5	0.0015	9.0	0.00012
1.6	0.202	4.1	0.017	6.6	0.0014	9.1	0.00011
1.7	0.183	4.2	0.015	6.7	0.0012	9.2	0.00010
1.8	0.165	4.3	0.014	6.8	0.0011	9.3	0.00009
1.9	0.150	4.4	0.012	6.9	0.0010	9.4	0.00008
2.0	0.135	4.5	0.011	7.0	0.0009	9.5	0.00008
2.1	0.122	4.6	0.010	7.1	0.0008	9.6	0.00007
2.2	0.111	4.7	0.009	7.2	0.0007	9.7	0.00006
2.3	0.100	4.8	0.008	7.3	0.0007	9.8	0.00006
2.4	0.091	4.9	0.007	7.4	0.0006	9.9	0.00005

TABLE XIV   Logarithms

N	0	1	2	3	4	5	6	7	8	9
10	0000	0043	0086	0128	0170	0212	0253	0294	0334	0374
11	0414	0453	0492	0531	0569	0607	0645	0682	0719	0755
12	0792	0828	0864	0899	0934	0969	1004	1038	1072	1106
13	1139	1173	1206	1239	1271	1303	1335	1367	1399	1430
14	1461	1492	1523	1553	1584	1614	1644	1673	1703	1732
15	1761	1790	1818	1847	1875	1903	1931	1959	1987	2014
16	2041	2068	2095	2122	2148	2175	2201	2227	2253	2279
17	2304	2330	2355	2380	2405	2430	2455	2480	2504	2529
18	2553	2577	2601	2625	2648	2672	2695	2718	2742	2765
19	2788	2810	2833	2856	2878	2900	2923	2945	2967	2989
20	3010	3032	3054	3075	3096	3118	3139	3160	3181	3201
21	3222	3243	3263	3284	3304	3324	3345	3365	3385	3404
22	3424	3444	3464	3483	3502	3522	3541	3560	3579	3598
23	3617	3636	3655	3674	3692	3711	3729	3747	3766	3784
24	3802	3820	3838	3856	3874	3892	3909	3927	3945	3962
25	3979	3997	4014	4031	4048	4065	4082	4099	4116	4133
26	4150	4166	4183	4200	4216	4232	4249	4265	4281	4298
27	4314	4330	4346	4362	4378	4393	4409	4425	4440	4456
28	4472	4487	4502	4518	4533	4548	4564	4579	4594	4609
29	4624	4639	4654	4669	4683	4698	4713	4728	4742	4757
30	4771	4786	4800	4814	4829	4843	4857	4871	4886	4900
31	4914	4928	4942	4955	4969	4983	4997	5011	5024	5038
32	5051	5065	5079	5092	5105	5119	5132	5145	5159	5172
33	5185	5198	5211	5224	5237	5250	5263	5276	5289	5302
34	5315	5328	5340	5353	5366	5378	5391	5403	5416	5428
35	5441	5453	5465	5478	5490	5502	5514	5527	5539	5551
36	5563	5575	5587	5599	5611	5623	5635	5647	5658	5670
37	5682	5694	5705	5717	5729	5740	5752	5763	5775	5786
38	5798	5809	5821	5832	5843	5855	5866	5877	5888	5899
39	5911	5922	5933	5944	5955	5966	5977	5988	5999	6010
40	6021	6031	6042	6053	6064	6075	6085	6096	6107	6117
41	6128	6138	6149	6160	6170	6180	6191	6201	6212	6222
42	6232	6243	6253	6263	6274	6284	6294	6304	6314	6325
43	6335	6345	6355	6365	6375	6385	6395	6405	6415	6425
44	6435	6444	6454	6464	6474	6484	6493	6503	6513	6522
45	6532	6542	6551	6561	6571	6580	6590	6599	6609	6618
46	6628	6637	6646	6656	6665	6675	6684	6693	6702	6712
47	6721	6730	6739	6749	6758	6767	6776	6785	6794	6803
48	6812	6821	6830	6839	6848	6857	6866	6875	6884	6893
49	6902	6911	6920	6928	6937	6946	6955	6964	6972	6981
50	6990	6998	7007	7016	7024	7033	7042	7050	7059	7067
51	7076	7084	7093	7101	7110	7118	7126	7135	7143	7152
52	7160	7168	7177	7185	7193	7202	7210	7218	7226	7235
53	7243	7251	7259	7267	7275	7284	7292	7300	7308	7316
54	7324	7332	7340	7348	7356	7364	7372	7380	7388	7396

TABLE XIV   Logarithms (*Continued*)

N	0	1	2	3	4	5	6	7	8	9
55	7404	7412	7419	7427	7435	7443	7451	7459	7466	7474
56	7482	7490	7497	7505	7513	7520	7528	7536	7543	7551
57	7559	7566	7574	7582	7589	7597	7604	7612	7619	7627
58	7634	7642	7649	7657	7664	7672	7679	7686	7694	7701
59	7709	7716	7723	7731	7738	7745	7752	7760	7767	7774
60	7782	7789	7796	7803	7810	7818	7825	7832	7839	7846
61	7853	7860	7868	7875	7882	7889	7896	7903	7910	7917
62	7924	7931	7938	7945	7952	7959	7966	7973	7980	7987
63	7993	8000	8007	8014	8021	8028	8035	8041	8048	8055
64	8062	8069	8075	8082	8089	8096	8102	8109	8116	8122
65	8129	8136	8142	8149	8156	8162	8169	8176	8182	8189
66	8195	8202	8209	8215	8222	8228	8235	8241	8248	8254
67	8261	8267	8274	8280	8287	8293	8299	8306	8312	8319
68	8325	8331	8338	8344	8351	8357	8363	8370	8376	8382
69	8388	8395	8401	8407	8414	8420	8426	8432	8439	8445
70	8451	8457	8463	8470	8476	8482	8488	8494	8500	8506
71	8513	8519	8525	8531	8537	8543	8549	8555	8561	8567
72	8573	8579	8585	8591	8597	8603	8609	8615	8621	8627
73	8633	8639	8645	8651	8657	8663	8669	8675	8681	8686
74	8692	8698	8704	8710	8716	8722	8727	8733	8739	8745
75	8751	8756	8762	8768	8774	8779	8785	8791	8797	8802
76	8808	8814	8820	8825	8831	8837	8842	8848	8854	8859
77	8865	8871	8876	8882	8887	8893	8899	8904	8910	8915
78	8921	8927	8932	8938	8943	8949	8954	8960	8965	8971
79	8976	8982	8987	8993	8998	9004	9009	9015	9020	9025
80	9031	9036	9042	9047	9053	9058	9063	9069	9074	9079
81	9085	9090	9096	9101	9106	9112	9117	9122	9128	9133
82	9138	9143	9149	9154	9159	9165	9170	9175	9180	9186
83	9191	9196	9201	9206	9212	9217	9222	9227	9232	9238
84	9243	9248	9253	9258	9263	9269	9274	9279	9284	9289
85	9294	9299	9304	9309	9315	9320	9325	9330	9335	9340
86	9345	9350	9355	9360	9365	9370	9375	9380	9385	9390
87	9395	9400	9405	9410	9415	9420	9425	9430	9435	9440
88	9445	9450	9455	9460	9465	9469	9474	9479	9484	9489
89	9494	9499	9504	9509	9513	9518	9523	9528	9533	9538
90	9542	9547	9552	9557	9562	9566	9571	9576	9581	9586
91	9590	9595	9600	9605	9609	9614	9619	9624	9628	9633
92	9638	9643	9647	9652	9657	9661	9666	9671	9675	9680
93	9685	9689	9694	9699	9703	9708	9713	9717	9722	9727
94	9731	9736	9741	9745	9750	9754	9759	9763	9768	9773
95	9777	9782	9786	9791	9795	9800	9805	9809	9814	9818
96	9823	9827	9832	9836	9341	9845	9850	9854	9859	9863
97	9868	9872	9877	9881	9886	9890	9894	9899	9903	9908
98	9912	9917	9921	9926	9930	9934	9939	9943	9948	9952
99	9956	9961	9965	9969	9974	9978	9983	9987	9991	9996

Table XV contains the square roots of the numbers from 1.00 to 9.99, and also the square roots of these numbers multiplied by 10, spaced at intervals of 0.01. The square roots are all rounded to four decimals. To find the square root of any positive number rounded to three significant digits, we use the following rule in deciding whether to take the entry of the $\sqrt{n}$ or $\sqrt{10n}$ column:

Move the decimal point an even number of places to the right or to the left until a number greater than or equal to 1 but less than 100 is reached. If the resulting number is less than 10 go to the $\sqrt{n}$ column; if it is 10 or more go to the $\sqrt{10n}$ column.

Thus, to find the square roots of 12,800, 379, and 0.0812, we go to the $\sqrt{n}$ column since the decimal point has to be moved, respectively, four places to the left, two places to the left, and two places to the right, to give 1.28, 3.79, and 8.12. Similarly, to find the square roots of 5,240, 0.281, and 0.0000259, we go to the $\sqrt{10n}$ column since the decimal point has to be moved, respectively, two places to the left, two places to the right, and six places to the right, to give 52.4, 28.1, and 25.9.

After we locate a square root in the appropriate column of Table XV, we must be sure to get the decimal point in the right place. Here it will help to use the following rule:

Having previously moved the decimal point an even number of places to the left or right to get a number greater than or equal to 1 but less than 100, move the decimal point of the entry of the appropriate column in Table XV half as many places in the opposite direction.

For example, to determine the square root of 12,800, we first note that the decimal point has to be moved *four places to the left* to give 1.28. We then take the entry of the $\sqrt{n}$ column corresponding to 1.28, move its decimal point *two places to the right*, and get $\sqrt{12,800} = 113.14$. Similarly, to find the square root of 0.0000259, we note that the decimal point has to be moved *six places to the right* to give 25.9. We thus take the entry of the $\sqrt{10n}$ column corresponding to 2.59, move the decimal point *three places to the left*, and get $\sqrt{0.0000259} = 0.0050892$. In actual practice, if a number whose square root we want to find is rounded, the square root will have to be rounded to as many significant digits as the original number.

TABLE XV Square Roots

n	$\sqrt{n}$	$\sqrt{10n}$	n	$\sqrt{n}$	$\sqrt{10n}$	n	$\sqrt{n}$	$\sqrt{10n}$
1.00	1.0000	3.1623	1.50	1.2247	3.8730	2.00	1.4142	4.4721
1.01	1.0050	3.1780	1.51	1.2288	3.8859	2.01	1.4177	4.4833
1.02	1.0100	3.1937	1.52	1.2329	3.8987	2.02	1.4213	4.4944
1.03	1.0149	3.2094	1.53	1.2369	3.9115	2.03	1.4248	4.5056
1.04	1.0198	3.2249	1.54	1.2410	3.9243	2.04	1.4283	4.5166
1.05	1.0247	3.2404	1.55	1.2450	3.9370	2.05	1.4318	4.5277
1.06	1.0296	3.2558	1.56	1.2490	3.9497	2.06	1.4353	4.5387
1.07	1.0344	3.2711	1.57	1.2530	3.9623	2.07	1.4387	4.5497
1.08	1.0392	3.2863	1.58	1.2570	3.9749	2.08	1.4422	4.5607
1.09	1.0440	3.3015	1.59	1.2610	3.9875	2.09	1.4457	4.5717
1.10	1.0488	3.3166	1.60	1.2649	4.0000	2.10	1.4491	4.5826
1.11	1.0536	3.3317	1.61	1.2689	4.0125	2.11	1.4526	4.5935
1.12	1.0583	3.3466	1.62	1.2728	4.0249	2.12	1.4560	4.6043
1.13	1.0630	3.3615	1.63	1.2767	4.0373	2.13	1.4595	4.6152
1.14	1.0677	3.3764	1.64	1.2806	4.0497	2.14	1.4629	4.6260
1.15	1.0724	3.3912	1.65	1.2845	4.0620	2.15	1.4663	4.6368
1.16	1.0770	3.4059	1.66	1.2884	4.0743	2.16	1.4697	4.6476
1.17	1.0817	3.4205	1.67	1.2923	4.0866	2.17	1.4731	4.6583
1.18	1.0863	3.4351	1.68	1.2961	4.0988	2.18	1.4765	4.6690
1.19	1.0909	3.4496	1.69	1.3000	4.1110	2.19	1.4799	4.6797
1.20	1.0954	3.4641	1.70	1.3038	4.1231	2.20	1.4832	4.6904
1.21	1.1000	3.4785	1.71	1.3077	4.1352	2.21	1.4866	4.7011
1.22	1.1045	3.4928	1.72	1.3115	4.1473	2.22	1.4900	4.7117
1.23	1.1091	3.5071	1.73	1.3153	4.1593	2.23	1.4933	4.7223
1.24	1.1136	3.5214	1.74	1.3191	4.1713	2.24	1.4967	4.7329
1.25	1.1180	3.5355	1.75	1.3229	4.1833	2.25	1.5000	4.7434
1.26	1.1225	3.5496	1.76	1.3266	4.1952	2.26	1.5033	4.7539
1.27	1.1269	3.5637	1.77	1.3304	4.2071	2.27	1.5067	4.7645
1.28	1.1314	3.5777	1.78	1.3342	4.2190	2.28	1.5100	4.7749
1.29	1.1358	3.5917	1.79	1.3379	4.2308	2.29	1.5133	4.7854
1.30	1.1402	3.6056	1.80	1.3416	4.2426	2.30	1.5166	4.7958
1.31	1.1446	3.6194	1.81	1.3454	4.2544	2.31	1.5199	4.8062
1.32	1.1489	3.6332	1.82	1.3491	4.2661	2.32	1.5232	4.8166
1.33	1.1533	3.6469	1.83	1.3528	4.2778	2.33	1.5264	4.8270
1.34	1.1576	3.6606	1.84	1.3565	4.2895	2.34	1.5297	4.8374
1.35	1.1619	3.6742	1.85	1.3601	4.3012	2.35	1.5330	4.8477
1.36	1.1662	3.6878	1.86	1.3638	4.3128	2.36	1.5362	4.8580
1.37	1.1705	3.7014	1.87	1.3675	4.3243	2.37	1.5395	4.8683
1.38	1.1747	3.7148	1.88	1.3711	4.3359	2.38	1.5427	4.8785
1.39	1.1790	3.7283	1.89	1.3748	4.3474	2.39	1.5460	4.8888
1.40	1.1832	3.7417	1.90	1.3784	4.3589	2.40	1.5492	4.8990
1.41	1.1874	3.7550	1.91	1.3820	4.3704	2.41	1.5524	4.9092
1.42	1.1916	3.7683	1.92	1.3856	4.3818	2.42	1.5556	4.9193
1.43	1.1958	3.7815	1.93	1.3892	4.3932	2.43	1.5588	4.9295
1.44	1.2000	3.7947	1.94	1.3928	4.4045	2.44	1.5620	4.9396
1.45	1.2042	3.8079	1.95	1.3964	4.4159	2.45	1.5652	4.9497
1.46	1.2083	3.8210	1.96	1.4000	4.4272	2.46	1.5684	4.9598
1.47	1.2124	3.8341	1.97	1.4036	4.4385	2.47	1.5716	4.9699
1.48	1.2166	3.8471	1.98	1.4071	4.4497	2.48	1.5748	4.9800
1.49	1.2207	3.8601	1.99	1.4107	4.4609	2.49	1.5780	4.9900

TABLE XV   Square Roots (*Continued*)

n	$\sqrt{n}$	$\sqrt{10n}$	n	$\sqrt{n}$	$\sqrt{10n}$	n	$\sqrt{n}$	$\sqrt{10n}$
2.50	1.5811	5.0000	3.00	1.7321	5.4772	3.50	1.8708	5.9161
2.51	1.5843	5.0100	3.01	1.7349	5.4863	3.51	1.8735	5.9245
2.52	1.5875	5.0200	3.02	1.7378	5.4955	3.52	1.8762	5.9330
2.53	1.5906	5.0299	3.03	1.7407	5.5045	3.53	1.8788	5.9414
2.54	1.5937	5.0398	3.04	1.7436	5.5136	3.54	1.8815	5.9498
2.55	1.5969	5.0498	3.05	1.7464	5.5227	3.55	1.8841	5.9582
2.56	1.6000	5.0596	3.06	1.7493	5.5317	3.56	1.8868	5.9666
2.57	1.6031	5.0695	3.07	1.7521	5.5408	3.57	1.8894	5.9749
2.58	1.6062	5.0794	3.08	1.7550	5.5498	3.58	1.8921	5.9833
2.59	1.6093	5.0892	3.09	1.7578	5.5588	3.59	1.8947	5.9917
2.60	1.6125	5.0990	3.10	1.7607	5.5678	3.60	1.8974	6.0000
2.61	1.6155	5.1088	3.11	1.7635	5.5767	3.61	1.9000	6.0083
2.62	1.6186	5.1186	3.12	1.7664	5.5857	3.62	1.9026	6.0166
2.63	1.6217	5.1284	3.13	1.7692	5.5946	3.63	1.9053	6.0249
2.64	1.6248	5.1381	3.14	1.7720	5.6036	3.64	1.9079	6.0332
2.65	1.6279	5.1478	3.15	1.7748	5.6125	3.65	1.9105	6.0415
2.66	1.6310	5.1575	3.16	1.7776	5.6214	3.66	1.9131	6.0498
2.67	1.6340	5.1672	3.17	1.7804	5.6303	3.67	1.9157	6.0581
2.68	1.6371	5.1769	3.18	1.7833	5.6391	3.68	1.9183	6.0663
2.69	1.6401	5.1865	3.19	1.7861	5.6480	3.69	1.9209	6.0745
2.70	1.6432	5.1962	3.20	1.7889	5.6569	3.70	1.9235	6.0828
2.71	1.6462	5.2058	3.21	1.7916	5.6657	3.71	1.9261	6.0910
2.72	1.6492	5.2154	3.22	1.7944	5.6745	3.72	1.9287	6.0992
2.73	1.6523	5.2249	3.23	1.7972	5.6833	3.73	1.9313	6.1074
2.74	1.6553	5.2345	3.24	1.8000	5.6921	3.74	1.9339	6.1156
2.75	1.6583	5.2440	3.25	1.8028	5.7009	3.75	1.9365	6.1237
2.76	1.6613	5.2536	3.26	1.8055	5.7096	3.76	1.9391	6.1319
2.77	1.6643	5.2631	3.27	1.8083	5.7184	3.77	1.9416	6.1400
2.78	1.6673	5.2726	3.28	1.8111	5.7271	3.78	1.9442	6.1482
2.79	1.6703	5.2820	3.29	1.8138	5.7359	3.79	1.9468	6.1563
2.80	1.6733	5.2915	3.30	1.8166	5.7446	3.80	1.9494	6.1644
2.81	1.6763	5.3009	3.31	1.8193	5.7533	3.81	1.9519	6.1725
2.82	1.6793	5.3104	3.32	1.8221	5.7619	3.82	1.9545	6.1806
2.83	1.6823	5.3198	3.33	1.8248	5.7706	3.83	1.9570	6.1887
2.84	1.6852	5.3292	3.34	1.8276	5.7793	3.84	1.9596	6.1968
2.85	1.6882	5.3385	3.35	1.8303	5.7879	3.85	1.9621	6.2048
2.86	1.6912	5.3479	3.36	1.8330	5.7966	3.86	1.9647	6.2129
2.87	1.6941	5.3572	3.37	1.8358	5.8052	3.87	1.9672	6.2209
2.88	1.6971	5.3666	3.38	1.8385	5.8138	3.88	1.9698	6.2290
2.89	1.7000	5.3759	3.39	1.8412	5.8224	3.89	1.9723	6.2370
2.90	1.7029	5.3852	3.40	1.8439	5.8310	3.90	1.9748	6.2450
2.91	1.7059	5.3944	3.41	1.8466	5.8395	3.91	1.9774	6.2530
2.92	1.7088	5.4037	3.42	1.8493	5.8481	3.92	1.9799	6.2610
2.93	1.7117	5.4129	3.43	1.8520	5.8566	3.93	1.9824	6.2690
2.94	1.7146	5.4222	3.44	1.8547	5.8652	3.94	1.9849	6.2769
2.95	1.7176	5.4314	3.45	1.8574	5.8737	3.95	1.9875	6.2849
2.96	1.7205	5.4406	3.46	1.8601	5.8822	3.96	1.9900	6.2929
2.97	1.7234	5.4498	3.47	1.8628	5.8907	3.97	1.9925	6.3008
2.98	1.7263	5.4589	3.48	1.8655	5.8992	3.98	1.9950	6.3087
2.99	1.7292	5.4681	3.49	1.8682	5.9076	3.99	1.9975	6.3166

TABLE XV   Square Roots (*Continued*)

n	$\sqrt{n}$	$\sqrt{10n}$	n	$\sqrt{n}$	$\sqrt{10n}$	n	$\sqrt{n}$	$\sqrt{10n}$
4.00	2.0000	6.3246	4.50	2.1213	6.7082	5.00	2.2361	7.0711
4.01	2.0025	6.3325	4.51	2.1237	6.7157	5.01	2.2383	7.0781
4.02	2.0050	6.3403	4.52	2.1260	6.7231	5.02	2.2405	7.0852
4.03	2.0075	6.3482	4.53	2.1284	6.7305	5.03	2.2428	7.0922
4.04	2.0100	6.3561	4.54	2.1307	6.7380	5.04	2.2450	7.0993
4.05	2.0125	6.3640	4.55	2.1331	6.7454	5.05	2.2472	7.1063
4.06	2.0149	6.3718	4.56	2.1354	6.7528	5.06	2.2494	7.1134
4.07	2.0174	6.3797	4.57	2.1378	6.7602	5.07	2.2517	7.1204
4.08	2.0199	6.3875	4.58	2.1401	6.7676	5.08	2.2539	7.1274
4.09	2.0224	6.3953	4.59	2.1424	6.7750	5.09	2.2561	7.1344
4.10	2.0248	6.4031	4.60	2.1448	6.7823	5.10	2.2583	7.1414
4.11	2.0273	6.4109	4.61	2.1471	6.7897	5.11	2.2605	7.1484
4.12	2.0298	6.4187	4.62	2.1494	6.7971	5.12	2.2627	7.1554
4.13	2.0322	6.4265	4.63	2.1517	6.8044	5.13	2.2650	7.1624
4.14	2.0347	6.4343	4.64	2.1541	6.8118	5.14	2.2672	7.1694
4.15	2.0372	6.4420	4.65	2.1564	6.8191	5.15	2.2694	7.1764
4.16	2.0396	6.4498	4.66	2.1587	6.8264	5.16	2.2716	7.1833
4.17	2.0421	6.4576	4.67	2.1610	6.8337	5.17	2.2738	7.1903
4.18	2.0445	6.4653	4.68	2.1633	6.8411	5.18	2.2760	7.1972
4.19	2.0469	6.4730	4.69	2.1656	6.8484	5.19	2.2782	7.2042
4.20	2.0494	6.4807	4.70	2.1679	6.8557	5.20	2.2804	7.2111
4.21	2.0518	6.4885	4.71	2.1703	6.8629	5.21	2.2825	7.2180
4.22	2.0543	6.4962	4.72	2.1726	6.8702	5.22	2.2847	7.2250
4.23	2.0567	6.5038	4.73	2.1749	6.8775	5.23	2.2869	7.2319
4.24	2.0591	6.5115	4.74	2.1772	6.8848	5.24	2.2891	7.2388
4.25	2.0616	6.5192	4.75	2.1794	6.8920	5.25	2.2913	7.2457
4.26	2.0640	6.5269	4.76	2.1817	6.8993	5.26	2.2935	7.2526
4.27	2.0664	6.5345	4.77	2.1840	6.9065	5.27	2.2956	7.2595
4.28	2.0688	6.5422	4.78	2.1863	6.9138	5.28	2.2978	7.2664
4.29	2.0712	6.5498	4.79	2.1886	6.9210	5.29	2.3000	7.2732
4.30	2.0736	6.5574	4.80	2.1909	6.9282	5.30	2.3022	7.2801
4.31	2.0761	6.5651	4.81	2.1932	6.9354	5.31	2.3043	7.2870
4.32	2.0785	6.5727	4.82	2.1954	6.9426	5.32	2.3065	7.2938
4.33	2.0809	6.5803	4.83	2.1977	6.9498	5.33	2.3087	7.3007
4.34	2.0833	6.5879	4.84	2.2000	6.9570	5.34	2.3108	7.3075
4.35	2.0857	6.5955	4.85	2.2023	6.9642	5.35	2.3130	7.3144
4.36	2.0881	6.6030	4.86	2.2045	6.9714	5.36	2.3152	7.3212
4.37	2.0905	6.6106	4.87	2.2068	6.9785	5.37	2.3173	7.3280
4.38	2.0928	6.6182	4.88	2.2091	6.9857	5.38	2.3195	7.3348
4.39	2.0952	6.6257	4.89	2.2113	6.9929	5.39	2.3216	7.3417
4.40	2.0976	6.6332	4.90	2.2136	7.0000	5.40	2.3238	7.3485
4.41	2.1000	6.6408	4.91	2.2159	7.0071	5.41	2.3259	7.3553
4.42	2.1024	6.6483	4.92	2.2181	7.0143	5.42	2.3281	7.3621
4.43	2.1048	6.6558	4.93	2.2204	7.0214	5.43	2.3302	7.3689
4.44	2.1071	6.6633	4.94	2.2226	7.0285	5.44	2.3324	7.3756
4.45	2.1095	6.6708	4.95	2.2249	7.0356	5.45	2.3345	7.3824
4.46	2.1119	6.6783	4.96	2.2271	7.0427	5.46	2.3367	7.3892
4.47	2.1142	6.6858	4.97	2.2293	7.0498	5.47	2.3388	7.3959
4.48	2.1166	6.6933	4.98	2.2316	7.0569	5.48	2.3409	7.4027
4.49	2.1190	6.7007	4.99	2.2338	7.0640	5.49	2.3431	7.4095

TABLE XV   Square Roots (*Continued*)

n	$\sqrt{n}$	$\sqrt{10n}$	n	$\sqrt{n}$	$\sqrt{10n}$	n	$\sqrt{n}$	$\sqrt{10n}$
5.50	2.3452	7.4162	6.00	2.4495	7.7460	6.50	2.5495	8.0623
5.51	2.3473	7.4229	6.01	2.4515	7.7524	6.51	2.5515	8.0685
5.52	2.3495	7.4297	6.02	2.4536	7.7589	6.52	2.5534	8.0747
5.53	2.3516	7.4364	6.03	2.4556	7.7653	6.53	2.5554	8.0808
5.54	2.3537	7.4431	6.04	2.4576	7.7717	6.54	2.5573	8.0870
5.55	2.3558	7.4498	6.05	2.4597	7.7782	6.55	2.5593	8.0932
5.56	2.3580	7.4565	6.06	2.4617	7.7846	6.56	2.5612	8.0994
5.57	2.3601	7.4632	6.07	2.4637	7.7910	6.57	2.5632	8.1056
5.58	2.3622	7.4699	6.08	2.4658	7.7974	6.58	2.5652	8.1117
5.59	2.3643	7.4766	6.09	2.4678	7.8038	6.59	2.5671	8.1179
5.60	2.3664	7.4833	6.10	2.4698	7.8102	6.60	2.5690	8.1240
5.61	2.3685	7.4900	6.11	2.4718	7.8166	6.61	2.5710	8.1302
5.62	2.3707	7.4967	6.12	2.4739	7.8230	6.62	2.5729	8.1363
5.63	2.3728	7.5033	6.13	2.4759	7.8294	6.63	2.5749	8.1425
5.64	2.3749	7.5100	6.14	2.4779	7.8358	6.64	2.5768	8.1486
5.65	2.3770	7.5166	6.15	2.4799	7.8422	6.65	2.5788	8.1548
5.66	2.3791	7.5233	6.16	2.4819	7.8486	6.66	2.5807	8.1609
5.67	2.3812	7.5299	6.17	2.4839	7.8549	6.67	2.5826	8.1670
5.68	2.3833	7.5366	6.18	2.4860	7.8613	6.68	2.5846	8.1731
5.69	2.3854	7.5432	6.19	2.4880	7.8677	6.69	2.5865	8.1792
5.70	2.3875	7.5498	6.20	2.4900	7.8740	6.70	2.5884	8.1854
5.71	2.3896	7.5565	6.21	2.4920	7.8804	6.71	2.5904	8.1915
5.72	2.3917	7.5631	6.22	2.4940	7.8867	6.72	2.5923	8.1976
5.73	2.3937	7.5697	6.23	2.4960	7.8930	6.73	2.5942	8.2037
5.74	2.3958	7.5763	6.24	2.4980	7.8994	6.74	2.5962	8.2098
5.75	2.3979	7.5829	6.25	2.5000	7.9057	6.75	2.5981	8.2158
5.76	2.4000	7.5895	6.26	2.5020	7.9120	6.76	2.6000	8.2219
5.77	2.4021	7.5961	6.27	2.5040	7.9183	6.77	2.6019	8.2280
5.78	2.4042	7.6026	6.28	2.5060	7.9246	6.78	2.6038	8.2341
5.79	2.4062	7.6092	6.29	2.5080	7.9310	6.79	2.6058	8.2401
5.80	2.4083	7.6158	6.30	2.5100	7.9373	6.80	2.6077	8.2462
5.81	2.4104	7.6223	6.31	2.5120	7.9436	6.81	2.6096	8.2523
5.82	2.4125	7.6289	6.32	2.5140	7.9498	6.82	2.6115	8.2583
5.83	2.4145	7.6354	6.33	2.5159	7.9561	6.83	2.6134	8.2644
5.84	2.4166	7.6420	6.34	2.5179	7.9624	6.84	2.6153	8.2704
5.85	2.4187	7.6485	6.35	2.5199	7.9687	6.85	2.6173	8.2765
5.86	2.4207	7.6551	6.36	2.5219	7.9750	6.86	2.6192	8.2825
5.87	2.4228	7.6616	6.37	2.5239	7.9812	6.87	2.6211	8.2885
5.88	2.4249	7.6681	6.38	2.5259	7.9875	6.88	2.6230	8.2946
5.89	2.4269	7.6746	6.39	2.5278	7.9937	6.89	2.6249	8.3006
5.90	2.4290	7.6811	6.40	2.5298	8.0000	6.90	2.6268	8.3066
5.91	2.4310	7.6877	6.41	2.5318	8.0062	6.91	2.6287	8.3126
5.92	2.4331	7.6942	6.42	2.5338	8.0125	6.92	2.6306	8.3187
5.93	2.4352	7.7006	6.43	2.5357	8.0187	6.93	2.6325	8.3247
5.94	2.4372	7.7071	6.44	2.5377	8.0250	6.94	2.6344	8.3307
5.95	2.4393	7.7136	6.45	2.5397	8.0312	6.95	2.6363	8.3367
5.96	2.4413	7.7201	6.46	2.5417	8.0374	6.96	2.6382	8.3427
5.97	2.4434	7.7266	6.47	2.5436	8.0436	6.97	2.6401	8.3487
5.98	2.4454	7.7330	6.48	2.5456	8.0498	6.98	2.6420	8.3546
5.99	2.4474	7.7395	6.49	2.5475	8.0561	6.99	2.6439	8.3606

TABLE XV   Square Roots (*Continued*)

n	$\sqrt{n}$	$\sqrt{10n}$	n	$\sqrt{n}$	$\sqrt{10n}$	n	$\sqrt{n}$	$\sqrt{10n}$
7.00	2.6458	8.3666	7.50	2.7386	8.6603	8.00	2.8284	8.9443
7.01	2.6476	8.3726	7.51	2.7404	8.6660	8.01	2.8302	8.9499
7.02	2.6495	8.3785	7.52	2.7423	8.6718	8.02	2.8320	8.9554
7.03	2.6514	8.3845	7.53	2.7441	8.6776	8.03	2.8337	8.9610
7.04	2.6533	8.3905	7.54	2.7459	8.6833	8.04	2.8355	8.9666
7.05	2.6552	8.3964	7.55	2.7477	8.6891	8.05	2.8373	8.9722
7.06	2.6571	8.4024	7.56	2.7495	8.6948	8.06	2.8390	8.9778
7.07	2.6589	8.4083	7.57	2.7514	8.7006	8.07	2.8408	8.9833
7.08	2.6608	8.4143	7.58	2.7532	8.7063	8.08	2.8425	8.9889
7.09	2.6627	8.4202	7.59	2.7550	8.7121	8.09	2.8443	8.9944
7.10	2.6646	8.4261	7.60	2.7568	8.7178	8.10	2.8460	9.0000
7.11	2.6665	8.4321	7.61	2.7586	8.7235	8.11	2.8478	9.0056
7.12	2.6683	8.4380	7.62	2.7604	8.7293	8.12	2.8496	9.0111
7.13	2.6702	8.4439	7.63	2.7622	8.7350	8.13	2.8513	9.0167
7.14	2.6721	8.4499	7.64	2.7641	8.7407	8.14	2.8531	9.0222
7.15	2.6739	8.4558	7.65	2.7659	8.7464	8.15	2.8548	9.0277
7.16	2.6758	8.4617	7.66	2.7677	8.7521	8.16	2.8566	9.0333
7.17	2.6777	8.4676	7.67	2.7695	8.7579	8.17	2.8583	9.0388
7.18	2.6796	8.4735	7.68	2.7713	8.7636	8.18	2.8601	9.0443
7.19	2.6814	8.4794	7.69	2.7731	8.7693	8.19	2.8618	9.0499
7.20	2.6833	8.4853	7.70	2.7749	8.7750	8.20	2.8636	9.0554
7.21	2.6851	8.4912	7.71	2.7767	8.7807	8.21	2.8653	9.0609
7.22	2.6870	8.4971	7.72	2.7785	8.7864	8.22	2.8671	9.0664
7.23	2.6889	8.5029	7.73	2.7803	8.7920	8.23	2.8688	9.0719
7.24	2.6907	8.5088	7.74	2.7821	8.7977	8.24	2.8705	9.0774
7.25	2.6926	8.5147	7.75	2.7839	8.8034	8.25	2.8723	9.0830
7.26	2.6944	8.5206	7.76	2.7857	8.8091	8.26	2.8740	9.0885
7.27	2.6963	8.5264	7.77	2.7875	8.8148	8.27	2.8758	9.0940
7.28	2.6981	8.5323	7.78	2.7893	8.8204	8.28	2.8775	9.0995
7.29	2.7000	8.5381	7.79	2.7911	8.8261	8.29	2.8792	9.1049
7.30	2.7019	8.5440	7.80	2.7928	8.8318	8.30	2.8810	9.1104
7.31	2.7037	8.5499	7.81	2.7946	8.8374	8.31	2.8827	9.1159
7.32	2.7055	8.5557	7.82	2.7964	8.8431	8.32	2.8844	9.1214
7.33	2.7074	8.5615	7.83	2.7982	8.8487	8.33	2.8862	9.1269
7.34	2.7092	8.5674	7.84	2.8000	8.8544	8.34	2.8879	9.1324
7.35	2.7111	8.5732	7.85	2.8018	8.8600	8.35	2.8896	9.1378
7.36	2.7129	8.5790	7.86	2.8036	8.8657	8.36	2.8914	9.1433
7.37	2.7148	8.5849	7.87	2.8054	8.8713	8.37	2.8931	9.1488
7.38	2.7166	8.5907	7.88	2.8071	8.8769	8.38	2.8948	9.1542
7.39	2.7185	8.5965	7.89	2.8089	8.8826	8.39	2.8965	9.1597
7.40	2.7203	8.6023	7.90	2.8107	8.8882	8.40	2.8983	9.1652
7.41	2.7221	8.6081	7.91	2.8125	8.8938	8.41	2.9000	9.1706
7.42	2.7240	8.6139	7.92	2.8142	8.8994	8.42	2.9017	9.1761
7.43	2.7258	8.6197	7.93	2.8160	8.9051	8.43	2.9034	9.1815
7.44	2.7276	8.6255	7.94	2.8178	8.9107	8.44	2.9052	9.1869
7.45	2.7295	8.6313	7.95	2.8196	8.9163	8.45	2.9069	9.1924
7.46	2.7313	8.6371	7.96	2.8213	8.9219	8.46	2.9086	9.1978
7.47	2.7331	8.6429	7.97	2.8231	8.9275	8.47	2.9103	9.2033
7.48	2.7350	8.6487	7.98	2.8249	8.9331	8.48	2.9120	9.2087
7.49	2.7368	8.6545	7.99	2.8267	8.9387	8.49	2.9138	9.2141

TABLE XV   Square Roots (*Continued*)

n	$\sqrt{n}$	$\sqrt{10n}$	n	$\sqrt{n}$	$\sqrt{10n}$	n	$\sqrt{n}$	$\sqrt{10n}$
8.50	2.9155	9.2195	9.00	3.0000	9.4868	9.50	3.0822	9.7468
8.51	2.9172	9.2250	9.01	3.0017	9.4921	9.51	3.0838	9.7519
8.52	2.9189	9.2304	9.02	3.0033	9.4974	9.52	3.0854	9.7570
8.53	2.9206	9.2358	9.03	3.0050	9.5026	9.53	3.0871	9.7622
8.54	2.9223	9.2412	9.04	3.0067	9.5079	9.54	3.0887	9.7673
8.55	2.9240	9.2466	9.05	3.0083	9.5131	9.55	3.0903	9.7724
8.56	2.9257	9.2520	9.06	3.0100	9.5184	9.56	3.0919	9.7775
8.57	2.9275	9.2574	9.07	3.0116	9.5237	9.57	3.0935	9.7826
8.58	2.9292	9.2628	9.08	3.0133	9.5289	9.58	3.0952	9.7877
8.59	2.9309	9.2682	9.09	3.0150	9.5341	9.59	3.0968	9.7929
8.60	2.9326	9.2736	9.10	3.0166	9.5394	9.60	3.0984	9.7980
8.61	2.9343	9.2790	9.11	3.0183	9.5446	9.61	3.1000	9.8031
8.62	2.9360	9.2844	9.12	3.0199	9.5499	9.62	3.1016	9.8082
8.63	2.9377	9.2898	9.13	3.0216	9.5551	9.63	3.1032	9.8133
8.64	2.9394	9.2952	9.14	3.0232	9.5603	9.64	3.1048	9.8184
8.65	2.9411	9.3005	9.15	3.0249	9.5656	9.65	3.1064	9.8234
8.66	2.9428	9.3059	9.16	3.0265	9.5708	9.66	3.1081	9.8285
8.67	2.9445	9.3113	9.17	3.0282	9.5760	9.67	3.1097	9.8336
8.68	2.9462	9.3167	9.18	3.0299	9.5812	9.68	3.1113	9.8387
8.69	2.9479	9.3220	9.19	3.0315	9.5864	9.69	3.1129	9.8438
8.70	2.9496	9.3274	9.20	3.0332	9.5917	9.70	3.1145	9.8489
8.71	2.9513	9.3327	9.21	3.0348	9.5969	9.71	3.1161	9.8539
8.72	2.9530	9.3381	9.22	3.0364	9.6021	9.72	3.1177	9.8590
8.73	2.9547	9.3434	9.23	3.0381	9.6073	9.73	3.1193	9.8641
8.74	2.9563	9.3488	9.24	3.0397	9.6125	9.74	3.1209	9.8691
8.75	2.9580	9.3541	9.25	3.0414	9.6177	9.75	3.1225	9.8742
8.76	2.9597	9.3595	9.26	3.0430	9.6229	9.76	3.1241	9.8793
8.77	2.9614	9.3648	9.27	3.0447	9.6281	9.77	3.1257	9.8843
8.78	2.9631	9.3702	9.28	3.0463	9.6333	9.78	3.1273	9.8894
8.79	2.9648	9.3755	9.29	3.0480	9.6385	9.79	3.1289	9.8944
8.80	2.9665	9.3808	9.30	3.0496	9.6437	9.80	3.1305	9.8995
8.81	2.9682	9.3862	9.31	3.0512	9.6488	9.81	3.1321	9.9045
8.82	2.9698	9.3915	9.32	3.0529	9.6540	9.82	3.1337	9.9096
8.83	2.9715	9.3968	9.33	3.0545	9.6592	9.83	3.1353	9.9146
8.84	2.9732	9.4021	9.34	3.0561	9.6644	9.84	3.1369	9.9197
8.85	2.9749	9.4074	9.35	3.0578	9.6695	9.85	3.1385	9.9247
8.86	2.9766	9.4128	9.36	3.0594	9.6747	9.86	3.1401	9.9298
8.87	2.9783	9.4181	9.37	3.0610	9.6799	9.87	3.1417	9.9348
8.88	2.9799	9.4234	9.38	3.0627	9.6850	9.88	3.1432	9.9398
8.89	2.9816	9.4287	9.39	3.0643	9.6902	9.89	3.1448	9.9448
8.90	2.9833	9.4340	9.40	3.0659	9.6954	9.90	3.1464	9.9499
8.91	2.9850	9.4393	9.41	3.0676	9.7005	9.91	3.1480	9.9549
8.92	2.9866	9.4446	9.42	3.0692	9.7057	9.92	3.1496	9.9599
8.93	2.9883	9.4499	9.43	3.0708	9.7108	9.93	3.1512	9.9649
8.94	2.9900	9.4552	9.44	3.0725	9.7160	9.94	3.1528	9.9700
8.95	2.9917	9.4604	9.45	3.0741	9.7211	9.95	3.1544	9.9750
8.96	2.9933	9.4657	9.46	3.0757	9.7263	9.96	3.1559	9.9800
8.97	2.9950	9.4710	9.47	3.0773	9.7314	9.97	3.1575	9.9850
8.98	2.9967	9.4763	9.48	3.0790	9.7365	9.98	3.1591	9.9900
8.99	2.9983	9.4816	9.49	3.0806	9.7417	9.99	3.1607	9.9950

ANSWERS TO ODD-NUMBERED EXERCISES

In exercises involving extensive calculations, the reader may well get answers differing somewhat from those given here due to rounding at various intermediate stages.

CHAPTER 1

1.1 The following are possibilities: (a) In a random sample of 200 women under thirty, 137 stated that they prefer perfume as a gift to either flowers or candy. At the 0.05 level of significance, does this refute the claim that 60 percent of all women under thirty prefer perfume as a gift to either flowers or candy? (b) In a random sample of 200 violinists, 137 stated that they started playing the violin before they were twelve years old. At the 0.05 level of significance, does this refute the claim that 60 percent of all violinists start playing the violin before they are twelve years old? (c) In a random sample of 200 cars coming to a certain intersection, 137 made right turns. At the 0.05 level of significance, does this refute the claim that 60 percent of all cars coming to this intersection make right turns?

1.3 (a) Persons coming out of the building housing the national headquarters of a political party are more likely to support that party; (b) December spending is not typical of spending throughout the year.

1.5 (a) Only 79 and 88 exceed 75, so that the conclusion can be obtained from the data; (b) If the student received the grades in the given order, the conclusion follows from the data; (c) Since there may be other reasons, the conclusion requires a generalization; (d) Since $88 - 46 = 42$, the conclusion follows from the data.

1.7 (a) Since $9 + 14 + 10 = 33$ exceeds $5 + 10 + 12 = 27$, the conclusion is obtained by descriptive methods; (b) Since 9 exceeds 5, 14 exceeds 10, and 10 is less than 12, the conclusion is obtained by descriptive methods; (c) Since there may be other reasons, the conclusion is a generalization; (d) The conclusion does not necessarily follow from the data, so that it is a generalization.

1.9 (a) Ordinal; (b) nominal; (c) ordinal; (d) nominal.

1.11 It would entail the assumption that the difference in intelligence between the second and third persons is three times the difference in intelligence between the first and second persons.

1.13 (a) This is a generalization, as it does not follow from the data; (b) Since 12,305 exceeds the other four attendance figures, the conclusion is a description; (c) It follows from the data and, hence, is a description; (d) Since there may be other reasons, the conclusion is a generalization.

1.15 (a) Since poor persons are less likely to have telephones, they are less likely to be included in the poll; (b) Referring to the practice as "unfair" introduces a bias.

1.17 (a) "Indian art" could be interpreted as art from India or as art of American Indians; (b) Few persons will give honest answers about their personal habits.

1.19 These ordinal data should not have been added; instead, we might compare the two golfers' performance by comparing their total prize money.

2.1 170–179, 180–189, 190–199, 200–209, 210–219, 220–229, 230–239, 240–249, 250–259, and 260–269.

2.3 (a) Yes;   (b) no;   (c) no;   (d) yes;   (e) no;   (f) no.

2.5 (a) 0, 20, 40, 60, 80, 100, 120, and 140; (b) 19, 39, 59, 79, 99, 119, 139, and 159; (c) 9.5, 29.5, 49.5, 69.5, 89.5, 109.5, 129.5, and 149.5; (d) 20.

2.7 (a) 11.5, 20.5, 29.5, 38.5, 47.5, 56.5, and 65.5; (b) 12–20, 21–29, 30–38, 39–47, 48–56, and 57–65.

2.9 The class frequencies are 1, 2, 2, 3, 12, 14, 12, and 4.

2.11 The cumulative "less than" frequencies are 0, 1, 3, 5, 8, 20, 34, 46, and 50.

2.13 3.3, 5.0, 16.7, 30.0, 26.7, 11.7, and 6.7 percent.

2.15 The class frequencies are 2, 3, 7, 8, 14, 5, and 1.

2.17
Car	20
Train	9
Plane	15
Bus	6

2.29
9	8
10	5 8 2 0 8 3 7 3
11	2 7 5 4
12	0 6
13	
14	3

2.31
1	67, 88, 95
2	55, 70, 91, 83, 17
3	05, 19, 34, 62
4	40, 08
5	12

2.33 (a)
16·	9
17*	0
17·	5
18*	1 3
18·	6 7 7
19*	4 4 0 2
19·	8 6 9
20*	4 3
20·	5 7
21*	2
21·	6 8
22*	3
22·	6

(b)
5·	7 9
6*	1 2 0 4 1 0
6·	8 5 9 8 6 6 5 7 7
7*	3 4 0 1 2 3 2
7·	6 8 6
8*	1 3
8·	5

2.35 The class frequencies are 7, 10, 20, 22, 10, 2, and 1.

2.37 The cumulative "less than" frequencies are 7, 17, 37, 69, 59, 71, and 72.

2.39 (a) 0.5, 2.5, 4.5, 6.5, and 8.5; (b) −0.5, 1.5, 3.5, 5.5, 7.5, and 9.5; (c) 2.

2.41
Never	8
Rarely	13
Occasionally	15
Frequently	4

2.43 (a)

6f	4
6s	6 6 7 6
6·	9 8 9 9 8 8
7*	1 0 0 1
7t	2 2
7f	5

(b) 235, 234, 234, 234, 235, 234, 236, 237, 236, 236, 239, 238, and 241.

2.45 19 is not accommodated and 34 falls into two classes.

2.47 $13,000–$15,999, $16,000–$18,999, $19,000–$21,999, $22,000–$24,999, and $25,000–$27,999.

2.49 6.0–7.9, 8.0–9.9, 10.0–11.9, 12.0–13.9, and 14.0–15.9.

CHAPTER 3

3.1 (a) The data would constitute a population if the dean wants to determine the average number of failing grades given by the faculty members in that academic year; (b) The data would constitute a sample if the dean wants to predict the total number of failing grades in future academic years.

3.3 58.1 years.

3.5 (a) 335.1 calories per serving; (b) 335.1 calories per serving.

3.7 (a) $2,430 < 3,000$ and the elevator is not overloaded; (b) $3,051 > 3,000$ and the elevator is overloaded.

3.9 (a) $\frac{65}{80} = 0.8125$; (b) $\frac{158}{250} = 0.632$.

3.11 (a) 8; (b) 9; (c) 10.

3.13 (a) 72; (b) 65.

3.15 (a) 146; (b) 146 and 149; (c) does not exist.

3.17 (a) The mean; (b) the median.

3.19 (a) The medians are 4, 5, 4, 3, 5, 2, 3, 5, 3, 2, 3, and 4; the means are 4, $4\frac{1}{3}$, 4, $3\frac{1}{3}$, 4, 2, $2\frac{2}{3}$, 4, $3\frac{1}{3}$, 3, 3, and 4;

(b)

Medians	Frequency		Means	Frequency
1.5–2.5	2		1.5–2.5	1
2.5–3.5	4		2.5–3.5	5
3.5–4.5	3		3.5–4.5	6
4.5–5.5	3		4.5–5.5	0

3.21 (a) 16; (b) 4; (c) 2; 96 and 192.

3.23 78.

3.25 9.05 percent.

3.27 $17,168.

3.29 41 percent.

3.31 12.

3.33 (a) 2.74; (b) 2.74.

3.35 (a) 3.77; (b) 3.77. If we subtract the same constant from each value, this will not affect the standard deviation.

3.37 The range is 27, $s = 8.81$, and $\dfrac{27}{8.81} = 3.06$ is very close to 3.

3.39 11.36.

3.41 (a) At least 98.4 percent; (b) at least 99.3 percent.

3.43 (a) 0.230 and 0.290;   (b) 0.220 and 0.300.

3.45 The lobster dinner is relatively most overpriced.

3.47 (a) The first university;   (b) the second university.

3.49 The first student is relatively more consistent.

3.51 $\bar{x} = 7.8$ and $s = 5.66$.

3.53 $\bar{x} = 56.45$ and $s = 20.62$.

3.55 -0.37.

3.57 (a) 27.23; (b) 22.57 and 32.74; (c) 21.50 and 34.13; (d) 14.97 and 16.84

3.59 (a) 56.83 and 31.08;   (b) 5.08 and 18.4 percent.

3.61 (a) 18.90;   (b) 5.66.

3.65 (a) $\displaystyle\sum_{i=1}^{5} z_i$;   (b) $\displaystyle\sum_{i=5}^{12} x_i$;   (c) $\displaystyle\sum_{i=1}^{6} x_i f_i$;   (d) $\displaystyle\sum_{i=1}^{3} y_i^2$;

   (e) $\displaystyle\sum_{i=1}^{7} 2x_i$;   (f) $\displaystyle\sum_{i=2}^{4} (x_i - y_i)$;   (g) $\displaystyle\sum_{i=2}^{5} (z_i + 3)$;   (h) $\displaystyle\sum_{i=1}^{4} x_i y_i f_i$.

3.67 (a) 25;   (b) 27;   (c) 131;   (d) 667.

3.69 (a) 7, 4, -2, and 10;   (b) 4, 8, and 7.

3.73 (a) 15.6;   (b) 15.5.

3.75 $2.88.

3.77 -0.35.

3.79 (a) 2;   (b) 1.

3.81 $\bar{x} = \$21,400$ (to nearest \$100), $\tilde{x} = \$21,300$, range $= \$19,600$, and $s = \$6,800$ (to nearest \$100).

3.83 2.22.

3.85 It is possible.

3.87 (a) \$68.80 and \$69.41;   (b) \$20.37;   (c) \$53.07 and \$85.00;   (d) 29.6 and 23.1 percent.

3.89 29.7, 29.5, and 28.

3.91 39.3 cents.

3.93

2	9
3	8 7
4	7 9
5	2 6 8
6	6 8 0 5 5 3 9 2 3 9 3 0
7	5 7 0 3 7 9 0 5 4 6 4 3 6 1
8	9 3 7 1 5 3 2 4 8 1 7 2
9	0 4 7 1

The median is 73.5.

4.7 30.

4.9 (a) 48; (b) 12; (c) 24.

4.11 32,768.

4.13 (a) True; (b) true; (c) false; (d) true.

4.15 175,560.

4.17 5,040.

4.19 24.

4.21 720.

4.23 (a) 1,260; (b) 20; (c) $\dfrac{n!}{r_1! \cdot r_2! \cdot r_3!}$; 90; (d) 83,160.

4.25 78.

4.27 (a) 6,435; (b) 3,003.

4.29 (a) 120; (b) 90; (c) 10.

4.31 1,200.

4.37 $\frac{1}{4}$, $\frac{1}{2}$, and $\frac{1}{4}$.

4.39 (a) $\frac{1}{3}$; (b) $\frac{1}{2}$.

4.41 (a) $\frac{77}{92}$; (b) $\frac{11}{69}$; (c) $\frac{1}{276}$.

4.43 $\dfrac{374}{1,015}$.

4.45 0.26.

4.47 0.85.

4.49 5.

4.51 0.64.

4.53 (a) True; (b) false; (c) false; (d) true.

4.57 9,765,625.

4.59 0.64.

4.61 220.

4.63 (a) 2; (b) 3.

5.1 (a) $\{a, c, d, f, g\}$ is the event that the scholarship is awarded to one of the female applicants; (b) $\{a, b, e, h\}$ is the event that the scholarship is awarded to Ms. Adam, Mr. Bean, Mr. Earl, or Mr. Hall; (c) $\{b, e\}$ is the event that the scholarship is awarded to Mr. Bean or Mr. Earl; (d) $\{b, c, d, e, g\}$ is the event that the scholarship is awarded to Mr. Bean, Miss Clark, Mrs. Daly, Mr. Earl, or Ms. Gardner; (e) $\{a, b, e, f, h\}$ is the event that the scholarship is awarded to Ms. Adam, Mr. Bean, Mr. Earl, Ms. Fuentes, or Mr. Hall; (f) $\{c, d, g\}$ is the event that the scholarship is awarded to Miss Clark, Mrs. Daly, or Ms. Gardner.

5.3 (a) $\{8\}$; (b) $\{1, 2, 3, 4, 5, 6, 7\}$; (c) $\{3, 4\}$; (d) $\{1, 2, 3, 4, 5\}$; (e) $\{1, 2, 3, 4, 5, 6, 7, 8\}$; (f) $\{1, 2, 3, 4, 5\}$.

5.5 (a) The two salespersons sell equally many of the two cars; (b) The first salesperson sells one of the two cars; (c) The first salesperson does not sell either of the two cars.

5.7 (a) (2, 0) and (0, 2); (b) (0, 1) and (1, 1); (c) (1, 0), (1, 1), and (2, 0).

5.9 T and W, U and V, and U and W are mutually exclusive.

5.11   (a) $\{A, D\}$;   (b) $\{C, E\}$;   (c) $\{B\}$.

5.13   (a) Not mutually exclusive since a driver can speed through a red light; (b) Mutually exclusive since by law the President of the United States cannot be foreign born; (c) Mutually exclusive since an at bat can have only one outcome; (d) Not mutually exclusive since the baseball player may get a walk and hit a home run in different at bats; (e) Not mutually exclusive since there can be rain and sunshine at different times of the day, or even at the same time.

5.15   (a) The driver has collision insurance; (b) The driver does not have liability insurance; (c) The driver has collision insurance, liability insurance, or both.

5.17   (a) The novel is well written; (b) The novel is not a financial success; (c) The novel is either not well written or not a financial success.

5.19   (a) The car needs an engine overhaul, transmission repairs, and new tires; (b) The car needs transmission repairs and new tires, but not an engine overhaul; (c) The car needs an engine overhaul but neither transmission repairs nor new tires; (d) The car needs an engine overhaul and new tires; (e) The car needs transmission repairs but not new tires; (f) The car does not need an engine overhaul.

5.23   (a) $P(F \cap Q)$;   (b) $P(T' \cap F)$;   (c) $P(Q \cup F)$;   (d) $P(F' \cap T')$.

5.25   (a) $P(A \cup B) \geq P(A)$;   (b) $P(A \cap B) \leq P(A)$.

5.27   (a) 0.42;   (b) 0.84;   (c) 0.16.

5.29   (a) The odds are 7 to 2 against rolling "7 or 11" with a pair of balanced dice; (b) The odds are 11 to 5 against getting three heads and three tails in six flips of a coin; (c) The odds are 3 to 2 that the last digit of a car's license plate is 2, 3, 4, 5, 6, or 7.

5.31   9 to 4.

5.33   No; $\frac{2}{3} + \frac{1}{5} + \frac{1}{10} \neq 1$.

5.35   The probability is greater than $\frac{5}{6}$.

5.37   The probabilities are consistent.

5.39   The odds are 11 to 9 that neither car will win.

5.43   (a) 0.43;   (b) 0.67;   (c) 0.11;   (d) 0.59.

5.45   (a) 0.45;   (b) 0.37;   (c) 0.53.

5.47   (a) $\frac{20}{35}$;   (b) $\frac{24}{35}$;   (c) $\frac{5}{35}$.

5.49 $\frac{1}{16}$, $\frac{4}{16}$, $\frac{6}{16}$, $\frac{4}{16}$, and $\frac{1}{16}$.

5.51   0.12.

5.53 $\frac{1}{16}$.

5.55   (a) $P(Q|W)$;   (b) $P(W'|Q)$;   (c) $P(Q'|W')$.

5.57   (a) $P(N|I)$;   (b) $P(I'|A')$;   (c) $P(I' \cap A'|N)$;   (d) $P(N'|I \cap A)$.

5.59   (a) $\frac{3}{5}$;   (b) $\frac{7}{10}$;   (c) $\frac{1}{5}$;   (d) $\frac{3}{10}$;   (e) $\frac{1}{3}$;   (f) $\frac{3}{7}$.

5.61   (a) $\frac{19}{58}$;   (b) $\frac{19}{25}$.

5.63   (a) 0.90;   (b) 0.15.

5.65   (a) $\frac{1}{16}$;   (b) $\frac{1}{17}$.

5.67   (a) $\frac{22}{145}$;   (b) $\frac{4}{25}$.

5.69   (a) $\frac{5}{36}$;   (b) $\frac{1}{9}$.

5.71 They are independent.

5.73 (a) $\frac{1}{22}$;   (b) $\frac{1}{91}$;   (c) $\frac{1}{12}$;   (d) 0.0288;   (e) 0.729 and 0.144.

5.75 $\frac{17}{38}$.

5.77 0.908.

5.79 0.805.

5.81 (a) 0.290;   (b) 0.387.

5.83 0.625.

5.85 (a) $P(Y) = 0.20 \cdot P(M) + 0.40$;   (b) 0.12.

5.89 (a) $\frac{1}{2}$;   (b) $\frac{2}{5}$.

5.91 0.0046.

5.93 They are independent.

5.95 (a) The probability that the horse will win is $\frac{5}{24}$; (b) The probability is $\frac{13}{16}$ that at most three of the cards will be red; (c) The probability is $\frac{40}{77}$ that one man and one woman will be selected.

5.97 0.68.

5.99 (a) 0.38;   (b) 0.42;   (c) 0.50.

5.101 (a) $\frac{15}{32}$;   (b) $\frac{13}{32}$;   (c) $\frac{5}{32}$;   (d) $\frac{23}{32}$;   (e) $\frac{8}{32}$;   (f) $\frac{9}{32}$.

5.103 (a) $\frac{94}{125}$;   (b) $\frac{4}{5}$;   (c) $\frac{4}{5}$;   (d) $\frac{40}{47}$;   (e) $\frac{11}{25}$;   (f) $\frac{20}{31}$.

5.105 0.224.

5.107 The probabilities are consistent.

5.109 0.269.

5.111 $\dfrac{1}{4,096}$.

5.113 (a) Person will visit friend in Portland but not friends in Bend or Eugene; (b) Person will visit friends in Bend and Eugene; (c) Person will visit friend in Eugene but not friend in Bend; (d) Person will visit friend in Portland, friend in Eugene, or both, but not friend in Bend; (e) Person will not visit friends in Bend and Eugene.

5.115 (a) 0.32;   (b) 0.68;   (c) 0.12;   (d) 0.60.

CHAPTER 6

6.1 20 cents.

6.3 $5.00.

6.5 40 cents; not worthwhile to spend 50 cents.

6.7 $1.00.

6.9 1.81.

6.11 The probability that her client will win is less than 0.20.

6.13 The probability is greater than 0.40.

6.15 $1,250.

6.17 (a) He should go to the construction site which is 18 miles from the lumberyard; (b) It makes no difference.

6.21 The retailer should stock 2.

6.23 (a) He should expand now; (b) Go to construction site which is 18 miles from the lumberyard.

6.25 (a) 2 miles; (b) $1,575,000; it would be worthwhile.

6.27 (a) $132; (b) 192.

6.29 (a) 18; (b) 19.

6.31 0, which does not entail any risk and will not yield any profit or loss; it is not reasonable to use the minimax criterion in a problem of this kind.

6.33 $E = 0.944$; it would not be worthwhile to pay $1.00.

6.35 The decision would not be the same.

6.37 40.

6.39 0.16.

6.41 (a) $p > 0.10$; (b) $p < 0.10$; (c) $p = 0.10$.

6.43 If the first player wins b dollars with probability p and loses a dollars with probability $1 - p$, his expectation is $bp - a(1 - p)$; equating this to 0 yields $p = \dfrac{a}{a + b}$.

CHAPTER 7

7.1 (a) No; (b) yes; (c) yes; (d) no.

7.3 0.211.

7.5 0.057.

7.7 (a) 0.115; (b) 0.115.

7.9 (a) 0.175; (b) 0.228; (c) 0.588.

7.11 (a) 0.203; (b) 0.316; (c) 0.050.

7.13 (a) 0.086; (b) 0.079; (c) 0.063.

7.15 0.509.

7.17 (a) 0.833; (b) 0.455; (c) 0.773.

7.19 (a) 0.266; (b) 0.265.

7.21 0.192.

7.23 0.879.

7.25 (a) 0.449; (b) 0.359; (c) 0.144; (d) 0.048.

7.27 0.117.

7.29 0.040.

7.31 $\sigma^2 = 1$.

7.33 1.37.

7.35 (a) 2.5; (b) 2.5.

7.37 $\mu = 7$ and $\sigma = 2.415$.

7.39 (a) $\mu = 18.983$ and $\sigma = 1.139$; (b) $\mu = 19$ and $\sigma = 0.975$.

7.41 (b) $\mu = 2.5$; (c) $\mu = 2.5$.

7.43 $\mu = 0.799$.

7.45 (a) $\mu = 2.501$ and $\sigma = 1.578$; (b) 2.501 and $(1.578)^2 = 2.490$ are both very close to 2.5.

7.47 At least 0.96.

7.49 (a) The probability is at least $\frac{143}{144}$ that there will be between 18 and 234 rainy days; (b) The probability is at least 0.91.

7.53 (a) 00–13, 14–40, 41–67, 68–85, 86–94, 95–98, and 99.

7.55 (a) 0000–2465, 2466–5917, 5918–8334, 8335–9462, 9463–9857, 9858–9968, 9969–9994, and 9995–9999.

7.57 (a) 0.279;   (b) 0.319.

7.59 (a) No;   (b) no;   (c) yes.

7.61 $\sigma = 1.17$.

7.63 (a) 0.396;   (b) 0.183.

7.65 0.0245.

7.67 0.140.

7.71 (b) $f(0) = \dfrac{729}{4,096}$, $f(1) = \dfrac{1,458}{4,096}$, $f(2) = \dfrac{1,215}{4,096}$, $f(3) = \dfrac{540}{4,096}$, $f(4) = \dfrac{135}{4,096}$, $f(5) = \dfrac{18}{4,096}$, and $f(6) = \dfrac{1}{4,096}$.

7.73 $\mu = 2.092$ and $\sigma^2 = 1.836$.

7.75 (a) $\mu = \frac{4}{9}$;   (b) $\mu = \frac{4}{9}$.

7.77 (a) 0.070;   (b) 0.902;   (c) 0.028.

7.79 $\mu = 162$ and $\sigma^2 = 81$.

CHAPTER 8

8.1 (a) $\frac{1}{4}$;   (b) $\frac{5}{8}$;   (c) 0.7625.

8.3 (a) 0.1598;   (b) 0.1068;   (c) 0.6444.

8.5 (a) 1.92; (b) 2.22; (c) 0.74; (d) -0.50; (e) 1.12; (f) 1.44 or -1.44; (g) 2.17 or -2.17.

8.9 $\sigma = 20$.

8.11 (a) 0.259, 0.121, and 0.223; (b) 0.593 and 0.053; (c) 0.670.

8.15 (a) Roughly, $\mu = 18.6$ and $\sigma = 5.4$; (b) Roughly, $\mu = 32$ and $\sigma = 5.8$.

8.17 (a) 0.0548;   (b) 0.2743;   (c) 0.7222;   (d) 0.3026.

8.19 (a) 1.54 percent;   (b) 30.71 percent.

8.23 15.08 seconds.

8.25 (a) 0.0723;   (b) 0.4364;   (c) 0.4649.

8.27 0.2128 compared to the tabular value of 0.209.

8.29 0.2266.

8.31 0.8365.

8.33 0.0808.

8.35 (a) 0.2358;   (b) 0.4908;   (c) 0.9556.

8.41 (a) 0.1200;   (b) 0.120.

8.43 (a) 0.4664;   (b) 0.9938;   (c) 0.7389;   (d) 0.1075;   (e) 0.2389.

8.45 (a) 2.05;   (b) 1.28.

8.47 0.0618.

8.49 (a) 0.0985;   (b) 0.1335;   (c) 0.6626.

8.51 (a) 0.0643;   (b) 0.0143.

8.53 (a) 0.3665;   (b) 0.4714;   (c) 0.0534;   (d) 0.1293;   (e) 0.8544.

8.55 (a) 0.53 or -0.53;   (b) -1.18;   (c) 1.83;   (d) 0.34 or -0.34.

9.1 (a) 15;   (b) 45;   (c) 300.

9.3 (a) $\dfrac{1}{495}$;   (b) $\dfrac{1}{26{,}334}$.

9.5 a and b, a and c, a and d, a and e, a and f, b and c, b and d, b and e, b and f, c and d, c and e, c and f, d and e, d and f, and e and f.

9.7 (a) $\frac{1}{15}$;   (b) $\frac{2}{3}$.

9.9 5190, 1250, 1377, 1312, 7648, 6367, 0796, 7900, 5223, 6686, 1543, 5872, 1586, 3357, and 5209.

9.11 077, 533, 338, 096, 160, 425, 085, and 541.

9.15 (a) 18, 11, 9, and 83; 23, 7, 8, and 28; 12, 11, 19, and 9; 10, 116, 25, and 13; 20, 34, 9, and 15; (b) 30.25, 16.50, 12.75, 41.00, and 19.50; their mean is 24 and so is the mean of all the data.

9.17 (a) 67,762,406,250;   (b) 136,446,750,000.

9.19 $n_1 = 20$, $n_2 = 48$, $n_3 = 8$, and $n_4 = 4$.

9.21 115 and 125, 115 and 135, 125 and 135, 185 and 195, 185 and 205, and 195 and 205; the means are 120, 125, 130, 190, 195, and 200, and they all differ from $\mu = 160$ by more than 5. The probability of getting a value close to the mean of the population is greatest for stratified sampling, then comes simple random sampling, and then cluster sampling.

9.23 (a) The means are 81 and 510, and their weighted mean is 216.47; (b) The mean of all the claims is 216.47.

9.25 (b) 5 and 6, 5 and 7, 5 and 8, 5 and 9, 5 and 10, 6 and 7, 6 and 8, 6 and 9, 6 and 10, 7 and 8, 7 and 9, 7 and 10, 8 and 9, 8 and 10, and 9 and 10; the means are 5.5, 6, 6.5, 7, 7.5, 6.5, 7, 7.5, 8, 7.5, 8, 8.5, 8.5, 9, and 9.5;   (c)

Mean	Probability
5.5	$\frac{1}{15}$
6	$\frac{1}{15}$
6.5	$\frac{2}{15}$
7	$\frac{2}{15}$
7.5	$\frac{3}{15}$
8	$\frac{2}{15}$
8.5	$\frac{2}{15}$
9	$\frac{1}{15}$
9.5	$\frac{1}{15}$

(d) $\mu_{\bar{x}} = 7.5$ and $\sigma_{\bar{x}} = \sqrt{7/6} = 1.08$.

9.27 (a) 5 and 5, 5 and 6, 5 and 7, 5 and 8, 5 and 9, 5 and 10, 6 and 5, 6 and 6, 6 and 7, 6 and 8, 6 and 9, 6 and 10, 7 and 5, 7 and 6, 7 and 7, 7 and 8, 7 and 9, 7 and 10, 8 and 5, 8 and 6, 8 and 7, 8 and 8, 8 and 9, 8 and 10, 9 and 5, 9 and 6, 9 and 7, 9 and

8, 9 and 9, 9 and 10, 10 and 5, 10 and 6, 10 and 7, 10 and 8, 10 and 9, and 10 and
10; the means are 5, 5.5, 6, 6.5, 7, 7.5, 5.5, 6, 6.5, 7, 7.5, 8, 6, 6.5, 7, 7.5, 8, 8.5, 6.5, 7,
7.5, 8, 8.5, 9, 7, 7.5, 8, 8.5, 9, 9.5, 7.5, 8, 8.5, 9, 9.5, and 10; (b)

Mean	Probability
5	$\frac{1}{36}$
5.5	$\frac{2}{36}$
6	$\frac{3}{36}$
6.5	$\frac{4}{36}$
7	$\frac{5}{36}$
7.5	$\frac{6}{36}$
8	$\frac{5}{36}$
8.5	$\frac{4}{36}$
9	$\frac{3}{36}$
9.5	$\frac{2}{36}$
10	$\frac{1}{36}$

(c) $\mu_{\bar{x}} = 7.5$ and $\sigma_{\bar{x}} = \sqrt{\frac{35}{24}} = 1.21$.

9.29 (a) Divided by 2;   (b) divided by 1.5;   (c) divided by 3;   (d) multiplied by 4.

9.31 (a) 1.74;   (b) 0.56.

9.35 $\sigma_{\bar{x}} = 35$ compared to 21.0 and 7.1.

9.37 (a) Probability is at least 0.84;   (b) 0.9876.

9.39 0.95.

9.41 0.0918.

9.45 (a) $\bar{x} = 3.64$ and $s = 1.41$;   (b) $\sigma_s = 1.265$.

9.53 (a) 300,105,000;   (b) 1,736,410,000.

9.55 $\dfrac{1}{82,160}$.

9.57 (a) 1 and 1, 1 and 3, 1 and 5, 1 and 7, 3 and 1, 3 and 3, 3 and 5, 3 and 7, 5 and 1, 5
and 3, 5 and 5, 5 and 7, 7 and 1, 7 and 3, 7 and 5, and 7 and 7; the means are 1, 2, 3,
4, 2, 3, 4, 5, 3, 4, 5, 6, 4, 5, 6, and 7; 1 and 7 have the probability $\frac{1}{16}$, 2 and 6 have the
probability $\frac{2}{16}$, 3 and 5 have the probability $\frac{3}{16}$, and 4 has the probability $\frac{4}{16}$;
(b) $\mu_{\bar{x}} = 4$ and $\sigma_{\bar{x}} = \sqrt{\frac{5}{2}} = 1.58$.

9.61 (a) 0.870;   (b) 0.981.

9.63 $n_1 = 27$, $n_2 = 36$, $n_3 = 18$, and $n_4 = 9$.

9.65 $\dfrac{1}{42,504}$.

CHAPTER 10

10.1 0.45 minute.

10.3 $51.7 < \mu < 53.9$.

10.5 (a) $E = 0.43$ minute;   (b) $9.12 < \mu < 9.78$.

10.7 (a) $104.08 < \mu < 107.44$;   (b) $E = \$1.18$.

10.9 99 percent.

10.11 (a) $533.3 < \mu < 592.7$;   (b) $76.3 < \mu < 79.9$.

10.13 355.

10.15 33.

10.17 $E = 1.37$ gallons.

10.19 $0.501 < \mu < 0.509$.

10.21 $11.12 < \mu < 14.88$.

10.23 (a) $E = 3.29$ fillings;   (b) $1.35 < \mu < 4.65$.

10.25 0.70.

10.27 Type I error; Type II error.

10.29 Null hypothesis: device is not effective.

10.31 (a) 0.0087;   (b) at most 0.0087.

10.35 (a) 0.04;   (b) 0.02, 0.15, 0.50, 0.845, 0.845, 0.50, 0.15, 0.02.

10.37 (a) $\mu < 20$ and make the modification only if the null hypothesis can be rejected; (b) $\mu > 20$ and make the modification unless the null hypothesis can be rejected.

10.39 (a) $\mu_2 > \mu_1$ and switch to the radial tires only if the null hypothesis can be rejected; (b) $\mu_2 < \mu_1$ and switch to the radial tires unless the null hypothesis can be rejected; (c) $\mu_2 \neq \mu_1$.

10.41 $z = 2.73$; the null hypothesis must be rejected.

10.43 $z = -1.44$; the null hypothesis cannot be rejected.

10.45 0.0202.

10.47 (a) $n = 126$;   (b) $n = 91$.

10.49 $t = 4.00$; the null hypothesis must be rejected.

10.51 $t = 3.08$; the null hypothesis must be rejected.

10.53 $t = 1.94$; the null hypothesis cannot be rejected.

10.55 $t = 2.11$; the null hypothesis cannot be rejected.

10.57 $z = -2.12$; the difference is not significant.

10.59 $t = 3.02$; the difference is significant.

10.61 $t = 1.11$; the difference is not significant.

10.63 (a) $z = 4.88$; the hypothesis $\delta = 0.5$ must be rejected; (b) $t = 4.17$; the hypothesis $\delta = 0.050$ must be rejected.

10.65 $t = 2.20$; the difference is not significant.

10.67 $z = -28.90$; we can conclude that their use in a tropical climate reduces the average useful life of such shirts.

10.69 $15.95 < \mu < 24.39$.

10.71 (a) $0.09 < \delta < 1.71$;   (b) $8.1 < \delta < 851.9$.

10.73 (a) 0.6543;   (b) $\mu_1 = 4721$ and $\sigma_1 = 109.1$;   (c) 0.9736.

10.75 $t = 4.06$; the exercises are effective in reducing weight.

10.77 (a) $E = 0.47$ pound;   (b) $18.92 < \mu < 20.48$.

10.79 $t = 0.73$; the difference is not significant.

10.81 0.88.

10.83 $n = 230$.

10.85 $32.7 < \mu < 36.9$.

11.1 $0.0027 < \sigma < 0.0077$.

11.3 $0.095 < \sigma < 0.331$.

11.5 $3.56 < \sigma^2 < 72.93$.

11.7 $3.67 < \sigma < 5.83$.

11.9 (a) 1.98 compared to 1.79;   (b) 0.172 compared to 0.158.

11.11 $\chi^2 = 6.66$; the null hypothesis cannot be rejected.

11.13 $\chi^2 = 3.56$; the null hypothesis cannot be rejected.

11.15 $z = -0.80$; the null hypothesis cannot be rejected.

11.17 $F = 2.86$; the null hypothesis must be rejected.

11.19 $F = 1.81$; the null hypothesis cannot be rejected.

11.21 $0.008 < \sigma < 0.017$.

11.23 (a) 4.80;   (b) 5.14.

11.25 $F = 2.57$; the null hypothesis cannot be rejected.

12.1 $0.471 < p < 0.609$.

12.3 $0.081 < p < 0.279$.

12.5 $0.617 < p < 0.743$.

12.7 $0.245 < p < 0.295$.

12.9 $0.642 < p < 0.758$.

12.11 $0.772 < p < 0.928$.

12.13 (a) $0.261 < p < 0.419$;   (b) $0.182 < p < 0.418$.

12.15 $n = 2{,}017$.

12.17 $n = 943$.

12.19 $n = 1{,}842$.

12.21 (a) 0.50, 0.45, and 0.05;   (b) 0.53, 0.44, and 0.03.

12.25 1 or less or 11 or more.

12.27 The probability of 11 or fewer spots is 0.017; the null hypothesis must be rejected.

12.29 The probability of 7 or fewer "successes" is 0.057; the null hypothesis cannot be rejected.

12.31 $z = -0.89$; the null hypothesis cannot be rejected.

12.33 $z = 1.49$; the null hypothesis cannot be rejected.

12.35 $z = -1.83$; the null hypothesis must be rejected.

12.37 $z = 0.81$; the null hypothesis cannot be rejected.

12.39 $z = 2.72$; the null hypothesis must be rejected.

12.41 $z = 3.01$; the null hypothesis must be rejected.

12.43 (a) $z = 1.81$; the null hypothesis must be rejected; (b) $z = 2.71$; the null hypothesis must be rejected.

12.45 $\chi^2 = 0.97$; the differences are not significant.

12.47 $\chi^2 = 16.55$; the null hypothesis must be rejected.

12.49 $z = 2.315$.

12.51 $\chi^2 = 20.13$; the therapy is effective.

12.53 $\chi^2 = 25.5$; the null hypothesis must be rejected.

12.55 $\chi^2 = 88.9$; there is a relationship.

12.59 $\chi^2 = 9.48$; the hypothesis that the die is balanced cannot be rejected.

12.61 $\chi^2 = 21.42$; the null hypothesis must be rejected.

12.63 (a) 0.1251, 0.2878, 0.3451, 0.1904, 0.0464, and 0.0052; (b) 10.0, 23.0, 27.6, 15.2, 3.7, and 0.4; (c) $\chi^2 = 1.77$; the null hypothesis cannot be rejected.

12.65 (a) $0.605 < p < 0.795$;   (b) $E = 0.124$.

12.67 (a) 0.073;   (b) 0.724;   (c) 0.194.

12.69 $\chi^2 = 8.3$; the difference is significant.

12.71 $0.895 < p < 0.965$.

12.73 $\chi^2 = 4.24$; the null hypothesis cannot be rejected.

12.75 $z = -7.75$; the claim must be rejected.

12.77 $z = -3.98$; the hypothesis that the probability is 0.70 must be rejected.

12.79 $\chi^2 = 52.8$; the null hypothesis must be rejected.

12.81 $E = 0.057$.

12.83 (a) 0 or 9 or more; 0.015; (b) 8 or more; 0.031; (c) 0 or 1; 0.048; (d) 0, 1, or 8 or more; 0.079.

12.85 $z = -1.59$; the null hypothesis cannot be rejected.

12.87 $\chi^2 = 28.3$; the null hypothesis (that the coins are balanced and randomly tossed) must be rejected.

CHAPTER 13   **13.1** (a) $F = 6.54$; the null hypothesis must be rejected; (b)

Source of variation	Degrees of freedom	Sum of squares	Mean square	F
Treatments	4	136	34	6.54
Error	15	78	5.2	
Total	19	214		

13.3 (a) We cannot differentiate between possible differences due to the designs and possible differences due to the different golf pros; (b) The results, whatever they may be, apply only to the particular driver; (c) Observed differences among designs may be due rather to fatigue, wear of equipment, or perhaps changes in the weather; (d) Observed differences among designs may be due instead to the use of a tee; (e) Observed differences among designs may be due to playing conditions at the different tees.

13.5 $F = 0.64$; the differences among the means are not significant.

13.7 $F = 10.79$; differences cannot be attributed to chance.

13.9 $F = 11.05$; differences are significant.

13.13 For the forms $F = 0.41$, so that the null hypothesis about the forms cannot be rejected; for the students $F = 23.8$, so that the null hypothesis about the students must be rejected.

13.15 For the salespersons $F = 0.18$, so that the null hypothesis about the salespersons cannot be rejected; for the weeks $F = 4.64$, so that the null hypothesis about the weeks must be rejected.

13.17 For the machines $F = 0.35$, so that the null hypothesis about the machines cannot be rejected; for the workmen $F = 7.07$, so that the null hypothesis about the workmen must be rejected.

13.19 160.

(a)

C	A	B
A	B	C
B	C	A

(b)

A	D	B	C
C	B	D	A
D	A	C	B
B	C	A	D

13.23 For the golf pros $F = 15.63$, so that the null hypothesis about the golf pros must be rejected; for the drivers $F = 0.82$, so that the null hypothesis about the drivers cannot be rejected; for the golf-ball designs $F = 45.80$, so that the null hypothesis about the golf-ball designs must be rejected.

13.25 4, 3, and 7 on Tuesday; 2, 5, and 7 on Thursday; and 5, 3, and 6 on Saturday.

13.27 (a) Each of the seven department heads serves together with each of the other department heads on two committees; (b) Griffith must be chairperson of the dramatics committee, Anderson must be chairperson of the discipline committee, Evans must be chairperson of the tenure committee and Fleming chairperson of the salaries committee, or vice versa.

13.29 For majors $F = 15.2$, so that the null hypothesis about the majors must be rejected; for instructors $F = 53.0$, so that the null hypothesis about the instructors must be rejected.

13.31 $F = 6.83$; the null hypothesis (that the different descriptions do not affect sales) must be rejected.

13.33 For the routes $F = 7.75$, so that the null hypothesis about the routes must be rejected; for the days of the week $F = 8.06$, so that the null hypothesis about the days of the week must be rejected.

CHAPTER 14

14.1 (a) $\hat{y} = 12.445 + 0.898x$; (b) $a = 12.447$ and $b = 0.898$.

14.3 (a) $\hat{y} = 1.900 - 0.086x$; (b) 1.470.

14.7 (a) $\hat{y} = 0.490 + 0.272x$; (b) 10.826.

14.9 (a) 4.46 million dollars; (b) $305.56.

14.11 $\hat{y} = 4.44x$.

14.13 $t = 2.84$; the null hypothesis must be rejected.

14.15 $t = -2.15$; the null hypothesis must be rejected.

14.17 $-18.05 < \beta < -10.25$.

14.19 $1.04 < \beta < 1.92$.

14.21 $0.34 < \alpha < 2.18$.

14.23 $69.8 < \mu_{y|45} < 76.6$.

14.25 (a) 12.48–26.72;   (b) 63.5–82.9.

14.27 $\hat{y} = 68.9(1.23)^x$; 197.

14.29 $\hat{y} = 27{,}200x^{-2.36}$; 77.2 thousand units.

14.31 $\hat{y} = 429.0 - 42.2x + 1.10x^2$; 81 thousand units.

14.33 $\hat{y} = -16{,}874 + 964x_1 + 2{,}980x_2$; \$32,642.

14.35 $\hat{y} = 69.73 + 2.975x_1 - 11.97x_2$, where x_1 and x_2 are the coded values.

14.39 $1.269 < \mu_{y|20} < 7.319$.

14.41 (a) $\hat{y} = -0.661 + 0.0973x_1 + 0.663x_2$;   (b) 29.05 thousand dollars.

14.43 $\hat{y} = -1.07 + 1.22x$.

14.45 $\hat{y} = 50.00 + 7.44x$; 236.

14.47 89.8–382.2.

CHAPTER 15

15.1 $r = -0.01$.

15.3 $r = 0.68$.

15.5 49.6 percent.

15.7 (a) 0.87;   (b) 76 percent;   (c) $0.20 < \rho < 0.98$.

15.9 We can draw a line through any two points. In (a) its slope is positive and in (b) it is negative.

15.11 Twice as strong.

15.13 (a) $t = 1.82$; not significant;   (b) $t = 3.35$; significant.

15.15 $z = -0.75$; the null hypothesis cannot be rejected.

15.17 $z = 2.42$; the null hypothesis cannot be rejected.

15.19 (a) $0.49 < \rho < 0.93$;   (b) $-0.54 < \rho < 0.15$;   (c) $0.51 < \rho < 0.74$.

15.21 (a) $r = 0.13$; (b) $r = 0.81$; (c) $r = 0.61$; (d) $r = -0.18$; (e) only those of parts (b) and (c) are significant; (f) effect is felt most strongly after two years and after that it diminishes gradually.

15.23 0.58.

15.25 0.96.

15.27 $r_{13.2} = -0.80$.

15.29 0.997.

15.31 $r = -0.92$.

15.33 6.25 percent.

15.35 0.68.

15.37 0.86.

15.39 (a) 0.21;   (b) 0.02;   (c) -0.22;   (d) -0.72;   (e) -0.46;   (f) only that of part (d) is significant;   (g) an eight-year cycle.

16.1 $x = 9$; the null hypothesis cannot be rejected.

16.3 $x = 9$; the null hypothesis cannot be rejected.

16.5 $x = 10$; the null hypothesis cannot be rejected.

16.7 $x = 7$; the null hypothesis cannot be rejected.

16.9 $z = 1.73$; the null hypothesis cannot be rejected.

16.11 $z = -2.04$; the null hypothesis must be rejected.

16.13 $U = 6$; the null hypothesis cannot be rejected.

16.15 $U_1 = 3$; the null hypothesis must be rejected.

16.17 $U_1 = 18$; the null hypothesis cannot be rejected.

16.19 $z = 2.12$; the null hypothesis must be rejected.

16.21 $z = -1.62$; the null hypothesis cannot be rejected.

16.23 $T = 20$; the null hypothesis cannot be rejected.

16.25 $T^+ = 10$; the null hypothesis must be rejected.

16.27 $T = 25.5$; the null hypothesis must be rejected.

16.29 $T^- = 13$; the null hypothesis must be rejected.

16.31 $z = -2.67$; the null hypothesis must be rejected.

16.33 $z = 0.26$; the null hypothesis cannot be rejected.

16.35 $H = 0.245$; the null hypothesis cannot be rejected.

16.37 $H = 4.51$; the null hypothesis cannot be rejected.

16.39 $u = 5$; the null hypothesis of randomness must be rejected.

16.41 $u = 5$; the null hypothesis of randomness cannot be rejected.

16.43 $z = -0.77$; the null hypothesis of randomness cannot be rejected.

16.47 $u = 5$; reject the null hypothesis of randomness.

16.49 $r_S = 0.65$.

16.51 $r_S = 0.57$.

16.53 $z = 3.00$; it is significant.

16.55 (a) Judges A and B;   (b) judges B and C.

16.57 $T = 19$; the null hypothesis must be rejected.

16.59 $H = 5.03$; the null hypothesis cannot be rejected.

16.61 $z = -2.14$; the null hypothesis must be rejected.

16.63 $x = 9$; the null hypothesis cannot be rejected.

16.65 $z = 1.34$; the null hypothesis cannot be rejected.

16.67 $z = -1.15$; the null hypothesis of randomness cannot be rejected.

16.69 $x = 2$; the null hypothesis must be rejected.

16.71 $u = 4$; the null hypothesis of randomness must be rejected.

INDEX

E

Effect:
  block, 381
  treatment, 366
Efficiency, 490
Elimination, rule of, 140
Empirical rule, 60
Empty set, 106
Equality of standard deviations, 319
Equitable game, 153
Error:
  experimental, 370
  probable, 257
  standard, 250
  Type I, 282
  Type II, 282
Error sum of squares:
  Latin square, 390
  one-way analysis of variance, 371
  two-way analysis of variance, 382
Estimate:
  Bayesian, 273
  interval, 269
  point, 265
Estimated regression coefficients, 414
Estimated regression line, 410
Event, 106
Events:
  dependent, 132
  independent, 132
  mutually exclusive, 107
Expectation, mathematical, 151
Expected frequencies, 341, 350, 353
Expected profit with perfect information, 162
Expected value of perfect information, 162
Experiment, 105
  two-factor, 381, 385
Experimental error, 370
Experimental sampling distribution, 245
Exploratory data analysis, 31
Exponential curve, fitting, 422
Exponential distribution, 215

F

Factorial notation, 87
Fair game, 153
Fair odds, 118
F distribution, 319
  degrees of freedom, 319
Finite population, 234
  correction factor, 251

  random sample from, 235
Finite sample space, 106
Fisher Z transformation, 445
Five-stem plot, 34
Fractiles, 69
Frequency:
  class, 16
  expected, 341, 350, 353
  observed, 341, 350, 353
Frequency distribution, 14
  bell-shaped, 59, 70
  categorical, 14
  class boundary, 17
  class frequency, 16
  class interval, 17
  class limit, 16
  class mark, 17
  cumulative, 18
  deciles, 69
  histogram, 22
  mean, 65
  median, 66, 68
  numerical, 14
  ogive, 23
  open class, 15
  percentage, 17
  percentiles, 69
  qualitative, 14
  quantitative, 14
  quartiles, 69
  real limits, 17
  skewness, 71
  standard deviation, 65
  symmetrical, 70
  U-shaped, 71
Frequency interpretation of probability, 96
Frequency polygon, 23
F statistic, 319, 368

G

Game:
  equitable, 153
  fair, 153
General addition rule, 126
Generalized addition rule, 122, 129
General multiplication rule, 133
Geometric distribution, 177
Geometric mean, 51
Goodness of fit, 353
Grand mean, 48, 366, 371
Grand total, 343
Graphical presentation, 22

coefficient of, 60
measures of, 53
Venn diagram, 108

W

Weight, 47
Weighted mean, 47

Wilcoxon signed-rank test, 473
Wilcoxon test, 465

Z

z-scores, 60, 209
Z transformation, 445